Structural Molecular Biology

Structural Molecular Biology

Edited by Erik Pierre

SYRAWOOD
PUBLISHING HOUSE

New York

Published by Syrawood Publishing House,
750 Third Avenue, 9th Floor,
New York, NY 10017, USA
www.syrawoodpublishinghouse.com

Structural Molecular Biology
Edited by Erik Pierre

International Standard Book Number: 978-1-68286-676-4 (Hardback)

Cataloging-in-Publication Data

Structural molecular biology / edited by Erik Pierre.
    p. cm.
Includes bibliographical references and index.
ISBN 978-1-68286-676-4
1. Molecular biology. 2. Molecular structure. 3. Biomolecules. 4. Systems biology.
I. Pierre, Erik.
QH506 .S77 2019
572.8--dc23

TABLE OF CONTENTS

PREFACE

Molecular biology is a broad field of biochemistry that is concerned with the study of biological activity between biomolecules as well as their interactions with DNA, RNA and proteins. It further strives to understand biosynthesis mechanisms occurring in cells. As a field of scientific study, structural biology is the sub-discipline of molecular biology that is concerned with the study of macromolecules and their structural development and functions. Recent research in structural and molecular biology has transcended into interdisciplinary areas of bioinformatics, computational biology, evolutionary biology, population genetics and many others. This book is compiled in such a manner, that it will provide in-depth knowledge about the theory and practice of this field. It elucidates new techniques and their applications in a multidisciplinary manner. Different approaches, evaluations, methodologies and advanced studies on molecular biology have been included herein. Those in search of information to further their knowledge will be greatly assisted by this book.

The researches compiled throughout the book are authentic and of high quality, combining several disciplines and from very diverse regions from around the world. Drawing on the contributions of many researchers from diverse countries, the book's objective is to provide the readers with the latest achievements in the area of research. This book will surely be a source of knowledge to all interested and researching the field.

In the end, I would like to express my deep sense of gratitude to all the authors for meeting the set deadlines in completing and submitting their research chapters. I would also like to thank the publisher for the support offered to us throughout the course of the book. Finally, I extend my sincere thanks to my family for being a constant source of inspiration and encouragement.

Editor

Molecular dynamics simulations on the Tre1 G protein-coupled receptor: exploring the role of the arginine of the NRY motif in Tre1 structure

Margaret M Pruitt[1], Monica H Lamm[2*] and Clark R Coffman[1*]

Abstract

Background: The arginine of the D/E/NRY motif in Rhodopsin family G protein-coupled receptors (GPCRs) is conserved in 96% of these proteins. In some GPCRs, this arginine in transmembrane 3 can form a salt bridge with an aspartic acid or glutamic acid in transmembrane 6. The *Drosophila melanogaster* GPCR Trapped in endoderm-1 (Tre1) is required for normal primordial germ cell migration. In a mutant form of the protein, Tre1[sctt], eight amino acids RYILIACH are missing, resulting in a severe disruption of primordial germ cell development. The impact of the loss of these amino acids on Tre1 structure is unknown. Since the missing amino acids in Tre1[sctt] include the arginine that is part of the D/E/NRY motif in Tre1, molecular dynamics simulations were performed to explore the hypothesis that these amino acids are involved in salt bridge formation and help maintain Tre1 structure.

Results: Structural predictions of wild type Tre1 (Tre1[+]) and Tre1[sctt] were subjected to over 250 ns of molecular dynamics simulations. The ability of the model systems to form a salt bridge between the arginine of the D/E/NRY motif and an aspartic acid residue in transmembrane 6 was analyzed. The results indicate that a stable salt bridge can form in the Tre1[+] systems and a weak salt bridge or no salt bridge, using an alternative arginine, is likely in the Tre1[sctt] systems.

Conclusions: The weak salt bridge or lack of a salt bridge in the Tre1[sctt] systems could be one possible explanation for the disrupted function of Tre1[sctt] in primordial germ cell migration. These results provide a framework for studying the importance of the arginine of the D/E/NRY motif in the structure and function of other GPCRs that are involved in cell migration, such as CXCR4 in the mouse, zebrafish, and chicken.

Keywords: GPCR, Tre1, Molecular dynamics, Germ cell migration, Salt bridge formation

Background

G protein-coupled receptors (GPCRs) are the largest class of membrane proteins, accounting for 2% of genes in the human genome [1-3]. In general, GPCRs are responsible for modulating signals from the extracellular environment and transducing these stimuli into intracellular signaling cascades and cellular responses. GPCRs are involved in a wide range of cellular processes including cell movement, neurotransmission and olfaction, and can also be involved in disease progression with roles in metastasis, angiogenesis, cell proliferation and inflammation [4,5]. Since GPCRs are involved in maintaining homeostasis as well as disease progression, GPCRs are an important group of proteins to study, informing basic cellular and molecular biology as well as pharmaceutical applications.

Although there are many different genes encoding for GPCRs, all GPCRs share a common structure. GPCRs have seven transmembrane α-helices (TM1-TM7) connected by three intracellular and three extracellular loops. There are five main families of human GPCRs (Rhodopsin, Secretin, Glutamate, Adhesion and Frizzled/Taste2) [1-3], and this classification holds true for GPCRs in other bilateral species [6].

GPCRs are inherently difficult to crystallize due to their transmembrane nature and the fact that individual

* Correspondence: mhlamm@iastate.edu; ccoffman@iastate.edu
[2]Department of Chemical and Biological Engineering, Iowa State University, Ames, IA 50011, USA
[1]Department of Genetics, Development and Cell Biology, Iowa State University, Ames, IA 50011, USA

GPCRs are typically expressed at low levels within cells. GPCRs, like other transmembrane proteins, require a membrane-like environment to remain in a properly folded conformation. The required presence of a membrane makes the overexpression and subsequent purification of GPCRs challenging. The first GPCR crystal structure, bovine rhodopsin, was determined in 2000 [7], with nearly seven years passing before a crystal structure for the second GPCR was published. To date there are 16 GPCRs crystallized, all representing the Rhodopsin family of GPCRs [8]. Additionally, six of these proteins (bovine opsin, bovine rhodopsin, human A_{2A}-adenosine receptor, turkey β_1-adrenergic receptor, human β_2-adrenergic receptor, and rat neurotensin receptor NTSR1) have been crystallized in active-like states [8].

Due to the difficulties of GPCR purification and crystallization, protein structure prediction programs and molecular dynamics (MD) simulations are frequently used to investigate the structures of GPCRs. There are currently three computational techniques available to generate a three-dimensional structural prediction of a protein: homology modeling, threading, and *ab initio* modeling. Homology modeling builds a three-dimensional structure by first identifying an evolutionarily related homologous protein with a known structure to use as a template. The program then aligns the amino acid sequence of the protein of interest to the amino acid sequence of the chosen template and finally builds the model [9-11]. The relatively low number of GPCR crystal structures is a major limitation to homology modeling. A lack of diverse structures means that a majority of GPCRs will still lack a homologous protein to use as a template. It is possible to build a highly accurate model when the template protein and the protein sequence of interest share 50% or more sequence identity [9,10]. However, when the sequence identity is below 30%, the protein structure prediction will likely more closely resemble the template structure than the native structure of the protein [12]. The sequence identity between crystallized GPCRs and other known GPCRs is often below 30% [13]. Due to the prevalence of low sequence identity, it is suggested that both sequence identity and structural information be used when choosing the template protein [13].

Threading, similar to homology modeling, is a template-based approach to structure prediction. The first step in threading is to search for evolutionary relatives to the protein sequence of interest. This is commonly accomplished with Position-Specific Iterative Basic Local Alignment Search Tool (PSI-BLAST) [14]. PSI-BLAST generates a sequence profile, which is used by a secondary structure predictor, like PSIPRED [15], to determine the secondary structure of the protein sequence of interest. Both the secondary structure and the sequence profile from PSI-

BLAST are used in a threading algorithm to identify template proteins from the Protein Data Bank that have similar protein folds to the sequence of interest. Templates used in threading may show no evolutionary relationship [11]. The use of multiple templates, creating a chimeric GPCR, has been shown to provide a more accurate model than using a single protein template [13,16,17]. Multiple templates can be used in both homology modeling and threading.

Ab initio modeling builds a three-dimensional protein model from sequence information alone, without using a template structure, based upon the assumption that the protein structure will assume the lowest free energy conformation [9]. *Ab initio* modeling can work well for proteins with less than 120 amino acids [11]. Although there are three different ways to build a protein structure prediction, some current modeling programs use a combination of approaches to predict a structure [11]. The accuracy of the final model is linked to the template(s) chosen, and some approaches to generating a protein structural prediction work better on certain proteins or parts of proteins than others [18,19].

With only 16 distinct GPCR proteins crystallized, it can be difficult to find a suitable template(s) to use with the modeling software. Part of this challenge has been alleviated by the availability of web servers specifically designed for modeling GPCRs, such as GPCR-ModSim [20] and GPCR-Iterative Threading ASSEmbly Refinement (GPCR-ITASSER) [21-23]. GPCR-ModSim is a server that allows investigators to model GPCRs using MODELLER [9,20,24] and GPCR-ModSim users have the option of choosing whether to align their GPCR sequence with inactive-like crystallized GPCRs or active-like crystallized GPCRs. GPCR-ModSim aligns the sequence and shows the percent identity with the available templates. The user can then choose which template to use and GPCR-ModSim generates a homology model using MODELLER. Once a homology model is generated, the user has the option of submitting it for MD simulations in a solvated 1-palmitoyl-2-oleoyl-*sn*-glycero-3-phosphocholine lipid bilayer [20].

GPCR-ITASSER is another web server that allows for protein structure prediction [21-23]. GPCR-ITASSER takes the initial GPCR sequence and identifies evolutionary relatives using PSI-BLAST and secondary structures using PSIPRED. The results from PSI-BLAST and PSIPRED are used by the Local Meta-Threading Server (LOMETS) to find potential templates in the Protein Data Bank. Any sequence without a matched template is modeled using an *ab initio* helix-modeling program. Additional restraints to the protein structure are incorporated through the use of the online database GPCRRD (GPCR Research Database), which contains experimental restraints from other GPCR databases and literature

[21-23]. The *ab initio* modeling, results from threading, and restraints from the GPCRRD are all used to assemble and build a structural model. This structural model is refined using Fragment-Guided Molecular Dynamics [22] to give the final model.

In this study, the GPCR-ModSim [20] and the GPCR-ITASSER [21] web servers were used to predict protein structures of the GPCR Trapped in endoderm-1 (Tre1). Tre1 is a Rhodopsin family GPCR required for proper *Drosophila melanogaster* primordial germ cell migration [25-27]. In a mutant form of the protein, Tre1[sctt], primordial germ cell migration is severely disrupted. The primordial germ cells scatter across the posterior half of the embryo rather than populating the two gonads. The molecular lesion in *tre1[sctt]* RNA is a point mutation that results in an in-frame deletion of eight amino acids, RYILIACH [27]. Two of these amino acids (RY) are part of the highly conserved D/E/NRY motif in TM3 of Rhodopsin family GPCRs. The D/E/NRY motif is thought to act as a micro-switch in the activation mechanism of Rhodopsin family GPCRs [3,28], and the arginine is conserved in 96% of Rhodopsin family GPCRs [29]. The arginine of the D/E/NRY motif (R3.50 following Ballesteros-Weinstein nomenclature [30]) can form a salt bridge with TM6 in numerous GPCRs [31-42], and while the exact role of the salt bridge is unknown [43], it is clear that the arginine is very important for Tre1[+] function [27]. The first position of the D/E/NRY motif is also highly conserved. It is an acidic residue (aspartic acid (D) or glutamic acid (E)) in 86% of GPCRs [29]. In some GPCRs, the aspartic acid or glutamic acid can interact with the neighboring arginine to form an intrahelical salt bridge in addition to the interaction with TM6 [36,44]. Since an acidic residue in the first position of the motif is not present in Tre1 (NRY motif), an intrahelical salt bridge with the arginine does not form. This could make the interhelical salt bridge with TM6 more important.

Protein structure predictions of putative inactive structures of both Tre1[+] and Tre1[sctt] were generated with GPCR-ModSim [20] and GPCR-ITASSER [21]. The NAMD simulation package [45] was used to perform MD simulations on both Tre1[+] and Tre1[sctt] embedded in a 1-palmitoyl-2-oleoyl-*sn*-glycero-3-phosphoethanolamine (POPE) lipid bilayer, and MD simulations were run on four different model systems for over 250 ns each. The proteins were embedded in a POPE lipid bilayer since phosphoethanolamine is the most abundant phospholipid in Drosophila cell membranes [46]. The NRY motif of Tre1 was studied by examining the possibility of salt bridge formation between the arginine of the NRY motif and an aspartic acid residue of TM6. One of the wild type Tre1 model systems shows potential for a strong salt bridge to form between R134 of the

NRY motif and D266 of TM6. The distances between the residues are favorable for salt bridge formation and could indicate that the salt bridge promotes interhelical stabilization of the Tre1 GPCR. The lack of similar interactions in the mutant model systems with an alternative arginine residue as well as *in vivo* data with Tre1[sctt] [27] suggests that the arginine of the NRY motif is important to the function and maintenance of an inactive structure that allows for subsequent activation of the Tre1 GPCR.

Results
Protein structure prediction
The amino acid sequences for Tre1[+] and Tre1[sctt] were used for protein structure predictions using GPCR-ModSim [20] and GPCR-ITASSER [21]. Both GPCR-ModSim and GPCR-ITASSER are web servers for GPCR protein structure prediction, however the web servers differ in the approach taken to generate a protein structure prediction. The GPCR-ModSim server automates the process of using the homology modeling program MODELLER to model GPCRs [20], while GPCR-ITASSER uses multiple threading programs as well as the GPCR Research Database to predict protein structures [21]. These two web servers were used in this study to generate four independent protein structure predictions, two each for Tre1[+] and Tre1[sctt]. While MODELLER is an established program that has been used to predict GPCR structures for simulations before, this is the first study to use structural predictions from GPCR-ITASSER.

Tre1[+] and Tre1[sctt] were modeled to the seven inactive-like GPCRs on the GPCR-ModSim web server. From the multiple sequence alignment generated, it was clear that any one of the seven available GPCR crystal structures could be used as a template to model Tre1[+] and Tre1[sctt]. However, GPCR-ModSim allows only one template to be chosen, and for both Tre1[+] and Tre1[sctt] the template chosen was squid rhodopsin (PDB ID: 2Z73). Not only did squid rhodopsin show the highest total sequence identity to Tre1[+] and Tre1[sctt] (17.4% for Tre1[+] and 16.7% for Tre1[sctt]) (Table 1), but also squid rhodopsin seemed to be the best choice from earlier work using the web server I-TASSER. The I-TASSER web server uses threading to generate a structural prediction of a protein and allows a user to submit a sequence with or without selecting a template to use [11]. The structural prediction from I-TASSER using Tre1[+] sequence and no selected template looked most similar to the structural prediction when squid rhodopsin was chosen as a template (data not shown). Therefore, squid rhodopsin was selected as the most appropriate template. Using squid rhodopsin as a template, ten models each for Tre1[+] and Tre1[sctt] were generated using MODELLER. The two models chosen for further study using

Table 1 Sequence alignments in GPCR-ModSim indicate squid rhodopsin has the greatest identity to Tre1$^+$ and Tre1^{sctt}

Template	Tre1$^+$	Tre1^{sctt}
	Total % identities	Total % identities
1U19 – bovine rhodopsin	14.3	14.1
2RH1 – human β_2-adrenergic receptor	15.1	14.6
2VT4 – turkey β_1-adrenergic receptor	14.8	14.3
2Z73 – squid rhodopsin *	17.4	16.7
3EML – human A_{2A}-adenosine receptor	14.0	13.5
3ODU – human chemokine receptor 4	16.3	14.8
3PBL – human D_3 dopamine receptor	16.1	15.9

* Denotes the template chosen for homology modeling.

MD simulations were selected based on the lowest Discrete Optimized Protein Energy (DOPE-HR) score [47] and are named mtre1 (Tre1$^+$) and msctt (Tre1^{sctt}).

The second set of independent protein structure predictions for Tre1$^+$ and Tre1^{sctt} were built using GPCR-ITASSER [21]. The amino acid sequences for Tre1$^+$ and Tre1^{sctt} were submitted to GPCR-ITASSER and used in the local threading server to find template proteins. Both Tre1$^+$ and Tre1^{sctt} were modeled to Substance P, human β_2-adrenergic receptor, bovine rhodopsin and human A_{2A} adenosine receptor. In addition, Tre1$^+$ was modeled to turkey β_1-adrenergic receptor and human β_2-adrenergic receptor-Gs protein complex, while Tre1^{sctt} was modeled to squid rhodopsin. Even though an active GPCR (human β_2-adrenergic receptor-Gs protein complex) was a template for the Tre1$^+$ model, it is thought that this Tre1$^+$ model reflects an inactive conformation of the protein as the resulting model from GPCR-ITASSER is a consensus of restraints from six templates, all but one being inactive. The two best models, based on GPCR-ITASSER confidence scores, were chosen for further study in MD simulations. The models generated by GPCR-ITASSER are named gtre1 (Tre1$^+$) and gsctt (Tre1^{sctt}).

The four models, mtre1, msctt, gtre1, and gsctt, selected for further study are shown in Figure 1A. At first glance, all four of the models look similar, but there are distinct differences. Namely, helices 5 and 6 (yellow and gold chains) are roughly the same length as the other five helices in mtre1 and msctt (arrowheads in Figure 1B). In gtre1 and gsctt, helices 5 and 6 are extended relative to the other five helices. Helical extensions of helices 5 and 6 are present in the crystal structure of squid rhodopsin (PDB ID: 2Z73) [48] which was used as a template structure for mtre1, msctt, and gsctt. These differences change the architecture of intracellular loop 3. Additionally, intracellular loop 2 (green) has different structures in mtre1 and gtre1 (arrows in Figure 1B). In mtre1, this loop region is unstructured. In contrast, there is a short helix in intracellular loop 2 in gtre1. These differences in intracellular loop structure

can be attributed to the template structure(s) used to generate mtre1 and gtre1. Intracellular loop 2 is of interest since it is the location of some of the residues missing in Tre1^{sctt}.

Building biologically relevant model systems

As GPCRs exist in a membrane environment, the four different protein structure predictions were inserted into a solvated POPE lipid bilayer using the Membrane Builder in the CHARMM-GUI [49,50]. The final solvated membrane systems are named the same as the structural predictions, mtre1, msctt, gtre1 and gsctt. Each system was subjected to over 250 ns of MD and an example of the mtre1 system after MD is shown in Figure 2.

Experimental work has shown that at 310 K a POPE lipid in a POPE lipid bilayer maintains a surface area of 59.75 – 60.75 Å^2 [51]. To confirm the correct surface areas per lipid were maintained in mtre1, msctt, gtre1 and gsctt, the Voronoi Tesselation and Monte Carlo (VTMC) integration method [52] was used. This method allows for calculation of the surface area per lipid in membrane-lipid systems. VTMC calculates the surface area per boundary and non-boundary lipids. Non-boundary lipids are described as those lipids not interacting with atoms of the protein. It is important to make the distinction between lipid types (boundary versus non-boundary) since lipids interacting with atoms of the protein will have a decreased surface area per lipid. The results of the VTMC analysis are shown in Table 2 and confirm that the non-boundary lipids in each of the model systems maintained the correct surface area per lipid, ranging from 59.0 – 60.2 Å^2.

Global movements of the model systems

Root mean squared deviation (RMSD) was computed over the course of the simulation for the Cα atoms of the transmembrane regions of the proteins to measure structural stability (Figure 3). As seen from the curves in Figure 3, the RMSD values did not change significantly from the starting structure and each curve began to

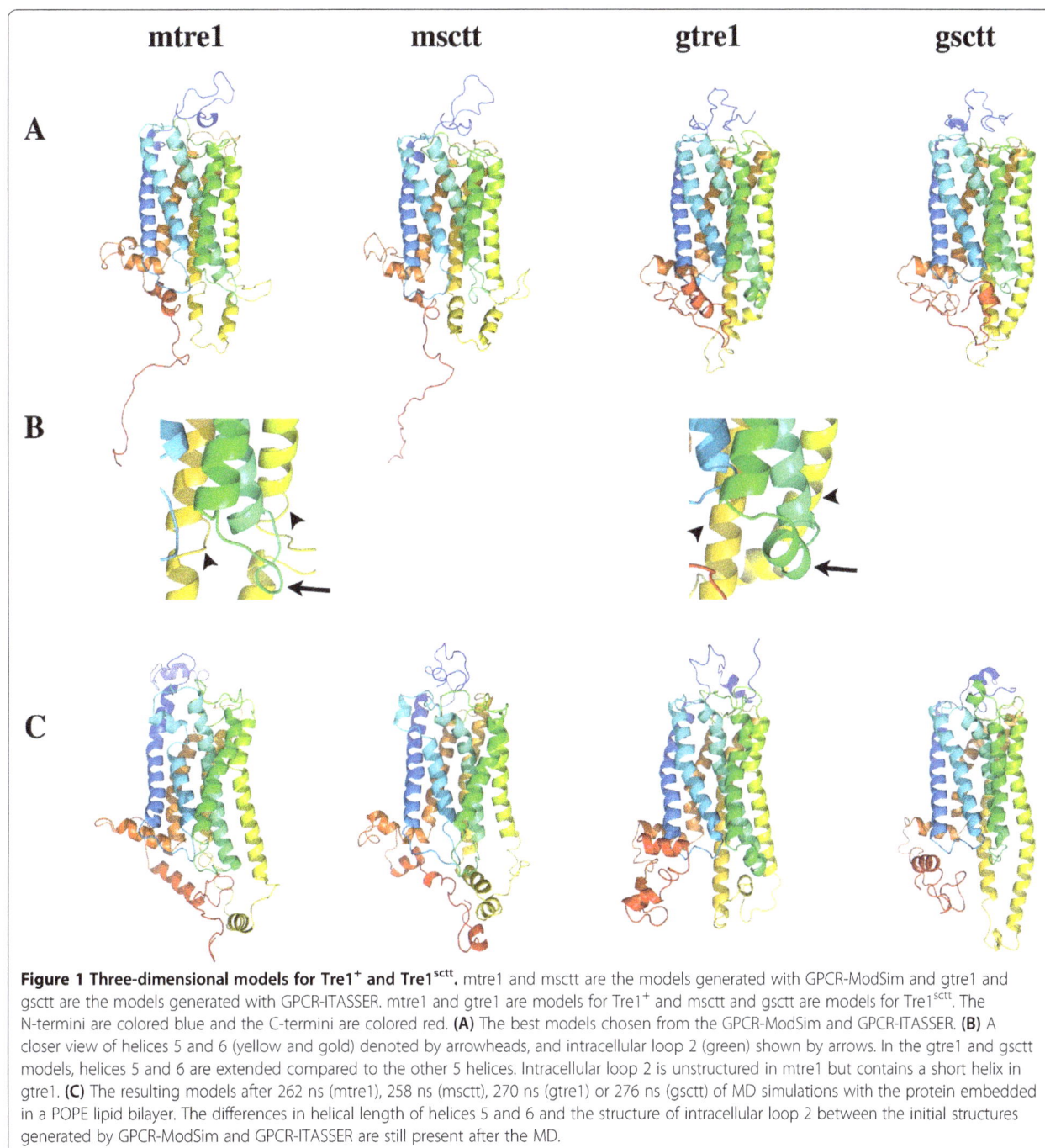

Figure 1 Three-dimensional models for Tre1⁺ and Tre1ˢᶜᵗᵗ. mtre1 and msctt are the models generated with GPCR-ModSim and gtre1 and gsctt are the models generated with GPCR-ITASSER. mtre1 and gtre1 are models for Tre1⁺ and msctt and gsctt are models for Tre1ˢᶜᵗᵗ. The N-termini are colored blue and the C-termini are colored red. **(A)** The best models chosen from the GPCR-ModSim and GPCR-ITASSER. **(B)** A closer view of helices 5 and 6 (yellow and gold) denoted by arrowheads, and intracellular loop 2 (green) shown by arrows. In the gtre1 and gsctt models, helices 5 and 6 are extended compared to the other 5 helices. Intracellular loop 2 is unstructured in mtre1 but contains a short helix in gtre1. **(C)** The resulting models after 262 ns (mtre1), 258 ns (msctt), 270 ns (gtre1) or 276 ns (gsctt) of MD simulations with the protein embedded in a POPE lipid bilayer. The differences in helical length of helices 5 and 6 and the structure of intracellular loop 2 between the initial structures generated by GPCR-ModSim and GPCR-ITASSER are still present after the MD.

stabilize after 150 ns of dynamics. This suggests that the systems have equilibrated. Further confirmation of structural stability after 150 ns can be seen from the curves of the RMSD values calculated for the complete proteins computed over the course of the simulation (Additional file 1: Figure S1).

Root mean squared fluctuation (RMSF) was also used as a way to qualitatively characterize the protein dynamics. Here, RMSF describes the fluctuations of each Cα atom of the amino acid residues in the proteins averaged over the simulation time, beginning at 150 ns (Figure 4). The general fluctuations of specific regions of the proteins are similar between each of the Tre1⁺ and Tre1ˢᶜᵗᵗ models. It is clear from all four plots that the regions of the protein with the least amount of movement are the transmembrane regions. Intracellular loop 3 shows the greatest fluctuations, which is expected since it is the longest loop in Tre1. Of all the model systems, intracellular loop 3 of gsctt has the highest RMSF values. The high degree of fluctuation seen in intracellular loop 3 of

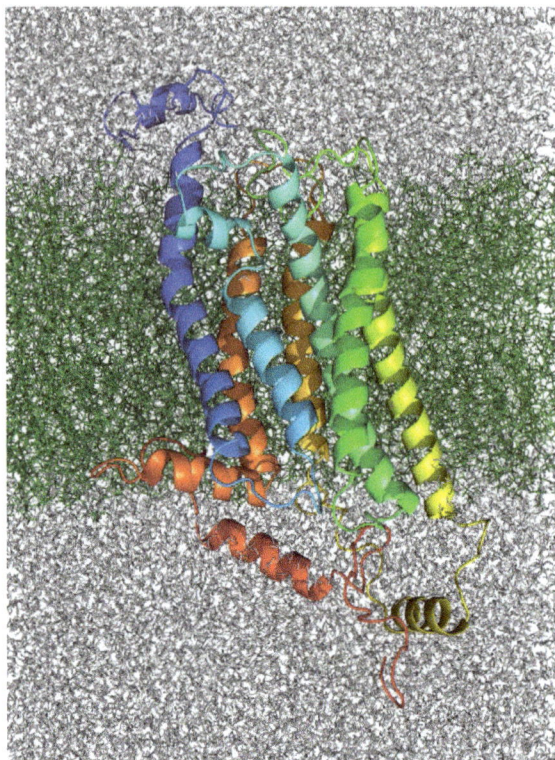

Figure 2 Protein structure prediction of mtre1 embedded in a POPE lipid bilayer. The mtre1 model system shown here is after 262 ns of molecular dynamics. The extracellular surface of the bilayer is on top and the intracellular surface of the bilayer is on the bottom. mtre1 is depicted as ribbons, with the N-terminus colored blue and the C-terminus colored red. The POPE bilayer is represented as sticks and is colored green. Water and ions are depicted as grey lines.

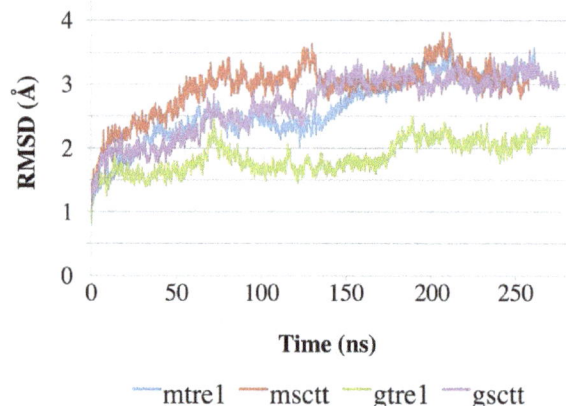

Figure 3 RMSD of the transmembrane regions shows the protein structures have equilibrated. Root mean squared deviation (RMSD) was calculated for the Cα atoms of the transmembrane domains and is plotted over simulation time. The transmembrane regions of the protein models do not change much compared to the starting structures over the course of simulations, as the final RMSD values are between 2.0 and 3.5 Å.

gsctt could be due to its architecture, as intracellular loop 3 is more helical in gsctt than it is in mtre1, msctt, and gtre1. In gsctt, intracellular loop 3 begins as a helical extension of TM5, and then it has an unstructured loop region, followed by another helical region that connects

Table 2 Voronoi Tessellation Monte Carlo integration method confirms the model systems have maintained a fluid-phase bilayer

Model system	Average surface area (Å²/lipid)	
	Boundary lipid	Non-boundary lipid
mtre1	45.2	59.8
msctt	45.1	59.0
gtre1	45.4	60.1
gsctt	47.9	60.2

Average surface areas per lipid were calculated using the Voronoi Tessellation Monte Carlo integration method [52] and calculated at the end of each simulation for all model systems. A boundary lipid is a lipid that contacts an atom from the protein and a non-boundary lipid is a lipid that makes no contact with protein atoms. Experimental values of surface area per non-boundary POPE lipid is 59.75 – 60.75 Å² [51]. Computationally determined surface area per non-boundary POPE lipid using CHARMM36 is 59.2 Å² [66].

with TM6. The high RMSF values in intracellular loop 3 in gsctt come from the residues in intracellular loop 3 that are in the unstructured loop. It is possible that the more rigid α-helical segments in intracellular loop 3 in gsctt prevent some of the other, unstructured residues in the loop from making important contacts with other parts of the protein. This could cause higher RMSF values.

A third qualitative assessment of the simulations was an all-to-all RMSD calculation on the transmembrane Cα atoms of the Tre1 protein, shown in Figure 5 as heat maps. Heat maps of all-to-all RMSD calculations show the number of different states the protein has visited during the course of the simulation. The darker diagonal blocks in each plot show when the Tre1 protein (Tre1[+] or Tre1[sctt]) explores conformations that are structurally very similar. Darker off-diagonal blocks suggest that the protein revisits a conformation over the course of the simulation, although in this case, low RMSD alone is not sufficient to guarantee that two, noncontiguous in time, structures are similar [53]. In the heat maps shown here, each simulation samples two or more conformational substates. The quality of sampling is comparable to the sampling obtained at 250 ns in previous MD simulations of GPCRs [53-55].

The final protein structures after the dynamics in a lipid bilayer are shown in Figure 1C (the lipids, water and ions are not shown). In general, the transmembrane regions of the protein structures did not change drastically from the initial structures. Some of the loops have changed considerably, but the distinctions between the initial structures generated by GPCR-ModSim and

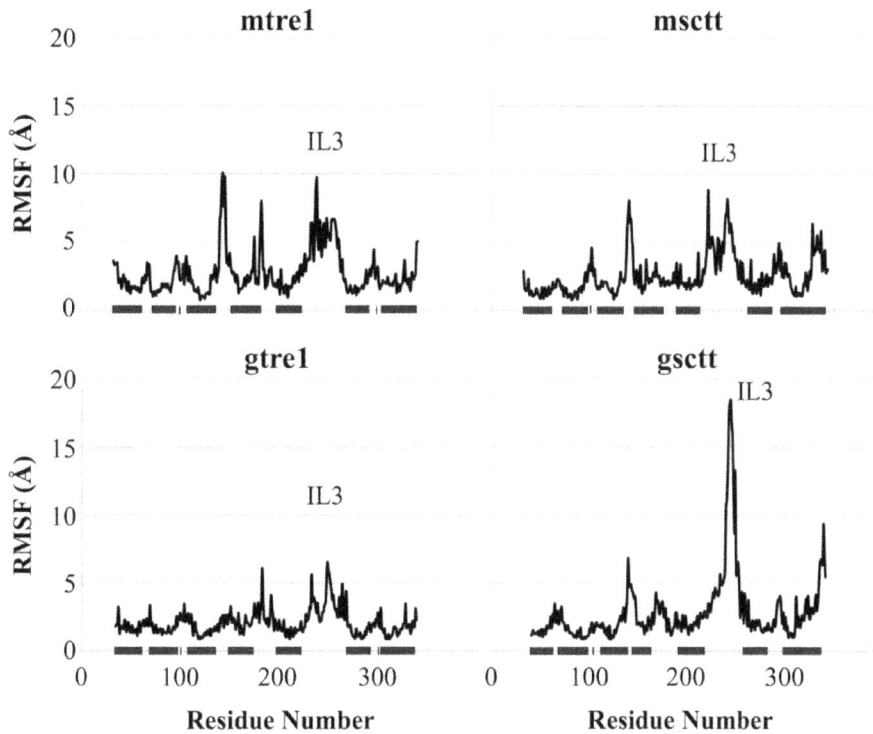

Figure 4 RMSF of each model system shows transmembrane regions move less than the loop regions. The root mean squared fluctuation (RMSF) for the mtre1, msctt, gtre1 and gsctt model systems beginning after 150 ns of dynamics are shown. The N- and C-termini are not included in the plots since the termini had high RMSF values and made it difficult to see the fluctuations in the other regions of the protein. The black bars denote the regions of the protein that are within the lipid bilayer. In most model systems, intracellular loop 3 (IL3) shows the greatest fluctuations.

GPCR-ITASSER noted previously are still present. To quantitatively determine changes in secondary structure over the course of the simulations, the Dictionary of Secondary Structure of Proteins [56,57] was used (Additional file 2: Figure S2). These plots confirm the seven transmembrane helices remain stable throughout the simulation, and the termini and loops are regions of change. While the significance of the differences between the GPCR-ModSim and GPCR-ITASSER models remain to be understood, it is interesting that even after over 250 ns of MD, some of the structural differences between the model systems were not resolved. This could mean that the structural models generated by the different modeling programs represent different protein conformations of Tre1.

Studies of the NRY motif of Tre1

From previous genetic studies, it is known that the arginine of the NRY motif (R3.50 following Ballesteros-Weinstein nomenclature [30]) in Tre1 is critical to the function of this GPCR [27]. Other than the critical nature of the arginine to Tre1 function, very little is known about the potential structural roles for this amino acid. It is possible that the arginine of the NRY motif in TM3 is involved in forming a salt bridge with an aspartic

acid residue in TM6. A similar salt bridge in other GPCRs is thought to be important for holding GPCRs in inactive or activated states [32-42]. If there is a salt bridge in Tre1 between the arginine of the NRY motif and an aspartic acid residue in TM6, loss of this arginine could remove this salt bridge and impair function, which would be consistent with experimental observations that germ cell migration requires this arginine in Tre1 for function [27]. It is also possible that an alternative arginine just downstream of the deleted amino acids in Tre1^{sctt} could be used to form a salt bridge with TM6. This alternative salt bridge could explain why the $tre1^{sctt}$ allele does not appear to be a complete loss-of-function allele of the $tre1$ gene. In Tre1$^+$ the sequence around the arginine is NRYILIACHSR*Y. In Tre1^{sctt}, the amino acids RYILIACH are missing and the remaining sequence is NSR*Y. The arginine of the NRY motif in Tre1$^+$ is numbered as R134. The arginine R* in Tre1^{sctt}, the alternative arginine, is numbered as R135, meaning this alternative arginine is located one residue from where the original arginine is located in Tre1$^+$. Therefore, this alternative arginine could be close enough to form a salt bridge in the Tre1^{sctt} protein. To test this hypothesis, the potential for salt bridge formation was evaluated in all model systems. Here, a salt bridge is

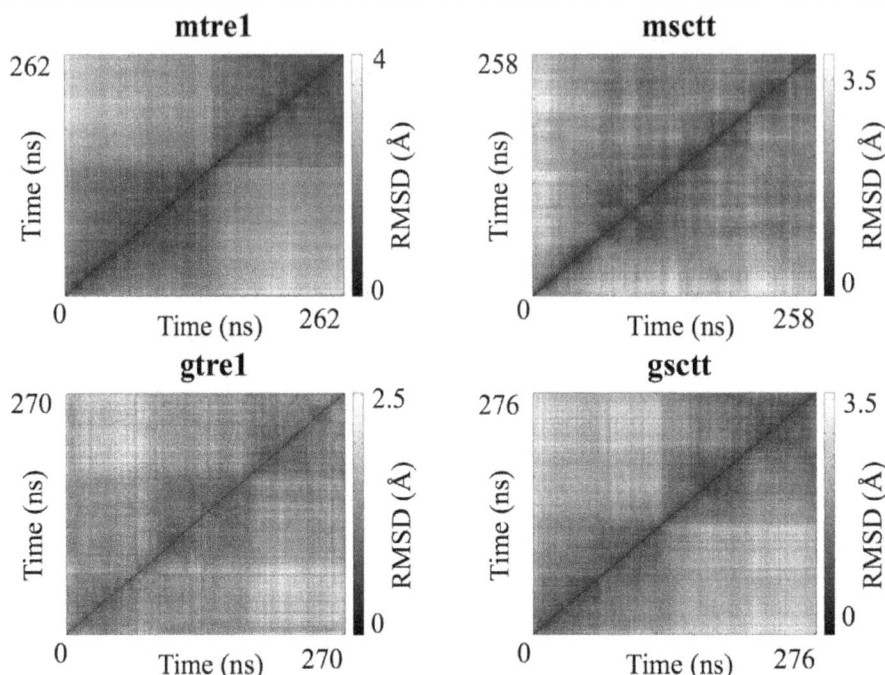

Figure 5 Heat maps of all-to-all RMSD calculations show sampling of conformational substates during MD simulations. Transmembrane Cα atoms in mtre1, msctt, gtre1 and gsctt model systems were used in these calculations. The all-to-all RMSD calculation computes the RMSD of all pairs of frames in a trajectory. This all-to-all RMSD calculation was performed using the VMD plug-in RMSD Trajectory Tool [71]. The RMSD scale is different for each model system and is located to the right of each plot.

defined as a noncovalent interaction between the carboxylate group of aspartic acid (D266 in Tre1$^+$ and D258 in Tre1^{sctt}) and the guanidium group of arginine (R134 in Tre1$^+$ and R135 in Tre1^{sctt}). As aspartic acid residues have two oxygen atoms that could be involved in a salt bridge, and arginine residues have two nitrogen atoms that could be involved in a salt bridge, the distance between both of the oxygen atoms of the aspartic acid in TM6 and both of the nitrogen atoms of the arginine in TM3 were calculated and plotted over the simulation time (Figure 6). Interatomic distances of 3.2 Å or less were considered favorable for salt bridge formation and such distances were seen in mtre1, gtre1, and msctt. The atoms studied in gsctt were never close enough to form a salt bridge. As shown in Figure 7B, the nitrogen atoms of R135 in gsctt are not oriented towards the oxygen atoms of the aspartic acid residue in TM6 as they are in the gtre1 system (Figure 7A). The MD simulation of gtre1 shows interatomic N-O distances of 3.2 Å or less for 3 of the 4 possible N-O pairs throughout most of the simulation (NH1-OD1, NH2-OD1, NH2-OD2). N-O distances in mtre1 were consistently greater than 3.2 Å except for ~75 ns towards the end of the simulation (NH2-OD2). The differences in interatomic N-O distances between mtre1 and gtre1 could be due to different conformations of Tre1 being represented or due to inherent

differences in the initial protein structure predictions. msctt does not appear to be able to form a stable salt bridge using the alternative arginine, R135, as correct interatomic N-O distances are only transiently seen over the course of the simulation. To ensure the differences in interatomic N-O distances were not due to differences in the distance between TM3 and TM6 carbon backbones, the Cα-Cα distances were measured between the arginine (R134 or R135) in TM3 and aspartic acid (D266 or D258) in TM6 in all systems. Plots of the Cα-Cα distances over the course of the simulation showed that the distances between the residues studied here were similar (Additional file 3: Figure S3).

Discussion

The Tre1 GPCR is an important component of primordial germ cell migration in Drosophila [25-27]. In a severe partial loss-of-function allele of the *tre1* gene, *tre1*^{sctt}, proper primordial germ cell migration is disrupted. The Tre1^{sctt} protein is missing eight amino acids, RYILIACH, from the junction of the third transmembrane domain and second intracellular loop [27]. This study was performed to analyze how the loss of the amino acids RYILIACH may affect Tre1 structure.

Protein structure predictions were generated for Tre1$^+$ and Tre1^{sctt} using GPCR-ModSim [20] and GPCR-

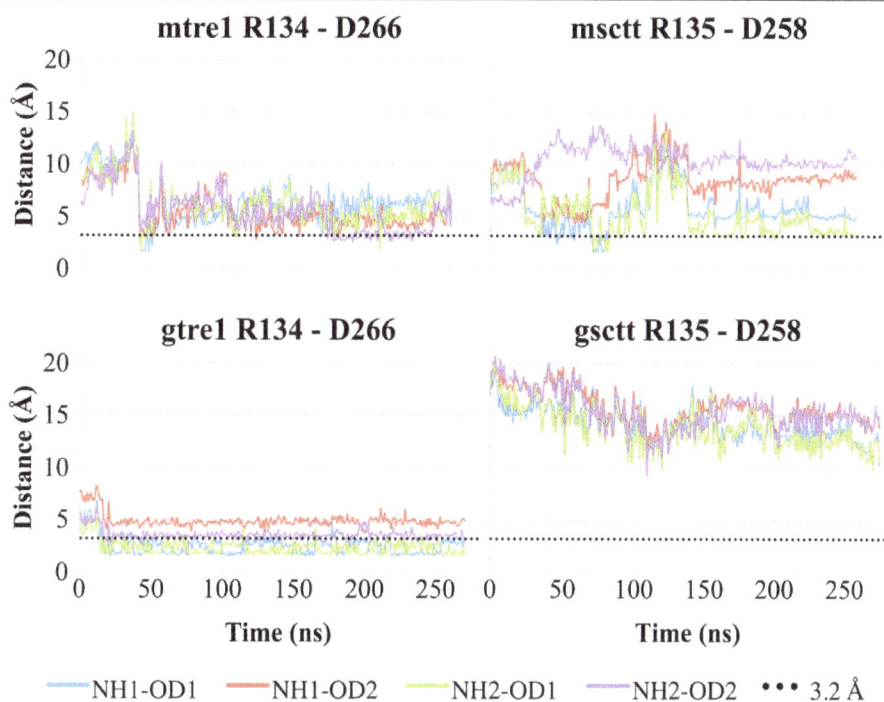

Figure 6 Distance calculations suggest a salt bridge could form in mtre1, msctt and gtre1 model systems. Distances between oxygen atoms of the carboxylate group of aspartic acid (D) (OD1, OD2) and nitrogen atoms of the guanidium group of arginine (R) (NH1, NH2) are plotted. N-O distances of 3.2 Å (dotted lines) or less are capable of forming a salt bridge. The N-O distances are close enough to form a salt bridge transiently in mtre1 and msctt. There is potential for a salt bridge to be consistently present in the gtre1 model system. The N-O distances are too great in gsctt to form a salt bridge at any point during the simulation.

ITASSER [21]. The four resulting structures were inserted into a POPE lipid bilayer and subjected to over 250 ns of MD simulations each. Interesting insights into the structures of Tre1$^+$ and Tre1^{sctt} were gained from this study. First, as shown by the RMSD versus time plots and the RMSF plots (Figures 3 and 4), Tre1$^+$ and Tre1^{sctt} behave similarly. The RMSD values for both Tre1$^+$ and Tre1^{sctt} protein structure predictions begin to stabilize around

150 ns of each MD run. Also, as shown by Figure 4, the general fluctuations of specific regions of the proteins are similar between Tre1$^+$ and Tre1^{sctt}.

The only primary sequence difference between Tre1$^+$ and Tre1^{sctt} is that Tre1^{sctt} is missing the eight amino acids RYILIACH. Of the eight amino acids, the arginine is the most conserved residue, being present in 96% of Rhodopsin family GPCRs [29]. The arginine and the

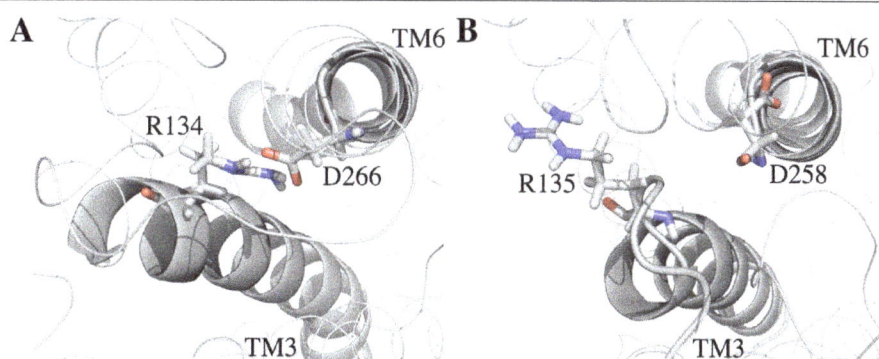

Figure 7 Orientation of the arginine in gsctt is not correct for salt bridge formation. Intracellular views of TM3 and TM6 are shown. The arginine and aspartic acid residues studied for formation of a salt bridge are shown as sticks, with the nitrogen atoms labeled blue and the oxygen atoms labeled red. TM3 and TM6 helices are represented as gray ribbons. **(A)** The gtre1 model system at 270 ns. In gtre1, R134 and D266 are close enough to form a salt bridge. Nitrogen – oxygen distances range from 2.4 – 4.9 Å. **(B)** The gsctt model system at 276 ns. The nitrogen atoms of R135 and the oxygen atoms of D266 are not facing each other, which prevents a salt bridge from forming between these residues. Nitrogen – oxygen distances range from 10.3 – 14.8 Å.

tyrosine are part of a highly conserved D/E/NRY motif in Rhodopsin family GPCRs [3,28]. The D/E/NRY motif is thought to have roles as a micro-switch, being involved in holding the GPCR in an active or inactive state by forming a salt bridge with an aspartic acid residue or a glutamic acid residue in TM6 [31-42]. Interestingly, the arginine is also the most critical residue of the amino acids RYILIACH for proper primordial germ cell migration in Drosophila. When the arginine is substituted by an alanine, a severe loss-of-function germ cell phenotype is observed [27]. The germ cell phenotype from the arginine substitution is indistinguishable from the phenotype when the amino acids RYILIACH are missing. Based upon this knowledge, it was hypothesized that a salt bridge involving the conserved D/E/NRY motif is present in Tre1[+] and absent in Tre1[sctt]. A salt bridge with the D/E/NRY motif could be important for maintaining Tre1[+] in a conformation required for efficient ligand binding. The lack of salt bridge in Tre1[sctt] could alter the protein conformation such that the ligand cannot recognize the receptor. It is also possible that an alternative salt bridge using a nearby arginine could be formed in Tre1[sctt]. If an alternative salt bridge forms, it could be involved in restoring some function of the GPCR.

The ability of the four model systems to form a salt bridge (Figure 6) was examined. Wild-type systems, mtre1 and gtre1, confirm that it is possible for a salt bridge to form between the arginine of the NRY motif (R134) and an aspartic acid (D266) in TM6. The salt bridge analysis using the mutant systems, msctt and gsctt, present a different picture. While it is possible for a salt bridge to form between the alternative arginine (R135) and the aspartic acid of TM6 (D258) in the msctt system, the salt bridge would not be very stable. Distances favorable for salt bridge formation were not consistently present during the simulation (Figure 6). It is clear from Figure 6 that no salt bridge would be expected to form in gsctt. It is possible that the salt bridge seen in gtre1 promotes interhelical stabilization of the protein, and this stabilization could be important for proper function of Tre1. The inability to form a stable salt bridge could disrupt Tre1[sctt] protein structure making it unable to properly receive its ligand, or could alter the confirmation of Tre1[sctt] such that it cannot bind interacting proteins. An alternative explanation for the salt bridge analysis results is that the systems have not been sufficiently sampled (Additional file 4: Table S1 and Additional file 5: Table S2), and a salt bridge could still form in the Tre1[sctt] forms of the protein.

With the salt bridge analysis there are significant differences between the independent model systems for Tre1[+] and Tre1[sctt]. While it is possible that the different model systems represent two different protein conformations of Tre1, it is also possible that these differences can be attributed to how the protein structure predictions were generated. mtre1 and msctt were built using GPCR-ModSim [20] which uses the homology modeling program MODELLER. gtre1 and gsctt were built using GPCR-ITASSER [21-23] which predicts protein structures through the use of threading. To further study Tre1 structure, a third independent structure of Tre1 could be built. For example, GPCR-Sequence Structure Feature Extractor (GPCR-SSFE) could be used to generate another starting structure. GPCR-SSFE is a database in addition to a homology modeling program that creates homology models of GPCRs using multiple templates and the program MODELLER [24,58]. The ability to use multiple templates is significant since the use of multiple templates with MODELLER has been shown to give more accurate homology models than using a single template [17].

Conclusions

In this study, the role of the arginine of the NRY motif in Tre1 was investigated. It is known from previous work that this arginine is critical to the proper function of the Tre1 GPCR in Drosophila primordial germ cell migration [27]. Whether or not it is important for Tre1 structure was unknown. The results presented here suggest that a salt bridge may form between this critical arginine and an aspartic acid in TM6 in Tre1.

GPCRs are a common class of proteins involved in cell migration. Similar to how Tre1 is involved in Drosophila primordial germ cell migration, another GPCR, CXCR4, is important for proper primordial germ cell migration in mouse, zebrafish and chicken [59-62]. Like Tre1, CXCR4 contains the highly conserved D/E/NRY motif. While a salt bridge with the arginine of the DRY motif was not present in the crystal structure of the human CXCR4 [63], it would be interesting to learn if the arginine of the DRY motif is important to the structure of mouse, zebrafish or chicken CXCR4 and what implications this would have on primordial germ cell migration. The importance of the arginine to both the function [27] and structure of Tre1 could also mean that the arginine of the DRY motif in CXCR4 is important for its structure and function.

Primordial germ cell migration is an important process to study as it serves as a model for cell migration. In many animals, the primordial germ cells are formed at a place distant to the presumptive gonads requiring the primordial germ cells to migrate to their target tissues. In order for the primordial germ cells to properly migrate to the presumptive gonads, the primordial germ cells are required to initiate migration, migrate through various tissues, evade or suppress cell death mechanisms and respond to directional cues. The study of primordial germ cell migration as a model for cell migration will help to better understand the mechanisms of cell movements, enabling the development of new techniques or approaches to treat cancer or other diseased states caused by improper cell migration.

Methods

Protein structure predictions

Protein structure predictions were generated for Tre1$^+$ (GenBank ID: AAF46059) and Tre1^{sctt} using GPCR-ModSim [20] and GPCR-ITASSER [21]. Using squid rhodopsin (PDB ID: 2Z73) as a template, ten homology models were generated for both Tre1$^+$ and Tre1^{sctt} by GPCR-ModSim. The best model (as judged by the lowest DOPE-HR score [47]) was chosen for further refinement of the loop regions. The DOPE-HR score is used to assess the quality of the models generated by examining the energy of the protein models. Five models were generated for each protein after loop refinement and again the best model was chosen based on DOPE-HR score.

Five models of both Tre1$^+$ and Tre1^{sctt} were generated by GPCR-ITASSER and the best model as judged by the confidence score (C-score) was chosen for this study. C-score is an estimation of the structural prediction and is based on the threading alignments from LOMETS and convergence during the structural refinements [11].

Building a system to reflect a Drosophila cellular membrane environment

The protein structure predictions (mtre1, msctt, gtre1 and gsctt) were embedded in a solvated (0.15 M NaCl) and pre-equilibrated POPE lipid bilayer using the Membrane Builder in the CHARMM-GUI [49,50]. For the mtre1 system, 101 Na$^+$ and 112 Cl$^-$ ions were added to neutralize the system and the system contained 37,557 water molecules. The upper and lower leaflets of the membrane contained 141 and 137 POPE lipids, respectively. The mtre1 system had a total of 153,870 atoms. The msctt system contained 110 Na$^+$ and 120 Cl$^-$ ions, 40,839 water molecules, 141 and 137 POPE lipids on upper and lower leaflets of the membrane, respectively, and 163,593 total atoms. The gtre1 system contained 69 Na$^+$ and 80 Cl$^-$ ions, 26,139 water molecules, 140 and 137 POPE lipids on upper and lower leaflets of the membrane, respectively, and 119,427 total atoms. The gsctt system contained 68 Na$^+$ and 78 Cl$^-$ ions, 26,102 water molecules, 139 and 140 POPE lipids on upper and lower leaflets of the membrane, respectively, and 119,423 total atoms.

Molecular dynamics simulations

MD simulations were performed using the NAMD 2.8 simulation package [45]. The CHARMM22 [64,65] and CHARMM36 [66] force fields were used for proteins and lipids, respectively, and water molecules were described

using TIP3P [67]. All systems were simulated at 310 K. Temperature and pressure were held constant with Langevin dynamics [45] and the Nose-Hoover Langevin piston [68,69]. Particle-mesh Ewald was used to calculate electrostatic interactions [70] and a 12 Å cut-off for van der Waals interactions was used. Each system was simulated on three compute nodes, each containing one Intel (R) Xeon(R) X5650 CPU (6 cores at 2.67 GHz), two Nvidia C2070 graphical processing units (GPUs) and 24 GB of RAM connected by QDR QLogic Infiniband.

After building the systems with the Membrane Builder in the CHARMM-GUI, six short (25 or 100 ps) equilibrium simulations were performed to gradually equilibrate the systems. Details for the equilibrium simulations can be found in reference [49]. Briefly, positional harmonic restraints were used on the protein backbone, protein side chains and ions. Additional harmonic restraints were used on the water molecules, to prevent water molecules from entering the hydrophobic region of the membrane, and the lipid head groups, to keep the lipid head groups level with the Z-axis. The restraints were reduced at each subsequent equilibrium simulation. The first two simulations used the NVT (constant volume and temperature) ensemble and the last four equilibrium simulations used the NPAT (constant pressure, area and temperature) ensemble. A timestep of 1 fs was used for the first three equilibrium simulations, which were 25 ps each. The last three equilibrium simulations used a 2 fs timestep and were run for 100 ps each [49]. Production runs began after the systems were equilibrated and used an NPT (constant pressure and temperature) ensemble and a 2 fs timestep. Harmonic restraints were not used in the production runs. Production runs were 262 ns, 258 ns, 270 ns, and 276 ns for mtre1, msctt, gtre1 and gsctt, respectively.

Data analysis

Visual Molecular Dynamics 1.9 (VMD) [71] was used to visualize the trajectories and to perform the all-to-all RMSD calculations and the salt bridge analysis. The Voronoi Tesselation and Monte Carlo (VTMC) integration method was used to calculate the surface area per lipid in all model systems [52] to ensure the systems maintained a biologically relevant, fluid phase lipid bilayer. The Dictionary of Secondary Structure in Proteins [56,57] with the do_dssp interface supplied by GROMACS [72] was used to calculate the evolution of secondary structures over time for each model.

Amino acid numbering

Amino acid residues are labeled using the single-letter code for the amino acid followed by the absolute sequence number. For example, arginine 134 is labeled R134. Tre1^{sctt} is missing eight amino acids compared to Tre1$^+$; however, the absolute sequence number of amino acid residues studied in this protein is still used. For example, an aspartic acid residue in TM6 is labeled D266 in Tre1$^+$ and D258 in Tre1^{sctt}.

Statistical analyses

Like any set of data, MD simulations are prone to statistical errors. The errors can be from inaccuracies in the model or inadequate sampling. For this reason, it is important to report the statistical uncertainty of values determined from simulations. In order to calculate the statistical uncertainty of different values in a simulation, the number of independent samples within a single simulation needs to be known. It has been suggested that estimation of a value of interest based on less than 20 statistically independent samples is considered unreliable [54]. To calculate the number of independent samples within a simulation, the decorrelation time must be calculated.

To calculate the decorrelation time, this study used the structural histogram analysis and automated effective sample size methods developed by the Zuckerman lab [73,74], as well as the block covariance overlap method (BCOM) from the Grossfield lab [53]. The effective sample size gives the degrees to which a simulation has sampled the conformational space of the protein and BCOM is a method used to measure the extent of convergence of a simulation. All of these tools are available through the LOOS (Lightweight Object-Oriented Structure) analysis library [75]. Only transmembrane Cα atoms were used in the decorrelation time calculations.

The decorrelation times as estimated by the structural histogram analysis and by the automated effective sample size are shown in Additional file 4: Table S1 and the decorrelation times calculated using the BCOM are shown in Additional file 5: Table S2. The results from the structural histogram analysis, the automated effective sample size calculation and the blocked covariance overlap method indicate that the systems have insufficient sampling and have not converged. This means that statistics generated from the data are not sufficient to draw statistically meaningful conclusions. This is not surprising. Microsecond simulations (or longer) with other GPCRs did not show convergence using these same methods [53]. Since the systems in this study have not converged, the values presented in this study represent a more qualitative assessment of the simulations.

Additional files

Additional file 1: Figure S1. RMSD of entire protein structures shows equilibration began at 150 ns. Description: Root mean squared deviation (RMSD) was calculated for each complete protein and is plotted over simulation time. The curves show the protein structures began to equilibrate after 150 ns.

Additional file 2: Figure S2. Evolution of protein secondary structure over time. Description: The secondary structure of the proteins in each of the model systems was calculated and plotted over the simulation time with the do_dssp interface supplied by GROMACS [72]. Residues with the same secondary structure are in the same color. These plots show that the transmembrane regions of the proteins (blue) remain stable throughout the simulations.

Additional file 3: Figure S3. Distances between Cα residues of TM3 and TM6 are similar in all model systems. Description: The distances between the Cα residues of R134 and D266 in Tre1$^+$ and R135 and D258 in Tre1^{sctt} were calculated and plotted over simulation time.

Additional file 4: Table S1. Approximate decorrelation times for the four different model systems. Description: $\tau_d 1$ is the decorrelation time as estimated from the plot of σ^2 (t) with step sizes 2, 4 and 5 [73]. $\tau_d 2$ is the decorrelation time from the automated effective sample size calculation [74]. Both calculations are part of the LOOS analysis library [75]. ^{a} 10 bins were used, ^{b} 20 bins were used.

Additional file 5: Table S2. Assessing convergence of the different model systems using the blocked covariance overlap method. Description: BCOM is the blocked covariance overlap method and BBCOM is the bootstrapped blocked covariance overlap. $t_1 - t_3$ are decorrelation times from fitting the BCOM/BBCOM curve to: $f(t) = k_1 e^{-t/t1} + k_2 e^{-t/t2} + k_3 e^{-t/t3} + 1$ [53]. The BCOM/BBCOM ratio decays to a final ratio of greater than 1 for each model system. This suggests that the systems have not yet converged. BCOM/BBCOM is part of the LOOS analysis library [75].

Abbreviations

GPCR: G protein-coupled receptor; Tre1: Trapped in endoderm-1; TM: Transmembrane; MD: Molecular dynamics; POPE: 1-palmitoyl-2-oleoyl-*sn*-glycero-3-phosphoethanolamine; GPCR-ITASSER: GPCR-Iterative Threading ASSEmbly Refinement; VTMC: Voronoi Tesselation and Monte Carlo integration method; RMSD: Root mean squared deviation; RMSF: Root mean squared fluctuation.

Competing interests

The authors declare that they have no competing interests.

Authors' contributions

MMP, MHL, and CRC conceived the study. MMP performed the simulations and data analysis. MMP, MHL, and CRC wrote the manuscript. All authors read and approved the final manuscript.

Acknowledgements

We thank Iowa State University for providing funds for the computer system used in this research. Support for MMP and CRC came from the Roy J. Carver Charitable Trust. We would also like to thank Dr. Spencer Pruitt for technical assistance and insightful discussions.

References

1. Fredriksson R, Lagerström MC, Lundin L-G, Schiöth HB: The G-protein-coupled receptors in the human genome form five main families. Phylogenetic analysis, paralogon groups, and fingerprints. *Mol Pharmacol* 2003, **63**(6):1256–1272.
2. Gloriam DE, Fredriksson R, Schiöth HB: The G protein-coupled receptor subset of the rat genome. *BMC Genomics* 2007, **8**:338.
3. Lagerström MC, Schiöth HB: Structural diversity of G protein-coupled receptors and significance for drug discovery. *Nat Rev Drug Discov* 2008, **7**(4):339–357.
4. Cotton M, Claing A: G protein-coupled receptors stimulation and the control of cell migration. *Cell Signal* 2009, **21**(7):1045–1053.
5. Dorsam RT, Gutkind JS: G-protein-coupled receptors and cancer. *Nat Rev Cancer* 2007, **7**(2):79–94.
6. Fredriksson R, Schiöth HB: The repertoire of G-protein-coupled receptors in fully sequenced genomes. *Mol Pharmacol* 2005, **67**(5):1414–1425.

7. Palczewski K, Kumasaka T, Hori T, Behnke C, Motoshima H, Fox B, Trong I, Teller D, Okada T, Stenkamp R, Yamamoto M, Miyano M: **Crystal structure of rhodopsin: a G protein-coupled receptor.** *Science* 2000, **289**:739–745.

8. Katritch V, Cherezov V, Stevens RC: **Structure-function of the G protein–coupled receptor superfamily.** *Annu Rev Pharmacol Toxicol* 2013, **53**(1):531–556.

9. Baker D, Sali A: **Protein structure prediction and structural genomics.** *Science* 2001, **294**(5540):93–96.

10. Bordoli L, Kiefer F, Arnold K, Benkert P, Battey J, Schwede T: **Protein structure homology modeling using SWISS-MODEL workspace.** *Nat Protoc* 2009, **4**(1):1–14.

11. Roy A, Kucukural A, Zhang Y: **I-TASSER: a unified platform for automated protein structure and function prediction.** *Nat Protoc* 2010, **5**(4):725–738.

12. Zhang Y, Devries ME, Skolnick J: **Structure modeling of all identified G protein-coupled receptors in the human genome.** *PLoS Comput Biol* 2006, **2**(2):e13.

13. Worth C, Kleinau G, Krause G: **Comparitive sequence and structural analyses of G-protein-coupled-receptor crystal structures and implications for molecular models.** *PLoS ONE* 2009, **4**(9):1–14.

14. Altschul SF, Madden TL, Schäffer AA, Zhang J, Zhang Z, Miller W, Lipman DJ: **Gapped BLAST and PSI-BLAST: a new generation of protein database search programs.** *Nucleic Acids Res* 1997, **25**(17):3389–3402.

15. Jones DT: **Protein secondary structure prediction based on position-specific scoring matrices.** *J Mol Biol* 1999, **292**(2):195–202.

16. Mobarec JC, Sanchez R, Filizola M: **Modern homology modeling of G-protein coupled receptors: which structural template to use?** *J Med Chem* 2009, **52**(16):5207–5216.

17. Sokkar P, Mohandass S, Ramachandran M: **Multiple templates-based homology modeling enhances structure quality of AT1 receptor: validation by molecular dynamics and antagonist docking.** *J Mol Model* 2010, **17**(7):1565–1577.

18. Zhang Y: **Progress and challenges in protein structure prediction.** *Curr Opin Struct Biol* 2008, **18**:342–348.

19. Zhang Y: **Protein structure prediction: when is it useful?** *Curr Opin Struct Biol* 2009, **19**(2):145–155.

20. Rodríguez D, Bello X, Gutiérrez-de-Terán H: **Molecular modelling of G protein-coupled receptors through the web.** *Molecular Informatics* 2012, **31**(5):334–341.

21. Zhang J, Zhang Y: **GPCRRD: G protein-coupled receptor spatial restraint database for 3D structure modeling and function annotation.** *Bioinformatics* 2010, **26**(23):3004–3005.

22. Zhang J, Liang Y, Zhang Y: **Atomic-level protein structure refinement using fragment-guided molecular dynamics conformation sampling.** *Structure* 2011, **19**(12):1784–1795.

23. Zhang J, Zhang Y: **A novel side-chain orientation dependent potential derived from random-walk reference state for protein fold selection and structure prediction.** *PLoS ONE* 2010, **5**(10):e15386.

24. Sali A, Blundell TL: **Comparative protein modelling by satisfaction of spatial restraints.** *J Mol Biol* 1993, **234**(3):779–815.

25. Kunwar P, Starz-Gaiano M, Bainton R, Heberlein U, Lehmann R: **Tre1, a GPCR, directs transepithelial migration of Drosophila germ cells.** *Plos Biol* 2003, **1**(3):372–384.

26. Kunwar PS, Sano H, Renault AD, Barbosa V, Fuse N, Lehmann R: **Tre1 GPCR initiates germ cell transepithelial migration by regulating Drosophila melanogaster E-cadherin.** *J Cell Biol* 2008, **183**:1–12.

27. Kamps AR, Pruitt MM, Herriges JC, Coffman CR: **An evolutionarily conserved arginine is essential for Tre1 G protein-coupled receptor function during germ cell migration in Drosophila melanogaster.** *PLoS ONE* 2010, **5**(7):e11839.

28. Nygaard R, Frimurer TM, Holst B, Rosenkilde MM, Schwartz TW: **Ligand binding and micro-switches in 7TM receptor structures.** *Trends Pharmacol Sci* 2009, **30**(5):249–259.

29. Mirzadegan T, Benkö G, Filipek S, Palczewski K: **Sequence analyses of G-protein-coupled receptors: similarities to rhodopsin.** *Biochemistry* 2003, **42**(10):2759–2767.

30. Ballesteros JA, Weinstein H: **Integrated methods for the construction of three-dimensional models and computational probing of structure-function relations in G protein-coupled receptors.** *Methods in Neuroscience* 1995, **25**:366–428.

31. Chien EYT, Liu W, Zhao Q, Katritch V, Han GW, Hanson MA, Shi L, Newman AH, Javitch JA, Cherezov V, Stevens RC: **Structure of the human dopamine D3 receptor in complex with a D2/D3 selective antagonist.** *Science* 2010, **330**(6007):1091–1095.

32. Aizaki Y, Maruyama K, Nakano-Tetsuka M, Saito Y: **Distinct roles of the DRY motif in rat melanin-concentrating hormone receptor 1 in signaling control.** *Peptides* 2009, **30**:974–981.

33. Ballesteros J, Kitanovic S, Guarnieri F, Davies P, Fromme BJ, Konvicka K, Chi L, Millar RP, Davidson JS, Weinstein H, Sealfon SC: **Functional microdomains in G-protein-coupled receptors. The conserved arginine-cage motif in the gonadotropin-releasing hormone receptor.** *J Biol Chem* 1998, **273**(17):10445–10453.

34. Ballesteros JA, Jensen AD, Liapakis G, Rasmussen SG, Shi L, Gether U, Javitch JA: **Activation of the beta 2-adrenergic receptor involves disruption of an ionic lock between the cytoplasmic ends of transmembrane segments 3 and 6.** *J Biol Chem* 2001, **276**(31):29171–29177.

35. Shapiro DA, Kristiansen K, Weiner DM, Kroeze WK, Roth BL: **Evidence for a model of agonist-induced activation of 5-hydroxytryptamine 2A serotonin receptors that involves the disruption of a strong ionic interaction between helices 3 and 6.** *J Biol Chem* 2002, **277**(13):11441–11449.

36. Vogel R, Mahalingam M, Ludeke S, Huber T, Siebert F, Sakmar T: **Functional Role of the "Ionic Lock" - an interhelical hydrogen-bond network in family a heptahelical receptors.** *J Mol Biol* 2008, **360**:648–655.

37. Zhu SZ, Wang SZ, Hu J, El-Fakahany EE: **An arginine residue conserved in most G protein-coupled receptors is essential for the function of the m1 muscarinic receptor.** *Mol Pharmacol* 1994, **45**(3):517–523.

38. Scheer A, Costa T, Fanelli F, DeBenedetti P, Mhaouty-Kodja S, Abuin L, Nenniger-Tosato M, Cotecchia S: **Mutational analysis of the highly conserved arginine within the Glu/Asp-Arg-Tyr Motif of the alpha1b-adrenergic receptor: effects on receptor isomerization and activation.** *Mol Pharmacol* 2000, **57**:219–231.

39. Scheer A, Fanelli F, Costa T, De Benedetti PG, Cotecchia S: **Constitutively active mutants of the alpha 1B-adrenergic receptor: role of highly conserved polar amino acids in receptor activation.** *EMBO J* 1996, **15**(14):3566–3578.

40. Greasley PJ, Fanelli F, Rossier O, Abuin L, Cotecchia S: **Mutagenesis and modelling of the alpha(1b)-adrenergic receptor highlight the role of the helix 3/helix 6 interface in receptor activation.** *Mol Pharmacol* 2002, **61**(5):1025–1032.

41. Capra V, Veltri A, Foglia C, Crimaldi L, Habib A, Parenti M, Rovati GE: **Mutational analysis of the highly conserved ERY motif of the thromboxane A2 receptor: alternative role in G protein-coupled receptor signaling.** *Mol Pharmacol* 2004, **66**(4):880–889.

42. Greasley PJ, Fanelli F, Scheer A, Abuin L, Nenniger-Tosato M, DeBenedetti P, Cotecchia S: **Mutational and computational analysis of the alpha1b-adrenergic receptor: involvement of basic and hydrophobic residues in receptor activation and G protein coupling.** *J Biol Chem* 2001, **276** (49):46485–46494.

43. Fanelli F, De Benedetti PG: **Computational modeling approaches to structure – function analysis of G protein-coupled receptors.** *Chem Rev* 2011, **111**(9):R438–PR535.

44. Rosenbaum DM, Cherezov V, Hanson MA, Rasmussen SGF, Thian FS, Kobilka TS, Choi HJ, Yao XJ, Weis WI, Stevens RC, Kobilka BK: **GPCR engineering yields high-resolution structural insights into 2-adrenergic receptor function.** *Science* 2007, **318**(5854):1266–1273.

45. Phillips J, Braun R, Wang W, Gumbart J, Tajkhorshid E, Villa E, Chipot C, Skeel R, Kale L, Schulten K: **Scalable molecular dynamics with NAMD.** *J Comput Chem* 2005, **26**(16):1781–1802.

46. Jones H, Harwood J, Bowen I, Griffiths G: **Lipid composition of subcellular membranes from larvae and prepupae of Drosophila melanogaster.** *LIPIDS* 1992, **27**(12):984–987.

47. Shen M-Y, Sali A: **Statistical potential for assessment and prediction of protein structures.** *Protein Sci* 2006, **15**:2507–2524.

48. Murakami M, Kouyama T: **Crystal structure of squid rhodopsin.** *Nature* 2008, **453**(7193):363–367.

49. Jo S, Kim T, Im W: **Automated builder and database of protein/membrane complexes for molecular dynamics simulations.** *PLoS ONE* 2007, **2**(9):e880.

50. Jo S, Lim JB, Klauda JB, Im W: **CHARMM-GUI membrane builder for mixed bilayers and its application to yeast membranes.** *Biophysj* 2009, **97**(1):50–58.

51. Rappolt M, Hickel A, Bringezu F, Lohner K: **Mechanism of the Lamellar: inverse hexagonal phase transition examined by high resolution x-ray diffraction.** *Biophys J* 2003, **84**:3111–3122.

52. Mori T, Ogushi F, Sugita Y: **Analysis of lipid surface area in protein-membrane systems combining Voronoi Tessellation and Monte Carlo integration methods.** *J Comput Chem* 2012, **33**(3):286–293.

53. Romo TD, Grossfield A: **Block covariance overlap method and convergence in molecular dynamics simulation.** *J Chem Theory Comput* 2011, **7**(8):2464–2472.

54. Grossfield A, Zuckerman DM: **Quantifying uncertainty and sampling quality in biomolecular simulations.** *Ann Rep Comput Chem* 2009, **5**:23–48.

55. Romo TD, Grossfield A, Pitman MC: **Concerted interconversion between ionic lock substates of the beta(2) adrenergic receptor revealed by microsecond timescale molecular dynamics.** *Biophys J* 2010, **98**(1):76–84.

56. Kabsch W, Sander C: **Dictionary of protein secondary structure: pattern recognition of hydrogen-bonded and geometrical features.** *Biopolym* 1983, **22**:2577–2637.

57. Joosten R, Te Beek T, Krieger E, Hekkelman M, Hooft R, Schneider R, Sander C, Vriend G: **A series of PDB related databases for everyday needs.** *Nucleic Acids Res* 2011, **39**:D411–D419.

58. Worth CL, Kreuchwig A, Kleinau G, Krause G: **GPCR-SSFE: a comprehensive database of G-protein-coupled receptor template predictions and homology models.** *BMC Bioinforma* 2011, **12**:185.

59. Ara T, Nakamura Y, Egawa T, Sugiyama T, Abe K, Kishimoto T, Matsui Y, Nagasawa T: **Impaired colonization of the gonads by primordial germ cells in mice lacking a chemokine, stromal cell-derived factor-1 (SDF-1).** *Proc Natl Acad Sci USA* 2003, **100**(9):5319–5323.

60. Doitsidou M, Reichman-Fried M, Stebler J, Koprunner M, Dorries J, Meyer D, Esguerra C, Leung T, Raz E: **Guidance of primordial germ cell migration by the chemokine SDF-1.** *Cell* 2002, **111**:647–659.

61. Knaut H, Werz C, Geisler R, Consortium TTS, Nüsslein-Volhard C: **A zebrafish homologue of the chemokine receptor Cxcr4 is a germ-cell guidance receptor.** *Nature* 2003, **421**(6920):279–282.

62. Molyneaux KA, Zinszer H, Kunwar P, Schaible K, Stebler J, Sunshine M, O'Brien W, Raz E, Littman D, Wylie C, Lehmann R: **The chemokine SDF1/CXCL12 and its receptor CXCR4 regulate mouse germ cell migration and survival.** *Development* 2003, **130**(18):4279–4286.

63. Wu B, Chien EYT, Mol CD, Fenalti G, Liu W, Katritch V, Abagyan R, Brooun A, Wells P, Bi FC, Hamel DJ, Kuhn P, Handel TM, Cherezov V, Stevens RC: **Structures of the CXCR4 Chemokine GPCR with Small-Molecule and Cyclic Peptide Antagonists.** *Science* 2010, **330**(6007):1066–1071.

64. MacKerell A, Bashford D, Bellott M, Dunbrack R, Evanseck J, Field M, Fischer S, Gao J, Guo H, Ha S, Joseph-McCarthy D, Kuchnir L, Kuczera K, Lau F, Mattos C, Micknick S, Ngo T, Nguyen D, Prodhom B, Reiher W, Roux B, Schlenkrich M, Smith J, Stote R, Straub J, Watanabe M, Wiorkiewicz-Kuezera J, Yin D, Karplus M: **All-Atom empirical potential for molecular modeling and dynamics studies of proteins.** *J Phys Chem B* 1998, **102**:3586–3616.

65. MacKerell AD Jr: **Empirical force fields for biological macromolecules: overview and issues.** *J Comput Chem* 2004, **25**(13):1584–1604.

66. Klauda JB, Venable RM, Freites JA, O'Connor JW, Tobias DJ, Mondragon-Ramirez C, Vorobyov I, MacKerell Jr AD, Pastor RW: **Update of the CHARMM all-atom additive force field for lipids: validation on six lipid types.** *J Phys Chem B* 2010, **114**:7830–7843.

67. Jorgensen W, Chandrasekhar J, Madura J, Impey R, Klein M: **Comparison of simple potential functions for simulating liquid water.** *J Chem Phys* 1983, **79**(2):926–935.

68. Feller S, Zhang Y, Pastor R, Brooks B: **Constant pressure molecular dynamics simulation: the Langevin piston method.** *J Chem Phys* 1995, **103**(11):4613–4621.

69. Nosé S: **A unified formulation of the constant temperature molecular dynamics methods.** *J Chem Phys* 1984, **81**(1):511.

70. Darden T, York D, Pedersen L: **Particle mesh Ewald: an Nlog(N) method for Ewald sums in large systems.** *J Chem Phys* 1992, **98**(12):10089–10092.

71. Humphry W, Dalke A, Schulten K: **VMD: Visual Molecular Dynamics.** *J Mol Graph* 1996, **14**:33–38.

72. Pronk S, Pall S, Schulz R, Larsson P, Bjelkmar P, Apostolov R, Shirts M, Smith J, Kasson P, Van der Spoel D, Hess B, Lindahl E: **GROMACS 4.5: a high-throughput and highly parallel open source molecular simulation toolkit.** *Bioinform* 2013, **29**(7):845–854.

73. Lyman E, Zuckerman DM: **On the structural convergence of biomolecular simulations by determination of the effective sample size.** *J Phys Chem B* 2007, **111**(44):12876–12882.

74. Zhang X, Bhatt D, Zuckerman DM: **Automated sampling assessment for molecular simulations using the effective sample size.** *J Chem Theory Comput* 2010, **6**(10):3048–3057.

75. Romo T, Grossfield A: **LOOS: an extensible platform for the structural analysis of simulations.** *Conf IEEE EMBS* 2009, **2009**:2332–2335.

2

Structural insight into the recognition of amino-acylated initiator tRNA by eIF5B in the 80S initiation complex

Bernhard Kuhle[*] and Ralf Ficner

Abstract

Background: From bacteria to eukarya, the specific recognition of the amino-acylated initiator tRNA by the universally conserved translational GTPase eIF5B/IF2 is one of the most central interactions in the process of translation initiation. However, the molecular details, particularly also in the context of ribosomal initiation complexes, are only partially understood.

Results: A reinterpretation of the 6.6 Å resolution cryo-electron microscopy (cryo-EM) structure of the eukaryal 80S initiation complex using the recently published crystal structure of eIF5B reveals that domain IV of eIF5B forms extensive interaction interfaces with the Met-tRNA$_i$, which, in contrast to the previous model, directly involve the methionylated 3' CCA-end of the acceptor stem. These contacts are mediated by a conserved surface area, which is homologous to the surface areas mediating the interactions between IF2 and fMet-tRNA^{fMet} as well as between domain II of EF-Tu and amino-acylated elongator tRNAs.

Conclusions: The reported observations provide novel direct structural insight into the specific recognition of the methionylated acceptor stem by eIF5B domain IV and demonstrate its universality among eIF5B/IF2 orthologs in the three domains of life.

Keywords: Ribosome, Translation initiation, Subunit joining, Initiator tRNA, eIF5B/IF2, Structure, Protein evolution

Background

The process of translation initiation results in the formation of an elongation-competent ribosome with the start codon of an mRNA in its P site, base paired to the amino-acylated initiator tRNA. In bacteria and eukarya this process follows significantly different mechanisms, highlighted by different numbers of auxiliary protein factors (initiation factors or IFs) that are employed by bacterial (three IFs) or eukaryal cells (at least 12 eIFs) for correct ribosome assembly [1]. Only two of these factors, a/eIF1A/IF1 and the translational GTPase a/eIF5B/IF2, are universally conserved in the three domains of life [2]. In bacteria, IF2 plays a critical role throughout the initiation pathway. In the early stages, IF2 binds to the 30S subunit in a GTP-dependent manner and stimulates the recruitment of the N-formylmethionylated initiator

tRNA (fMet-tRNA^{fMet}) to the P site of the 30S ribosomal subunit to form the 30S pre-initiation complex (pre-IC). Finally, IF2·GTP catalyzes the joining of the 50S ribosomal subunit to form the elongation-competent ribosome [3,4]. Speed and accuracy of both processes depend of the specific recognition of the αNH-blocked methionine esterified to the 3' CCA-end of tRNA^{fMet} [5-8]. Biochemical studies showed that all determinants required for this interaction are located in domain IV of IF2, which consists of a six-stranded β barrel [9-12]. Domain IV of IF2 exhibits a marked structural homology to domain II of EF-Tu that, together with the G domain, forms the universally conserved structural core among translational GTPases [13] and in EF-Tu constitutes part of the binding pocket for the amino-acylated acceptor arm of elongator tRNAs [12,14,15]. Based on this observation it was suggested that IF2 domain IV and EF-Tu domain II use similar interfaces for their interactions with the tRNA [12]. This assumption is at least partially corroborated by mutational and NMR spectroscopy

* Correspondence: bkuhle@gwdg.de
Abteilung für Molekulare Strukturbiologie, Institut für Mikrobiologie und Genetik, Göttinger Zentrum für Molekulare Biowissenschaften, Georg-August-Universität Göttingen, D-37077 Göttingen, Germany

analyses [11]. Cryo-EM structures of bacterial 30S pre-ICs and 70S IC containing GTP/GDPNP-bound IF2 show how this interaction mutually stabilizes fMet-tRNA^{fMet} and IF2 in conformations that allow the efficient association of the 50S subunit [16,17]. However, none of these structures were determined at sufficiently high resolution to give any detailed insight into the interaction that would allow a correlation with the biochemical data.

In contrast to bacterial IF2, the role of a/eIF5B in eukarya and archaea seems to be confined to the GTP-dependent promotion of subunit joining, the last step of the initiation process [18-20]. The recruitment of the charged initiator tRNA (Met-tRNA$_i$) to the small ribosomal subunit is carried out by the heterotrimeric a/eIF2, a specialized EF-Tu paralog that has no counterpart in bacteria [21]. Accordingly, a/eIF5B·GTP binds to the small ribosomal subunit already containing the P site-bound Met-tRNA$_i$, which invokes the question whether a/eIF5B still has to interact with Met-tRNA$_i$ to promote joining of the large ribosomal subunit, and whether this interaction would involve a specific recognition of the methionylated acceptor end, similar to the recognition of the fMet-tRNA^{fMet} by IF2. Genetic, biochemical and structural studies point toward essentially the same mechanisms for eIF5B and IF2 catalyzed subunit joining [18,19,22-25]. Crystal structures of aIF5B and eIF5B revealed a six-stranded β barrel fold for domain IV, homologous to domain IV in IF2 [22,24]. Indirect biochemical assays showed that a/eIF5B binds Met-tRNA$_i$ in solution, however, with very low affinity and specificity for the methionyl moiety in case of eIF5B [22,26]. NMR studies revealed a weak but specific interaction between methionine-ethyl ester (mimicking the ester bond between tRNA and the methionly moiety) and eIF5B domain IV in the area corresponding to the surface on IF2 that is affected by N-formylmethionine binding [27]. Finally, the recently determined 6.6 Å resolution cryo-EM structure of the yeast 80S IC (EM-Databank: EMD-2422) demonstrates that, like IF2 in the corresponding bacterial 70S complex [16], eIF5B and the P site-bound Met-tRNA$_i$ stabilize each other in their subunit joining-competent conformations through the direct contact between domain IV and acceptor stem [25]. Surprisingly however, according to the structural model this contact does not involve the methionylated 3′ CCA-end of the tRNA [25]. Instead, the CCA-end points away from domain IV, placing the methionyl moiety ∼ 23 Å from the protein. Thus, this model is clearly at odds with the observations from biochemical studies [26,27] and fails to explain why deacylation of the initiator tRNA results in the loss of its ability to stabilize eIF5B [25].

Here, we provide an analysis and reinterpretation for the cryo-EM density of the yeast 80S IC [25] for domain IV and its contact to the initiator tRNA. We show that the original structural model for this region, based on the fit of the archaeal aIF5B ortholog, is only partially consistent with the available density. Fitting of the recently determined structure of eIF5B domain IV from C. thermophilum, which shows a significantly higher degree of sequence similarity to the S. cerevisiae ortholog, allows a reinterpretation of the 6.6 Å resolution density. The resulting model demonstrates a direct contact between the methionylated CCA-end of the tRNA and a conserved surface area of domain IV that directly corresponds to the binding sites for the tRNA acceptor arm on domain IV of IF2 or domain II of EF-Tu [11,12,14]. Thus, we show that the high-quality cryo-EM density of the 80S complex not only provides the first direct structural indications for the EF-Tu-like interactions between eIF5B/IF2 domain IV and the initiator tRNA but also for their universality among a/eIF5B/IF2 orthologs in the three domains of life. Finally, we use our observations to propose a possible scenario for the evolution of the translational β barrel fold in eIF5B/IF2 and EF-Tu and its interactions with tRNAs.

Results and discussion

Model of eIF5B domain IV and the acceptor end of Met-tRNA$_i$ in the 80S IC

Recently, we were able to solve two structures of eIF5B domain IV from the fungus C. thermophilum [24]. It consists of six antiparallel β strands (β1-β6) forming a closed β barrel that is followed by two α helices (Figure 1A). At its top and bottom, the β barrel is closed by an additional short β strand (β^{L4}) and a one-turn α helix (α^{L5}), respectively. Despite relatively low sequence similarity, it is structurally very similar to domain IV of aIF5B from the archaeon M. thermoautotrophicum (rmsd of 2.2 Å with ∼ 20% sequence identity and ∼ 30% similarity). However, in the cteIF5B ortholog the β hairpin formed by β strands 3 and 4 and the loop following strand β5 (L5) contain 9 and 5 additional amino acids, respectively (Figure 1A/B). Further differences can be found in the organization of the two C-terminal α helices that are rotated by ∼ 25° with respect to the β barrel.

Compared to the archaeal ortholog, domain IV from cteIF5B shows a relatively high sequence similarity to the yeast ortholog (19% sequence identity and 30% similarity for mtaIF5B compared to 49% identity and 65% similarity for cteIF5B) including the β3-β4 hairpin, L5 and the two C-terminal α helices. Based on this observation, we assumed that the cteIF5B structure allows a better fit to the recently determined cryo-EM density of the yeast 80S IC with initiator tRNA and eIF5B·GDPCP than obtained with the mtaIF5B structure [25] (Figure 1C). Rigid-body fitting of cteIF5B domain IV (cross-correlation coefficient (CCC) of 73%) results in an improved correlation between structural model and density (Figure 1D):

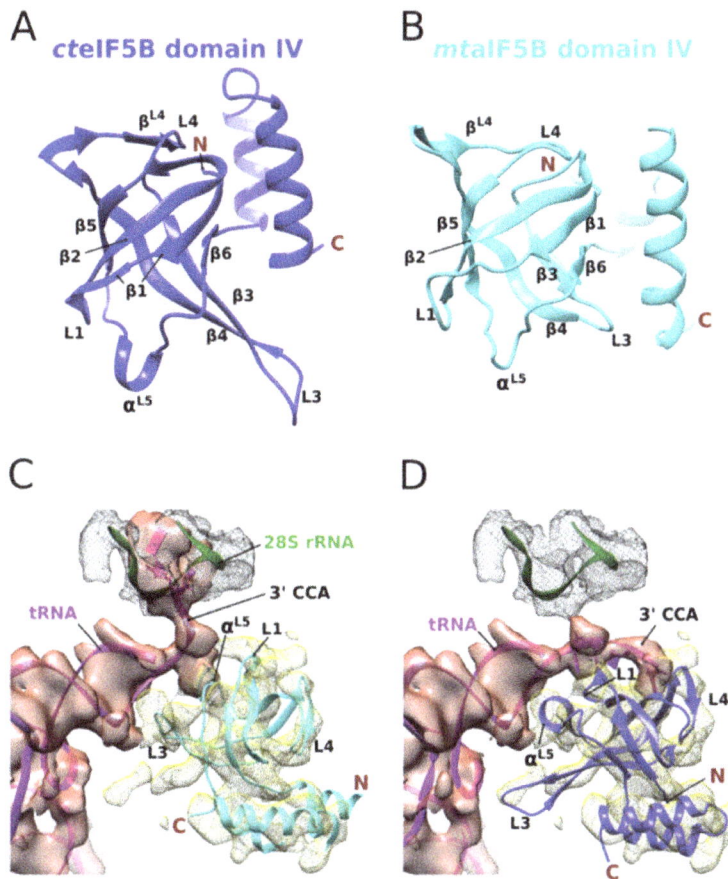

Figure 1 Model for the interactions between eIF5B domain IV and Met-tRNA_i in the 80S IC. A) Crystal structure of cteIF5B domain IV (PDB: 4N3G). The most marked differences to domain IV of aIF5B from the archaeon *M. thermoautotrophicum* (PDB: 1G7T) **(B)** are found in the lengths of the β3-β4 hairpin and loop L5 as well as in the arrangement of the two C-terminal α helices. **C)** Original model for the interactions between domain IV (cyan) and Met-tRNA_i (purple), fitted into the cryo-EM density of the 80S IC (EMD-2422) [25]. **D)** New model for the interactions between domain IV (blue) and Met-tRNA_i (purple), based on the rigid-body fitting of the crystal structure of domain IV from cteIF5B. The 3' CCA-end now forms a direct contact with the surface of domain IV, and α^{L5} occupies a position in the major groove of the acceptor stem.

In contrast to the original fit of *mt*aIF5B [25], no clashes occur between the ribosomal RNA and *ct*eIF5B domain IV, as the loop between β strands 1 and 2 (L1) is moved away from C2284-U2286 and now lies next to the acceptor stem of the tRNA. Compared to the original model [25] the β barrel is rotated by ~ 30°, causing the conserved helix α^{L5} at the bottom of the β barrel to displace the β3-β4 hairpin in the major groove of the initiator tRNA acceptor stem. In turn, the long, poorly conserved β3-β4 hairpin now occupies previously unexplained density close to the C-terminus of the last α helix and forms apparently no direct contacts to the tRNA (Figure 1D).

An interesting consequence of this reorganization of domain IV is the emergence of a well defined but unexplained density packed alongside the β barrel, directly opposite to the C-terminal α helices (Figure 1D). This density starts next to the very C-terminus of β strand 4 and the following loop (L4) and runs across strands β5, β2 and finally β1 where it directly leads into the continuous

density of the phosphate backbone of the initiator tRNA at A73. Interestingly, this same position (A73) also marks the starting point for the distortion of the following 3' CCA-end in the original model that is markedly different from its canonical conformation [25]. For the following reasons, it is unlikely that this original model gives the correct conformation for the tRNA acceptor arm in the 80S pre-IC: First, C75 and A76 clash extensively with the ribosomal RNA between G2615 and C2625 (Figure 1C). Second, the CCA-end is oriented away from eIF5B domain IV, resulting in a distance of ~ 23 Å between the ribose of A76 (which carries the methionyl moiety) and the nearest parts of domain IV. This is clearly inconsistent with the observation that deacylation of the tRNA results in the loss of its contact to eIF5B in the 80S complex [25] and is at odds with the expected direct contact between the methionyl moiety and domain IV [19,26,27]. Remodeling of the 3' CCA-end into the vacant density next to the β barrel of eIF5B avoids the clashes with the rRNA and, moreover, allows a direct recognition of the aminoacyl group by the protein by placing the 3' end of the tRNA directly on top of the conserved Ala1056 of strand β5 (Figure 1D). It is important to note, that there is no alternative density present that could accommodate the entire CCA-end without causing a sterical conflict with the rRNA. Independent support for this new placement of the CCA-end is provided by the just recently published lower resolution (8–9 Å) cryo-EM model of the mammalian 80S pre-IC with eIF5B bound on HCV-IRES RNA, which suggests a direct contact between the acceptor end of the Met-tRNA$_i$ and domain IV of eIF5B [28].

As reported previously for *mt*aIF5B [22], the β barrel of eIF5B is structurally homologous to domain IV (C2 domain) of bacterial IF2 [12] and domain II of EF-Tu homologs, despite an overall low sequence similarity (Figures 2 and 3). Using site directed mutagenesis and NMR spectroscopy, it was shown that IF2 interacts with the αNH-formylmethionylated CCA-end of fMet-tRNA^{fMet} through a surface of domain IV that overlaps with that used by EF-Tu domain II to interact with the acceptor end of elongator tRNAs [11,12,14,15]. The superposition of domain IV of *cte*IF5B with domain IV of IF2 from *Bacillus*

stearothermophilus [12] and domain II of EF-Tu from *Thermus aquaticus* in complex with Phe-tRNA^{Phe} [14] reveals that these surface areas coincide perfectly with those occupied by the CCA-end in our cryo-EM density-based model (Figure 2). Consistently, a structure-based sequence alignment reveals the highest degree of sequence conservation between the eIF5B orthologs and EF-Tu domain II in those residues implicated in tRNA binding in IF2 and EF-Tu (Figure 3).

The analysis of the surface area in *cte*IF5B reveals two pockets next to the modeled 3' end of the tRNA (Figures 2 and 4). The first formed by Val999, Gly1037, Glu1039 and the aliphatic part of Lys1058, corresponding to the EF-Tu residues Val237, Gly269, Glu271 and Leu289, respectively, which accommodate the base of A76 in Phe-tRNA^{Phe} [14]. A similar pocket is found on the surface of IF2, whose residues are directly affected by fMet-tRNA^{fMet} binding [11,12]. The second pocket is separated from the first by the methyl group of Ala1056 (corresponding to the conserved Gly287 in EF-Tu and Gly715 in *bs*IF2) and is formed on the one side by the hydrophobic Val989, Ala990, Phe992, Gly1001 and Ala1054 and on the other by the peptide backbone of Glu1039 to His1042. The

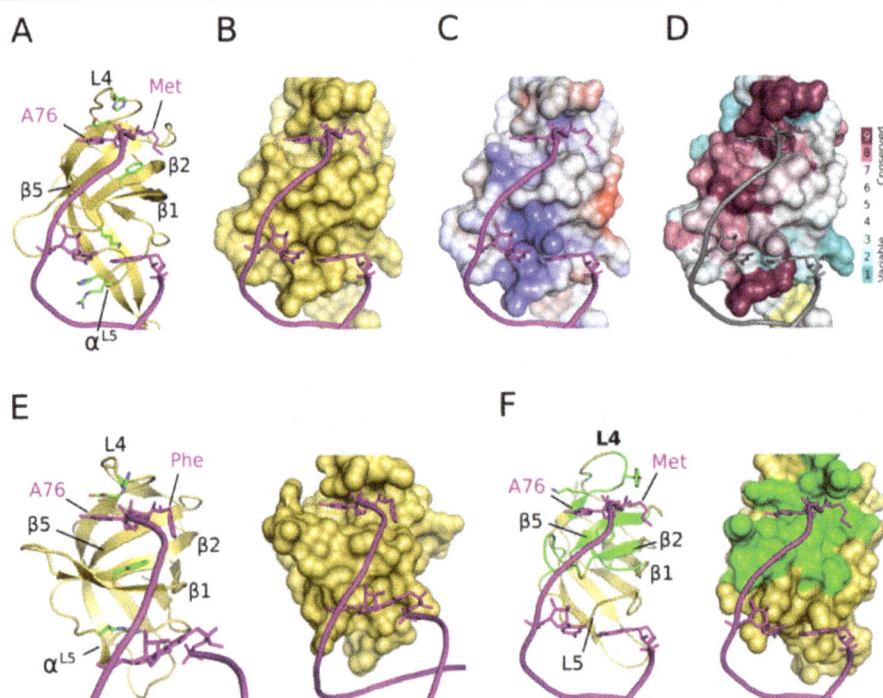

Figure 2 The interaction interface between the acceptor stem of Met-tRNA$_i$ and domain IV of eIF5B. Interactions between Met-tRNA$_i$ (purple) and eIF5B domain IV **(A-D)**, Phe-tRNA^{Phe} and domain II of EF-Tu (PDB: 1TTT) **(E)** and fMet-tRNA^{fMet} and IF2 domain IV **(F)**. **A)** Domain IV of eIF5B in ribbon presentation (yellow) with residues potentially involved in interactions with the tRNA as green sticks. **B)** Surface presentation of domain IV, revealing the two well defined pockets below loop L4 that are also visible on the surfaces of EF-Tu domain II (E, right) and IF2 domain IV (F, right), and might accommodate A76 and the methionyl moiety of the 3' CCA-end. **C)** Electrostatic surface potential of domain IV. **D)** Conservation of residues lying in the proposed interaction interface to the acceptor stem. **E)** Position of the acceptor stem of Phe-tRNA^{Phe} (purple) on the surface of domain II of EF-Tu (yellow; PDB: 1TTT). **F)** Model of domain IV of IF2 (PDB: 1D1N) and the initiator tRNA positioned as in **A)**. The green surfaces indicate residues of IF2 that were shown to interact with fMet-tRNA^{fMet} [11,12].

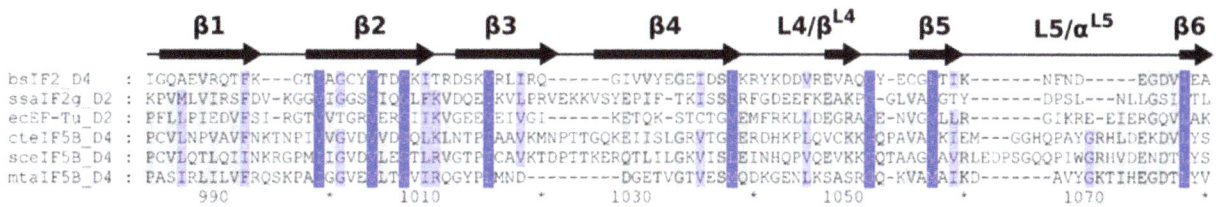

Figure 3 Partial structure-based sequence alignment of the β barrel fold of domain IV (D4). The aligned sequences are from *B. stearothermophilus* IF2, *M. thermoautotrophicum* aIF5B, *C. thermophilum* eIF5B and *S. cerevisiae* eIF5B with domain II (D2) from *E. coli* EF-Tu and *S. solfataricus* aIF2γ (GenBank: CAA27987, AAB84765, EGS21143, AAC04996, CAA40370, AAK40740, respectively). Highly conserved residues are highlighted in dark blue, conserved residues in light blue and similar residues in grey. Sequence numbering and secondary structure elements correspond to the *cte*IF5B structure (PDB: 4N3G). As there is no structure of *sce*IF5B domain IV available so far, its sequence was aligned directly with that of *cte*IF5B.

position of this pocket corresponds to the localization of the aminoacyl groups in ternary complexes of EF-Tu, and residues of this area were found to interact specifically with N-formylmethionine in IF2 [11] and methionine-ethyl ester in eIF5B [27]. Consistently, this second pocket is compatible with the binding of a methionyl moiety in size as well as electrostatic surface properties (Figures 2 and 4). Notably, in both available crystal structures of *cte*IF5B domain IV, this pocket is occupied by a large additional electron density. Due to the absence of alternatives in the crystallization condition (100 mM MES, 12% PEG

20000, 10 mM Na-lactate; ethylene glycol was used for cryo protection), this density was originally assigned to a lactate molecule [24]. However, refinement with the lactate molecule still results in positive difference electron density. A simulated annealing omit map for this area gives a density too large for a lactate (Figure 4B). Thus, the density would be compatible with the size of a methionine or other similarly large amino-acids whose α-carboxylate and α-amino groups form hydrogen bonds to the amide proton of Asp1041 and the main chain CO of Ala1054, respectively, corresponding to His273 and

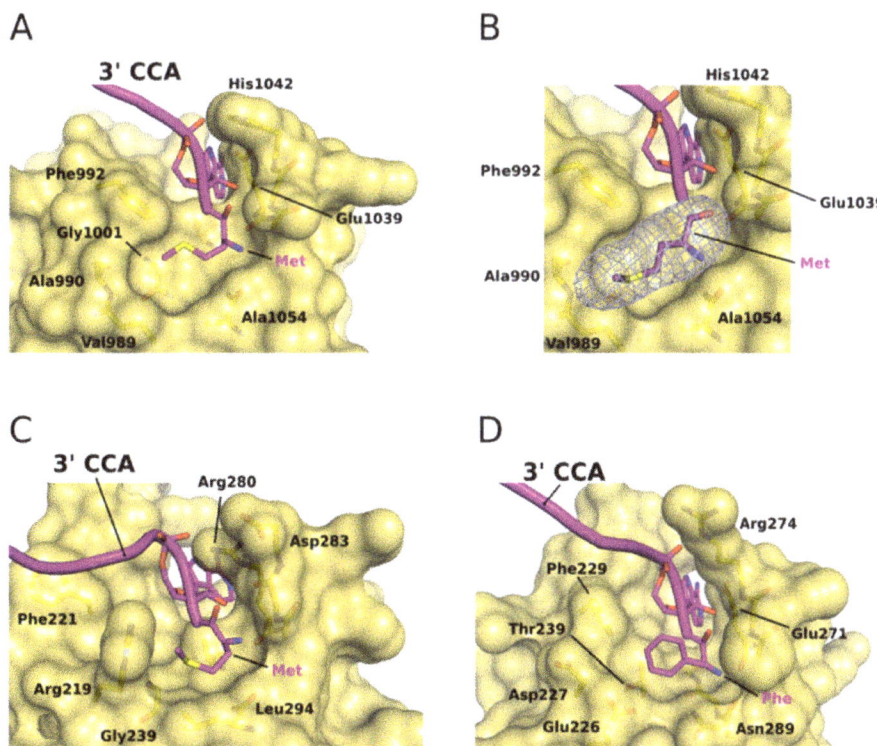

Figure 4 Recognition of the methionylated 3' CCA-end by eIF5B domain IV. A) CCA arm, A76 and the methionly moiety of Met-tRNA$_i$ modeled into the two surface pockets on eIF5B domain IV according to the cryo-EM density. Residues implicated in the interactions with the 3' end are indicated. **B)** Simulated annealing f$_o$-f$_c$ omit map for the putative methionyl-binding pocket in eIF5B domain IV (blue mesh; contoured at 3σ). **C)** Met-tRNA$_i$ bound to aIF2γ (PDB: 3V11). **D)** Phe-tRNA^{Phe} bound to EF-Tu (PDB: 1TTT).

Asn285 that form similar contacts to the aminoacyl group in ternary complexes of EF-Tu [14,29] (Figure 4D). However, the resolution of the structures (2.75 and 3.02 Å) necessarily does not allow an unambiguous assignment of the densities to a certain ligand, and a possible origin for a putative amino acid in this position remains elusive, as the weak binding between IF2 and fMet or eIF5B and methionine-ethyl ester [11,27] makes a co-purification unlikely. The critical point, however, is the observation that the described pocket is evidently suited to accommodate organic molecules of a size similar to that of methionine and could thus accommodate the methionyl moiety of the Met-tRNA$_i$ in a way analogous to domains II in EF-Tu and aIF2γ (Figure 4).

In ternary complexes of EF-Tu, the binding site for the 3′ CCA-end on domain II is complemented by the conserved Phe229 in strand β1 that stacks against C4 and C5 of the A76 ribose and by Arg274 (sometimes Gln or Lys) in the flexible loop following strand β4 that interacts with the phosphate of A76 [14,29] (Figure 4C). The density assigned to the CCA-end of the Met-tRNA$_i$ suggests similar interactions for the conserved Phe992 (in few cases Tyr or Ile) and His1042 in cteIF5B (Figure 4A). According to the model, the rest of the acceptor stem of the tRNA adopts a slightly different orientation relative to the β barrel than observed for aa-tRNA bound to EF-Tu. In good agreement with the predictions made for IF2 [11], C75 and C74 seem to be rotated ~ 20° toward the L1 loop (Figure 2). Interestingly, the orientation of the β barrel would allow several positively charged residues to interact directly with the acceptor stem. Lys994 in the L1 loop could contact the initiator tRNA specific A1:U72 base pair. The conserved Arg1070 and His1071 in helix $α^{L5}$, positioned in a well defined density in the cryo-EM map (Figure 1D), are within contact distance to the phosphate backbone at G68 and C69 in the major groove. Notably, EF-Tu domain II contains a corresponding short helix $α^{L5}$ in which the conserved Arg295 as well forms a contact to the acceptor arm of the bound tRNA [14,29] (Figure 2E). Based on the comparison with EF-Tu it was previously assumed that a similar contact might be formed between Lys725 and Glu726 of bsIF2 and fMet-tRNA^{fMet}. However, such an interaction could not be observed by NMR spectroscopy in solution [11]. It is therefore conceivable that these interactions are formed only in the context of the ribosomal pre-IC, where the tRNA is stabilized in a specific orientation relative to domain IV of eIF5B/IF2.

Biological relevance of this domain IV-tRNA interaction lies in the mutual stabilization of initiator tRNA and eIF5B in conformations that allow efficient recruitment of the large ribosomal subunit and insertion of the acceptor arm into the peptidyl-transferase center (PTC). In the 80S complex domain IV stabilizes the tRNA in a non-canonical P/I orientation [25] that according to our model places the 3′ end ~ 20 Å from the PTC without inducing a major distortion of the CCA-end from its canonical conformation (Figure 5). The following GTP hydrolysis and dissociation of eIF5B would thus allow the acceptor stem to rotate into the PTC while the overall tRNA relaxes into its canonical P site conformation. Through the specificity of domain IV for the methionylated acceptor arm, which might be more pronounced in the context of the preassembled 40S·Met-tRNA$_i$ complex than in solution [26], this interaction would mark a final checkpoint in the initiation process that allows subunit joining only on correctly assembled 48S pre-ICs with a charged initiator tRNA bound in the P site [24,30].

Implications for the evolution of the translational β barrel fold

As reported previously, domain IV of IF2 and the structurally homologous domain II of EF-Tu use similar surface areas to interact with amino-acylated tRNAs [11,12] (Figure 2E/F). The structure of the ternary complex of aIF2 shows the same interface for the interactions between domain II of aIF2γ (a paralog of EF-Tu) and the Met-tRNA$_i$ [31] (Figure 4C). Our observations provide structural evidence that this also applies to domain IV of eIF5B. This common binding interface for the 3′ CCA-end on the translational β barrel fold is centered on β strands 1, 2 and 5 and framed by the flexible loops L1

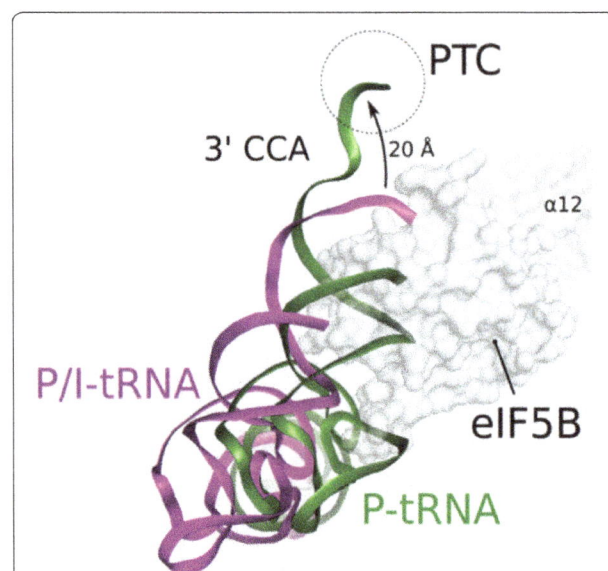

Figure 5 Conformational rearrangement of Met-tRNA$_i$ on the initiation complex. eIF5B stabilizes the initiator tRNA (purple) in a non-canonical P/I conformation [25] with the 3′ CCA-end outside of the PTC. Upon GTP hydrolysis in eIF5B and the release of the 3′ CCA-end from its contacts to domain IV, the initiator tRNA rearranges into the canonical P site conformation, involving a 20 Å repositioning of the 3′ end into the PTC.

and L4 (Figures 2 and 4). In all cases additional interactions are made by the short capping α-helix that provides positively charged residues for contacts to the phosphate backbone of the acceptor stem, while at the same time allowing substantially different overall orientations of the tRNA relative to the β barrel, irrespective of an identical polarity of the bound CCA-end (Figure 6). Despite the low average sequence identity over the various β barrel folds (Figure 3), the significant structural and functional parallels in their interactions with amino-acylated tRNAs clearly point toward a common evolutionary origin. As eIF5B/IF2 and EF-Tu are both universally conserved in the three domains of life, their divergence and thus the origin for their respective tRNA-binding domains most likely lies long before the onset of speciation; this raises the interesting question of potential homologies to other ancient RNA-binding protein folds.

A central question for the problem of cellular evolution is the appearance of the basic protein-folding types and of domains as functional building blocks for proteins. Folded proteins adopt only a limited number of folding structures; however, whether these folds emerged by divergent evolution from a single ancestor or independently by convergent evolution from different lineages is unclear. In this context, it is interesting that the characteristic features of tRNA binding by the translational β barrel fold show significant parallels to those between OB-fold domains and single-stranded nucleic acids (Figure 7). The OB-fold is a five-stranded mixed β

barrel, capped on one end by an α-helix [32]. Most known OB-fold domains are involved in interactions with single-stranded RNA or DNA [33]. Despite very low sequence similarity among its members, the OB-fold superfamily is thought to be an ancient domain structure that derived by divergent evolution from a common ancestral protein – an assumption that is based on the common features of their fold-related ligand-binding interface [33,34]. Despite a different overall topology of the OB-fold (Figure 7C/D) and a different classification in the SCOP (Structural Classification of Proteins) database, this interface, composed of β1-L1-β2-β3-L3/α^{L3}-β4-L4 (Figure 7B), shows an intriguing structural and functional correspondence to the identically arranged but differently connected building blocks of β1-L1-β2 and β4-L4-β5-L5/α^{L5}-β6 in the translational β barrel fold that are responsible for its interactions with tRNA (Figure 7A).

These similarities might merely be a functional analogy between both protein families that arose by convergent evolution from two distinct starting points. However, by the argument of a common descent based on a fold-related ligand-binding interface, the evident similarities might as well be indicative of a common evolutionary origin for the two equally ancient protein folds. For this hypothesis, two previously proposed theories are of particular interest: i) The emergence of domain folds by polyphyletic evolution from self-assembling short peptide ancestors, whose remnants (in sequence, structure or function) still exist in extant proteins [36]; and ii) the theory of a chemoautotrophic origin of life on volcanic iron-sulfur surfaces, according to which protein domains emerged from functional peptides that used metal ions as folding determinants or formed surface-bonded β-sheets that finally detached from the stabilizing surfaces (e.g. to form β-barrel domains) in the course of progressing cellularization [37-39]. In both theories, the transition from the peptide- to the independently folding protein-domain proceeds concomitant to the refinement of the genetic machinery that allows the synthesis of increasingly long polypeptides with sufficiently high fidelity [36,38,39].

In light of these hypotheses, we suggest a possible common polyphyletic origin of both fold-related RNA binding interfaces discussed above. At the earliest stages of cellular evolution, when the fidelity of the primordial translation apparatus allowed the synthesis only of short peptides, nucleic acid-peptide interactions most likely played an essential role, particularly for the genetic machinery. In this context, it would be conceivable that the common ligand-binding interface in the ancient lineages of OB-fold proteins and translational GTPases has arisen as an ancient structural entity formed by individually synthesized peptides, associating with single-stranded nucleic acids as folding determinants, similar to metallo-

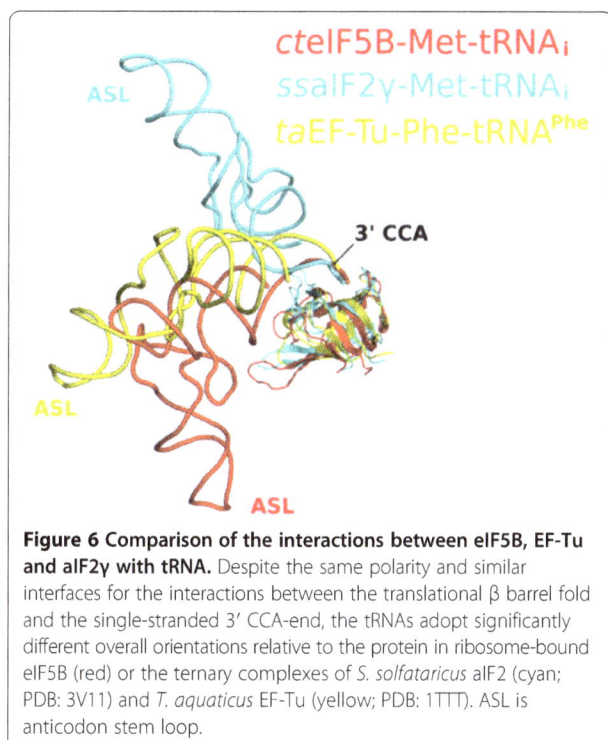

Figure 6 Comparison of the interactions between eIF5B, EF-Tu and aIF2γ with tRNA. Despite the same polarity and similar interfaces for the interactions between the translational β barrel fold and the single-stranded 3' CCA-end, the tRNAs adopt significantly different overall orientations relative to the protein in ribosome-bound eIF5B (red) or the ternary complexes of *S. solfataricus* aIF2 (cyan; PDB: 3V11) and *T. aquaticus* EF-Tu (yellow; PDB: 1TTT). ASL is anticodon stem loop.

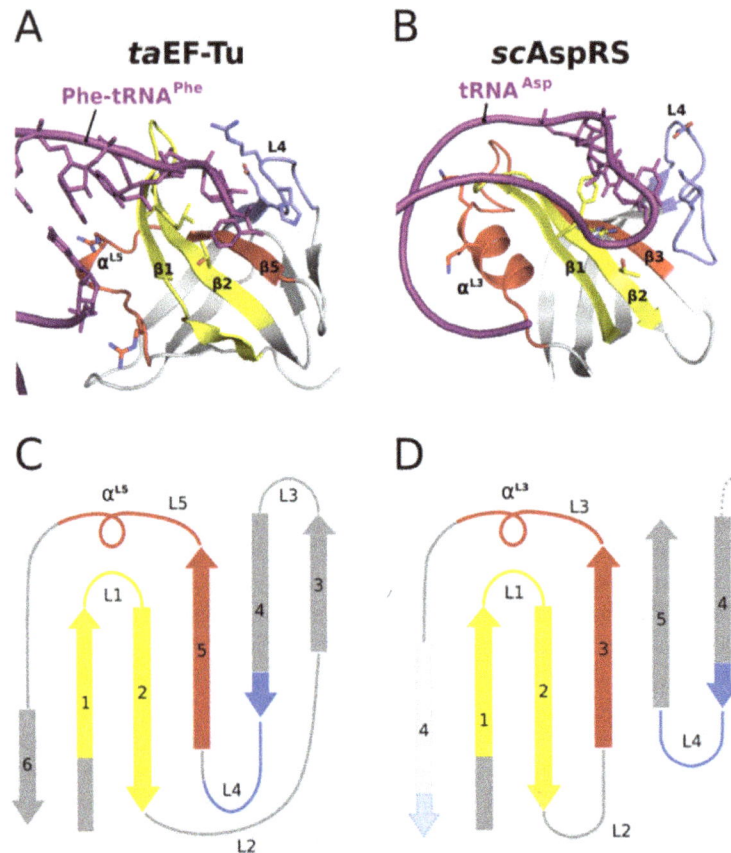

Figure 7 Translational β barrel fold and OB-fold share the same fold related ligand-binding interface. A and **C)** The interaction interface between the 3' CCA-end of the tRNA and the translational β barrel (here Phe-tRNA[Phe] (purple) bound to *T. aquaticus* EF-Tu domain II (PDB: 1TTT)) centers on β strands 1, 2 and 5 and is augmented by loops L1, L4 and L5, containing helix α[L5]. **B** and **D)** A related ligand-binding interface is found in single-stranded nucleic acid binding OB-folds (here the anticodon binding domain of the aspartyl-tRNA synthetase (PDB: 1ASZ) from *S. cerevisiae* [35]). Similar to the interactions observed for EF-Tu and despite a different topology, the bases of the anticodon stem loop (purple) point toward the surface of the β barrel, centered on β strands 1, 2 and 3, while the flexible loops L1, L4 and L3 with α[L3] form additional contacts to the phosphate backbone.

peptides as precursors for metallo-proteins [37-39]; during the gradual replacement of peptides by their fusion into independently folding proteins, this would ensure the conservation of the nucleic acid-binding interface, while at the same time allowing a substantially different connectivity of the individual building blocks in the emerging protein families.

Conclusions

In this study, we used the recently reported medium resolution cryo-EM density of the yeast 80S IC [25] and high resolution crystal structures of eIF5B from *C. thermophilum* to propose a new model for the interactions between eIF5B domain IV and the Met-tRNA$_i$ in the context of the ribosome (Figure 1). According to this model, domain IV forms direct interactions with the phosphate backbone in the major groove of the acceptor arm, the initiator tRNA specific A1:U72 base pair and – most importantly – with the methionylated 3' CCA-end.

The relevance of these findings lies in the novel insight into the specific recognition of the amino-acylated initiator tRNA by eIF5B/IF2 in the context of pre-initiation complexes, which, as a final checkpoint for ribosomal subunit joining, is one of the central interactions in the process translation initiation. Finally, the identified binding interface between eIF5B and Met-tRNA$_i$ directly corresponds to that reported earlier for the interaction between the homologous domains in IF2, EF-Tu and aIF2γ with their respective tRNA ligands [12,14] and exhibits a striking structural and functional similarity to the fold-related ligand-binding interface of OB-fold domains, possibly reflecting a common evolutionary origin of the two ancient domain folds.

Methods
Model building

Rigid-body fitting of *ct*eIF5B domain IV (residues 382–1116; PDB codes 4N3N and 4N3G) was performed using

UCSF Chimera [40]. Despite entirely different sets of crystal contacts for domain IV in the two X-ray structures they are very similar to each other (rmsd of 0.34), indicating a high degree of structural rigidity. Thus, although domain IV might undergo minor conformational changes upon interacting with ribosome and tRNA, particularly in the loop regions, we decided not to include any flexible fitting procedures to avoid overfitting of the model. Manual rebuilding of the acceptor stem of the Met-tRNA$_i$ between bases G70 and A76 into the density next to domain IV was done in COOT [41]. Figures were prepared using UCSF Chimera [40] or Pymol (http://www.pymol.org).

We would like to mention here that in our hands the isolated domain IV of the *mt*aIF5B crystal structure (PDB: 1G7R) is fitted to the cryo-EM density in the same way as the *cte*IF5B structure (with a CCC of 67%), supporting the newly proposed fit shown in Figure 1D. However, the *mt*aIF5B domain IV of the cryo-EM based model (PDB: 4BYX) is fitted as presented in [25] with a CCC of 78.7% (Figure 1C), most likely as the result of a combination of rigid-body and flexible fitting procedures [25], which gave rise to rmsds of 5.8 Å and 7.5 Å (over 106 C$_\alpha$ atoms) relative to the crystal structures of *mte*IF5B and *cte*IF5B, respectively, while the two crystal structures themselves differ only by an rmsd of 2.2 Å. Thus, the higher CCC for the cryo-EM-based model is most likely due to its distortion from the original rigid structure of *mt*aIF5B and is therefore not comparable to the CCC values obtained for our rigid-body fit.

Sequence alignments

Multiple sequence alignments were done using the iterative alignment program MUSCLE [42]. Structural sequence alignments were done using the DALI server [43].

Abbreviations

IF: Initiation factor; eIF: Eukaryal initiation factor; aIF: Archaeal initiation factor; cryo-EM: Cryo-electron microscopy; pre-IC: Pre-initiation complex; EMD: EM-Databank; EF-Tu: Elongation factor-Tu; ct: *Chaetomium thermophilum*; mt: *Methanococcus thermoautotrophicum*; sc: *Saccharomyces cerevisiae*; bs: *Bacillus stearothermophilus*; PDB: Protein data bank; CCC: Cross-correlation coefficient; SCOP: Structural Classification of Proteins; OB-fold: Oligo-nucleotide/oligo-saccharide binding fold; rmsd: Root mean square deviation; PTC: Peptidyl-transferase center; ASL: Anticodon stem loop.

Competing interests

The authors declare that they have no competing interests.

Authors' contributions

BK designed the study, analyzed structures and cryo-EM density, performed the structural modeling and sequence alignments, analyzed and interpreted the data and wrote the manuscript. RF analyzed the data and helped to draft the manuscript. Both authors read and approved the final manuscript.

Acknowledgments

We thank L. K. Dörfel for critical reading of the manuscript.

References

1. Marintchev A, Wagner G: Translation initiation: structures, mechanisms and evolution. *Q Rev Biophys* 2004, 37(3–4):197–284.
2. Kyrpides NC, Woese CR: Universally conserved translation initiation factors. *Proc Natl Acad Sci U S A* 1998, 95(1):224–228.
3. Gualerzi CO, Brandi L, Caserta E, Garofalo C, Lammi M, La Teana A, Petrelli D, Spurio R, Tomsic J, Pon CL: Initiation factors in the early events of mRNA translation in bacteria. *Cold Spring Harb Symp Quant Biol* 2001, 66:363–376.
4. Milon P, Rodnina MV: Kinetic control of translation initiation in bacteria. *Crit Rev Biochem Mol Biol* 2012, 47(4):334–348.
5. Sundari RM, Stringer EA, Schulman LH, Maitra U: Interaction of bacterial initiation factor 2 with initiator tRNA. *J Biol Chem* 1976, 251(11):3338–3345.
6. Petersen HU, Roll T, Grunberg-Manago M, Clark BF: Specific interaction of initiation factor IF2 of E. coli with formylmethionyl-tRNA f Met. *Biochem Biophys Res Commun* 1979, 91(3):1068–1074.
7. Boelens R, Gualerzi CO: Structure and function of bacterial initiation factors. *Curr Protein Pept Sci* 2002, 3(1):107–119.
8. Antoun A, Pavlov MY, Lovmar M, Ehrenberg M: How initiation factors maximize the accuracy of tRNA selection in initiation of bacterial protein synthesis. *Mol Cell* 2006, 23(2):183–193.
9. Krafft C, Diehl A, Laettig S, Behlke J, Heinemann U, Pon CL, Gualerzi CO, Welfe H: Interaction of fMet-tRNA(fMet) with the C-terminal domain of translational initiation factor IF2 from Bacillus stearothermophilus. *FEBS Lett* 2000, 471(2–3):128–132.
10. Spurio R, Brandi L, Caserta E, Pon CL, Gualerzi CO, Misselwitz R, Krafft C, Welfle K, Welfle H: The C-terminal subdomain (IF2 C-2) contains the entire fMet-tRNA binding site of initiation factor IF2. *J Biol Chem* 2000, 275(4):2447–2454.
11. Guenneugues M, Caserta E, Brandi L, Spurio R, Meunier S, Pon CL, Boelens R, Gualerzi CO: Mapping the fMet-tRNA(f)(Met) binding site of initiation factor IF2. *EMBO J* 2000, 19(19):5233–5240.
12. Meunier S, Spurio R, Czisch M, Wechselberger R, Guenneugues M, Gualerzi CO, Boelens R: Structure of the fMet-tRNA(fMet)-binding domain of B. stearothermophilus initiation factor IF2. *EMBO J* 2000, 19(8):1918–1926.
13. Avarsson A: Structure-based sequence alignment of elongation factors Tu and G with related GTPases involved in translation. *J Mol Evol* 1995, 41(6):1096–1104.
14. Nissen P, Kjeldgaard M, Thirup S, Polekhina G, Reshetnikova L, Clark BF, Nyborg J: Crystal structure of the ternary complex of Phe-tRNAPhe, EF-Tu, and a GTP analog. *Science* 1995, 270(5241):1464–1472.
15. Berchtold H, Reshetnikova L, Reiser CO, Schirmer NK, Sprinzl M, Hilgenfeld R: Crystal structure of active elongation factor Tu reveals major domain rearrangements. *Nature* 1993, 365(6442):126–132.
16. Allen GS, Zavialov A, Gursky R, Ehrenberg M, Frank J: The cryo-EM structure of a translation initiation complex from Escherichia coli. *Cell* 2005, 121(5):703–712.
17. Simonetti A, Marzi S, Myasnikov AG, Fabbretti A, Yusupov M, Gualerzi CO, Klaholz BP: Structure of the 30S translation initiation complex. *Nature* 2008, 455(7211):416–420.
18. Pestova TV, Lomakin IB, Lee JH, Choi SK, Dever TE, Hellen CU: The joining of ribosomal subunits in eukaryotes requires eIF5B. *Nature* 2000, 403(6767):332–335.
19. Choi SK, Lee JH, Zoll WL, Merrick WC, Dever TE: Promotion of met-tRNAiMet binding to ribosomes by yIF2, a bacterial IF2 homolog in yeast. *Science* 1998, 280(5370):1757–1760.
20. Lee JH, Pestova TV, Shin BS, Cao C, Choi SK, Dever TE: Initiation factor eIF5B catalyzes second GTP-dependent step in eukaryotic translation initiation. *Proc Natl Acad Sci U S A* 2002, 99(26):16689–16694.
21. Pain VM: Initiation of protein synthesis in eukaryotic cells. *Eur J Biochem* 1996, 236(3):747–771.
22. Roll-Mecak A, Cao C, Dever TE, Burley SK: X-Ray structures of the universal translation initiation factor IF2/eIF5B: conformational changes on GDP and GTP binding. *Cell* 2000, 103(5):781–792.
23. Pavlov MY, Zorzet A, Andersson DI, Ehrenberg M: Activation of initiation factor 2 by ligands and mutations for rapid docking of ribosomal subunits. *EMBO J* 2011, 30(2):289–301.
24. Kuhle B, Ficner R: eIF5B employs a novel domain release mechanism to catalyze ribosomal subunit joining. *EMBO J* 2014.
25. Fernandez IS, Bai XC, Hussain T, Kelley AC, Lorsch JR, Ramakrishnan V, Scheres SH: Molecular architecture of a eukaryotic translational initiation complex. *Science* 2013, 342(6160):1240585.

26. Guillon L, Schmitt E, Blanquet S, Mechulam Y: **Initiator tRNA binding by e/aIF5B, the eukaryotic/archaeal homologue of bacterial initiation factor IF2.** *Biochemistry* 2005, **44**(47):15594–15601.

27. Marintchev A, Kolupaeva VG, Pestova TV, Wagner G: **Mapping the binding interface between human eukaryotic initiation factors 1A and 5B: a new interaction between old partners.** *Proc Natl Acad Sci U S A* 2003, **100**(4):1535–1540.

28. Yamamoto H, Unbehaun A, Loerke J, Behrmann E, Collier M, Burger J, Mielke T, Spahn CM: **Structure of the mammalian 80S initiation complex with initiation factor 5B on HCV-IRES RNA.** *Nat Struct Mol Biol* 2014, **21**(8):721–727.

29. Nissen P, Thirup S, Kjeldgaard M, Nyborg J: **The crystal structure of Cys-tRNACys-EF-Tu-GDPNP reveals general and specific features in the ternary complex and in tRNA.** *Structure* 1999, **7**(2):143–156.

30. Acker MG, Shin BS, Nanda JS, Saini AK, Dever TE, Lorsch JR: **Kinetic analysis of late steps of eukaryotic translation initiation.** *J Mol Biol* 2009, **385**(2):491–506.

31. Schmitt E, Panvert M, Lazennec-Schurdevin C, Coureux PD, Perez J, Thompson A, Mechulam Y: **Structure of the ternary initiation complex aIF2-GDPNP-methionylated initiator tRNA.** *Nat Struct Mol Biol* 2012, **19**(4):450–454.

32. Murzin AG: **OB(oligonucleotide/oligosaccharide binding)-fold: common structural and functional solution for non-homologous sequences.** *EMBO J* 1993, **12**(3):861–867.

33. Theobald DL, Mitton-Fry RM, Wuttke DS: **Nucleic acid recognition by OB-fold proteins.** *Annu Rev Biophys Biomol Struct* 2003, **32**:115–133.

34. Arcus V: **OB-fold domains: a snapshot of the evolution of sequence, structure and function.** *Curr Opin Struct Biol* 2002, **12**(6):794–801.

35. Cavarelli J, Eriani G, Rees B, Ruff M, Boeglin M, Mitschler A, Martin F, Gangloff J, Thierry JC, Moras D: **The active site of yeast aspartyl-tRNA synthetase: structural and functional aspects of the aminoacylation reaction.** *EMBO J* 1994, **13**(2):327–337.

36. Lupas AN, Ponting CP, Russell RB: **On the evolution of protein folds: are similar motifs in different protein folds the result of convergence, insertion, or relics of an ancient peptide world?** *J Struct Biol* 2001, **134**(2–3):191–203.

37. Huber C, Wachtershauser G: **Peptides by activation of amino acids with CO on (Ni, Fe)S surfaces: implications for the origin of life.** *Science* 1998, **281**(5377):670–672.

38. Wachtershauser G: **Before enzymes and templates: theory of surface metabolism.** *Microbiol Rev* 1988, **52**(4):452–484.

39. Wachtershauser G: **On the chemistry and evolution of the pioneer organism.** *Chem Biodivers* 2007, **4**(4):584–602.

40. Pettersen EF, Goddard TD, Huang CC, Couch GS, Greenblatt DM, Meng EC, Ferrin TE: **UCSF Chimera–a visualization system for exploratory research and analysis.** *J Comput Chem* 2004, **25**(13):1605–1612.

41. Emsley P, Lohkamp B, Scott WG, Cowtan K: **Features and development of Coot.** *Acta Crystallogr D Biol Crystallogr* 2010, **66**(Pt 4):486–501.

42. Edgar RC: **MUSCLE: a multiple sequence alignment method with reduced time and space complexity.** *BMC Bioinformatics* 2004, **5**:113.

43. Holm L, Sander C: **Protein structure comparison by alignment of distance matrices.** *J Mol Biol* 1993, **233**(1):123–138.

3

Structural insights into Noonan/LEOPARD syndrome-related mutants of protein-tyrosine phosphatase SHP2 (*PTPN11*)

Wei Qiu[1], Xiaonan Wang[1,2], Vladimir Romanov[1], Ashley Hutchinson[1], Andrés Lin[1], Maxim Ruzanov[1], Kevin P Battaile[3], Emil F Pai[1,4], Benjamin G Neel[1,2*] and Nickolay Y Chirgadze[1,5*]

Abstract

Background: The ubiquitous non-receptor protein tyrosine phosphatase SHP2 (encoded by *PTPN11*) plays a key role in RAS/ERK signaling downstream of most, if not all growth factors, cytokines and integrins, although its major substrates remain controversial. Mutations in *PTPN11* lead to several distinct human diseases. Germ-line *PTPN11* mutations cause about 50% of Noonan Syndrome (NS), which is among the most common autosomal dominant disorders. LEOPARD Syndrome (LS) is an acronym for its major syndromic manifestations: multiple Lentigines, Electrocardiographic abnormalities, Ocular hypertelorism, Pulmonary stenosis, Abnormalities of genitalia, Retardation of growth, and sensorineural Deafness. Frequently, LS patients have hypertrophic cardiomyopathy, and they might also have an increased risk of neuroblastoma (NS) and acute myeloid leukemia (AML). Consistent with the distinct pathogenesis of NS and LS, different types of *PTPN11* mutations cause these disorders.

Results: Although multiple studies have reported the biochemical and biological consequences of NS- and LS-associated *PTPN11* mutations, their structural consequences have not been analyzed fully. Here we report the crystal structures of WT SHP2 and five NS/LS-associated SHP2 mutants. These findings enable direct structural comparisons of the local conformational changes caused by each mutation.

Conclusions: Our structural analysis agrees with, and provides additional mechanistic insight into, the previously reported catalytic properties of these mutants. The results of our research provide new information regarding the structure-function relationship of this medically important target, and should serve as a solid foundation for structure-based drug discovery programs.

Background

The ubiquitous non-receptor protein tyrosine phosphatase SHP2 (encoded by *PTPN11*) plays a key role in RAS/ERK signaling downstream of most, if not all growth factors, cytokines and integrins, although its major substrates remain controversial [1,2]. SHP2 contains two N-terminal SH2 domains, a catalytic (PTP) domain, a C-terminal tail with two tyrosine phosphorylation sites and a proline-rich domain [2-5], and is regulated by an elegant molecular switch mechanism that couples appropriate cellular localization to catalytic activation [3,5]. In the absence of a tyrosine-phosphorylated binding partner for its SH2 domains (basal state), SHP2 assumes a "closed" conformation wherein the N-terminal SH2 (N-SH2) domain is wedged into the PTP domain, blocking substrate access (Figure 1a). Upon agonist stimulation, recruitment of the N-SH2 domain to specific phosphotyrosyl (pTyr-) peptides disrupts this self-locking conformation, freeing the PTP domain for catalysis [3,5,6].

Mutations in *PTPN11* cause several human diseases. Germ-line *PTPN11* mutations cause ~50% of Noonan Syndrome (NS), which is among the most common autosomal dominant disorders [7,8]. Gain-of-function mutations in other RAS-RAF-MEK-ERK pathway members, including *SOS1* [9,10], *KRAS* [11], *NRAS* [12], *SHOC2* [13], and *RAF1* [14,15], are responsible for most remaining NS cases. With an estimated incidence of 1/2,000 live

* Correspondence: bneel@uhnresearch.ca; nchirgadze@gmail.com
[1]Princess Margaret Cancer Center, University Health Network, Toronto, Ontario, M5G 2C4, Canada
[5]Department of Pharmacology and Toxicology, University of Toronto, Toronto, Ontario, M5S 1A8, Canada

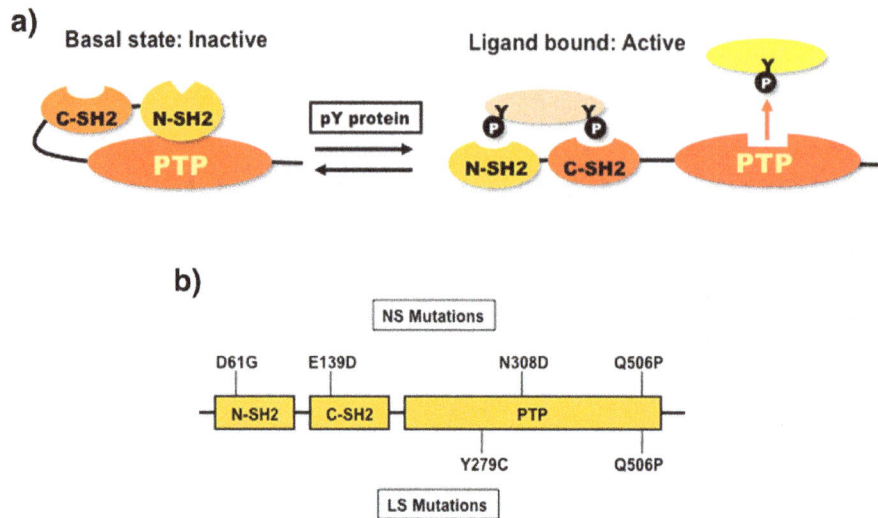

Figure 1 SHP2 regulation and disease-associated mutants. a) SHP2 is regulated via a "self-locking" mechanism: in the absence of pTyr- proteins (pY), SHP2 exists in a closed conformation with the N-SH2 domain bound to the PTP domain, blocking the catalytic site. Upon binding of the appropriate pTyr-proteins, the closed confirmation is disrupted, opening up SHP2 so that substrates can bind to the active site. **b)** Positions of human disease-associated mutants used in this study.

births [16], NS is characterized by facial dysmorphism, proportional short stature, cardiac anomalies, and various less penetrant phenotypes, such as webbed neck, deafness, and motor delay. Many (20-50%) NS patients develop some type of myeloproliferative disorder (MPD), which is typically mild and self-limited [17]. Rare NS patients progress to Juvenile Myelomonocytic Leukemia (JMML), which is fatal if not treated by bone marrow transplantation, somatic *PTPN11* mutations are the single most common cause of sporadic JMML [7,18-20]. LEOPARD Syndrome (LS), a much less common autosomal dominant disorder, is almost always caused by *PTPN11* mutations, and is related to, but distinguishable from, NS [7,16,21]. LEOPARD is an acronym for its major syndromic manifestations: multiple Lentigines, Electrocardiographic abnormalities, Ocular hypertelorism, Pulmonary stenosis, Abnormalities of genitalia, Retardation of growth, and Deafness [22]. These patients often have hypertrophic cardiomyopathy (HCM), and might also have an increased risk of neuroblastoma (NS) and acute myeloid leukemia (AML) [23,24]. Knock-in mouse models have been generated for NS and LS alleles of *Ptpn11* and generally reproduce the phenotypes seen in the cognate human syndromes [25-27].

Consistent with the distinct pathogenesis of NS and LS, different types of *PTPN11* mutations cause these disorders. Most NS-associated *PTPN11* mutations alter residues that reside at the interface between the N-SH2 and PTP domains [16], resulting in elevated enzymatic activity and enhanced RAS/ERK activation [27-31]. These data suggest that NS mutations disrupt the intramolecular interaction between the N-SH2 and PTP domains, shifting the equilibrium between the closed and open

conformations and lowering the activation threshold for SHP2. By contrast, LS mutations typically affect PTP domain residues, result in markedly decreased catalytic activity, and lower RAS/ERK activation in transient transfection assays [31-33]. Studies of the LS Y279C mouse model also indicate that LS mutants may have dominant negative effects in at least some tissues *in vivo* [25]. Whereas NS phenotypes arise from enhanced MEK/ERK activation and can be prevented or reversed by MEK inhibition [34-36], LS-associated HCM is caused by enhanced PI3K/AKT/mTORC1 activity and can be reversed by rapamycin [25].

Although multiple studies have reported the biochemical and biological consequences of NS- and LS-associated *PTPN11* mutations, their structural consequences have not been analyzed. Here, we report the X-ray structures of five NS/LS SHP2 mutants and discuss how these mutations affect the interaction between different SHP2 domains and its catalytic activity.

Methods

Cloning

A wild type (WT) SHP2 expression construct 1–539 (comprising the N + C SH2 and PTP domains) was PCR-amplified from *PTPN11* cDNA [37] with a set of custom-designed primers (see Additional file 1: Description S1). The resultant PCR fragment was cloned into a modified version of the plasmid pET28b (Novagen) that generates a fusion protein with an N-terminal hexahistidine tag. The SHP2 catalytic domain expression construct (a.a. 221–524) was cloned into pGEX4T, which introduces a GST-tag at its N-terminus. Mutations were introduced into these expression constructs by site directed-mutagenesis with specifically designed primers bearing one substitution each (see Additional file 1: Description S1). Pfu Ultra II high fidelity DNA polymerase (Stratagene) was used for PCR, with an extension temperature of 68°C over 10 minutes. To remove any traces of the original cDNA, all reactions were subjected to digestion with *DpnI* (New England Biolabs) for 1 hour at 37°C. Reaction mixtures were transformed into DH5α cells, and the genetic content of all constructs was verified by Sanger sequencing.

Protein expression & purification

Vectors encoding full length versions of SHP2 mutants were transferred into *E. coli* BL21(DE3). Cells were grown in Terrific Broth containing kanamycin (50 mg/l) in 1 L Tunair flasks at 37°C to an OD_{600} of 3–5, after which the temperature was lowered to 16°C, and isopropyl-1-thio-β-D-galactopyranoside (IPTG) was added to 0.2 mM. Expression was allowed to proceed overnight, then cells were harvested by centrifugation, flash-frozen in liquid nitrogen, and stored at −80°C. Due to the low level of expression of the Q506P construct, cells expressing this mutant were washed using the osmotic shock technique [38] prior to freezing. Unless stated otherwise, all purification procedures were carried out at 4°C. Cells were thawed on ice and resuspended in Binding Buffer (see Additional file 2: Table S1 for detailed buffer components), supplemented with phenylmethylsulfonylfluoride and benzamidine. After disruption by sonication and centrifugation at 60,000 g for 40 min, cell-free extracts were passed through a DE-52 column (2.6 × 7 cm) that had been pre-equilibrated with the same buffer, and loaded by gravity flow onto a Ni-nitrilotriacetic acid (NTA) column (Qiagen, Germantown, MD). The latter column was washed with 20–25 volumes of Wash Buffer A, followed by 20–25 volumes of Wash Buffer B and finally with Elution Buffer. N308D and E139D eluted in Elution Buffer, whereas the other three mutants eluted with the Wash buffers. For the latter proteins, the wash fractions were diluted 15-fold and reloaded on fresh Ni-NTA columns. After washing with 10 column volumes of Binding Buffer, N308D and E139D proteins were eluted in elution buffer. These samples were concentrated using a VIVASpin unit (Sartorius NA, Edgewood, NY), and loaded onto a 2.6 × 60 cm Superdex 200 column (GE Healthcare), equilibrated with Gel Filtration buffer. Elution was performed at a flow rate of 3 ml/min at 8°C, with the SHP2 proteins behaving as apparent monomers. Final protein samples were concentrated to 20–40 mg/ml, divided into 1.5 mg aliquots, flash-frozen and stored at −80°C.

SHP2 catalytic domain mutants were transformed into *Escherichia coli* strain BL21(DE3). A 25 ml aliquot of an overnight culture from a single colony was added to 500 ml of LB/ampicillin (50 μg/ml) and grown at 37°C

to A_{600} = 0.8. IPTG was added to a final concentration of 0.1 mM, and the bacteria were maintained for 16 h at 25°C with shaking, then centrifuged at 6,000 × g for 10 min. at 4°C. Pellets were resuspended in 12.5 ml of a buffer containing 50 mM Tris–HCl, pH 7.5, 150 mM NaCl, 5 mM MgCl2, 1% Triton X-100, 10% glycerol, 5 mM dithiothreitol, 2 μg/ml aprotinin, 10 μg/ml leupeptin, 1 μg/ml antipain, 1 μg/ml pepstatin A, 0.5 mg/ml lysozyme and 1 mg/mL DNase I. Suspensions were incubated on ice for 30 minutes, and then sonicated for 10 seconds on ice. Lysates were centrifuged at 14,000 × g for 30 min. at 4°C, and supernatants were transferred to a fresh 15-ml polypropylene tube containing 0.5 ml of glutathione-Sepharose 4B (GE Healthcare Life Sciences). This suspension was rotated end-over-end overnight at 4°C, and then centrifuged at 1000 × g for 1 min. at 4°C. The supernatants were discarded, and the beads were washed 3 times for 5 min. each at 4°C with 10 ml of wash buffer (25 mM Tris–HCl, pH 7.5, 150 mM NaCl, 5 mM MgCl2, 1% Triton X-100, 10% glycerol, 5 mM dithiothreitol), and then once with PTP assay buffer (25 mM Hepes, pH 7.5, 100 mM NaCl, 2 mM EDTA and 5 mM dithiothreitol). Bound GST fusion proteins were resuspended 1:1 in PTP assay buffer. A 20 uL aliquot of slurry for each mutant was separated on a 10% SDS-polyacrylamide gel, together with different amounts of BSA. The gel was washed in water for 10 minutes, and stained with Colloidal Coomassie Blue for 1 hour at room temperature. Bands were quantified using a LI-COR Odyssey.

PTP assays

To determine kinetic parameters, fixed amounts of purified GST-WT or -mutant SHP2 catalytic domains (1.6 pmol of WT and N308D, 115pmol of Y279C and 16.3 pmol of Q506P) were incubated with variable concentrations of substrate peptide (R-R-L-I-E-D-A-E-pY-A-A-R-G, Millipore #12-217; Kit #12-217) in PTP assay buffer in a total volume of 50 uL. Reactions were carried out for 10 minutes at 25°C, and phosphate release was quantified by adding Malachite Green (Millipore #17-125) to the supernatants, measuring absorbance at 620 nm, and comparing values to a standard curve generated with varying amounts of KH_2PO_4. All reactions fell within the linear range. Phosphatase activity is expressed in pmol Pi released/min/pmol enzyme.

Crystallization

Mutant SHP2 proteins were crystallized under conditions similar to those reported previously [3]. In order to obtain the best diffracting crystals, 0.1 M LiCl was added to the literature crystallization conditions for D61G, 5% glycerol for N308D, and 10% glycerol and 0.3 M cycohexyl-methyl-β-D-maltoside for Q506P. The other two mutant proteins and the WT protein were crystallized under literature conditions with optimized precipitant concentrations. Crystals appeared overnight, and reached their full size of about 300 × 300 × 30 microns in one week at room temperature. The stacked plate crystals were separated and flash frozen in liquid nitrogen, using paratone-N oil (Hampton Research Inc.) as a cryo-protectant.

Data collection

Data were collected at 100 K with a wavelength of 1.0 Å on the Industrial Macromolecular Crystallography Association (IMCA-CAT) beam line at the Advanced Photon Source (Argonne National Laboratory, IL USA). The data were indexed, integrated, and scaled with *XDS* and *XSCALE* [39].

Structure determination

The first mutant structure of N308D was determined by molecular replacement, using the previously solved structure of SHP2 (PDB access code 2SHP) [3] as the search model. The wild type SHP2 and D61G, E139D, Y279C and Q506P mutant structures were determined by molecular replacement, using the structure of the N308D SHP2 mutant as a search model. Following the initial rigid body refinement, interactive cycles of model building and refinement were performed by using *COOT* [40] and *Buster-TNT* [41] software. Special attention was paid to the mutation sites, which were initially replaced with alanine to reduce model bias and later positioned based on the $2mF_o$-DF_c and difference Fourier electron density maps after a few rounds of refinement to confirm that those amino acids had, indeed, been mutated. Data collection and refinement statistics are shown in Table 1. The atomic coordinates have been deposited in the RCSB Protein Data Bank under accession numbers 4NXD, 4H10, 4NWG, 4GWF, 4NWF, and 4H34. All figures were produced using *PyMOL* (http://www.pymol.org).

Table 1 Summary of crystallographic data and refinement statistics

Parameters	Wild type	D61G	E139D	Y279C	N308D	Q506P
Data collection:						
Resolution, (Å)	2.75	2.20	2.45	2.10	2.10	2.70
Outermost resolution shell, (Å)	(2.85-2.75)	(2.30-2.20)	(2.55-2.45)	(2.20-2.10)	(2.20-2.10)	(2.80-2.70)
Space group	P2$_1$	P2$_1$2$_1$2	P2$_1$2$_1$2$_1$	P2$_1$	P2$_1$2$_1$2$_1$	P2$_1$2$_1$2
Unit cell parameters						
a, (Å)	55.7	55.0	56.3	55.7	55.9	54.8
b, (Å)	211.7	220.3	212.4	212.0	211.2	202.4
c, (Å)	91.2	41.7	92.2	46.0	91.6	44.5
β, (°)	89.97			96.6		
Molecules per asymmetric unit	4	1	2	2	2	1
Unique reflections	53,849	26,689	41,625	61,515	64,401	14,342
Multiplicity	3.1 (3.4)	6.3 (6.1)	6.5 (6.3)	3.5 (3.5)	7.0 (7.2)	6.5 (6.3)
Average I/σ (I)	5.5 (2.2)	11.1 (2.7)	9.2 (2.3)	6.9 (1.9)	11.0 (3.2)	12.0 (2.8)
R_{merge} (%)	18.9 (46.0)	10.6 (52.0)	11.4 (57.1)	10.4 (49.7)	9.4 (43.0)	13.9 (56.0)
Completeness, (%)	95.9 (98.5)	99.3 (96.2)	99.7 (98.1)	99.9 (99.6)	99.9 (100)	99.9 (100)
Refinement and structure statistics						
R_{work} (%)	25.6	21.1	21.3	21.1	24.2	22.7
R_{free} (%)	28.5	23.3	25.4	24.3	28.7	24.2
RMSD from ideal geometry						
Bond lengths, (Å)	0.007	0.007	0.010	0.008	0.009	0.007
Bond angles, (°)	0.91	0.99	1.24	1.00	1.10	0.91
Numbers of atoms						
Protein (non-hydrogen)	15,560	4,021	8,214	8,013	8,100	3,951
Water oxygen atoms	784	134	365	261	679	54
Ligand's atoms		20		90		36
PDB ID	4NXD	4H10	4NWG	4GWF	4NWF	4H34

$R_{merge} = \Sigma_{hkl}|I - \langle I \rangle|/\Sigma_{hkl}I$, where I is the intensity of the individual reflections.
$R_{work} = \Sigma |F_{obs} - F_{cal}|/\Sigma|F_{obs}|$, where F_{obs} and F_{calc} are the observed and the calculated structure factors, respectively.
R_{free} was calculated using 5% of total reflections randomly chosen and excluded from the crystallographic refinement.

Results and discussion

We determined the crystal structures of WT SHP2 (residues 1–539), as well as five mutants (D61G, E139D, Y279C, N308D, and Q506P), chosen to represent the spectrum of disease-associated *PTPN11* mutations. Mutants D61G, E139D, and N308D are found in NS, Y279C is a canonical LS mutation [16,21,42], and Q506P has been reported in both disorders [16], although it is unclear whether this reflects misdiagnosis or true bipotentiality of this allele. The D61G mutation affects the N-SH2 domain, E139D lies within the C-SH2 domain and the other three mutations alter the PTP domain (Figure 1b). The enzymatic properties of the full-length versions of these mutants (including the C-terminal tail, which is missing in our crystallization constructs) were characterized previously by our group [29,33] (Additional file 3: Figure S1), and range from strongly activated (D61G), to mildly activated (N308D), to catalytically impaired (Y279C). Q506P shows altered specificity for some substrates [29].

The SHP2 structure published by Hof *et al.* (PDB accession code: 2SHP; hereafter termed "2SHP") has three mutations (T2K, F41L and F513S) and is in complex with a detergent molecule (CTAB). We corrected these mutations, and crystallized WT SHP2 under detergent-free conditions. In our WT structure, the SH2 domains and the PTP domain assume a "closed" conformation with the N-SH2 domain locked into the PTP catalytic site, similar to the 2SHP structure (Figure 2). Superimposition of our *bona fide* WT structure with the earlier "WT" SHP2 structure revealed an overall root mean square deviation (*rmsd*) of 0.59 Å over 487 aligned residues. The N-SH2 domain had a smaller *rmsd* of 0.35 Å, whereas the *rmsd* for the C-SH2 was larger (0.74 Å), implying that the C-SH2 domain might have a higher degree of flexibility than the other two domains in the closed conformation. The three mutations and the CTAB binding site in the 2SHP structure are distant from the C-SH2 domain, so we do not think it likely that they affect C-SH2 domain flexibility directly. In the PTP domain, the major deviations between the two structures were seen in residues 425–436 of the αF helix (average *rmsd* =1.33 Å), corresponding to the CTAB binding site in the 2SHP structure (Figure 2). Although the overall structural difference between WT and 2SHP was small, we used our WT structure as the reference for comparison with the mutant structures to exclude any potential structural changes induced by the three mutations and the detergent molecule (CTAB) in 2SHP.

D61G

In the WT structure, Asp61 was located on the surface of the N-SH2 domain. The side chain of Asp61 formed hydrogen bonds with Ser460 from the catalytic P-loop (residues 458–464), a water-mediated hydrogen bond with the catalytic cysteinyl residue, Cys459, two water-mediated hydrogen bonds with Arg465, and another water-mediated hydrogen bond with Asp425 (Figure 3a). Consequently, Asp61 plays an important role in the N-SH2 and PTP domain interaction. In the D61G mutant structure, these hydrogen bonds were abolished. The change from aspartate to glycine also altered the surface charge from very negative to neutral (Figure 3b). Opposite D61G on the interface surface, the PTP domain presents a predominantly positively charged pocket (Figure 3c). Thus, the D61G mutation greatly loosened the interactions between N-SH2 and PTP domains. These data are consistent with previous publications that observed increased basal activity for this mutant against artificial (e.g., pNPP) and pTyr-peptide substrates [29,33].

E139D

Residue Glu139 was located on the surface of the C-SH2 domain. The overall crystal structure of the E139D mutant was very similar to that of WT SHP2, with an *rmsd* of 0.4 Å (Figure 4a). Glu139 was about 40 Å away from the catalytic site, with the N-SH2 domain interposed between these domains. E139 is, however, located in the vicinity of the phosphate group of the pTyr peptide-binding site of the C-SH2 domain, and the E139D mutation stabilizes the conformation of the 139–147 loop that plays an essential role in pTyr peptide-binding. The mutant structure has well defined electron density for this loop, whereas this loop is disordered in the WT structure. Compared with WT SHP2, there also were

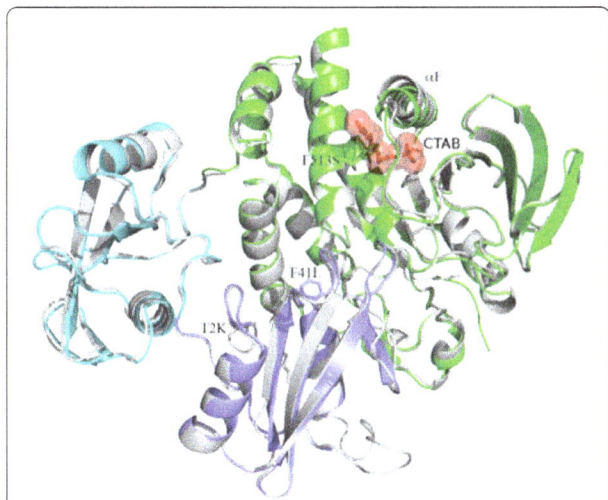

Figure 2 Comparison of "true" WT SHP2 structure (gray) with previously determined "WT" structure (PDB accession code: 2SHP). Note that the earlier structure has three mutations (T2K, F41L and F513S, presented in red sticks) and also contains a molecule of the detergent CTAB, (presented as red sticks and transparent sphere). The F513S mutation creates a cavity that binds the detergent molecule, which alters the orientation of the αF helix.

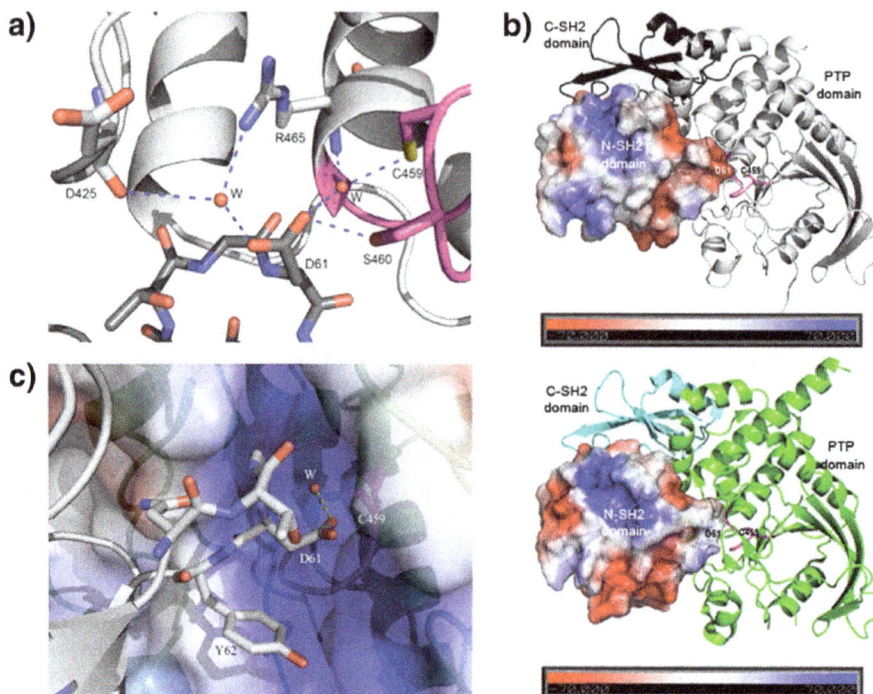

Figure 3 Crystal structure of the D61G mutant. a) In WT SHP2, the side chain of Asp61 (in the N-SH2 domain) forms a direct hydrogen bond with Ser460, a water-mediated hydrogen bond with the catalytic cysteinyl residue Cys459 and two water-mediated hydrogen bonds with Arg465 in the PTP domain. **b)** The D61G mutation alters the electrostatic surface charge on the N-SH2 domain along its interaction interface with the PTP domain catalytic pocket. The N-SH2 domain is rendered in electrostatic surface representation, the C-SH2 domain is colored dark grey (WT structure; top panel) or cyan (D61G mutant structure; bottom panel), and the PTP domain is shown in grey (WT) or green (D61G). The catalytic P-loop (458–464) is shown in magenta. **c)** The N-SH2/PTP domain interface near the catalytic site. The PTP domain is rendered in electrostatic surface presentation, and shows a mostly positively charged catalytic site opposite to Asp61. A conserved water molecule mediates a hydrogen bond between Asp61 and Cys459.

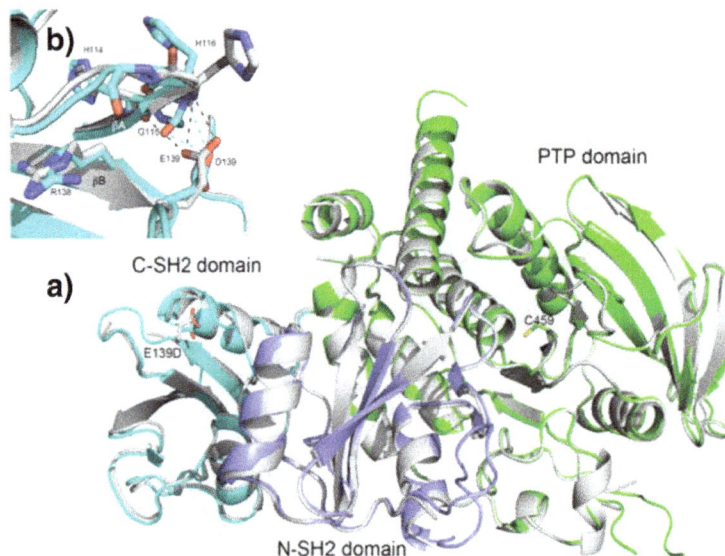

Figure 4 Crystal structure of the E139D mutant. a) Crystal structure of the E139D mutant does not display any obvious conformational changes in the C-SH2 (shown in cyan) or PTP domains (shown in green), superimposed on the WT SHP2 structure (colored in grey). Note that the mutant residue E139D is far away from the catalytic cysteine, Cys459 (~40 Å). The N-SH2 domain is shown in blue. **b)** In the E139D structure, the side chain of Asp139 forms only two hydrogen bonds with the main chain atoms of G115 and His116 (cyan dash lines), whereas in the WT structure, Glu139 forms three hydrogen bonds with His114 and His116 (grey dash lines). The mutation could loosen the connection between the βA and βB strand in C-SH2 domain, thereby exposing the side chain of Arg138, a key residue for pTyr-peptide binding.

some local structural rearrangements in the E139D mutant, with the most noticeable difference being a conformational change of His116 (Figure 4b). The E139D mutant shows a small increase in basal activity, but when it is assayed with a pTyr peptide (pTyr1172 from IRS-1) that can bind both SH2 domains and be dephosphorylated by the catalytic domain, the activity of this mutant is more than 5 times higher than that of WT SHP2 [29]. Although the side chain charge remained unchanged when glutamic acid was changed to aspartic acid, the size of the side chain was reduced. This change could alter the surface of the adjacent residue, Arg138, a key residue for pTyr binding [29]. As shown in Figure 4b, aspartic acid 139 (located in the βB strand) contributes less hydrogen bonding to residues His114-His116 of the βA strand than does glutamic acid. This subtle change might loosen the connection between the βA and βB strands, helping to expose the side chain of Arg138, and thus enhancing the affinity for pTyr-peptide binding. The binding of pTyr -peptides to the C-SH2 domain also could affect the interaction between the N-SH2 and C-SH2 that is critical for enzymatic activation [43]. The mutant also could facilitate the binding of the C-SH2 domain to certain physiologically important binding partners; e.g., IRS-1, in which pTyr-1222 binds to the C-SH2 domain, while pTyr-1172 binds to the N-SH2 domain. The E139D mutation could indirectly increase the binding affinity of the N-SH2 domain for SHP2 substrates (for the reasons discussed above), and therefore increase catalytic activity. In concert, these effects likely explain why E139D is activated by pTyr-peptide binding more than WT SHP2.

Y279C

Y279C is a catalytically impaired mutant associated with LS [16,21,42]. In the crystal structure of this mutant, Tyr279 was located in the long pTyr loop (residues 277–288) with two small α-helices, αC (residues 265–269) and αD (residues 271–276), at its upstream end and one small β-strand, βB (residues 289–292), at its downstream end. In general, because the intervening region (residues 262–288) lacks a structurally stable long α-helix or β-strand, it is likely to be flexible. It also contained a large number of positively charged side chains pointing toward the surface of the PTP domain that might interact with solvent molecules or other binding partners and thus increase the mobility of the pTyr loop. In the WT structure, the side chain of Tyr279 makes *van der Waals* contacts with Ser460 and Ala461 of the catalytic P-loop, as well as with Tyr62 and Lys70 from the N-SH2 domain (Figure 5). The –OH group of Tyr62 interacts with the π-electrons of the Tyr279's aromatic ring. Together with Q506 in the Q-loop (residues 501–507), Tyr279 is believed to play a key role in binding the

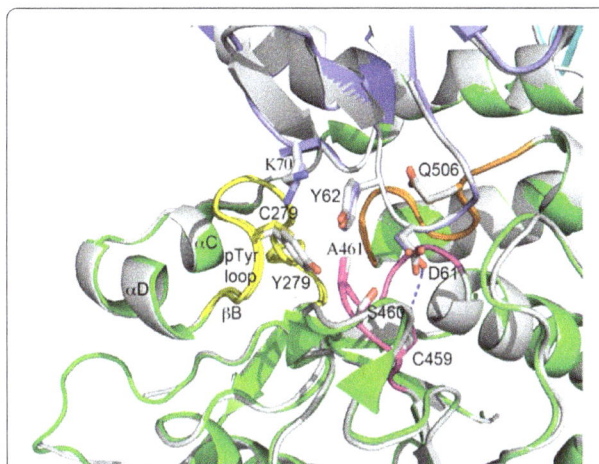

Figure 5 Superimposition of the Y279C mutant and WT SHP2 (grey) structures. In WT SHP2, Tyr279 lies in proximity to Cys459 and also interacts with Tyr62 of the N-SH2 domain (shown in blue and dark grey, respectively). Mutation from Tyr to Cys decreases these interactions. The P-loop (458–464) and pTyr loop (277–288) are highlighted in magenta and yellow, respectively.

tyrosine side chains of substrate proteins/peptides during catalysis. In the PTP1B structure [44], Tyr46 is the residue equivalent to Tyr279 in the SHP2 structure [3,45]. The stacking interaction of three residues, Tyr46, the substrate pTyr, and Phe182 from the "WPDF" loop, help to properly position the substrate for catalysis. In the Y279C structure, the interactions of Tyr279 with the P-loop and the N-SH2 domain were disrupted, due to the significantly shorter side chain of cysteine compared with that of tyrosine. SHP2 has a WPDH loop that corresponds to WPDF in PTP1B. The Y279C mutation would have less stacking interactions with Tyr279, a bound pTyr substrate, and His426 (the equivalent of Phe182 in PTP1B). The mutation also distorts the pTyr substrate/SHP2 interaction, and thus would be expected to disrupt catalysis significantly. At the same time, the Y279C mutation results in loss of the Tyr279/Tyr62 interaction, diminishing the strength of intramolecular binding between the PTP and N-SH2 domains. This would be expected to facilitate the "open" conformation, and can explain the enhanced interaction of this mutant with binding partners (e.g., GAB1) observed previously [33,45,46]. While this manuscript was in preparation, Yu *et al.* [46] reported a very similar model of Y279C (PDB accession code: 4DGX), as well as WT SHP2 (PDB accession code: 4DGP). The Cα carbon atom comparison for these structures revealed RMSD of 0.42 and 0.47 Å for the corresponding mutant and wild-type crystal structure pairs correspondingly. Both published crystal structures belong to the P2$_1$2$_1$2 space group, which is different from the space group (P2$_1$) for our Y279C mutant and WT structures. Importantly, the residues surrounding the mutant residue are in a similar conformation.

Our WT structure has two disordered regions (perhaps due to their flexibility) lacking electron density (88–95 and 156–164). From both Y279C structures, we can see that replacing tyrosine with cysteine at position 279 does not block the accessibility to the PTP catalytic site; instead, it could facilitate local conformational changes that lead to the release of the N-SH2 domain, and thereby open the conformation of the substrate-binding site.

To test the predictions of these structural studies, we assayed the enzymatic activity of isolated SHP2 catalytic domains (present as part of GST-fusion proteins). Compared with WT GST-SHP2, the Y279C mutant had ~80 times lower k_{cat} and 7 times higher K_m towards a pY peptide (Figure 5, Table 2). These results are quite similar to those were reported by Yu *et al.* [46]. Yu *et al.* also provided hydrogen/deuterium exchange mass spectrometry experiments and molecular dynamics simulations showing that the N-SH2 and PTP domain interaction was decreased in the Y279C mutant. Moreover, they found that the Y279C mutant displayed higher affinity for, and was preferentially activated by, a non-hydrolyzable N-SH2 ligand.

N308D

Residue 308 was located in the βC strand of the PTP domain, and was not involved in direct interactions with the N-SH2 domain. However, the Oδ1 atom from the side chain of Asn308 formed a strong hydrogen bond with the side chain of the conserved Arg501. Arg 501 also made direct hydrogen bonds with the main chains of the P-loop residues Ala461 and Gly462 (Figure 6a). The Nδ2 atom of Asn308 formed two hydrogen bonds with the main chain of Phe285 from the pTyr-loop. In the N308D mutant, the charge of the side chain changes from neutral to negative, whereas the side chain polarity changes from polar to acidic polar. Compared with the WT Asn residue, Asp308 formed more hydrogen bonds with pTyr-loop residues, and an especially strong one (2.5 Å) with Thr288 (Figure 6b). Consequently, this mutation could make the pTyr- and P-loops less flexible, locking the enzyme in a more favorable position for

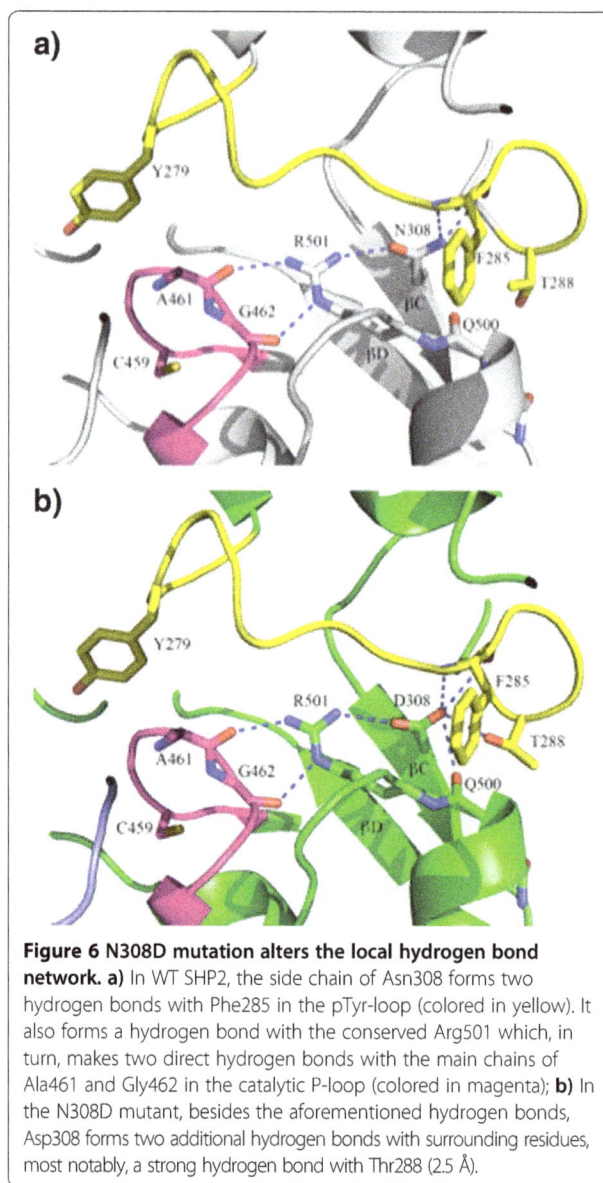

Figure 6 N308D mutation alters the local hydrogen bond network. a) In WT SHP2, the side chain of Asn308 forms two hydrogen bonds with Phe285 in the pTyr-loop (colored in yellow). It also forms a hydrogen bond with the conserved Arg501 which, in turn, makes two direct hydrogen bonds with the main chains of Ala461 and Gly462 in the catalytic P-loop (colored in magenta); **b)** In the N308D mutant, besides the aforementioned hydrogen bonds, Asp308 forms two additional hydrogen bonds with surrounding residues, most notably, a strong hydrogen bond with Thr288 (2.5 Å).

Table 2 Catalytic activities of the indicated SHP2 catalytic domain (221–524) GST fusion proteins were measured using the Malachite Green assay in the presence of different concentrations of PTP-1B peptide R-R-L-I-E-D-A-E-pY-A-A-R-G. K_m and k_{cat} calculated by: $1/V = (K_m/V_{max}) / [pY] + 1/V_{max}$

	k_{cat} (s^{-1})	K_m (mM)	k_{cat} / K_m (s^{-1} mM^{-1})
WT	9.26	1.31	7.07
N308D	12.82	1.92	6.68
Y279C	0.12	9.14	0.01
Q506P	0.61	2.83	0.22

catalysis. The greater rigidity of the pTyr and P loops makes it more difficult for the N-SH2 domain to close back on the PTP domain once it opens. Since the open and closed forms are in equilibrium, this could mean that the ability (*i.e.*, rate constant) for closing back is significantly diminished, hence favoring the open form, leading to the increased activity (basally and in response to pTyr peptide) of the N308D mutant as a full-length enzyme. The residue Asn308 is a hot spot for NS mutations, with N308D accounting for 25% of NS cases. Previous studies showed that this mutant (as a full length protein) has a 3-fold higher basal activity than WT [29]. Our PTP assay showed that catalytic domain of this mutation had slightly higher k_{cat} and K_m values when compared with the WT PTP domain (Figure 1, Table 2).

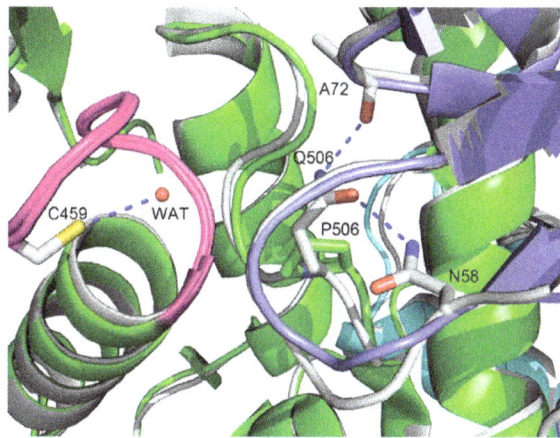

Figure 7 Q506 is necessary to position and activate a water molecule for hydrolysis of the phospho-enzyme intermediate (wild type). The Q506P mutant loses this functionality. Q506 also forms two important hydrogen bonds with Asn58 and Ala72 from the N-SH2 domain (colored in blue), which connect the N-SH2 domain to the PTP domain (colored in green). In the Q506P structure, these connections no longer are present. The P-loop (458–464) is highlighted in magenta.

Q506P

Residue 506 was located in the interface between the PTP and N-SH2 domains. In the WT structure, Q506 formed two important hydrogen bonds, with the main chain of Ala72 and with the side chain of Asn58 in the N-SH2 domain. Together with Tyr279, Gln506 also plays an important role in PTP catalysis by binding the tyrosine side chain of the substrate [44] and by helping to properly position a water molecule for hydrolysis of the thiophosphate intermediate (Figure 7). In the Q506P mutant structure, the proline mutation abolished the two hydrogen bonds between the PTP and N-SH2 domain, which predicts that this mutant should be more "open" than WT SHP2. However, this mutation also disrupts the C459-H_2O-Q506 interaction. As a result, the water molecule needed in the second step of catalysis probably cannot be positioned properly. Consistent with this notion, basal PTP activity and PTP activity in the presence of an N-SH2 domain binding pTyr peptide are slightly lower (when measured against the artificial substrate pNPP) than in WT SHP2 [29]. We also monitored the activity of the isolated catalytic domain of the Q506P mutant. In accord with our structural data, the k_{cat} of Q506P was ~15–fold lower and the K_m about 2× higher than in WT SHP2 (Table 2).

Conclusion

SHP2 is regulated by a molecular switch mechanism that controls its catalytic activity. Upon binding to a tyrosine-phosphorylated binding partner for its SH2 domains, the N-terminal SH2 domain is released from the PTP

domain, activating the enzyme. This elegant mechanism ensures that PTP activity is delivered to the right place in the cell at the right time. Remarkably, germ line mutations that disrupt this regulatory machinery in different ways result in distinct disease syndromes. The crystal structures of "true" WT SHP2 and five NS/LS-associated SHP2 mutants reported herein provide direct comparisons of the local conformational changes caused by each mutation. Our structural observations are in agreement with, and can provide mechanistic insight into, the previously reported catalytic properties of these mutants. For example, mutation of D61G in the N-SH2 domain significantly impacts SHP2 activity because this residue is located at the N-SH2/PTP domain interface and its alteration weakens key interactions between the two domains. On the other hand, our data suggest that the C-SH2 domain mutation E139D might interfere with SHP2 binding to tryrosine-phosphorylated ligands. The other three mutants, Y279C, N308D and Q506P, are located in PTP domain, and the local conformational changes induced by each mutation provide insight into their abnormal catalytic properties. The results of our research provide structural insights into this medically important target and could aid in future structure-based drug discovery programs.

Additional files

Additional file 1: Description S1. SHP2 cloning and mutagenesis.

Additional file 2: Table S1. Buffers used for purification of SHP2 mutants.

Additional file 3: Figure S1. Activities of full-length WT and mutant SHP2 studied in this manuscript (from references [29,33]). The in vitro catalytic activities of the indicated GST-SHP2-FLAG proteins were measured using the artificial substrate ^{32}P-labeled reduced carboxamido-methylated and –maleylated lysozyme (^{32}P-RCML) in the absence or presence of an insulin receptor substrate-1-derived peptide containing phospho-tyrosine-1172 (pY1172) (100 μM). The pY1172 peptide binds to the N-SH2 domain, which in turn "opens up" the enzyme.

Competing interest
The authors declare that they have no competing interests.

Authors' contribution
QW carried out crystal structure determination, crystallographic refinement, structure analysis, and drafted the manuscript. WX performed assay studies and participated in the drafting of the manuscript. RV and HA performed protein sample preparation and structure analysis. LA carried out crystallization studies. RM performed construct design and cloning experiments for the mutants and wild type structures. BKP performed X-ray diffraction synchrotron experiments and participated in structure analysis. EFP performed structure analysis and participated in drafting the manuscript. BGN, NYC conceived of the study, participated in its design and coordination,

and drafted the manuscript. All authors read and approved the final manuscript.

Acknowledgements

These studies were supported by the Ontario Research and Development Challenge Fund (99-SEP-0512) and R37 CA49152 (to BGN). BGN and EFP are Canada Research Chairs, Tier 1, and work in their laboratories is partially supported by the Ontario Ministry of Health and Long Term Care and The Princess Margaret Cancer Foundation. Use of the IMCA-CAT beamline 17-ID at the Advanced Photon Source was supported by the companies of the Industrial Macromolecular Crystallography Association through a contract with the Hauptman-Woodward Medical Research Institute. Use of the Advanced Photon Source was supported by the U.S. Department of Energy, Office of Science, Office of Basic Energy Sciences, under Contract No. DE-AC02-06CH11357. We thank Aiping Dong for providing technical support and the Structural Genomics Consortium, University of Toronto, for the use of their X-ray facilities.

Author details

[1]Princess Margaret Cancer Center, University Health Network, Toronto, Ontario, M5G 2C4, Canada. [2]Department of Medical Biophysics, University of Toronto, Toronto, Ontario, M5S 1A8, Canada. [3]Hauptman-Woodward Medical Research Institute, IMCA-CAT, Advanced Photon Source, Argonne National Laboratory, Argonne, Illinois 60439, USA. [4]Departments of Biochemistry, Molecular Genetics, and Medical Biophysics, University of Toronto, Toronto, Ontario, M5S 1A8, Canada. [5]Department of Pharmacology and Toxicology, University of Toronto, Toronto, Ontario, M5S 1A8, Canada.

References

1. Neel BG, Gu H, Pao L: The 'Shp'ing news: SH2 domain-containing tyrosine phosphatases in cell signaling. *Trends Biochem Sci* 2003, **28**(6):284–293.
2. Feng GS: Shp-2 tyrosine phosphatase: signaling one cell or many. *Exp Cell Res* 1999, **253**(1):47–54.
3. Hof P, Pluskey S, Dhe-Paganon S, Eck MJ, Shoelson SE: Crystal structure of the tyrosine phosphatase SHP-2. *Cell* 1998, **92**(4):441–450.
4. Tonks NK, Neel BG: Combinatorial control of the specificity of protein tyrosine phosphatases. *Curr Opin Cell Biol* 2001, **13**(2):182–195.
5. Barford D, Neel BG: Revealing mechanisms for SH2 domain mediated regulation of the protein tyrosine phosphatase SHP-2. *Structure* 1998, **6**(3):249–254.
6. O'Reilly AM, Neel BG: Structural determinants of SHP-2 function and specificity in Xenopus mesoderm induction. *Mol Cell Biol* 1998, **18**(1):161–177.
7. Tartaglia M, Gelb BD: Germ-line and somatic PTPN11 mutations in human disease. *Eur J Med Genet* 2005, **48**(2):81–96.
8. Tartaglia M, Mehler EL, Goldberg R, Zampino G, Brunner HG, Kremer H, van der Burgt I, Crosby AH, Ion A, Jeffery S, Kalidas K, Patton MA, Kucherlapati RS, Gelb BD: Mutations in PTPN11, encoding the protein tyrosine phosphatase SHP-2, cause Noonan syndrome. *Nature genetics* 2001, **29**(4):465–468.
9. Tartaglia M, Pennacchio LA, Zhao C, Yadav KK, Fodale V, Sarkozy A, Pandit B, Oishi K, Martinelli S, Schackwitz W, Ustaszewska A, Martin J, Bristow J, Carta C, Lepri F, Neri C, Vasta I, Gibson K, Curry CJ, Siguero JP, Digilio MC, Zampino G, Dallapiccola B, Bar-Sagi D, Gelb BD: Gain-of-function SOS1 mutations cause a distinctive form of Noonan syndrome. *Nat Genet* 2007, **39**(1):75–79.
10. Roberts AE, Araki T, Swanson KD, Montgomery KT, Schiripo TA, Joshi VA, Li L, Yassin Y, Tamburino AM, Neel BG, Kucherlapati RS: Germline gain-of-function mutations in SOS1 cause Noonan syndrome. *Nat Genet* 2007, **39**(1):70–74.
11. Schubert S, Zenker M, Rowe SL, Boll S, Klein C, Bollag G, van der Burgt I, Musante L, Kalscheuer V, Wehner LE, Nguyen H, West B, Zhang KY, Sistermans E, Rauch A, Niemeyer CM, Shannon K, Kratz CP: Germline KRAS mutations cause Noonan syndrome. *Nat Genet* 2006, **38**(3):331–336.
12. Cirstea IC, Kutsche K, Dvorsky R, Gremer L, Carta C, Horn D, Roberts AE, Lepri F, Merbitz-Zahradnik T, Konig R, Kratz CP, Pantaleoni F, Dentici ML, Joshi VA, Kucherlapati RS, Mazzanti L, Mundlos S, Patton MA, Silengo MC, Rossi C, Zampino G, Digilio C, Stuppia L, Seemanova E, Pennacchio LA, Gelb BD, Dallapiccola B, Wittinghofer A, Ahmadian MR, Tartaglia M, Zenker M: A restricted spectrum of NRAS mutations causes Noonan syndrome. *Nat Genet* 2010, **42**(1):27–29.
13. Cordeddu V, Di Schiavi E, Pennacchio LA, Ma'ayan A, Sarkozy A, Fodale V, Cecchetti S, Cardinale A, Martin J, Schackwitz W, Lipzen A, Zampino G, Mazzanti L, Digilio MC, Martinelli S, Flex E, Lepri F, Bartholdi D, Kutsche K, Ferrero GB, Anichini C, Selicorni A, Rossi C, Tenconi R, Zenker M, Merlo D, Dallapiccola B, Iyengar R, Bazzicalupo P, Gelb BD, Tartaglia M: Mutation of SHOC2 promotes aberrant protein N-myristoylation and causes Noonan-like syndrome with loose anagen hair. *Nat Genet* 2009, **41**(9):1022–1026.
14. Pandit B, Sarkozy A, Pennacchio LA, Carta C, Oishi K, Martinelli S, Pogna EA, Schackwitz W, Ustaszewska A, Landstrom A, Bos JM, Ommen SR, Esposito G, Lepri F, Faul C, Mundel P, López Siguero JP, Tenconi R, Selicorni A, Rossi C, Mazzanti L, Torrente I, Marino B, Digilio MC, Zampino G, Ackerman MJ, Dallapiccola B, Tartaglia M, Gelb BD: Gain-of-function RAF1 mutations cause Noonan and LEOPARD syndromes with hypertrophic cardiomyopathy. *Nat Genet* 2007, **39**(8):1007–1012.
15. Razzaque MA, Nishizawa T, Komoike Y, Yagi H, Furutani M, Amo R, Kamisago M, Momma K, Katayama H, Nakagawa M, Fujiwara Y, Matsushima M, Mizuno K, Tokuyama M, Hirota H, Muneuchi J, Higashinakagawa T, Matsuoka R: Germline gain-of-function mutations in RAF1 cause Noonan syndrome. *Nat Genet* 2007, **39**(8):1013–1017.
16. Tartaglia M, Gelb BD: Noonan syndrome and related disorders: genetics and pathogenesis. *Annu Rev Genomics Hum Genet* 2005, **6**:45–68.
17. Bader-Meunier B, Tchernia G, Mielot F, Fontaine JL, Thomas C, Lyonnet S, Lavergne JM, Dommergues JP: Occurrence of myeloproliferative disorder in patients with Noonan syndrome. *J Pediatr* 1997, **130**(6):885–889.
18. Tartaglia M, Niemeyer CM, Shannon KM, Loh ML: SHP-2 and myeloid malignancies. *Curr Opin Hematol* 2004, **11**(1):44–50.
19. Loh ML, Vattikuti S, Schubbert S, Reynolds MG, Carlson E, Lieuw KH, Cheng JW, Lee CM, Stokoe D, Bonifas JM, Curtiss NP, Gotlib J, Meshinchi S, Le Beau MM, Emanuel PD, Shannon KM: Mutations in PTPN11 implicate the SHP-2 phosphatase in leukemogenesis. *Blood* 2004, **103**(6):2325–2331.
20. Chan G, Kalaitzidis D, Neel BG: The tyrosine phosphatase Shp2 (PTPN11) in cancer. *Cancer Metastasis Rev* 2008, **27**(2):179–192.
21. Legius E, Schrander-Stumpel C, Schollen E, Pulles-Heintzberger C, Gewillig M, Fryns JP: PTPN11 mutations in LEOPARD syndrome. *J Med Genet* 2002, **39**(8):571–574.
22. Gorlin RJ, Anderson RC, Moller JH: The leopard (multiple lentigines) syndrome revisited. *Laryngoscope* 1971, **81**(10):1674–1681.
23. Merks JH, Caron HN, Hennekam RC: High incidence of malformation syndromes in a series of 1,073 children with cancer. *Am J Med Genet A* 2005, **134A**(2):132–143.
24. Ucar C, Calyskan U, Martini S, Heinritz W: Acute myelomonocytic leukemia in a boy with LEOPARD syndrome (PTPN11 gene mutation positive). *J Pediatr Hematol Oncol* 2006, **28**(3):123–125.
25. Marin TM, Keith K, Davies B, Conner DA, Guha P, Kalaitzidis D, Wu X, Lauriol J, Wang B, Bauer M, Bronson R, Franchini KG, Neel BG, Kontaridis MI: Rapamycin reverses hypertrophic cardiomyopathy in a mouse model of LEOPARD syndrome-associated PTPN11 mutation. *The Journal of clinical investigation* 2011, **121**(3):1026–1043.
26. Krenz M, Yutzey KE, Robbins J: Noonan syndrome mutation Q79R in Shp2 increases proliferation of valve primordia mesenchymal cells via extracellular signal-regulated kinase 1/2 signaling. *Circ Res* 2005, **97**(8):813–820.
27. Araki T, Mohi MG, Ismat FA, Bronson RT, Williams IR, Kutok JL, Yang W, Pao LI, Gilliland DG, Epstein JA, Neel BG: Mouse model of Noonan syndrome reveals cell type- and gene dosage-dependent effects of Ptpn11 mutation. *Nat Med* 2004, **10**(8):849–857.
28. Fragale A, Tartaglia M, Wu J, Gelb BD: Noonan syndrome-associated SHP2/PTPN11 mutants cause EGF-dependent prolonged GAB1 binding and sustained ERK2/MAPK1 activation. *Hum Mutat* 2004, **23**(3):267–277.
29. Keilhack H, David FS, McGregor M, Cantley LC, Neel BG: Diverse biochemical properties of Shp2 mutants. Implications for disease phenotypes. *J Biol Chem* 2005, **280**(35):30984–30993.
30. Niihori T, Aoki Y, Ohashi H, Kurosawa K, Kondoh T, Ishikiriyama S, Kawame H, Kamasaki H, Yamanaka T, Takada F, Nishio K, Sakurai M, Tamai H, Nagashima T, Suzuki Y, Kure S, Fujii K, Imaizumi M, Matsubara Y: Functional analysis of PTPN11/SHP-2 mutants identified in Noonan syndrome and childhood leukemia. *J Hum Genet* 2005, **50**(4):192–202.
31. Tartaglia M, Martinelli S, Stella L, Bocchinfuso G, Flex E, Cordeddu V, Zampino G, Burgt I, Palleschi A, Petrucci TC, Alsorcini M, Schoch C, Foa R, Emanuel PD, Gelb BD: Diversity and functional consequences of germline and somatic PTPN11 mutations in human disease. *Am J Hum Genet* 2006, **78**(2):279–290.

32. Hanna N, Montagner A, Lee WH, Miteva M, Vidal M, Vidaud M, Parfait B, Raynal P: **Reduced phosphatase activity of SHP-2 in LEOPARD syndrome: consequences for PI3K binding on Gab1.** *FEBS Lett* 2006, **580**(10):2477–2482.

33. Kontaridis MI, Swanson KD, David FS, Barford D, Neel BG: **PTPN11 (Shp2) mutations in LEOPARD syndrome have dominant negative, not activating, effects.** *J Biol Chem* 2006, **281**(10):6785–6792.

34. Krenz M, Gulick J, Osinska HE, Colbert MC, Molkentin JD, Robbins J: **Role of ERK1/2 signaling in congenital valve malformations in Noonan syndrome.** *Proc Natl Acad Sci U S A* 2008, **105**(48):18930–18935.

35. Wu X, Simpson J, Hong JH, Kim KH, Thavarajah NK, Backx PH, Neel BG, Araki T: **MEK-ERK pathway modulation ameliorates disease phenotypes in a mouse model of Noonan syndrome associated with the Raf1(L613V) mutation.** *J Clin Invest* 2011, **121**(3):1009–1025.

36. Lauriol J, Kontaridis MI: **PTPN11-associated mutations in the heart: has LEOPARD changed Its RASpots?** *Trends Cardiovasc Med* 2011, **21**(4):97–104.

37. Freeman RM Jr, Plutzky J, Neel BG: **Identification of a human src homology 2-containing protein-tyrosine-phosphatase: a putative homolog of Drosophila corkscrew.** *Proc Natl Acad Sci U S A* 1992, **89**(23):11239–11243.

38. Magnusdottir A, Johansson I, Dahlgren LG, Nordlund P, Berglund H: **Enabling IMAC purification of low abundance recombinant proteins from E. coli lysates.** *Nature methods* 2009, **6**(7):477–478.

39. Kabsch W: **Integration, scaling, space-group assignment and post-refinement.** *Acta crystallographica Section D, Biol Crystallogr* 2010, **66**(Pt 2):133–144.

40. Emsley P, Lohkamp B, Scott W, Cowtan K: **Features and development of Coot.** *Acta Crystallogr D Biol Crystallogr* 2010, **66**(4):486–501.

41. Bricogne GBE, Brandl M, Flensburg C, Keller P, Paciorek W, Roversi P, Sharff A, Smart OS, Vonrhein C, Womack TO: **Buster 2.10.0.** In *2.10.0 edn*. Cambridge: United Kingdom: Global Phasing Ltd; 2011.

42. Digilio MC, Conti E, Sarkozy A, Mingarelli R, Dottorini T, Marino B, Pizzuti A, Dallapiccola B: **Grouping of multiple-lentigines/LEOPARD and Noonan syndromes on the PTPN11 gene.** *Am J Hum Genet* 2002, **71**(2):389–394.

43. Eck MJ, Pluskey S, Trub T, Harrison SC, Shoelson SE: **Spatial constraints on the recognition of phosphoproteins by the tandem SH2 domains of the phosphatase SH-PTP2.** *Nature* 1996, **379**(6562):277–280.

44. Tonks NK: **PTP1B: from the sidelines to the front lines!** *FEBS Lett* 2003, **546**(1):140–148.

45. Edouard T, Combier JP, Nedelec A, Bel-Vialar S, Metrich M, Conte-Auriol F, Lyonnet S, Parfait B, Tauber M, Salles JP, Lezoualc'h F, Yart A, Raynal P: **Functional effects of PTPN11 (SHP2) mutations causing LEOPARD syndrome on epidermal growth factor-induced phosphoinositide 3-kinase/AKT/ glycogen synthase kinase 3beta signaling.** *Mol Cell Biol* 2010, **30**(10):2498–2507.

46. Yu ZH, Xu J, Walls CD, Chen L, Zhang S, Zhang R, Wu L, Wang L, Liu S, Zhang ZY: **Structural and mechanistic insights into LEOPARD syndrome-associated SHP2 mutations.** *J Biol Chem* 2013, **288**(15):10472–10482.

4

High resolution structure of cleaved Serpin 42 Da from *Drosophila melanogaster*

Andrew M Ellisdon[1], Qingwei Zhang[1], Michelle A Henstridge[2], Travis K Johnson[1,2], Coral G Warr[2], Ruby HP Law[1] and James C Whisstock[1*]

Abstract

Background: The *Drosophila melanogaster Serpin 42 Da* gene (previously *Serpin 4*) encodes a serine protease inhibitor that is capable of remarkable functional diversity through the alternative splicing of four different reactive centre loop exons. Eight protein isoforms of Serpin 42 Da have been identified to date, targeting the protease inhibitor to both different proteases and cellular locations. Biochemical and genetic studies suggest that Serpin 42 Da inhibits target proteases through the classical serpin 'suicide' inhibition mechanism, however the crystal structure of a representative Serpin 42 Da isoform remains to be determined.

Results: We report two high-resolution crystal structures of Serpin 42 Da representing the A/B isoforms in the cleaved conformation, belonging to two different space-groups and diffracting to 1.7 Å and 1.8 Å. Structural analysis reveals the archetypal serpin fold, with the major elements of secondary structure displaying significant homology to the vertebrate serpin, neuroserpin. Key residues known to have central roles in the serpin inhibitory mechanism are conserved in both the hinge and shutter regions of Serpin 42 Da. Furthermore, these structures identify important conserved interactions that appear to be of crucial importance in allowing the Serpin 42 Da fold to act as a versatile template for multiple reactive centre loops that have different sequences and protease specificities.

Conclusions: In combination with previous biochemical and genetic studies, these structures confirm for the first time that the Serpin 42 Da isoforms are typical inhibitory serpin family members with the conserved serpin fold and inhibitory mechanism. Additionally, these data reveal the remarkable structural plasticity of serpins, whereby the basic fold is harnessed as a template for inhibition of a large spectrum of proteases by reactive centre loop exon 'switching'. This is the first structure of a *Drosophila* serpin reported to date, and will provide a platform for future mutational studies in *Drosophila* to ascertain the functional role of each of the Serpin 42 Da isoforms.

Keywords: Serpin 42Da, Serpin 4, Serine protease inhibitor, Neuroserpin, Drosophila, Furin

Background

Serpins are a large superfamily of protease inhibitors that were originally identified as *ser*ine *p*rotease *in*hibitors, but now encompass proteins that inhibit cysteine proteases or have non-inhibitory roles [1,2]. The serpin superfamily is represented in all branches of life with over 1500 serpins identified to date [1-3]. As such, serpins have a remarkably wide array of functions that include roles in immune defence, the blood coagulation pathway, and in hormone regulation and transport [1-3]. Within *Drosophila* there are over 20 inhibitory serpins,

many of which modulate the innate immune response [4]. Of these *Drosophila* serpins, eight are coded by the single *Serpin 42 Da* (*Spn42Da*) gene with each isoform formed through alternative splicing of different reactive centre loop (RCL) exons or signalling peptides [4].

Serpin structures are typified by a meta-stable native state, with a solvent exposed RCL that serves as 'bait' to bind and inhibit the target protease [1]. Specific recognition of the RCL by the target protease is primarily defined by the sequence of the RCL from the P15 to P3' positions, albeit studies have also shown a role for other exosites in determining protease-inhibitor recognition [1-3]. The peptide bond between the P1 and P1' residues is severed upon proteolytic attack by the target protease. Subsequently, the metastable serpin native

* Correspondence: james.whisstock@monash.edu
[1]Department of Biochemistry and Molecular Biology, Monash University, Clayton, VIC 3800, Australia
Full list of author information is available at the end of the article

state undergoes a large conformational change translocating the protease to the other pole of the serpin. The serpin-protease complex remains covalently bound, forming an ester bond between the catalytic residue of the protease and the main chain carbonyl of the P1 position. Thus, the protease is inhibited at the acyl-enzyme intermediate stage of the enzymatic cleavage reaction [2,3]. The resultant serpin-protease complex is highly stable, and effectively inhibits both the protease and serpin, leading to the description of serpins as 'suicide' inhibitors [3]. Within the final inhibitory complex, the serpin is in a hyper-stable conformation with the 'hinge' region of RCL forming the top of the central 4th strand of β-sheet A. This conformation can also spontaneously occur upon cleavage of the RCL loop, forming the stable 'cleaved' serpin conformation [1].

The *Drosophila melanogaster Spn42Da* gene, previously known as *Serpin 4*, encodes for eight different protein isoforms. Spn42Da isoforms B, D, E/F and I have N-terminal signal peptides and contain different RCL sequences due to the alternative splicing of four RCL encoding exons within the *Spn42Da* gene (Figure 1). The remaining four isoforms (A, G, H/K/L and J) have the same RCL splicing pattern, but do not have identifiable signal peptides and are thought to function within the cytosol [5,6]. A similar RCL splicing pattern, that generates multiple serpin isoforms from single genes, has been identified in nematodes and urochodates [5,7-9]. Spn42Da-A was the first isoform

to be characterised in *Drosophila*, with effective inhibition of proprotein convertases (PC), including human furin and *Drosophila* PC2, amontillado. Upon inhibition, Spn42Da-A forms a SDS-stable complex and has a stoichiometry of inhibition characteristic of other serpins [10-12]. Further biochemical studies have identified a diverse range of putative target protease families for the different Spn42Da isoforms, including serine proteases of the subtilase family, papain-like cysteine proteases, and members of the chymotrypsin family [13]. Therefore, through alternative splicing of the RCL, the *Spn42Da* gene is able to produce a wide range of intracellular and extracellular protease inhibitors that are targeted towards a remarkably diverse range of protease families. This has led to the hypothesis that in addition to a role in inhibition of PCs, Spn42Da isoforms may be essential for immune defence by inhibiting a large spectrum of pathogenic proteolytic enzymes [13]. However, further work is still required to characterise the potential diverse range of functions of the eight Spn42Da isoforms within *Drosophila*.

Although evidence suggests that the Spn42Da isoforms function as *bona fide* serpins and can form a covalent complex with target inhibitors, there are no reported structures of any Spn42Da isoform. Furthermore, it is currently unclear how the structure of the Spn42Da core can act as a versatile template to accommodate the switching of various RCL sequences whilst maintaining its function. In order to gain insight into these questions we have expressed and crystallised Spn42Da bearing the RCL from isoforms A and B to a high resolution. The Spn42Da-A/B structure (referred herein as Spn42Da) is in the cleaved conformation with a high degree of structural homology to the vertebrate serpin, neuroserpin. The structures illustrate the plasticity of the Spn42Da fold, and begin to describe how this fold is able to accommodate a wide variety of sequences through the use of alternatively spliced RCLs.

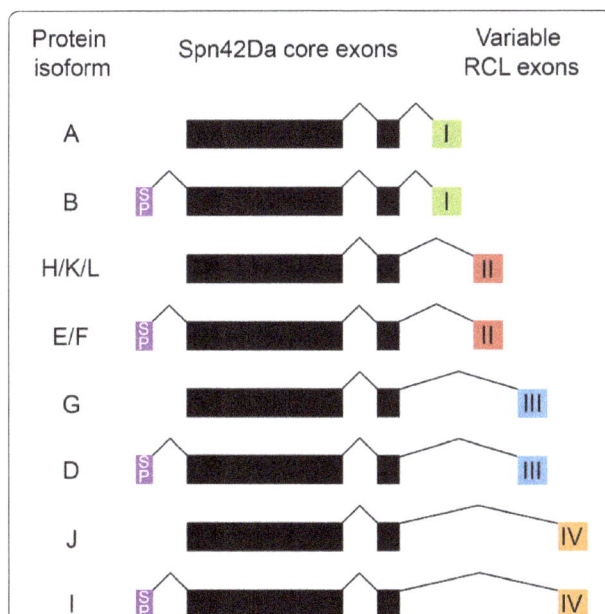

Figure 1 *Spn42Da* **encodes eight protein isoforms.** Two invariable core exons are followed by four alternatively spliced exons (I-IV) encoding different reactive centre loops (RCLs). For each of these, there is an alternative N-terminal exon encoding a secretion signal peptide (SP). Isoform names are according to FlyBase annotation (release 5.55) [6].

Results and discussion
Crystallisation and data quality
Crystals of Spn42Da grew after 6months to 1year at 20°C in two crystal forms, each with a single Spn42Da molecule in the asymmetric unit. The first crystal form, designated Spn42Da-1, crystallised in the $C222_1$ spacegroup (a = 59.74 Å, b = 125.95 Å, c = 119.94 Å) and diffracted to a resolution of 1.7 Å. The second crystal form, designated Spn42Da-2, crystallised in the I222 spacegroup (a = 87.22 Å, b = 109.77 Å, c = 140.58 Å) and diffracted to a resolution of 1.8 Å. The structure of Spn42Da-1 was determined by molecular replacement with cleaved neuroserpin as the search model (PDB code 3F02) [14]. The Spn42Da-2 structure was solved using the Spn42Da-1 structure as the search model for molecular replacement.

The structures are of high quality; the Spn42Da-1 structure refined to a R_{work}/R_{free} of 16.51% and 18.81% respectively, and the Spn42Da-2 structure to a R_{work}/R_{free} of 16.58% and 18.54% respectively. Within the Ramachandran plot, the Spn42Da-1 structure contains 98.9% of residues in the favoured region and no residues in the outlier regions. The Spn42Da-1 structure scored in the 100th percentile of structures in MolProbity, with a clashscore of 2.39 for all atoms and an overall MolProbity score of 1.02 [15]. The Spn42Da-2 structure also has excellent geometry, with 98.6% of amino acids in the favoured region of the Ramachandran plot and no outliers. The Spn42Da-2 structure scored in the 100th percentile in MolProbity with a clashscore of 1.02 and an overall MolProbity score of 0.80 [15]. The complete data collection statistics for the two crystal forms are reported in Table 1.

Spn42Da-1 has 369 amino acids from residue number 4 to 387 of the molecule, with three loops (D86-Q88, D191-R194, K367) and seven residues of the RCL (A343-E349) missing due to poor electron density - most likely reflecting protein cleavage or their mobility within the protein. The Spn42Da-2 structure has 371 residues, from residue 4 to 381 of the Spn42Da protein. Clear electron density is present for all loops within the structure except for seven residues of the RCL (R342-

E348). The last four residues of the Spn42Da protein, corresponding to a likely ER retention signal, are not present in the density of either structure.

The cleaved Spn42Da crystal structure

Spn42Da has a typical serpin fold comprising a mixed α-β secondary structure with an N-terminal helical region and a C-terminal β-barrel fold [2] (Figure 2A and B). The major elements of secondary structure characteristic of serpins are present, with a total of 3 β-sheets and 9 α-helices. The two crystal forms of Spn42Da are structurally homologous with no large conformational differences and an RMSD of 0.35 Å across 363 aligned residues (Figure 2C). As such, except for the presence of an extra six C-terminal amino acids within the Spn42Da-1 structure there are no major regions of difference between the two crystal forms. As Spn42Da-2 has the most complete electron density of internal loops between the two structures, it has been used for figures and comparative analysis unless otherwise stated.

The Spn42Da RCL sequence from P4 to P1 (R-R-K-R) corresponds to a classic furin-like consensus recognition sequence, with experimental evidence suggesting that cleavage occurs after the P1 R342 [10,12]. In the Spn42Da-1 crystal structure, clear electron density ends after the P1 Arg residue, suggesting that the RCL was cleaved after this

Table 1 Data collection and refinement statistics (highest resolution shell in brackets)

	Spn42Da-1	Spn42Da-2
Data collection statistics		
Beamline	Australian Synchrotron MX2	Australian Synchrotron MX2
Oscillation range (°)	1	1
Space group	C 2 2 21	I 2 2 2
Cell parameters	a = 59.74, b = 125.95, c = 119.94. α = 90, β = 90, γ = 90.	a = 87.22, b = 109.77, c = 140.58. α = 90, β = 90, γ = 90.
Resolution Range	27.88- 1.7 (1.79 - 1.7)	40.0 - 1.80 (1.9 - 1.8)
Observed reflections	378728	398398
Unique reflections	50081 (7222)	62423 (8962)
Completeness (%)	99.9 (100.00)	99.6 (98.9)
R_{merge}	0.100 (0.672)	0.128 (0.935)
I/σ (I)	12.7 (3.1)	7.3 (1.5)
Refinement statistics		
Resolution range (Å)	27.88 - 1.7 (1.761 - 1.7)	37.06 - 1.80 (1.864 - 1.8)
No. of protein atoms	2927	2946
No. of water atoms	344	377
R_{work}/R_{free}	0.1651/0.1881 (0.2341/0.2497)	0.1658/0.1854 (0.2774/0.3012)
Rmsd from ideal bond length (Å)	0.006	0.008
angles (°)	1.03	1.09
Ramachandran plot (%)		
Favoured region	99	99
Outlier region	0	0

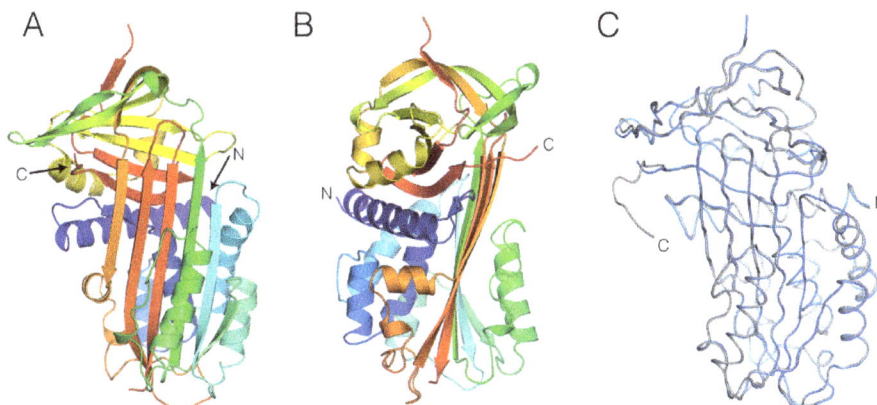

Figure 2 Crystal structure of cleaved Spn42Da. A, Overview of the Spn42Da-2 crystal structure in cartoon format with rainbow colouring from the amino (blue) to the carboxy (red) terminus. **B,** As per **(A)** rotated 90 degrees about the y-axis. **C,** Ribbon representation of the two Spn42Da models (Spn42Da-1 grey, Spn42Da-2 blue) superimposed.

consensus recognition site between the P1 Arg and the P1' Ala residues. Whereby in the Spn42Da-2 structure, clear electron density is only present up to the P2 Lys residue. Spn42Da is in the highly stable 'cleaved' conformation, with the cleaved RCL inserted into β-sheet A to form the 4th strand in the six stranded central β-sheet (Figure 3A) [1]. No density corresponding to residues P1' to P7' is present within the crystal structures, thus confirming they represent the cleaved confirmation and not the latent conformation whereby the intact RCL is inserted into β-sheet A [3]. Spn42Da was purified in the native conformation, with cleavage of the RCL most likely occurring during the extended crystallisation time. It is common for the RCL to be cleaved within the crystallisation solution over time, thereby allowing the protein to readily crystalize in the stable 'cleaved' conformation. This is the first reported crystal structure of a *Drosophila* serpin, and confirms that the Spn42Da isoforms have a protein fold typical of

serpins. Furthermore, the ability of the RCL to insert into β-sheet A confirms the capability of the Spn42Da protein to transit from the metastable native state to a stable RCL inserted confirmation typical of inhibitory serpins. Combined with the biochemical and genetic evidence provided by previous studies [10-12], these structures confirm that the Spn42Da isoforms indeed act as typical serpins, capable of inhibiting diverse families of proteases by the classic serpin 'suicide' inhibition mechanism.

The serpin inhibitory mechanism is conserved in Spn42Da

Serpin structures from other insects, including serpin 1K from *Manduca sexta* and SPN48 from *Tenebrio molitor*, have been solved in their native conformation [16,17]. These structures have the typical fold of other native-state inhibitory serpins, whereby the RCL extends into the solvent to act as bait for the target protease. Of the vertebrate serpins, Spn42Da is most closely related to

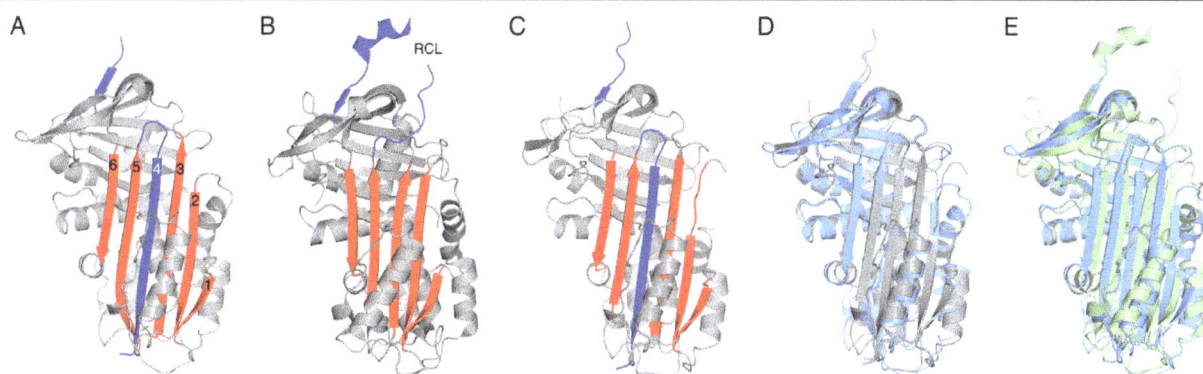

Figure 3 Conserved homology between Spn42Da and neuroserpin. A, Cartoon representation of cleaved Spn42Da highlighting the β-sheet A in red and the inserted RCL in blue with the strands numbered. **B,** Crystal structure of native neuroserpin (PDB code 3FGQ [18]) coloured as per **(A). C,** Crystal structure of cleaved neuroserpin (PDB code 3F02 [14]) coloured as per **(A). D,** Cartoon representation of cleaved neuroserpin (grey; PDB code 3F02 [14]) superimposed with cleaved Spn42Da (blue). **E,** Ribbon representation of native neuroserpin (green; PDB code 3FGQ [18]) superimposed with cleaved Spn42Da (blue).

mammalian neuroserpin with 34% sequence similarity and shared functional characteristics [11]. Indeed, the sequence similarity between Spn42Da and neuroserpin is very close to that shared between Spn42Da and the other insect serpins solved to date (~35%). Here, we have exploited the availability of highly characterised structures of both the native and cleaved conformations of neuroserpin, in addition to the similar functionality and high sequence similarity to Spn42Da. This has allowed us to gain insight into the conserved inhibitory mechanism of Spn42Da, by directly comparing our Spn42Da structure with the two (native and cleaved) neuroserpin structures [14,18].

Previous studies have shown that the RCL of native neuroserpin extends into the solvent with partial α-helical structure (PDB 3FGQ) [18] (Figure 3B). Upon cleavage, the RCL inserts into the molecule forming a single strand (S4A) of the central β-sheet A (PDB 3F02) (Figure 3C) [14]. This conformational change between native and cleaved neuroserpin is representative of the structural transition of the majority of other characterised inhibitory serpins [1,19]. However differences in the initial RCL position are apparent in some serpins, including mammalian antithrombin and insect SPN48, where the RCL hinge is partially inserted into the 'breach' region of β-sheet A forming a short β-strand 4A [17,20]. Attempts are ongoing to crystallise Spn42Da in its native conformation in order to determine the precise conformation of the RCL (extended or partially inserted) and the implications of this orientation for Spn42Da activity.

Cleaved Spn42Da and cleaved neuroserpin (PDB 3F02) are highly homologous with an RMSD of 1.25 Å across 343 aligned residues [14] (Figure 3D). The major elements of secondary structure align: both Spn42Da and neuroserpin are in the typical cleaved conformation with the inserted RCL forming one (S4A) of the 6 strands within the central β-sheet A (Figure 3D). The major difference between the two structures occurs in the position and length of the connecting loop regions. Spn42Da and native neuroserpin superpose with a RMSD of 2.2 Å across 327 aligned residues (Figure 3E). The β-sheet A is smaller by a single strand in native neuroserpin which gives the molecule a more compact fold. The C^α positions of Spn42Da helices B, D, E, and F and β-sheet A strands 1–3 undergo the greatest displacement upon cleavage and subsequent insertion of the RCL into β-sheet A (Figure 3E). This secondary structural movement is consistent with models of RCL cleavage and insertion for neuroserpin and other serpins [3,18].

The shutter region is a conserved cluster of amino acids that provide key interactions for controlling the opening of the central β-sheet A and insertion of the RCL [18,21]. This region is therefore of critical importance for the serpin inhibitory mechanism and protein stability. Sequence alignment of Spn42Da and neuroserpin identified conserved residues in the shutter region between the two proteins, suggesting a uniform inhibitory mechanism (Figure 4A). Specifically, five residues that have previously been identified to play a key role in the hydrogen bonding network of the shutter region, are highly conserved between neuroserpin and Spn42Da thus allowing us to analyse the likely changes that occur in Spn42Da upon protease inhibition and RCL insertion. Indeed, these residues are highly conserved across the entire serpin superfamily, with sequence alignment of over 200 serpins highlighting that S36, S39, N166, and H317 (Spn42Da numbering) are the most common residues to be found at these positions within the shutter region [21].

In native neuroserpin, the shutter region is composed of a hydrogen bonding network between the central H338 and S340 on strand S5A, residues S49 and S52 at the top of helix B, and N182 on strand S3A [18] (Figure 4B). In comparison, the shutter region of cleaved Spn42Da between S5A and S3A has opened to accommodate the RCL which forms strand S4A of β-sheet A (Figure 4C). Residue N166 of strand S3A and S39 on helix B are heavily displaced when T334 of the RCL inserts into β-sheet A as it opens to form the cleaved conformation. Despite this major change in secondary structure, a hydrogen bonding network within the shutter region is retained in the cleaved conformation. RCL residue T334 likely forms multiple hydrogen bonds with surrounding key conserved amino acids including N166, S39 and H317. Therefore, the residues within the shutter region that are required to accommodate the large conformational changes that occur in the transition from the native state to the hyperstable cleaved conformation, are conserved in Spn42Da. These data suggest that the classic serpin inhibitory mechanisms are conserved in the Spn42Da isoforms.

Structural basis for gene splicing and RCL switching

The most remarkable aspect of the *Spn42Da* gene is the capacity of the RCL exons to be variably spliced, with the eight protein isoforms exhibiting differential cellular localisation and protease targets [5,10-13]. Sequence alignment of the Spn42Da isoforms and neuroserpin reveals sequence conservation in the critical hinge region and in two clusters of the variably spliced RCL (Figure 5A). Each of the isoforms display clear variability across the consensus recognition sequence upstream of the critical P1 position. This variability affects the activity of the isoforms for different protease families. Previous studies identified Spn42Da-A/B as a potent inhibitor of furins, and the remaining variants as inhibitors of cathepsins, chymotrypsin and elastases [10-13].

The cleaved Spn42Da structure provides our first structural insight into how RCL switching is accommodated by

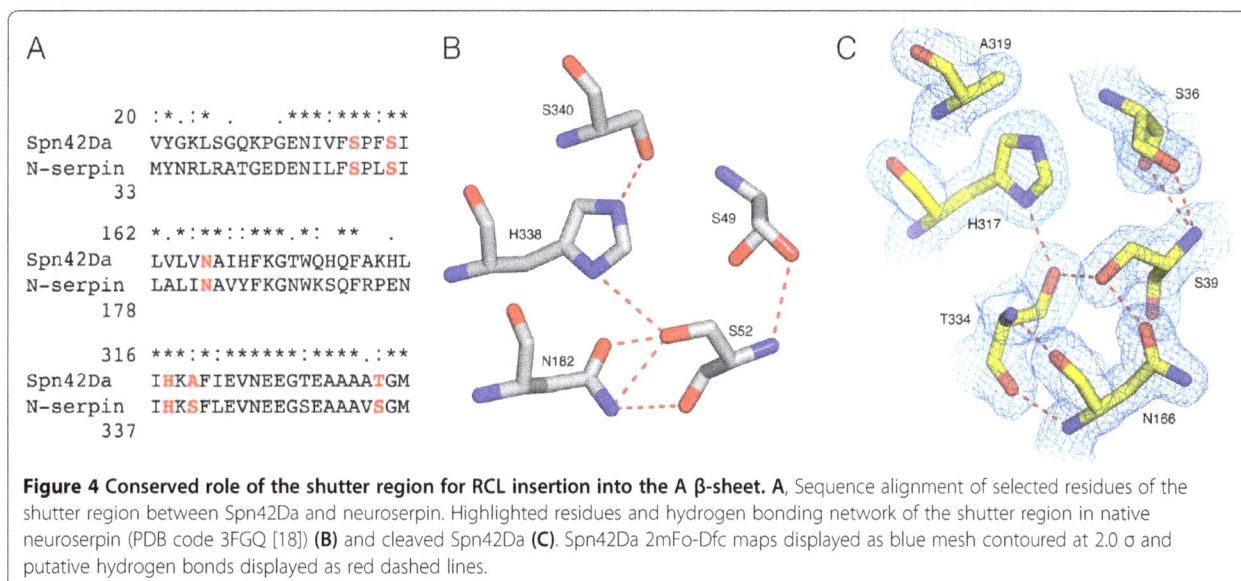

Figure 4 Conserved role of the shutter region for RCL insertion into the A β-sheet. **A**, Sequence alignment of selected residues of the shutter region between Spn42Da and neuroserpin. Highlighted residues and hydrogen bonding network of the shutter region in native neuroserpin (PDB code 3FGQ [18]) **(B)** and cleaved Spn42Da **(C)**. Spn42Da 2mFo-Dfc maps displayed as blue mesh contoured at 2.0 σ and putative hydrogen bonds displayed as red dashed lines.

the Spn42Da protein fold (Figure 5B). Spn42Da is cleaved after the P2 K341, with residues R342-E348 of the variably spliced region RCL not evident in the structure. These missing residues are either mobile in the crystal structure or missing due to the result of further proteolysis (Figure 5B). Clear electron density is present for the highly conserved hinge region comprised of relatively small amino acids, with residues N324 to T334 forming the S5A-S4A loop and the top of S4A within the central β-sheet A (Figure 5B). This hinge region is highly conserved between Spn42Da and neuroserpin, highlighting the recognised importance of this critical flexible region in the serpin inhibitory mechanism [2,3]. Residues G335 to V338 of the variably spliced region

are accommodated into the bottom of strand S4A, with the larger residues of the consensus recognition sequence positioned into the solvent at the bottom of strand S4A, making minimal important interactions. Within the variably spliced region there are only two clusters of amino acids that are highly conserved and appear crucial to incorporating the variably spliced RCL exons into the serpin scaffold to produce active protease inhibitors. Residues H357, P358, and F359 are completely conserved between the Spn42Da isoforms and neuroserpin, and form essential interactions in the turn leading into, and the beginning of strand S4B. Conserved residues F372 and G374 also appear vital in maintaining interactions between the interface of β-sheet B and β-sheet

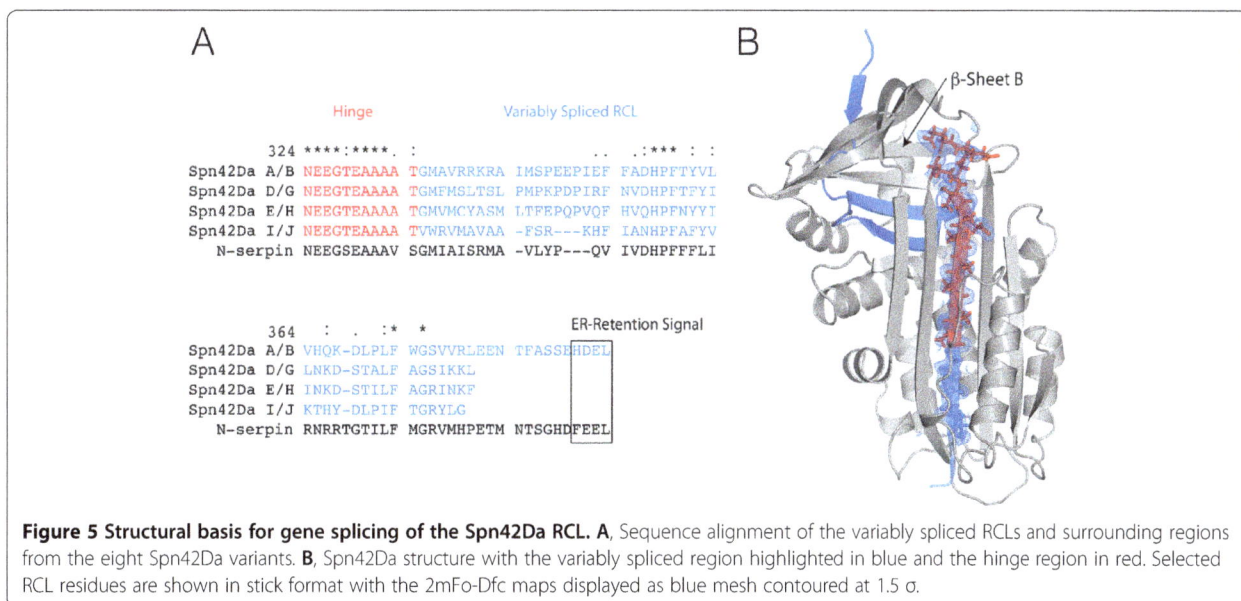

Figure 5 Structural basis for gene splicing of the Spn42Da RCL. **A**, Sequence alignment of the variably spliced RCLs and surrounding regions from the eight Spn42Da variants. **B**, Spn42Da structure with the variably spliced region highlighted in blue and the hinge region in red. Selected RCL residues are shown in stick format with the 2mFo-Dfc maps displayed as blue mesh contoured at 1.5 σ.

A and the packing between strands of the β-sheet B, respectively.

These data highlight the remarkable ability of the serpin protein fold to act as an accommodating scaffold for variable RCL sequences. Strict conservation appears only required in two clusters of the variably spliced RCL, in order to maintain the structural integrity of the top β-sheet B. After cleavage, the P1 residue and consensus recognition sequence are positioned at the bottom of strand S4A, where they make few critical interactions and therefore show high sequence variability. As such, the Spn42Da protein fold can accommodate the high sequence variability across the Spn42Da isoforms, allowing for extensive versatility in targeting a range of protease families.

Conclusions

We have solved the first crystal structures of Spn42Da from *Drosophila melanogaster* in the cleaved conformation, with two different crystal forms diffracting to 1.7 and 1.8 Å resolution. These data confirm for the first time that Spn42Da is a *bona fide* serpin, with the typical protein fold of this family of protease inhibitors. The Spn42Da structure has a high degree of homology to the mammalian neural serpin, neuroserpin, albeit with minor differences in the loop regions connecting the major elements of secondary structure. Structural comparison between Spn42Da and neuroserpin defines a likely conserved inhibitory mechanism, with sequence conservation of critical residues in both the hinge and shutter regions of Spn42Da. Importantly, the Spn42Da structure illustrates the structural features of the Spn42Da protein fold that are crucial in allowing it to act as a template that can be directed to inhibit diverse protease families through RCL switching. Furthermore, the Spn42Da structure provides the basis for future mutational targeting of Spn42Da isoforms to understand their various roles in *Drosophila*.

Methods

Cloning, expression, purification

PCR amplified Spn42Da cDNA was cloned into a pET3a vector (Novagen) as an untagged protein corresponding to residues 1–392. Spn42Da was expressed overnight in Rosetta2 (DE3) pLysS cells (Novagen) by IPTG induction at 16°C. The cells were lysed by sonication in 50 mM Tris–HCl (pH 8.0), 150 mM NaCl, 5 mM β-mercaptoethanol, and a complete EDTA-free protease inhibitor tablet (Roche). The lysate was clarified by centrifugation, filtered through a 0.45 μm membrane, and diluted at a 1:1 ratio in buffer containing 50 mM Tris–HCl (pH 8.0), and 5 mM β-mercaptoethanol. The supernatent was loaded onto a 5 ml Hitrap Q HP column (GE Healthcare), and eluted with a gradient from 50 mM Tris–HCl (pH 8.0), 50 mM NaCl, and 5 mM β-mercaptoethanol to 50 mM Tris–HCl (pH 8.0), 1 M NaCl, and 5 mM β-mercaptoethanol. Fractions containing Spn42Da were combined and dialysed against buffer containing 50 mM Tris–HCl (pH 8.0), 20 mM NaCl, and 5 mM β-mercaptoethanol. The Hitrap Q purification stage was repeated two times to increase the purity of Spn42Da. The resultant fractions were applied to a Superdex 75 16/60 prep-grade column (GE Healthcare) and eluted in 25 mM Tris–HCl (pH 8.0), 75 mM NaCl, 5 mM β-mercaptoethanol, and 0.02% (w/v) NaN$_3$.

Crystallisation, and structure determination

Spn42Da-1 crystals were grown at 20°C by hanging drop vapour diffusion in 0.2M lithium chloride, 20% (w/v) PEG3350. Spn42Da-2 crystals were grown at 20°C by hanging drop vapour diffusion in 0.2M ammonium phosphate monobasic, 20% (w/v) PEG3350. Spn42Da crystals were flash cooled in liquid nitrogen in mother liquor containing 20% ethylene glycol. Crystallographic data were collected at the Australian Synchrotron at the MX2 beamline [22]. Data were processed and scaled using iMOSFLM and programs within the CCP4 suite [23]. The Spn42Da-1 structure in the C222$_1$ spacegroup was solved by obtaining initial phases by molecular replacement using a cleaved neuroserpin structure [14] (PDB code 3F02) as the search model in PHENIX using the Phaser program [24]. The structure was automatically built in ArpWarp and iterative cycles of refinement were carried out using PHENIX Refine and REFMAC5 [24-26]. Local rebuilding was performed in Coot [27], resulting in a model with an *R*-factor of 16.51% (R_{free} of 18.81%) and excellent geometry (Table 1). The Spn4A-2 structure in the I222 spacegroup was determined using the Spn42Da-1 structure as the molecular replacement search model with refinement and rebuilding carried out as per the Spn42Da-1 structure. The resultant model has an *R*-factor of 16.58% (R_{free} of 18.54%) with excellent geometry (Table 1). All figures were made using PyMol and coordinates have been deposited at the Protein Data Bank with accession codes 4P0F and 4P0O for the Spn42Da-1 and Spn42Da-2 structures respectively.

Abbreviations
Spn42Da: Serpin 42Da; RCL: Reactive centre loop; PC: Proprotein convertases.

Competing interests
The author's declare that they have no competing interests.

Author's contributions
AME determined the structures and wrote the manuscript. QZ, MAH, TKJ and RHPL supervised and carried out the cloning, protein expression and crystallization. AME, TKJ, RHPL, CGW and JCW conceived the study, helped in its design, coordination, and drafted the manuscript. All authors read and approved the final manuscript.

Acknowledgements
Data were collected on the MX2 beamline at the Australian Synchrotron, Victoria, Australia.

Author details
[1]Department of Biochemistry and Molecular Biology, Monash University, Clayton, VIC 3800, Australia. [2]School of Biological Sciences, Monash University, Clayton, VIC 3800, Australia.

References

1. Law RHP, Zhang Q, McGowan S, Buckle AM, Silverman GA, Wong W, Rosado CJ, Langendorf CG, Pike RN, Bird PI, Whisstock JC: An overview of the serpin superfamily. *Genome Biol* 2006, **7**:216.

2. Huntington JA: Serpin structure, function and dysfunction. *J Thromb Haemost* 2011, **9**:26–34.

3. Whisstock JC, Silverman GA, Bird PI, Bottomley SP, Kaiserman D, Luke CJ, Pak SC, Reichhart J-M, Huntington JA: Serpins flex their muscle: II. Structural insights into target peptidase recognition, polymerization, and transport functions. *J Biol Chem* 2010, **285**:24307–24312.

4. Reichhart J-M, Gubb D, Leclerc V: The Drosophila Serpins. In *Methods in Enzymology. Volume 499*. Cambridge, MA, USA: Elsevier; 2011:205–225.

5. Krüger O, Ladewig J, Köster K, Ragg H: Widespread occurrence of serpin genes with multiple reactive centre-containing exon cassettes in insects and nematodes. *Gene* 2002, **293**:97–105.

6. St Pierre SE, Ponting L, Stefancsik R, McQuilton P: FlyBase Consortium: FlyBase 102–advanced approaches to interrogating FlyBase. *Nucleic Acids Res* 2014, **42**:D780–D788.

7. Pak SC, Pak SC, Kumar V, Tsu C, Luke CJ, Askew YS, Askew DJ, Mills DR, Brömme D, Silverman GA: SRP-2 Is a Cross-class Inhibitor That Participates in Postembryonic Development of the Nematode Caenorhabditis elegans: INITIAL CHARACTERIZATION OF THE CLADE L SERPINS. *J Biol Chem* 2004, **279**:15448–15459.

8. Ragg H: The role of serpins in the surveillance of the secretory pathway. *Cell Mol Life Sci* 2007, **64**:2763–2770.

9. Börner S, Ragg H: Functional diversification of a protease inhibitor gene in the genus Drosophila and its molecular basis. *Gene* 2008, **415**:23–31.

10. Oley M, Letzel MC, Ragg H: Inhibition of furin by serpin Spn4A from Drosophila melanogaster. *FEBS Lett* 2004, **577**:165–169.

11. Osterwalder T, Kuhnen A, Leiserson WM, Kim Y-S, Keshishian H: Drosophila serpin 4 functions as a neuroserpin-like inhibitor of subtilisin-like proprotein convertases. *J Neurosci* 2004, **24**:5482–5491.

12. Richer MJ, Keays CA, Waterhouse J, Minhas J, Hashimoto C, Jean F: The Spn4 gene of Drosophila encodes a potent furin-directed secretory pathway serpin. *Proc Natl Acad Sci U S A* 2004, **101**:10560–10565.

13. Brüning M, Lummer M, Bentele C, Smolenaars MMW, Rodenburg KW, Ragg H: The Spn4 gene from Drosophila melanogaster is a multipurpose defence tool directed against proteases from three different peptidase families. *Biochem J* 2007, **401**:325–331.

14. Ricagno S, Caccia S, Sorrentino G, Antonini G, Bolognesi M: Human neuroserpin: structure and time-dependent inhibition. *J Mol Biol* 2009, **388**:109–121.

15. Chen VB, Arendall WB, Arendall WB III, Headd JJ, Keedy DA, Immormino RM, Kapral GJ, Murray LW, Richardson JS, Richardson DC: MolProbity: all-atom structure validation for macromolecular crystallography. *Acta Crystallogr D Biol Crystallogr* 2009, **66**:12–21.

16. Li J, Wang Z, Canagarajah B, Jiang H, Kanost M, Goldsmith EJ: The structure of active serpin 1K from Manduca sexta. *Structure* 1999, **7**:103–109.

17. Park SH, Jiang R, Piao S, Zhang B, Kim E-H, Kwon H-M, Jin XL, Lee BL, Ha N-C: Structural and functional characterization of a highly specific serpin in the insect innate immunity. *J Biol Chem* 2011, **286**:1567–1575.

18. Takehara S, Onda M, Zhang J, Nishiyama M, Yang X, Mikami B, Lomas DA: The 2.1-A crystal structure of native neuroserpin reveals unique structural elements that contribute to conformational instability. *J Mol Biol* 2009, **388**:11–20.

19. Whisstock JC, Bottomley SP: Molecular gymnastics: serpin structure, folding and misfolding. *Curr Opin Struct Biol* 2006, **16**:761–768.

20. Carrell RW, Stein PE, Fermi G, Wardell MR: Biological implications of a 3 A structure of dimeric antithrombin. *Structure* 1994, **2**:257–270.

21. Irving JA, Pike RN, Lesk AM, Whisstock JC: Phylogeny of the serpin superfamily: implications of patterns of amino acid conservation for structure and function. *Genome Res* 2000, **10**:1845–1864.

22. McPhillips TM, McPhillips SE, Chiu H-J, Cohen AE, Deacon AM, Ellis PJ, Garman E, Gonzalez A, Sauter NK, Phizackerley RP, Soltis SM, Kuhn P: Blu-Ice and the Distributed Control System: software for data acquisition and instrument control at macromolecular crystallography beamlines. *J Synchrotron Radiat* 2002, **9**:401–406.

23. Collaborative Computational Project, Number 4: The CCP4 suite: programs for protein crystallography. *Acta Crystallogr D Biol Crystallogr* 1994, **50**:760–763.

24. Adams PD, Afonine PV, Bunkóczi G, Chen VB, Davis IW, Echols N, Headd JJ, Hung L-W, Kapral GJ, Grosse-Kunstleve RW, McCoy AJ, Moriarty NW, Oeffner R, Read RJ, Richardson DC, Richardson JS, Terwilliger TC, Zwart PH: PHENIX: a comprehensive Python-based system for macromolecular structure solution. *Acta Crystallogr D Biol Crystallogr* 2010, **66**:213–221.

25. Langer G, Cohen SX, Lamzin VS, Perrakis A: Automated macromolecular model building for X-ray crystallography using ARP/wARP version 7. *Nat Protoc* 2008, **3**:1171–1179.

26. Murshudov GN, Skubák P, Lebedev AA, Pannu NS, Steiner RA, Nicholls RA, Winn MD, Long F, Vagin AA: REFMAC5 for the refinement of macromolecular crystal structures. *Acta Crystallogr D Biol Crystallogr* 2011, **67**:355–367.

27. Emsley P, Cowtan K: Coot: model-building tools for molecular graphics. *Acta Crystallogr D Biol Crystallogr* 2004, **60**:2126–2132.

5

A simple method for finding a protein's ligand-binding pockets

Seyed Majid Saberi Fathi[1] and Jack A Tuszynski[2*]

Abstract

Background: This paper provides a simple and rapid method for a protein-clustering strategy. The basic idea implemented here is to use computational geometry methods to predict and characterize ligand-binding pockets of a given protein structure. In addition to geometrical characteristics of the protein structure, we consider some simple biochemical properties that help recognize the best candidates for pockets in a protein's active site.

Results: Our results are shown to produce good agreement with known empirical results.

Conclusions: The method presented in this paper is a low-cost rapid computational method that could be used to classify proteins and other biomolecules, and furthermore could be useful in reducing the cost and time of drug discovery.

Keywords: Protein structure, Ligand-binding pockets, Computational methods

Background

Essential information regarding protein function is generally dependent on the protein's tertiary structure. This includes the enzymatic function of a protein, and also the binding of ligands, such as small molecule inhibitors [1]. Methods developed for predicting an enzymatic function of a protein by identifying catalytic residues include: finding local characteristics of functional residues [2,3], applying known templates of active sites [4,5] or identifying the surface shape of active sites [6-10].

In order to predict ligand binding (sites, poses and affinities), we first need to determine a 3-dimensional structure of the protein in question, which can be done using several experimental or computational methods [11,12]. Structure-based pocket prediction employs geometrical algorithms or probes mapping/docking algorithms [13]. Comparing these two kinds of methods, it can be said that the geometrical algorithms have low computational costs in contrast to the mapping/docking and scoring of molecular fragments, but the latter algorithms have a greater physical meaning. Geometrical algorithms analyze protein surfaces, and once a structure has been determined, a number of algorithms may be used to predict binding pockets on the protein surface

[14-19]. One such example, SURFNET [15], fits spheres into the spaces between protein atoms and finds gap regions. The results obtained this way correspond to the cavities and keys of a given protein. An algorithm based on geometric hashing called VISGRID [20] uses the visibility of constituent atoms to identify cavities. "Active site points" are identified by PASS [19]. In this method the protein surface is coated with a layer of spherical probes, then those that clash with the protein or which are not sufficiently buried are filtered out. The active site points are identified from the final probes. Another method is LIGSITE [14,21], which is an improvement of the POCKET algorithm [22]. This algorithm puts protein-occupied space in a grid and identifies clefts by scanning areas that are enclosed on both sides by the protein's atoms. An alpha-shape algorithm is used by CAST [17] and APROPOS [18]. DRUGSITE [13] and POCKET-FINDER [23], in addition to the protein's shape, also consider physicochemical properties for identification of ligand binding pockets. Further geometrical algorithms are TRAVEL DEPTH [24], VOIDOO [25], and CAVITY SEARCH [26]. QSITEFINDER [16] uses interaction energy computation between the protein and a van der Waals probe to find favorable binding sites. Some methods using mapping/docking and scoring of molecular fragment concepts are described by Dennis et al. [27], Kortvelyesi et al. [28], Ruppert et al. [29], and

* Correspondence: jackt@ualberta.ca
[2]Department of Physics, University of Alberta, Edmonton, Alberta, Canada
Full list of author information is available at the end of the article

Verdonk et al. [30]. There are also several docking based methods that use ligands to probe the proteins for binding sites [31-34].

Computer-aided drug design often applies protein–ligand docking methods, most commonly structure-based methods. These methods provide support to the rational design and optimization of novel drug candidates [35]. Many structure-based protein–ligand docking methods have been reported in the literature [36-41]. These methods generally rely on first identifying a ligand-binding pocket in the protein structure.

Finding a comprehensive, fast and automated method that can accurately predict ligand-binding pockets on protein surfaces is a major challenge in virtual screening biophysics. This goal leads us to introduce a new method for finding putative ligand-binding pockets on a protein surface, and for identifying the most important characteristics of these pockets: surface area, volume, and potential interacting atoms. This information could be used to cluster protein pockets into similarity classes, and could be a valuable resource leading to a significant decrease in the cost and time expended in the drug discovery process.

The method we present in this paper is based on computational geometry and voxelization concepts. In this method we do not use Delaunay tessellation, the vision criterion, or fitting spheres between atoms, in contrast to some of the methods mentioned above. The CASTp method has used the Delaunay triangulation and the Voronoi concepts to find putative pockets and voids. This method triangulates the surface atoms and clusters triangles by merging small triangles to neighboring large triangles [14,17]. In our work we simply use the convex hull concept and generate a pocket by a grid box formed by the extreme points of a triangle. Then, we consider only the atoms closest to the triangle in the formed pocket. The distance to the convex hull is used for choosing the surface atoms. Thus, our method is not iterative and does not require a flow through all points, hence the computational cost is relatively low. We also take only a given number of empty voxel neighbors for each atom. Voxelization of space for finding putative pockets does not have an essential role for finding surface atoms, unlike VISGRID or grid-based methods, which are based on searching for empty voxels in different directions. We also use voxelization for finding the positions of possible ligands and also to determine physical properties of the pockets.

Comparative modeling methods use fold assignment and template selection for comparing the target protein to a set of proteins with known structures and to search for homologous proteins that have approximately similar structures. Some of these methods are BLAST [42,43], PSI-BLAST [44] and HHpred [45]. I-TASSER [46] is a composite approach of comparative modeling and threading methods

[47]. A summary of comparative modeling is given by [48]. In our method we also consider *some* biochemical properties of the protein's atoms and residues as is explained below. Hence, the proposed method is not purely geometrical. We demonstrate that the results obtained using this method are in good agreement with empirically known results. Hence developing it further may offer even more accurate and reliable results.

Methods

We first voxelize the volume of a box defined to contain the extreme points of the protein's atomic positions. Then, we use the convex hull concept to obtain the smallest convex polyhedron containing all of the protein's atoms. In 3-dimensional space, a convex hull surface is formed by triangles, as shown in Figure 1. In the present context, each of these triangles can define a pocket, as illustrated in Figure 2. To define a specific pocket, we consider the volume generated by the extreme positions of the triangle vertices as follows: each triangle contains three vertex points,

$$r_i \equiv (x_i, y_i, z_i), (i = 1, 2, 3),$$

which we should consider as

$$(\text{extreme}(x_i), \text{extreme}(y_i), \text{extreme}(z_i)),$$

where "extreme" indicates either a minimum or a maximum value. Figure 2 shows a given triangle on a convex hull. We have made the grids with a length of 1 Å between $([x_{min} : x_{max}], ([y_{min} : y_{max}], ([z_{min} : z_{max}])$ in each axis.

Normally, some parts of this rectangular cube are out of the convex hull, but we do not concern ourselves with

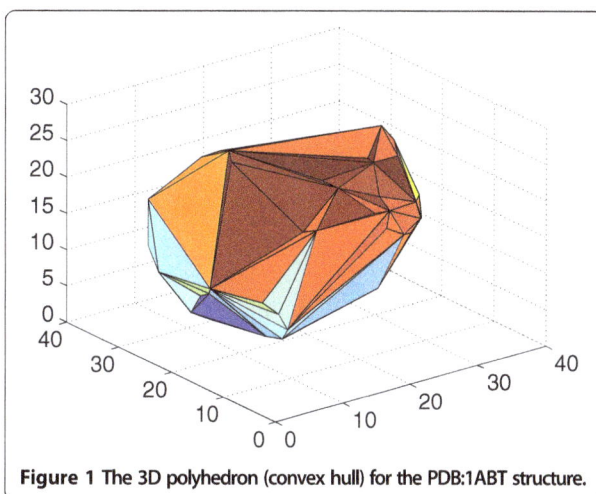

Figure 1 The 3D polyhedron (convex hull) for the PDB:1ABT structure.

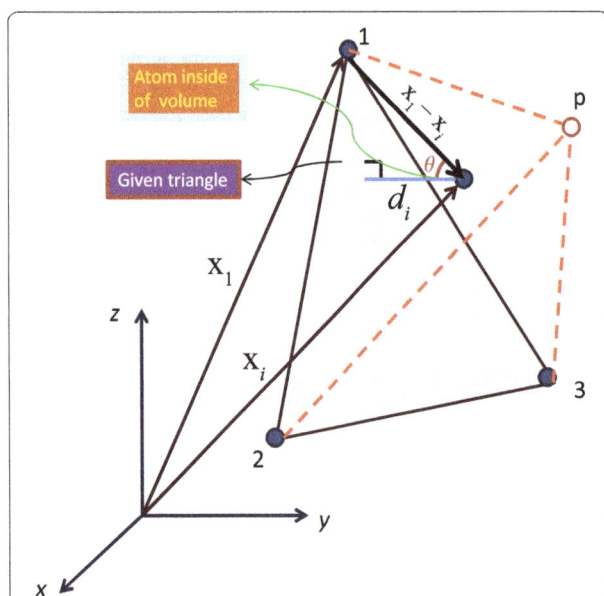

Figure 2 A given triangle on the convex hull for the PDB:1ABT structure. The three vertices are labeled as 1, 2, and 3. The point p is determined by the extreme values of x, y, and z of these three vertices. The distance of atom i to the triangle is obtained as follows: first obtain the normal vector to the triangle, N, $\mathbf{N} = (\mathbf{x_2} - \mathbf{x_1}) \times (\mathbf{x_3} - \mathbf{x_1})$, where x_1, x_2, and x_3 are the vectors from the origin of the systems of Cartesian coordinates to the three vertices. Then, calculate the angle between the normal vector and the line passing through atom i and one of the vertices of this triangle using the following relation: $\cos\theta = \frac{(\mathbf{x_1} - \mathbf{x_i}) \cdot \mathbf{N}}{|\mathbf{x_1} - \mathbf{x_i}||\mathbf{N}|}$. Finally, we compute this distance by $d_i = |\mathbf{x_1} - \mathbf{x_i}| \cos\theta$, where x_i is a vector joining the origin and a given point in this volume.

color represents the atoms and the red represents the empty grid points).

Then, we obtain the voxels, which are contained within this generated volume, and separate the voxels into those that contain protein atoms and those which do not. Next, we identify the nearest empty voxels with respect to these protein atoms. These empty voxels give us the possible positions of ligand atoms for this particular protein pocket. At this step, we have found a large number of "pocket" envelopes and all the atoms belonging to these pockets are the "protein's surface atoms".

In some cases, the entire space (or part thereof) under a triangle is common with another space so we say that these spaces overlap with each other. The overlap is defined by the number of atoms in common between the two pockets divided by the total number of atoms in a pocket, which means the overlap is also dependent on the size of a pocket, so that the overlap between two pockets is not symmetric. Figure 4 shows the overlap between two pockets in 2-dimensional space. As we can see in this figure, the overlap size of the common site (determined by the number of common atoms) divided by the size of the pocket (the total number of atoms in the pocket) for each pocket is different.

If all atoms contained in a set of the pocket atoms exist in the other pocket, it has an overlap of 100%. However, the second pocket may have more atoms than the first one, i.e. it has all atoms of the first pocket plus other atoms. For example, the overlap between pockets #1 and #2 might be 100% while the overlap between pockets #2 and #1 is only 50%, because the number of atoms in pocket #2 is twice as large as the number of atoms in pocket #1, and all atoms belonging to pocket #1 are also contained in pocket #2, but only a half of the atoms in pocket #2 are also in pocket #1. Accumulating all pockets with a given overlap between them as new pockets is the next step.

them because they will be eliminated by another criterion, namely to keep only a given number of empty voxels near each protein atom in a pocket. Figure 3 shows only the inside of a convex hull part of a pocket in 2-dimensions and its grid is shown by points (the blue

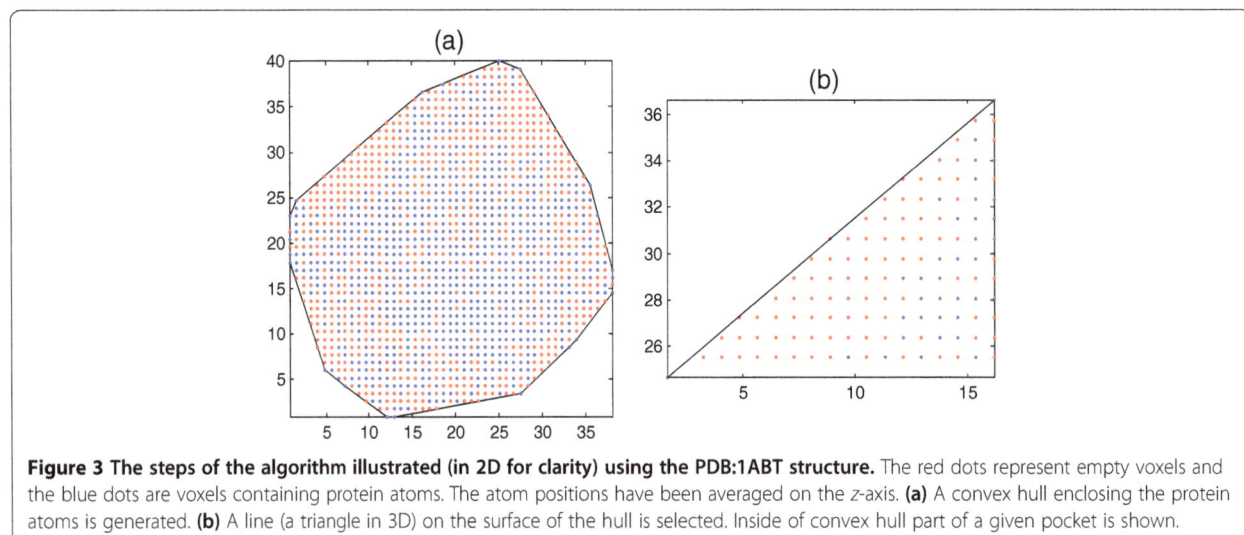

Figure 3 The steps of the algorithm illustrated (in 2D for clarity) using the PDB:1ABT structure. The red dots represent empty voxels and the blue dots are voxels containing protein atoms. The atom positions have been averaged on the z-axis. **(a)** A convex hull enclosing the protein atoms is generated. **(b)** A line (a triangle in 3D) on the surface of the hull is selected. Inside of convex hull part of a given pocket is shown.

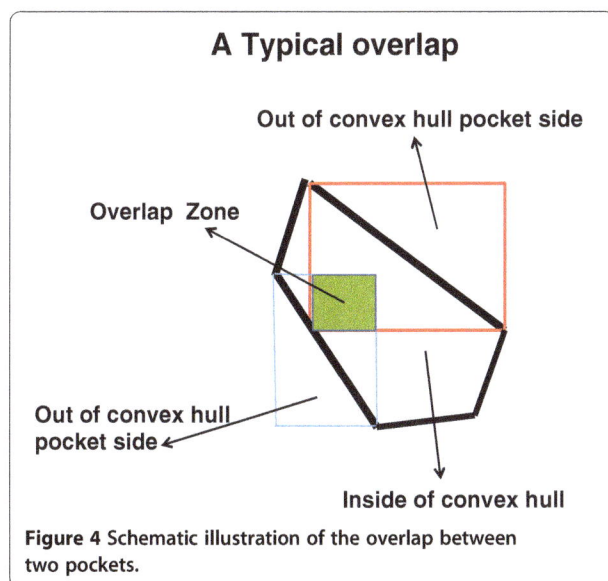

Figure 4 Schematic illustration of the overlap between two pockets.

Table 1 Main biochemical interactions of atoms and residues in the proteins [49,51,52]

Residue Name	Interactions
THR	**HBD**: OG1 (OH)
SER	**HBD**: OG (OH)
GLN	**HBA**: NE2 (NH₂)
ASN	**HBA**: ND2 (NH₂)
TYR	**HBA**: O – **HBD**: N, OH – **CR**: CE1, CE2, CD1, CD2, CZ, CG
CYS	**Sul**: SG (SH)
MET	**vdW**: CE (CH₃) – **Sul**: SD (S-CH₃)
ALA	**vdW**: CB (CH₃)
PRO	**vdW**: CB (CH₂), CD (CH₂), CG (CH₂)
LEU	**vdW**: CD1 (CH₃), CD2 (CH₃), CG (CH)
VAL	**vdW**: CG1 (CH₃), CG2 (CH₃), CB (CH)
ILE	**vdW**: CD1 (CH₃)
ASP	**HBA**: OD1(C = O) – **Ion(−)**: OD2 (OH)
GLU	**HBA**: OE1(C = O) – **Ion(−)**: OE2 (OH)
LYS	**Ion(+)**: NZ (NH₃)
ARG	**Ion(+)**: NH1 (NH₂) *trans*, NH2 (NH₂) *cis*
HIS	**Ion(+)**: NE1 (NH₂) *trans*, NE2 (NH₂) *cis* – **CR**: CD1, CE1, CD2, CE2, CG
PHE	**CR**: CG, CD1, CE1, CZ, CE2, CD2
TRP	**HBD**: NE1 (NH) – **CR**: CD2, CE2, CZ2, CH2, CZ3, CE3
TYR	**HBD**: OH – **CR**: CD1, CE1, CE2, CZ, CD2, CG
GLY	No participation

Abbreviations used: HBA: Hydrogen bond acceptor, HBD: Hydrogen bond donor, vdW: van der Waals interaction, Ion: Ionic interaction, Sul: Sulfur interaction.

The final step is related to biochemical and physical criteria such as hydrophobicity, hydrogen bonding, ionic and van der Waals interactions, and also the depth, surface area and volume comparisons between a given pocket and a ligand. By using biochemical conditions, we can find which atoms and which corresponding residues could potentially participate in an interaction with the ligand's atoms. Tables 1 and 2 propose a set of simple biochemical conditions. It should be noted that to find an active site, more accurate conditions should lead to more accurate results. In this step we can also compute the size of pockets.

A detailed description of the algorithm is given in the following:

The algorithm

1. Input protein atom position data, and define a box by using the extreme positions of the atoms.
2. Voxelize the box by considering the voxel with 1 Å in length, width and height.
3. Compute the convex hull surrounding the protein atoms and obtain the volume of the convex hull and the surface area of atoms.
4. Separate empty voxels (possible ligand atom positions) from voxels filled by the protein atoms in the convex hull.
5. Define the pockets by the volume generated by the vertices of each triangle on the convex hull.
6. Compute the overlap between two neighboring pockets and assemble the pockets with an overlap greater than a minimum value (reconstruct new pockets).
7. Find the physical properties of the pockets such as depth, surface and volume.
8. Find the residues corresponding to the pocket atoms.
9. Assess the biochemical conditions [49,50] as introduced in Table 1 (we use the IUPAC nomenclature [51] and the PDB format [52]). In this step we can find the atoms and residues that participate in the potential active site.
10. Compare physical and biochemical properties between ligand atoms (Table 2) and the atoms of a given pocket, such as: the size of pockets (depth, surface and volume) with ligand size, the number of

Table 2 Ligand biochemistry

C-Ring in ligand	C or N atoms in ligand recognizing by connection information in the PDB
Unprotonated atoms in ligand	1) O has a connection with N, P or Zn
	2) O only has a connection with C
Protonated atoms in ligand	1) Ca
	2) N has only two connection with C

The bond list is given in the PDB file CONECT lines.

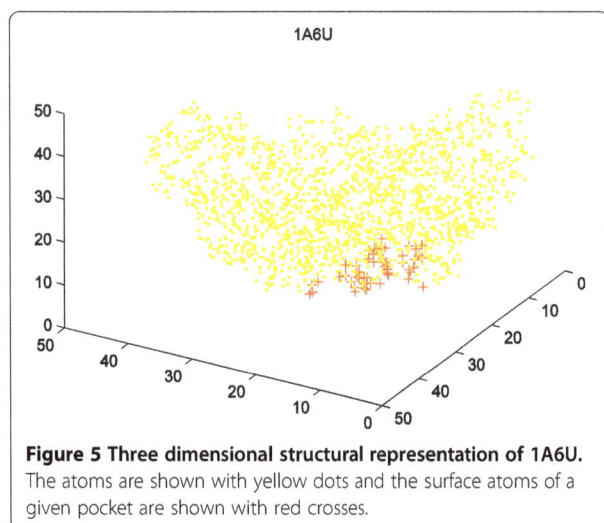

Figure 5 Three dimensional structural representation of 1A6U. The atoms are shown with yellow dots and the surface atoms of a given pocket are shown with red crosses.

hydrogen donor/acceptor atoms, possible rings, or van der Waals interactions, etc.

Supplementary steps to compare our results with known active sites

11. Compute the number of correct residues predicted in each pocket of the unliganded protein and divide it by the number of residues in an "active site" of the liganded protein as reported in the PDB, i.e.

cf = correct fraction

$$= \frac{\text{number of correct residues predicted in pocket}}{\text{number of residues in active site}}$$

12. *Optional step.* Compute the minimum distance between the ligand atoms and each residue atoms in the pocket. Then, filter residues of a pocket with the minimum distance greater than the given values, for example 3.50 Å.

In Figure 3, we illustrate these steps in 2-dimensional space for better clarity. Here, we need to use a line instead of a triangle to define a pocket. Figure 5 uses the example of the protein labeled 1A6U in the PDB. It shows 3-dimensional atomic positions of the protein and the atoms that belong to a pocket.

Results and discussion

In reality, the geometrical criteria give initial information about physical properties for the possible protein-protein or protein-ligand docking, determining shapes, sizes, etc. For docking to occur, the recognized geometrical protein

Table 3 Pockets and their characteristics recognized by our method for 1A6U protein atoms

Pocket Num.[*]	Num. of Atoms	Num. of Empty voxels	Surface of Pocket	Depth of Pocket	NoA[**] HA[a] Bonds	NoA HD[b] Bonds	NoA vdW[***] Bonds	NoA Ionic Bonds	NoA Sulfur Bonds	NoA C-Ring	cf of the 1st AS, HAP[c]	cf of the 2nd AS, AC1[c]
1	63	401	116.25	28.40	5	8	0	1	0	20	0.31	0.33
5	80	481	21.83	38.66	2	3	10	2	0	2	0	0
18	101	648	187.27	25.83	5	7	6	2	0	14	0.12	0.11
19	67	411	84.36	19.35	1	2	5	0	0	2	0	0
38	44	266	138.90	20.63	1	4	1	0	0	6	0	0
39	85	499	82.58	28.26	3	5	2	0	0	14	0.31	0.22
40	21	127	77.97	14.53	2	3	0	0	0	4	0.06	0
58	118	765	340.90	29.83	5	4	7	3	0	3	0	0
59	86	529	253.20	26.72	4	4	4	2	0	6	0.06	0
85	226	1360	370.14	36.18	7	7	26	3	1	27	0	0
89	21	141	212.35	21.47	0	1	4	1	0	4	0	0
90	92	573	293.28	28.54	4	2	15	2	0	11	0	0
112	44	241	36.33	27.39	1	2	1	0	0	6	0.06	0
117	38	215	76.66	17.42	1	3	0	0	0	8	0	0
137	15	99	127.57	17.53	2	4	0	0	0	3	0.25	0.33
143	55	354	259.10	24.24	4	8	0	1	0	20	0.43	0.55

[*]Pocket number indicates the number in the protein's atomic positions convex hull surface rows, and they correspond to three vertices of triangles.
[**]NoA means the number of atoms.
[***]vdW means van der Waals.
[a]HA means hydrogen bond acceptor.
[b]HD means hydrogen bond donor.
[c]These are the *cf*-values (ratio of the number of correct residues to the total number of residues in the active site). For 1A6W in PDB two active sites (AS) are reported as HAP and AC1.

pocket should be a protein's active site. Finding active sites is very complicated for both *in vitro* and *in silico* methods. There are many computer programs that find active sites [13-23] but they have high computational cost associated with them and also they do not typically determine physical properties of the active site which means that we need to find a ligand in spite of lacking some important information. Therefore, it is imperative to use mixed geometrical and biochemical methods to find possible pockets in a protein. This paper has introduced a method to find protein pockets with a higher probability of interactions than based on exclusively biochemical methods. This method offers a speed-up of the drug discovery process by allowing clustering of both the protein pockets and ligands.

We first demonstrate our method by describing an example, namely a pair of unliganded and liganded proteins, 1A6U and 1A6W. We have used only non-water atoms of 1A6U to find its pockets. These pockets are reported in Table 3. To verify these results, we check the SITE REMARK lines for the PDB file of its liganded pair structure, i.e. 1A6W, and we compare the residues of each active sites of the PDB file 1A6W with the residues obtained in each computed pocket. Then, we obtain the *cf* –value for each active site. The last two columns of Table 3 report these values.

Here, we give a summary discussion regarding the properties of the unliganded protein structure 1A6U. It has 1737 atoms and its box has $43 \times 49 \times 41$ voxels. The convex hull completely surrounded by triangles involves 148 triangles, which means the 1A6U structure can have at most 148 possible pockets. However, only 81 pockets remain with a 0.8 overlap cutoff between pockets. By using biochemical conditions, only 20 pockets remain and then by using physical conditions of depth and surface, only 16 pockets remain. These remaining pockets are listed in Table 3. Finally, only four pockets are left with a *cf* of 25% correctly predicted residues as shown in Table 4. The liganded protein reported in the PDB is 1A6W (1774 non-water atoms), and has the NIP ligand, which has 17 atoms with an 8.97 Å length and a 20.87 Å² surface area. Thus, the protein pockets should have values of depth and surface area greater than these. The minimum distance between the atoms of ARG 350H in 1A6U with the atoms of the active sites in 1A6W is 2.89 Å. Table 4 shows the pockets' residues and their minimum residue distances for 1A6U to the ligand atoms of NIP reported in the heterogenic atom lines in the PDB file of 1A6W.

Table 3 gives all pockets of 1A6U, where only the two last columns are obtained by the comparison of the results with the binding sites HAP and AC1 of 1A6W (the corresponding liganded protein of 1A6U). In Table 3 the pockets are numbered and ordered arbitrarily. This

Table 4 1A6U best pockets with residues in common with the 2 active sites, HAP and AC1

POCKET # 1, *cf* = 0.31 & 0.33		
ASN 354H (11.61)	SER 331H (10.79)	TYR 34 L (4.27)
ASP 352H (7.07)	THR 328H (14.41)	TYR 332H (8.34)
ILE 351H (6.25)	THR 330H (12.29)	TYR 401H (2.92)
SER 32 L (6.81)	TRP 333H (1.734)	TYR 402H (5.75)
POCKET # 39, *cf* = 0.31 & 0.22		
ALA 2 L (15.1365)	HIS 97 L (6.8477)	THR 26 L (15.7431)
ARG 350H (2.89)	ILE 348H (9.34)	TRP 98 L (3.24)
ASN 96 L (7.12)	LYS 359H (5.38)	TRP 347H (4.78)
ASN 361H (9.75)	LYS 365H (14.84)	TYR 94 L (7.84)
GLU 362H (12.30)	PHE 364H (13.46)	TYR 360H (8.34)
GLY 349H (6.45)	SER 366H (17.38)	VAL 99 L (9.69)
POCKET # 137, *cf* = 0.25 & 0.33		
ASP 400H (5.44)	THR 31 L (8.29)	TYR 401H (2.92)
SER 405H (3.65)	TYR 34 L (4.27)	TYR 402H (5.75)
POCKET # 143, *cf* = 0.44 & 0.56		
ARG 350H (2.89)	SER 95 L (5.42)	TYR 332H (8.34)
ASN 354H (11.61)	SER 331H (10.79)	TYR 401H (2.92)
ASP 352H (7.07)	TRP 93 L (3.36)	TYR 402H (5.75)
ILE 351H (6.25)	TRP 333H (1.73)	
SER 32 L (6.81)	TYR 34 L (4.27)	

There are four predicted pockets with more than 25% of residues in common between the pockets and the active sites. The values in parentheses are the minimum residue distances for 1A6U to the ligand atoms of NIP reported in the heterogenic atom lines in the PDB file of 1A6W.

table and all results were produced independently of the final answer.

As can be seen in Figure 6, which is shown in the PDB website for the 1A6W protein, only five residues – TYR 399H, ARG 350H, TRP 93 L, TYR 401H and TRP 98 L –

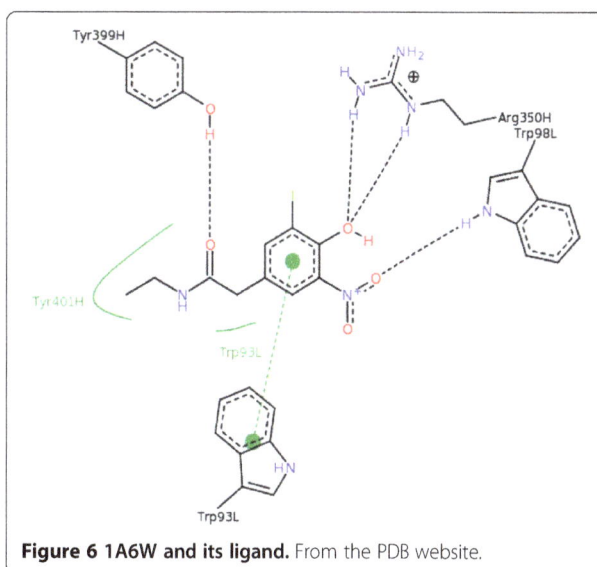

Figure 6 1A6W and its ligand. From the PDB website.

participate in the interaction with the NIP ligand, while in the PDB file of 1AW6 two active sites with 16 and 10 residues are reported (using the SITE REMARK lines in the PDB file). This shows that a maximum of 50% of the active site residues reported in the PDB for 1A6W participate in the interaction with the NIP ligand (a *cf* equal to 0.5). In our computation, for example, in the unliganded protein 1A6U the best pocket has a *cf* equal to 0.43 and to 0.55 for the first and second active site of the liganded protein 1A6W, respectively.

For illustration purposes we have taken the set of 48 and 86 "liganded and unliganded proteins", respectively, listed in the supplementary material of Li et al. [20] and downloaded the files from the PDB site (see Additional file 1 for a list of the PDB files). We found the pockets of the *unliganded proteins*, and then we compared these pockets with the known active sites reported in the PDB files of the corresponding *liganded proteins*.

The correct fraction, *cf*, of residues predicted in a given pocket is computed and the histograms of maximum *cf* in each protein's pockets are reported in Figures 7 and 8. These results are obtained for a 0.8 overlap cutoff between pockets, and they show that 76% of the pockets predicted by our algorithm in the 86-element data set have at least half of their residues belonging to an active site in the liganded protein; for the 48-element data set the corresponding number is 50%. By using instead a 0.5 overlap cutoff, the results are 78% and 54% for the 86-element and the 48-element data set, respectively. Note that not all

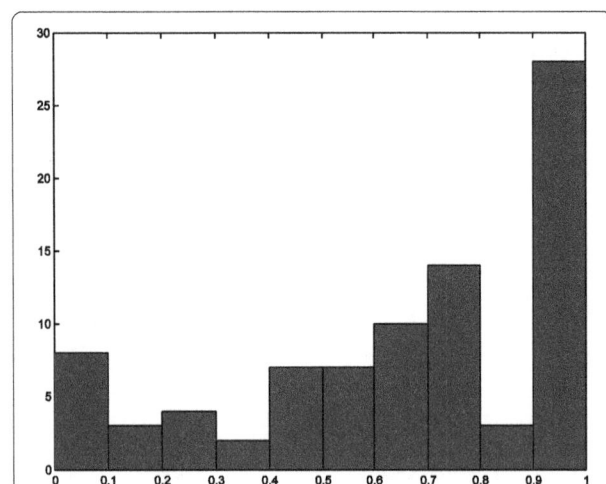

Figure 8 Histogram of the 48-element data set. The horizontal axis is the percentage of correct prediction of residues. The vertical axis is the number of proteins. The number of proteins with predicted pockets including more than half of the active site residues is 24 proteins (50% of the data set). Overlap threshold between pockets is 0.8.

residues in the active sites reported in the PDB participate in protein-ligand interactions.

In Table 5 we compare the performance of our method with the other methods CASTp, LIGSITE, PASS, SURFNET and VISGRID. This table shows that our method with an overlap cutoff of 0.8 has comparable performance with the other methods. We should also note that the low computational cost of our method is a major advantage. In Additional file 2, full pockets of the 48-element set with a *cf* (ratio of the number of correct residues to the total number of residues in the active site) of more than 25% are reported. Additional file 2 also gives the minimum distance between each residue of the protein and ligand atoms.

We have also chosen another 130 pairs of unliganded and liganded protein structures of (listed in Additonal file 3). In Figure 9 the histograms of the maximum *cf* in each protein's pockets are reported (with a 0.8 overlap).

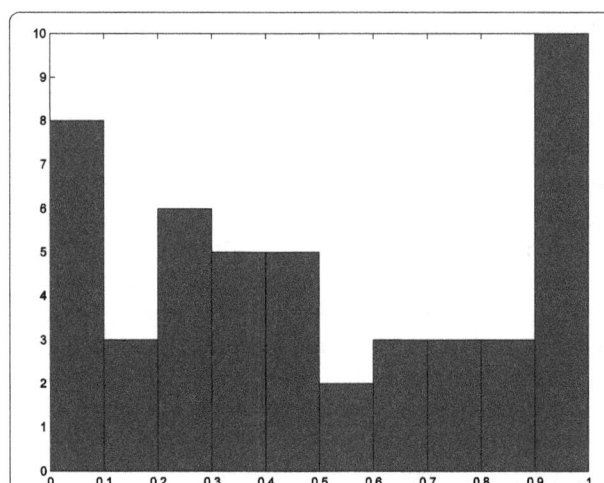

Figure 7 Histogram of the 86-element data set. Due to the RAM memory limits the protein number 55 in the 86-element data set list (PDB structures 2NGR and 1KZ7) was not included. The results are reported for the 85-element data set. The horizontal axis is the percentage of correct prediction of residues. The vertical axis is the number of proteins. The number of proteins with predicted pockets including more than half of the active site residues is 66 proteins (78% of the data set). Overlap threshold between pockets is 0.8.

Table 5 Performance comparison of our results with the other methods CASTp, LIGSITE, PASS, SURFNET and VISGRID

	48 Unbound structures (Top 1)	86 Unbound structures (Top 1)
CAST	31 (64.6%)	66 (76.7%)
LIGSITE	36 (75.0%)	69 (80.2%)
PASS	27 (56.3%)	54 (62.8%)
SURFNET	19 (39.6%)	63 (73.3%)
VISGRID: Top 0.8% voxels	34 (70.8%)	55 (64.0%)
Our method: Overlap 0.8	24 (50%)	66 (78%)

The other results reported in Table III of Li et al. [20].

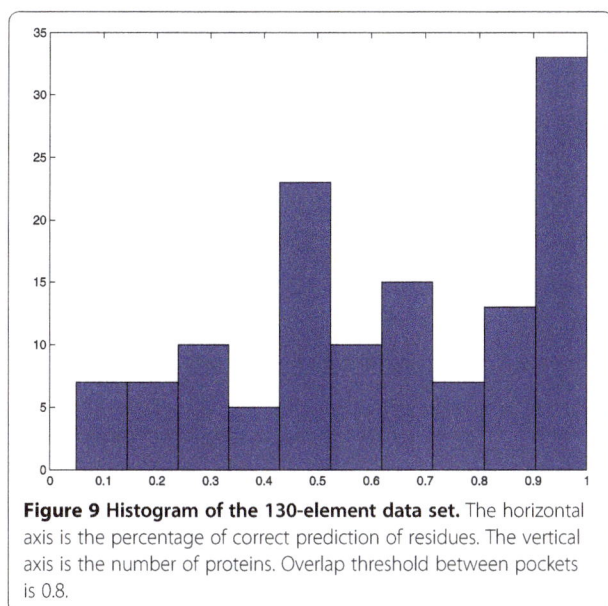

Figure 9 Histogram of the 130-element data set. The horizontal axis is the percentage of correct prediction of residues. The vertical axis is the number of proteins. Overlap threshold between pockets is 0.8.

It shows that 73.8% of the pockets predicted by our algorithm in the 130-element data set have at least half of their residues belonging to an active site in the liganded protein, i.e. $cf \geq 0.5$.

An important step which allows a decrease of the time and effort for the drug discovery process is to find suitable ligands through *in silico* methods using, for example, the virtual screening techniques. Our algorithm is easy to use and the cost of computation is approximately between 10 seconds for small proteins and up to 320 seconds for large proteins. The program was implemented in Matlab. The computer used for these computations is a laptop with an Intel Core i7 CPU and 8 GB RAM. The program usually uses 13% of the CPU time, but sometimes for a while it uses up to 50%. The program also while occupied in computation usually required less than 0.5 GB of RAM memory, but it was observed for some proteins to go up to 2 GB. The execution time for the 130 pair dataset is given in Additional file 3.

Conclusions

In this paper, we have introduced a new simple method for predicting putative ligand-binding protein pockets. For each pocket, we can identify possible interacting protein atoms and residues, surface atoms, and also determine the size of a pocket (volume, surface area and depth). This information can help us verify possible ligands having a shape and size that is geometrically compatible with the pocket, and thus could be docked to the protein. We have used some biochemical properties to find the possible interacting atoms and residues in the pockets. Our method is a low cost computational method which voxelizes the protein space, and uses the convex hull concept commonly employed in computational geometry. This method could

be used to classify proteins by the geometric properties of their pockets and also by their biochemical properties. An application of this method could be useful in reducing the cost and time of drug discovery.

Additional files

Additional file 1: Table with pairs (bound and unbound) of PDB files in the 48 element set and in the 86 element set. For each pair, the RMSD (in angstroms) is given.

Additional file 2: List of the full pockets for each unliganded structure in the 48-element set with a *cf* (ratio of the number of correct residues to the total number of residues in an active site) of more than 25%. For each pocket the *cf* for each active site ("AC") is given after the label "Res. in common with *N* AC:". Residues are named in the form "<resname > <resid > <chain>". For each residue in a pocket, the minimum distance between the residue and the ligand atoms of the corresponding liganded structure is given.

Additional file 3: Tab-delimited text file. Table with 130 pairs of unliganded (unlig) and liganded (lig) PDB files. For each pair, the *cf* and the time of execution (in seconds) is given.

Competing interests
The authors declare that they have no competing interests.

Authors' contributions
SMSF and JAT conceived of the study, and participated in its design and coordination and helped to draft the manuscript. Both authors read and approved the final manuscript.

Acknowledgements
SMSF acknowledges grant number 2/21897 from Ferdowsi University of Mashhad. JAT gratefully acknowledges research support received from the National Science and Engineering Research Council of Canada, the Canadian Breast Cancer Foundation, and the Allard Foundation.

Author details
[1]Department of Physics, Ferdowsi University of Mashhad, Mashhad, Iran.
[2]Department of Physics, University of Alberta, Edmonton, Alberta, Canada.

References
1. Polgár L: **The catalytic triad of serine peptidases.** *Cell Mol Life Sci CMLS* 2005, **62**:2161–2172.
2. Mooney SD, Liang MH-P, DeConde R, Altman RB: **Structural characterization of proteins using residue environments.** *Proteins* 2005, **61**:741–747.
3. Shulman-Peleg A, Nussinov R, Wolfson HJ: **Recognition of functional sites in protein structures.** *J Mol Biol* 2004, **339**:607–633.
4. Fetrow JS, Godzik A, Skolnick J: **Functional analysis of the Escherichia coli genome using the sequence-to-structure-to-function paradigm: identification of proteins exhibiting the glutaredoxin/thioredoxin disulfide oxidoreductase activity.** *J Mol Biol* 1998, **282**:703–711.
5. Wallace AC, Borkakoti N, Thornton JM: **TESS: a geometric hashing algorithm for deriving 3D coordinate templates for searching structural databases. Application to enzyme active sites.** *Protein Sci Publ Protein Soc* 1997, **6**:2308–2323.
6. Connolly ML: **Solvent-accessible surfaces of proteins and nucleic acids.** *Science* 1983, **221**:709–713.
7. Goldman BB, Wipke WT: **QSD quadratic shape descriptors. 2. Molecular docking using quadratic shape descriptors (QSDock).** *Proteins* 2000, **38**:79–94.
8. Duncan BS, Olson AJ: **Approximation and characterization of molecular surfaces.** *Biopolymers* 1993, **33**:219–229.

9. Exner TE, Keil M, Brickmann J: **Pattern recognition strategies for molecular surfaces. I. Pattern generation using fuzzy set theory.** *J Comput Chem* 2002, **23**:1176–1187.

10. Kinoshita K, Nakamura H: **Identification of protein biochemical functions by similarity search using the molecular surface database eF-site.** *Protein Sci Publ Protein Soc* 2003, **12**:1589–1595.

11. Rupp B, Wang J: **Predictive models for protein crystallization.** *Methods San Diego Calif* 2004, **34**:390–407.

12. Arnold K, Bordoli L, Kopp J, Schwede T: **The SWISS-MODEL workspace: a web-based environment for protein structure homology modelling.** *Bioinforma Oxf Engl* 2006, **22**:195–201.

13. An J, Totrov M, Abagyan R: **Comprehensive identification of "druggable" protein ligand binding sites.** *Genome Inform Int Conf Genome Inform* 2004, **15**:31–41.

14. Huang B, Schroeder M: **LIGSITEcsc: predicting ligand binding sites using the connolly surface and degree of conservation.** *BMC Struct Biol* 2006, **6**:19.

15. Laskowski RA: **SURFNET: a program for visualizing molecular surfaces, cavities, and intermolecular interactions.** *J Mol Graph* 1995, **13**:323–330. 307–308.

16. Laurie ATR, Jackson RM: **Q-SiteFinder: an energy-based method for the prediction of protein-ligand binding sites.** *Bioinforma Oxf Engl* 2005, **21**:1908–1916.

17. Liang J, Edelsbrunner H, Woodward C: **Anatomy of protein pockets and cavities: measurement of binding site geometry and implications for ligand design.** *Protein Sci Publ Protein Soc* 1998, **7**:1884–1897.

18. Peters KP, Fauck J, Frömmel C: **The automatic search for ligand binding sites in proteins of known three-dimensional structure using only geometric criteria.** *J Mol Biol* 1996, **256**:201–213.

19. Brady GP Jr, Stouten PF: **Fast prediction and visualization of protein binding pockets with PASS.** *J Comput Aided Mol Des* 2000, **14**:383–401.

20. Li B, Turuvekere S, Agrawal M, La D, Ramani K, Kihara D: **Characterization of local geometry of protein surfaces with the visibility criterion.** *Proteins* 2008, **71**:670–683.

21. Hendlich M, Rippmann F, Barnickel G: **LIGSITE: automatic and efficient detection of potential small molecule-binding sites in proteins.** *J Mol Graph Model* 1997, **15**:359–363.

22. Levitt DG, Banaszak LJ: **POCKET: a computer graphics method for identifying and displaying protein cavities and their surrounding amino acids.** *J Mol Graph* 1992, **10**:229–234.

23. An J, Totrov M, Abagyan R: **Pocketome via comprehensive identification and classification of ligand binding envelopes.** *Mol Cell Proteomics MCP* 2005, **4**:752–761.

24. Coleman RG, Sharp KA: **Travel depth, a new shape descriptor for macromolecules: application to ligand binding.** *J Mol Biol* 2006, **362**:441–458.

25. Kleywegt GJ, Jones TA: **Detection, delineation, measurement and display of cavities in macromolecular structures.** *Acta Crystallogr D Biol Crystallogr* 1994, **50**(Pt 2):178–185.

26. Ho CM, Marshall GR: **Cavity search: an algorithm for the isolation and display of cavity-like binding regions.** *J Comput Aided Mol Des* 1990, **4**:337–354.

27. Dennis S, Kortvelyesi T, Vajda S: **Computational mapping identifies the binding sites of organic solvents on proteins.** *Proc Natl Acad Sci U S A* 2002, **99**:4290–4295.

28. Kortvelyesi T, Silberstein M, Dennis S, Vajda S: **Improved mapping of protein binding sites.** *J Comput Aided Mol Des* 2003, **17**:173–186.

29. Ruppert J, Welch W, Jain AN: **Automatic identification and representation of protein binding sites for molecular docking.** *Protein Sci Publ Protein Soc* 1997, **6**:524–533.

30. Verdonk ML, Cole JC, Watson P, Gillet V, Willett P: **SuperStar: improved knowledge-based interaction fields for protein binding sites.** *J Mol Biol* 2001, **307**:841–859.

31. Bliznyuk AA, Gready JE: **Simple method for locating possible ligand binding sites on protein surfaces.** *J Comput Chem* 1999, **20**:983–988.

32. Campbell SJ, Gold ND, Jackson RM, Westhead DR: **Ligand binding: functional site location, similarity and docking.** *Curr Opin Struct Biol* 2003, **13**:389–395.

33. Glick M, Robinson DD, Grant GH, Richards WG: **Identification of ligand binding sites on proteins using a multi-scale approach.** *J Am Chem Soc* 2002, **124**:2337–2344.

34. Sotriffer C, Klebe G: **Identification and mapping of small-molecule binding sites in proteins: computational tools for structure-based drug design.** *Farm Soc Chim Ital 1989* 2002, **57**:243–251.

35. Andricopulo AD, Salum LB, Abraham DJ: **Structure-based drug design strategies in medicinal chemistry.** *Curr Top Med Chem* 2009, **9**:771–790.

36. Waszkowycz B, Clark DE, Gancia E: **Outstanding challenges in protein-ligand docking and structure-based virtual screening.** *Wiley Interdiscip Rev Comput Mol Sci* 2011, **1**:229–259.

37. Cavasotto CN, Orry AJW: **Ligand docking and structure-based virtual screening in drug discovery.** *Curr Top Med Chem* 2007, **7**:1006–1014.

38. Des Jarlais RL, Cummings MD, Gibbs AC: **Virtual docking: how are we doing and how can we improve?** *Front Drug Des Discov Struct-Based Drug Des 21st Century* 2007, **3**:81–103.

39. Moitessier N, Englebienne P, Lee D, Lawandi J, Corbeil CR: **Towards the development of universal, fast and highly accurate docking/scoring methods: a long way to go.** *Br J Pharmacol* 2008, **153**(Suppl 1):S7–S26.

40. Kontoyianni M, Madhav P, Suchanek E, Seibel W: **Theoretical and practical considerations in virtual screening: a beaten field?** *Curr Med Chem* 2008, **15**:107–116.

41. Tuccinardi T: **Docking-based virtual screening: recent developments.** *Comb Chem High Throughput Screen* 2009, **12**:303–314.

42. Altschul SF, Gish W, Miller W, Myers EW, Lipman DJ: **Basic local alignment search tool.** *J Mol Biol* 1990, **215**:403–410.

43. Pearson WR: **Rapid and sensitive sequence comparison with FASTP and FASTA.** *Methods Enzymol* 1990, **183**:63–98.

44. Altschul SF, Madden TL, Schäffer AA, Zhang J, Zhang Z, Miller W, Lipman DJ: **Gapped BLAST and PSI-BLAST: a new generation of protein database search programs.** *Nucleic Acids Res* 1997, **25**:3389–3402.

45. Söding J, Biegert A, Lupas AN: **The HHpred interactive server for protein homology detection and structure prediction.** *Nucleic Acids Res* 2005, **33**(Web Server issue):W244–W248.

46. Zhang Y: **Template-based modeling and free modeling by I-TASSER in CASP7.** *Proteins* 2007, **69**(Suppl 8):108–117.

47. Roy A, Kucukural A, Zhang Y: **I-TASSER: a unified platform for automated protein structure and function prediction.** *Nat Protoc* 2010, **5**:725–738.

48. Liu TW, Tang G, Capriotti E: **Comparative modeling: the state of the art and protein drug target structure prediction.** *Comb Chem High Throughput Screen* 2011, **14**:532–547.

49. Nelson DL, Cox MM, Lehninger AL: *Principles of Biochemistry.* New York: Freeman; 2004.

50. Murray RK: *Harper's Illustrated Biochemistry.* New York: McGraw-Hill; 2003.

51. Markley JL, Bax A, Arata Y, Hilbers CW, Kaptein R, Sykes BD, Wright PE, Wüthrich K: **Recommendations for the presentation of NMR structures of proteins and nucleic acids (IUPAC Recommendations 1998).** *Pure Appl Chem* 1998, **70**:117–142.

52. **Atomic coordinate entry format version 3.3.** http://www.wwpdb.org/documentation/format33/v3.3.html.

6

PcrG protects the two long helical oligomerization domains of PcrV, by an interaction mediated by the intramolecular coiled-coil region of PcrG

Abhishek Basu, Urmisha Das, Supratim Dey and Saumen Datta[*]

Abstract

Background: PcrV is a hydrophilic translocator of type three secretion system (TTSS) and a structural component of the functional translocon. C-terminal helix of PcrV is essential for its oligomerization at the needle tip. Conformational changes within PcrV regulate the effector translocation. PcrG is a cytoplasmic regulator of TTSS and forms a high affinity complex with PcrV. C-terminal residues of PcrG control the effector secretion.

Result: Both PcrV and PcrG-PcrV complex exhibit elongated conformation like their close homologs LcrV and LcrG-LcrV complex. The homology model of PcrV depicts a dumbbell shaped structure with N and C-terminal globular domains. The grip of the dumbbell is formed by two long helices (helix-7 and 12), which show high level of conservation both structurally and evolutionary. PcrG specifically protects a region of PcrV extending from helix-12 to helix-7, and encompassing the C-terminal globular domain. This fragment $\Delta PcrV_{(128-294)}$ interacts with PcrG with high affinity, comparable to the wild type interaction. Deletion of N-terminal globular domain leads to the oligomerization of PcrV, but PcrG restores the monomeric state of PcrV by forming a heterodimeric complex. The N-terminal globular domain ($\Delta PcrV_{(1-127)}$) does not interact with PcrG but maintains its monomeric state. Interaction affinities of various domains of PcrV with PcrG illustrates that helix-12 is the key mediator of PcrG-PcrV interaction, supported by helix-7. Bioinformatic analysis and study with our deletion mutant $\Delta PcrG_{(13-72)}$ revealed that the first predicted intramolecular coiled-coil domain of PcrG contains the PcrV interaction site. However, 12 N-terminal amino acids of PcrG play an indirect role in PcrG-PcrV interaction, as their deletion causes 40-fold reduction in binding affinity and changes the kinetic parameters of interaction. $\Delta PcrG_{(13-72)}$ fits within the groove formed between the two globular domains of PcrV, through hydrophobic interaction.

Conclusion: PcrG interacts with PcrV through its intramolecular coiled-coil region and masks the domains responsible for oligomerization of PcrV at the needle tip. Also, PcrG could restore the monomeric state of oligomeric PcrV. Therefore, PcrG prevents the premature oligomerization of PcrV and maintains its functional state within the bacterial cytoplasm, which is a pre-requisite for formation of the functional translocon.

Keywords: Regulation of TTSS, Functional translocon, Dynamic light scattering and elongated conformation, Homology model, Protease protected fragment, MS/MS sequence analysis, Reversal of oligomerization, Intramolecular coiled-coil, Deletion mutants, Surface plasmon resonance and protein-protein interaction, Molecular docking

* Correspondence: saumen_datta@iicb.res.in
Structural Biology and Bioinformatics division, Indian Institute of Chemical
Biology, 4 Raja S.C. Mullick Road, Kolkata 700032 West Bengal, India

Background

The Gram negative bacterium *Pseudomonas aeruginosa* is an opportunistic pathogen which causes acute infections in immune-compromised individuals. It is the causative agent of nosocomial pneumonia and other infections associated with burns, wounds, urinary tract, and cystic fibrosis [1-3]. *P. aeruginosa* possesses a TTSS, which uses an injectisome for delivery of bacterial toxic effector proteins within the host cell. The injectisome comprises of a basal structure and a needle complex. At the tip of the needle a translocon is formed by a set of three translocator proteins. This structure is essential for transport of the effector proteins and regulation of TTSS [3-7]. Two of the translocators are hydrophobic (like PopB, PopD from *Pseudomonas sp.* or YopB, YopD from *Yersinia sp.*) and form pores in the host cell membrane [6-10]. While the hydrophilic translocators (like PcrV from *Pseudomonas sp.* or LcrV from *Yersinia sp.*), also known as V-antigen, form a platform for the assembly of the hydrophobic translocators [6,7,11]. The hydrophilic translocators act as protective antigens against infection and are targets for the development of vaccine [12].

PcrV is a regulator of TTSS. It is chaperoned by PcrG, which itself is a cytoplasmic regulator of TTSS. Although, PcrV is a secretory protein, PcrG is completely cytoplasmic and both of these proteins regulate the TTSS independently [13,14]. PcrV oligomerizes at the needle tip for formation of the functional translocon. The C-terminal helix of PcrV is essential for its oligomerization. For the regulation of TTSS, PcrV exists in various conformational states. It alters the structure of translocation apparatus hence, controlling secretion of the effectors [15,16]. PcrV belongs to the Ysc family of translocators, which includes LcrV, AcrV (*Aeromonas sp.*). Other important family of translocators is Inv-Mxi, which consists of IpaD, SipD and BipD proteins. Translocators belonging to Inv-Mxi family possess two distinct domains. The N-terminal domain shows similarity with the common chaperones, suggesting a self-chaperoning function of these translocators. This is contrary to the behaviour of the translocators PcrV and LcrV, which utilize their cognate chaperones PcrG and LcrG in the bacterial cytoplasm [17-20]. The crystal structure of LcrV depicts dumbbell shape with two globular domains. The grip of the dumbbell is formed by two long helices [21]. PcrG and PcrV form a 1:1 high affinity complex and the deletion of the 24 C-terminal amino acids of PcrG does not alter the affinity of the complex formation [14]. Comparatively, little is known regarding the mechanism of regulation of TTSS by PcrG, apart from the fact that it is a negative regulator of TTSS, and C-terminal residues of PcrG regulate the effector secretion. Formation of PcrG-PcrV complex is not essential for the regulation of TTSS, but it confers stability to both the proteins within the bacterial cytoplasm and prevents the misfolding of PcrV [13-15].

The existing literature mainly focuses on the function of PcrV with respect to the regulation of TTSS. In this study, we have emphasized on the structural aspects of PcrG-PcrV interaction and conclusively proposed a model for the formation of PcrG-PcrV complex.

Results and discussion

PcrV retains the elongated conformation in the complex with PcrG, but with structural alteration

Based on Far UV CD spectrum and thermal denaturation curve, it was proposed that PcrV imparts structural stability to PcrG. Also, it was established that PcrG-PcrV interaction imparts stability to both the proteins [14]. The near UV CD spectrum showed the absence of tertiary structure signal for PcrG (Figure 1A). PcrV showed a negative signal at 285 nm (Figure 1A), which was almost in compliance with the previously reported minimum at 284 nm [15]. However, there was a shift of the minimum to 287 nm in case of PcrG-PcrV complex and the negative signal was more prominent in case of the complex structure (Figure 1A). This indicated towards a structural alteration of PcrV in presence of PcrG, in the complex. When PcrG was incubated with 8-anilinonapthalene-1-sulfonate (ANS), it showed almost similar spectrum to that of ANS (Figure 1B). This observation can be attributed to the lack of solvent exposed hydrophobic domains within PcrG, required for ANS binding. However, in the ANS-binding experiment, both PcrV and PcrG-PcrV complex showed significant blue shift in the λ_{max} to 484 nm and 476 nm, respectively. Moreover, the summation of PcrG-ANS and PcrV-ANS spectra was significantly different from PcrG-PcrV-ANS spectrum, revealing the differential structural moulding of both PcrG and PcrV within the complex (Figure 1B).

Dynamic light scattering (DLS) profile of PcrV and PcrG-PcrV revealed single predominant peaks corresponding to hydrodynamic diameters of 6.503 nm and 7.531 nm, respectively (Figure 1C). The estimated molecular weights of 53.1 kDa for PcrV and 74.9 kDa for PcrG-PcrV, clearly revealed their extended conformation. PcrV exists in a monomeric physiological state with a molecular weight of ~33 kDa. While PcrG-PcrV forms a 1:1 complex with a molecular weight of ~46 KDa [14]. Similar to PcrV, the elongated conformation was also observed in the crystal structure of its close homologue LcrV [21]. DLS experiments revealed that the hydrodynamic diameter of LcrV and LcrG-LcrV were 7.75 nm and 8.71 nm, respectively, in the solution state. This indicated towards the elongated conformation of LcrV in individual form and complex form (Figure 1C). PcrG and PcrV interaction is not essential for regulation of TTSS by these proteins. However, the formation of PcrG-PcrV complex provides structural stability to both the proteins and aids in proper folding and export of PcrV [13-15].

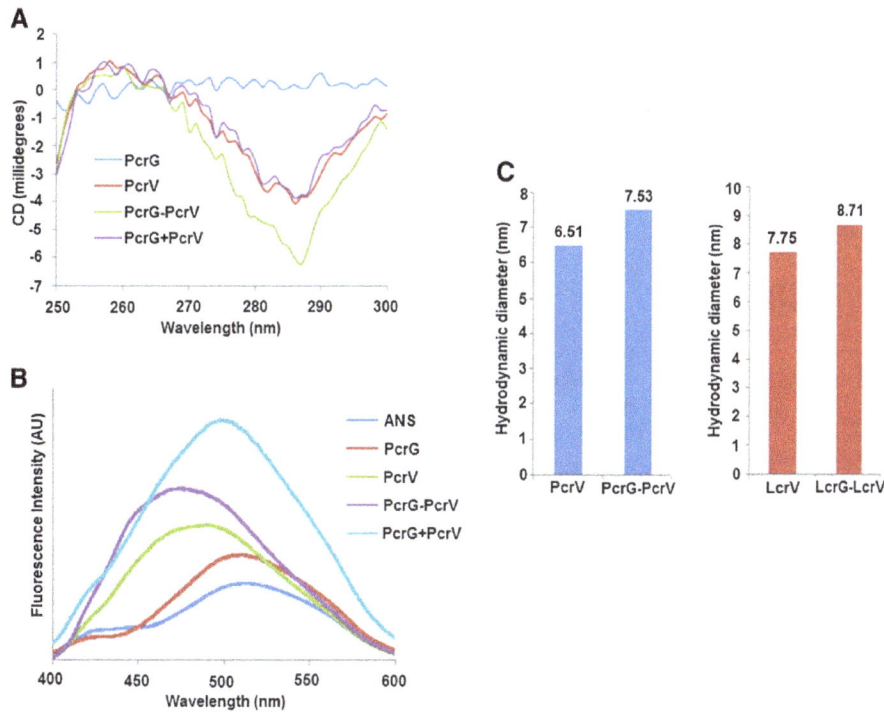

Figure 1 Near UV CD, ANS-fluorescence spectra, and DLS profile show structural stabilization and elongated conformation of the proteins.
A. Near UV CD signal recorded from 300 nm to 250 nm, shows negative signal for PcrV and PcrG-PcrV at 285 nm and 287 nm, respectively. PcrG shows absence of tertiary structure signal. **B**. ANS-binding profile of proteins was scanned from 400 nm to 600 nm. Both PcrV and PcrG-PcrV exhibit a blue-shift in the ANS-binding spectra, showing the presence of solvent exposed hydrophobic patches in the proteins. However, the spectrum of PcrG was similar to that of ANS. **C**. DLS profiles of PcrV, PcrG-PcrV and LcrV, LcrG-LcrV with corresponding hydrodynamic diameters, confirm an elongated conformation of the proteins.

A homology model of PcrV shows the elongated dumbbell shaped conformation

The homology model of PcrV was built by I-Tasser using LcrV [PDB ID: 1R6F] as template, which has 37% sequence identity with PcrV. The homology model was validated by PROCHECK (Additional file 1, Additional file 2) [22-24]. This model has a C-score of −0.07, TM Score of 0.70 ± 0.12 and RMSD of 6.3 ± 3.8 Å. The cartoon representation of the dumbbell shaped model of PcrV depicted the predominance of α-helical structures, interspersed by coiled regions, and few β-sheets. Out of the 12 α-helices in the structure, helix-7 (128–158) and helix-12 (251–293) are the longest. These two helices run anti-parallel to each other and form the grip of the dumbbell. Two β-sheets are localized between the 2nd and 3rd, and 4th and 5th helices, respectively. There is a short helix and a long coil region extending from residues 198–233 in the C-terminal, which falls within the protective epitope region of PcrV [19]. It is also evident from the orientation of PcrV on the needle tip, that aforesaid region must be exposed to the outer environment [7,19,25]. The N-terminal globular domain is formed by α-1, α-2, β-1, α-3, α-4, β-2, α-5, α-6. This domain is predicted to interact with the needle forming protein of *P. aeruginosa* [7,25]. The C-terminal globular

domain forming the outer part of the needle tip comprises of a short helix followed by α-8, α-9, a long coiled region and α-10, α-11. The N and C-terminal globular domains are structurally flexible due to their probable fate towards insertions (Figure 2A) [16,19,20,25]. The helix-12 is preceded by a loop in the model which might provide flexibility to the helix for attaining different conformations [15,21]. The elongated conformation proposed by the DLS data also corroborates with the dimensions of the dumbbell shaped model, which is 8.1 nm in length and 4.68 nm in width.

The homology model was used to generate ConSurf prediction model. This model is based on phylogenetic relations and evolutionary changes between homologous sequences [26]. The ConSurf model specified structurally and functionally conserved residues in a graded fashion. The helix-7, helix-12, the short helix, and the coil region preceding helix-12 showed maximum conservation. Broadly, the grip of the dumbbell is highly conserved both structurally and functionally (Figure 2B). From the multiple sequence alignment (MSA) file, a high sequence identity could be noticed between PcrV and its orthologs like AcrV, LssV and LcrV, specifically within helix-7 (residues 137–157) and helix-12 (residue 250–287) (Additional file 3) [27]. Sequence Logos of these two helices were generated

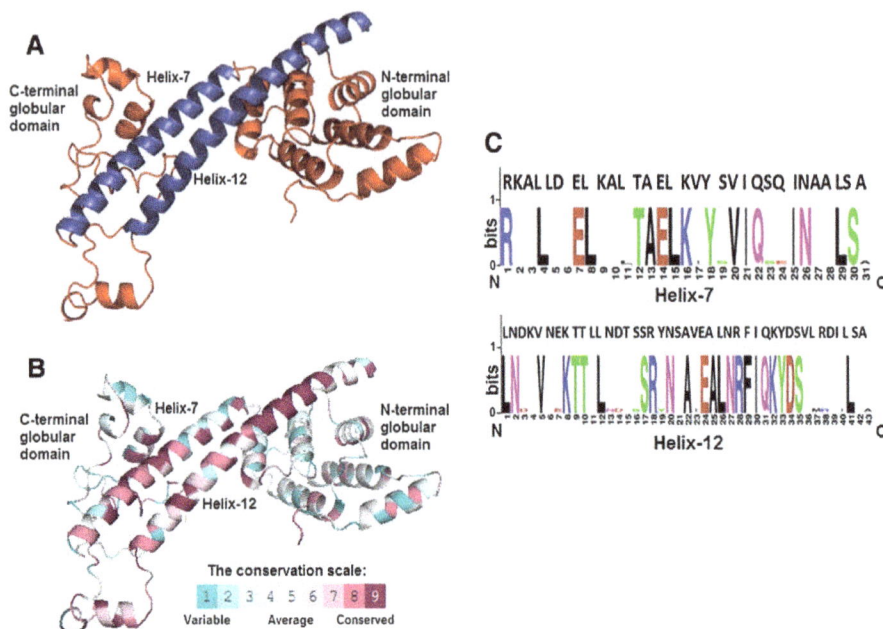

Figure 2 Homology model of PcrV and its analysis indicated the structurally and functionally conserved regions. A. Cartoon representation of homology model of PcrV depicts a dumbbell shaped structure with N and C-terminal globular domains. Helices 7 and 12 form the grip of the dumbbell. **B**. ConSurf predicted the structurally and functionally conserved residues within the homology model of PcrV, in a graded manner as shown by the colour code. **C**. WebLogo generated sequence Logos of helix-7 and 12 of V-antigens. Sequences of helix-7 and 12 of PcrV were aligned with the sequence Logos to determine the conserved positions.

using the proteins belonging to the LcrV family. The sequences of helix-7 and helix-12 of PcrV exhibited high level of conservation when compared to the consensus sequences of the Logos. In case of helix-7 (31 residues long) and helix-12 (43 residues long), 17 and 24 residues were conserved, respectively (Figure 2C) [28].

Proteolytic digestion identified a specific region of PcrV protected by PcrG

Proteolytic digestion of PcrV was carried out at different time points with 1:500 dilution of α-chymotrypsin. The digestion profile showed two predominant bands (fragments) existing till 50 minutes. One band was close to the 17 kDa marker and MS/MS sequence analysis showed that this fragment approximately extended from helix-7 up to helix-12 in the C-terminal of PcrV (Figure 3A, Additional file 4). Another band was present between 10 kDa and 17 kDa. MS/MS sequence analysis revealed that this fragment consists of bulk portion of the N-terminal, and completely includes the helix-7 of PcrV (Figure 3A, Additional file 5). Similar proteolytic digestion of PcrG-PcrV (as done for PcrV) revealed the presence of an extra band in between 17 and 26 kDa, corresponding to 19487.0064 Dalton in addition to the two aforesaid bands, as observed from the mass spectrometry profile (Figure 3B & 3C, Additional file 6). So, it can be concluded that PcrG specifically protects certain interacting region of PcrV. MS/MS sequence analysis revealed that this

protected region encompasses the entire C-terminal of PcrV comprising of helix-12, C-terminal globular domain and extending up to major portions of helix-7 (Additional file 7).

PcrG restores the monomeric state of oligomeric ΔPcrV$_{(128-294)}$ by forming a high affinity heterodimeric PcrG-ΔPcrV$_{(128-294)}$ complex

Based on the results of proteolytic digestion, showing a specific region of PcrV protected by PcrG, and bioinformatic analysis showing the conservation at helix-12 and 7, we designed a deletion mutant of PcrV comprising of helix-7, the C-terminal globular domain and helix-12 (ΔPcrV$_{(128-294)}$) (Figure 2B & 2C, Figure 3B & 3C, Additional file 6, Additional file 7). The complementary fragment of PcrV was also designed containing only the N-terminal globular domain (ΔPcrV$_{(1-127)}$). When both ΔPcrV$_{(1-127)}$ and ΔPcrV$_{(128-294)}$ were incubated with PcrG and purified by affinity chromatography, the interaction studies showed that only ΔPcrV$_{(128-294)}$ interacted with PcrG (Figure 4A). In Figure 4B, the region corresponding to ΔPcrV$_{(128-294)}$ is highlighted in deep blue colour in the homology model of PcrV. In order to check the affinity and kinetics of interaction of ΔPcrV$_{(128-294)}$ and PcrG, we used surface plasmon resonance (SPR). K$_D$ value of 2.43×10^{-8} M confirmed the formation of a high affinity complex between ΔPcrV$_{(128-294)}$ and PcrG. Although there was a slight reduction in the affinity of the complex

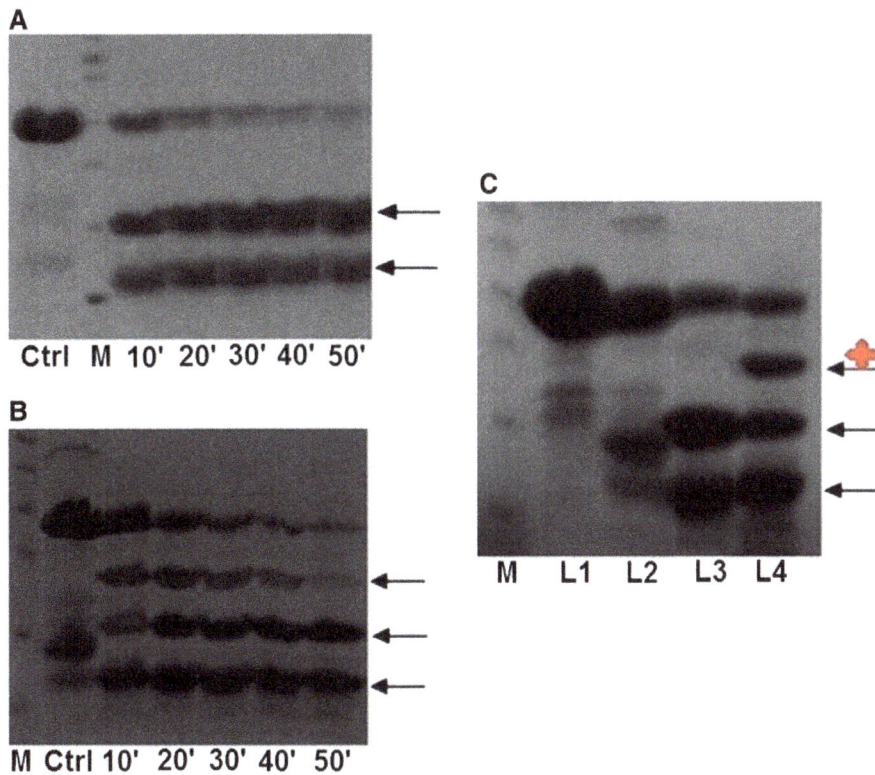

Figure 3 Proteolytic digestion identified a specifically protected region of PcrV in presence of PcrG. A. and **B**. Proteolytic digestion profiles of PcrV and PcrG-PcrV, respectively, with α-chymotrypsin from 10 to 50 minutes. Black arrows indicate the stable fragments generated after proteolysis. **C**. The digestion patterns of PcrV and PcrG-PcrV shows the presence of a specifically protected region of PcrV in the PcrG-PcrV complex, which is highlighted by a black arrow and a red star. L1 and L2 denote native PcrV and PcrG-PcrV. L3 and L4 denote PcrV and PcrG-PcrV, respectively, cleaved by α-chymotrypsin after 30 minutes. M is the protein molecular weight marker (10, 17, 26, 34, 43, 55 kDa bands from bottom to top).

formation (compared to PcrG-PcrV), a marked change in the association and dissociation rates of the reaction was observed (Figure 4C, Table 1) [14]. Size exclusion chromatography (SEC) profile showed that $\Delta PcrV_{(128-294)}$ exists as an oligomeric species eluting at 60 ml, which corresponds to a molecular weight ~193 kDa. We could not assign a proper state to the oligomer due to the elongated conformation of PcrV. The previous report also emphasized on the propensity of PcrV towards oligomerization on deletion of the N-terminal globular domain [20]. Gebus et al. [15], observed dimeric to hexameric states of PcrV on oligomerization, and concluded from further studies that V-antigens of *Y. pestis* and *P. aeruginosa* could oligomerize into higher order ring like structures with molecular weights greater than 130 kDa. However, in the complex form, PcrG-$\Delta PcrV_{(128-294)}$ showed a shift in the elution volume to 77 ml corresponding to a molecular weight of 41 kDa (actual mass of 1:1 PcrG-$\Delta PcrV_{(128-294)}$ complex is ~32 kDa), implying on the restoration of the monomeric state of $\Delta PcrV_{(128-294)}$ in the complex. PcrG elutes at 83 ml corresponding to a molecular weight of ~31 kDa, indicating towards a dimeric

state (actual mass of PcrG dimer is ~25 kDa) (Figure 4D). Further, DLS was performed to corroborate the observation of SEC and to check the existence of any aggregation state of $\Delta PcrV_{(128-294)}$. DLS studies showed that $\Delta PcrV_{(128-294)}$ forms oligomeric species with a hydrodynamic diameter of 11.8 nm, suggesting a molecular weight greater than 200 kDa (Figure 4E). However, when equimolar concentration of PcrG is incubated with $\Delta PcrV_{(128-294)}$, the higher order oligomeric species attains a lower hydrodynamic diameter of 7.3 nm, which corresponds to an elongated 1:1 heterodimeric complex of PcrG-$\Delta PcrV_{(128-294)}$ (Figure 4E). This reversion of oligomeric state of $\Delta PcrV_{(128-294)}$ could be visualized by native PAGE (Additional file 8). The 1:1 heterodimeric complex could also be seen by chemically crosslinking PcrG and $\Delta PcrV_{(128-294)}$ by ethylene glycol bis [succinimidyl succinate] (EGS)-sulfonate (Figure 4 F). The results depicted that the N-terminal globular domain regulates the physiological state of PcrV. The N-terminal globular domain does not interact with PcrG, but SEC profile showed that it maintains a monomeric state and it elutes at 88 ml corresponding to a ~21 kDa species (actual mass of $\Delta PcrV_{(1-127)}$ is ~15 kDa) (Figure 4G).

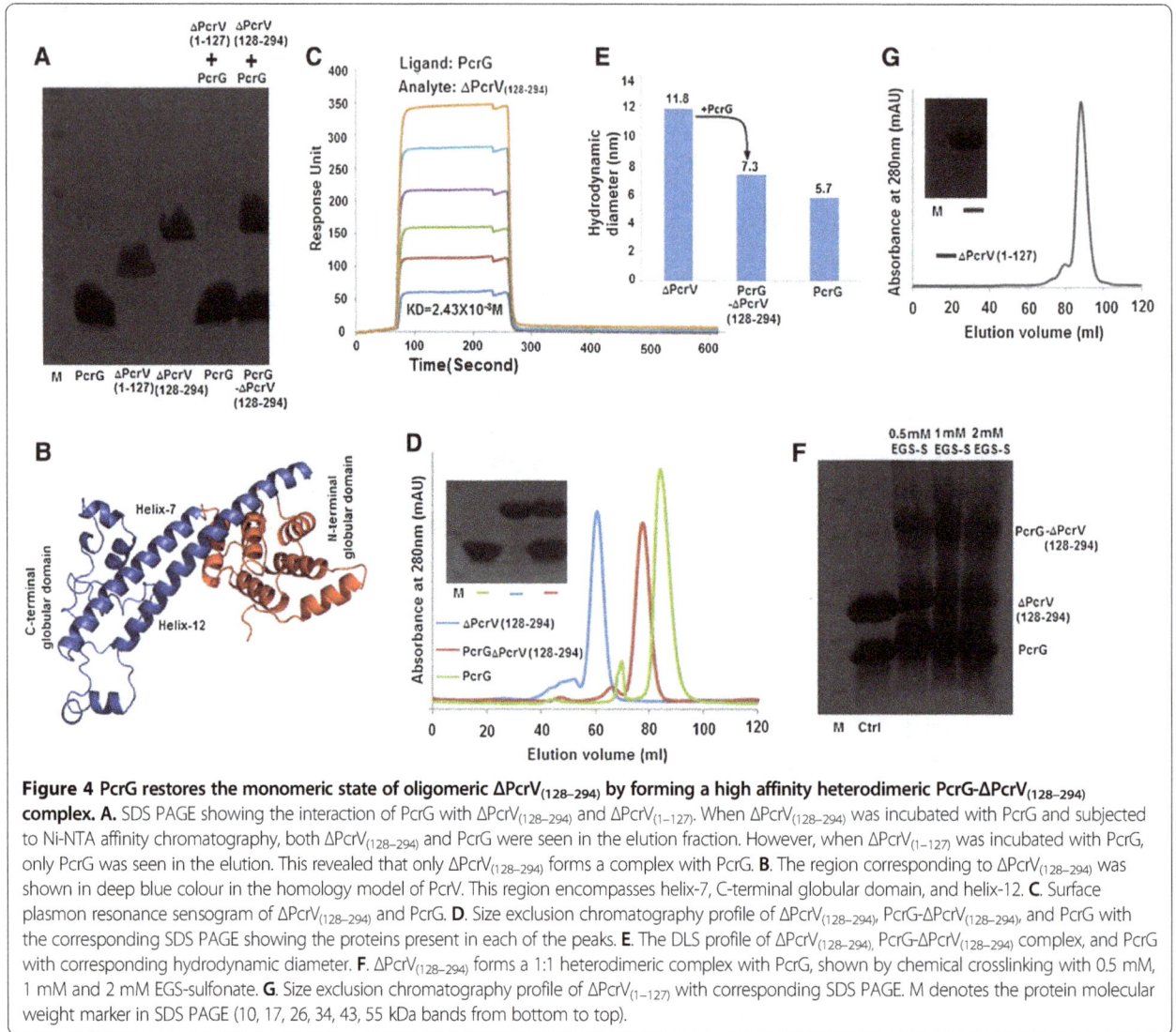

Figure 4 PcrG restores the monomeric state of oligomeric ΔPcrV(128–294) by forming a high affinity heterodimeric PcrG-ΔPcrV(128–294) complex. A. SDS PAGE showing the interaction of PcrG with ΔPcrV(128–294) and ΔPcrV(1–127). When ΔPcrV(128–294) was incubated with PcrG and subjected to Ni-NTA affinity chromatography, both ΔPcrV(128–294) and PcrG were seen in the elution fraction. However, when ΔPcrV(1–127) was incubated with PcrG, only PcrG was seen in the elution. This revealed that only ΔPcrV(128–294) forms a complex with PcrG. **B.** The region corresponding to ΔPcrV(128–294) was shown in deep blue colour in the homology model of PcrV. This region encompasses helix-7, C-terminal globular domain, and helix-12. **C.** Surface plasmon resonance sensogram of ΔPcrV(128–294) and PcrG. **D.** Size exclusion chromatography profile of ΔPcrV(128–294), PcrG-ΔPcrV(128–294), and PcrG with the corresponding SDS PAGE showing the proteins present in each of the peaks. **E.** The DLS profile of ΔPcrV(128–294), PcrG-ΔPcrV(128–294) complex, and PcrG with corresponding hydrodynamic diameter. **F.** ΔPcrV(128–294) forms a 1:1 heterodimeric complex with PcrG, shown by chemical crosslinking with 0.5 mM, 1 mM and 2 mM EGS-sulfonate. **G.** Size exclusion chromatography profile of ΔPcrV(1–127) with corresponding SDS PAGE. M denotes the protein molecular weight marker in SDS PAGE (10, 17, 26, 34, 43, 55 kDa bands from bottom to top).

The high affinity interaction between ΔPcrV(128–294) and PcrG indicates that the location of PcrG-binding site could be within the two long helices of PcrV. Oligomerization of PcrV subunits at the tip of the needle is proposed to occur by intramolecular exchange of helix-7 and helix-12 by "domain swapping" mechanism and the last 41 amino acids forming helix-12 is essential for oligomerization. Deletion of the C-terminal helix inhibits the multimerization of PcrV and has post-secretory implications, abolishing the bacterial cytotoxicity [15]. So, protection or masking of these helices by PcrG, prevented the oligomerization and misfolding of PcrV within the bacterial cytoplasm. It also helped to maintain the functional form of PcrV by formation of the heterodimeric complex of PcrG-PcrV in the cytoplasm. The result established the neutral role of the N-terminal domain of PcrV in interaction with PcrG.

Table 1 Kinetic parameters of interaction between deletion mutants of PcrG and PcrV, were determined by SPR

Kinetic parameter	PcrG-PcrV	ΔPcrG(1–74)-PcrV	PcrG-ΔPcrV(128–294)	ΔPcrG(13–72)-PcrV	ΔPcrG(13–72)-ΔPcrV(128–294)
K_A (1/M)	6.4×10^7	6.4×10^7	4.11×10^7	1.75×10^6	1.61×10^6
K_D (M)	1.56×10^{-8}	1.56×10^{-8}	2.43×10^{-8}	5.7×10^{-7}	6.21×10^{-7}
K_a (1/MS)	4.45×10^5	3.16×10^5	5.11×10^2	1.71×10^4	2.72×10^1
K_d (1/S)	6.94×10^{-3}	4.91×10^{-3}	1.24×10^{-5}	9.78×10^{-3}	1.69×10^{-5}

Table 1 Binding affinities, and association and dissociation rates of interaction between deletion mutants of PcrG and PcrV, were determined by SPR. Kinetic parameters of PcrG-PcrV interaction and ΔPcrG(1–74)-PcrV interaction were adapted from Nanao *et al.* [14].

Therefore, the N-terminal is responsible for the maintenance of the non-oligomeric state of PcrV and regulates the secretion of PcrV through the injectisome. Specifically, deletion of 3–20 amino acids leads to defects in the secretion of the PcrV [13]. Furthermore, the N-terminal globular domain of PcrV interacts with the needle forming protein PscF. Also, this domain of LcrV is essential for recruitment of YopB in target cell membrane. These observations prompted us to propose a chaperoning activity associated with this domain, as seen in Inv-Mxi family of hydrophilic translocators like IpaD, SipD, and BipD [3,7,19,25].

Helix-12 is the key mediator for PcrG-PcrV interaction and helix-7 might support the interaction

From the proteolytic cleavage data, it was established that PcrG protects the C-terminal helix-12 of PcrV and major part of helix-7 (Figure 3B & 3C, Additional file 6, Additional file 7). Lee et al. [13], reported that mutations in C-terminal helix of PcrV affect PcrG-PcrV interaction. The F279R mutation in helix-12 has a deleterious effect towards PcrG interaction and L262D mutation abolished PcrG binding. However, the entire region of PcrV involved in PcrG-interaction is still unknown. To specifically assign the PcrG-interaction domain of PcrV, we have dissected PcrV into various fragments, comprising of a combination of four domains (i.e. N-terminal globular domain, helix-7, C-terminal globular domain, and helix-12).

Apart from the stable fragment of PcrV protected by PcrG, which is $\Delta PcrV_{(128-294)}$, we designed four more deletion fragments of PcrV to understand the role of various domains of PcrV in PcrG-PcrV interaction. $\Delta PcrV_{(1-158)}$ comprises of the N-terminal globular domain and helix-7; $\Delta PcrV_{(1-250)}$ contains the N-terminal globular domain, helix-7 and the C-terminal globular domain; $\Delta PcrV_{(128-250)}$ comprises of helix-7 and the C-terminal globular domain, and $\Delta PcrV_{(159-294)}$ contains the C-terminal globular domain and helix-12. Molecular masses of all these deletion mutants were checked by native mass spectrometry, and their experimental masses were extremely close to their theoretical masses. Quantitative binding of PcrG with each of the deletion mutants of PcrV, was analyzed by SPR. We have found that PcrG-$\Delta PcrV_{(159-294)}$ has K_D of 3.66×10^{-8} M. So, we observed a 1.5 fold reduction in the binding affinity on deletion of helix-7, compared to K_D of PcrG-$\Delta PcrV_{(128-294)}$ complex, however, the order of interaction remains 10^{-8} M (Figure 5). Interactions of PcrG with fragments of PcrV devoid of helix-12- $\Delta PcrV_{(1-158)}$, $\Delta PcrV_{(1-250)}$, $\Delta PcrV_{(128-250)}$ showed K_D values 2.03×10^{-7} M, 1.79×10^{-7} M and 2.47×10^{-7} M, respectively (Figure 5). These results emphasized that deletion of helix-12 leads to greater than 10 fold reduction in binding affinity, when the binding affinities of $\Delta PcrV_{(128-294)}$ and $\Delta PcrV_{(128-250)}$ were compared (Figure 5). By comparing the binding affinity of $\Delta PcrV_{(1-158)}$ and $\Delta PcrV_{(1-250)}$,

we conclude an insignificant role of the C-terminal globular domain in PcrG-PcrV interaction (Figure 5).

The above experimental evidences suggest that the main PcrG-interaction domain is localized within helix-12. However, helix-7 also enables PcrG-PcrV complex formation, so it might stabilize the PcrG-PcrV interaction due to the close packing of the helices, as observed in the model. Lawton et al. [29], identified certain residues within helix-7 of LcrV, involved in LcrG-LcrV interaction. These residues occupy crucial positions of the heptad repeats in helix-7; hence, they are critical for the formation of intramolecular coiled-coil and α7-α12 interaction. Mutagenesis of these residues affects the binding between LcrV and LcrG [21,29,30]. Insertional and point mutagenesis identified residues in helix-12 crucial for LcrG-LcrV interaction [30]. Therefore, in LcrV, helix-7 as well as helix-12 possesses residues involved in LcrG-LcrV interaction. Interestingly, mutation of the residues in α7 of LcrV (involved in LcrG interaction) affects multimerization of LcrV, which might be due to disruption of intramolecular coiled-coil and α7-α12 interaction, but it allows the formation of the complex between LcrG-LcrV [29]. Mutations in C-terminal helix-12 of PcrV perturb the formation of the intramolecular coiled-coil and abolish the oligomerization of PcrV, revealing that formation of the coiled-coil is essential for oligomerization of the V-antigen [15]. Also, PcrG could prevent the oligomerization of PcrV, giving an indication that PcrG binding might affect the intramolecular coiled-coil structure of PcrV. Based on the crystal structure of LcrV and mechanism of LcrG-LcrV interaction mediated through the heptad repeats of α7 of LcrV, Gebus et al. [15], proposed that V-antigen could exist in a closed and an open state. The closed state corresponds to the monomeric shape of the V-antigen, while the open state renders α7 and α12 free for interaction with its partners [15,21,29]. Therefore, the formation of intramolecular coiled-coil is not absolutely essential for the interaction of V-antigen with its partners. These facts corroborate our SPR analysis, where we have observed that the presence of either of the two helices (7 or 12) in PcrV enables the complex formation with PcrG, because both the helices contain residues for PcrG interaction. Structural stabilization by intramolecular coiled-coil structure is not an absolute necessity for PcrG-PcrV interaction. However, the affinity of the interaction is reduced due to the changes in the local structure in absence of one of the helices. Single mutation of L262D in helix-12 of PcrV, abolishes PcrG-PcrV interaction [13]. The corresponding residue L291 in LcrV is not only involved in LcrG interaction, but also present within the zipper motif and takes part in α7-α12 interaction [30]. Importantly, L262D mutation inhibits the oligomerization of PcrV and had a profound effect on bacterial cytotoxicity [13,15]. Therefore, under the influence of this mutation V-antigen attains a monomeric shape

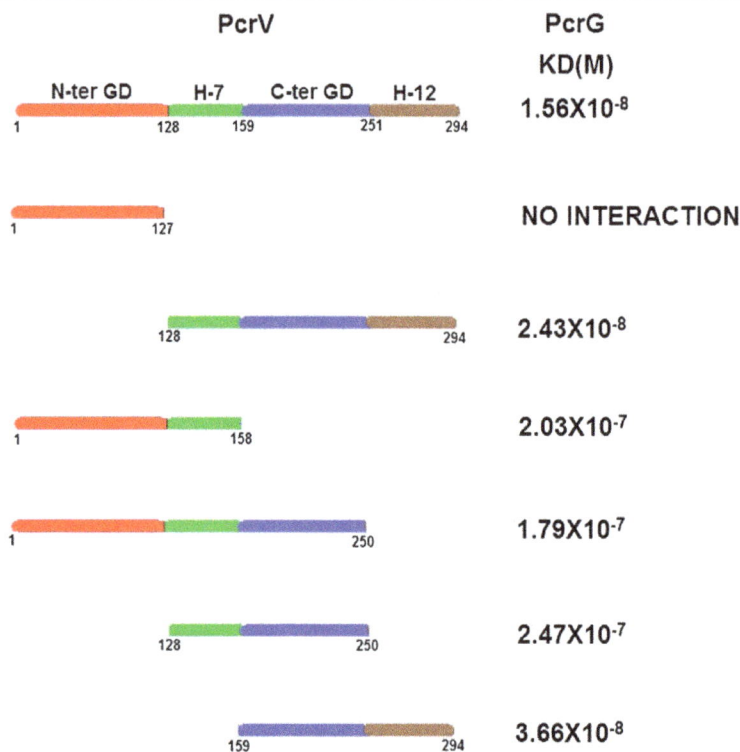

Figure 5 Interaction of various domains of PcrV with PcrG. K_D values estimated by SPR, reveal the binding affinities of various deletion mutants of PcrV with PcrG. Higher affinity of interaction was observed when helix-12 is present in the deletion mutant, but presence of helix-7 also allows the interaction.

corresponding to its "closed conformation", which is not favourable for the interaction of V-antigen with its binding partners [15]. Changing the hydrophobic leucine to charged aspartic acid residue might enforce a local structural change within both the helices due to their close packing. So, it would be interesting to test whether other single amino acid substitution in the helix-12 leading to oligomerization defects, could also abolish PcrG-PcrV interaction.

Intramolecular coiled-coil region of PcrG contains the PcrV interaction site with an indirect role of N-terminal residues of PcrG in the interaction

24 C-terminal amino acids of PcrG, exhibit disorder and absence of secondary structure [14]. Further analysis of PcrG with various disorder prediction servers revealed that a few amino acids in the N-terminal and a patch in the C-terminal are disordered. PrDOS predicted disorder region of 1–13 amino acids in the N-terminal and 73–98 in the C-terminal (Figure 6A) [31]. DisEMBL 1.5 and Disopred version-2.0 provided almost similar predictions (Additional file 9) [32,33]. Controlled proteolytic digestion with elastase produced a fragment of PcrG with molecular weight of 9782.1182 Dalton, with an intact C-terminal, as revealed by mass spectrometry and MS/MS sequence analysis (Figure 6B, Additional file 10, Additional file 11).

The C-terminal region is an essential part of PcrG since, deletion of this region leads to deregulation of effector secretion [13]. Also, MSA showed that there is substantial homology and conservation in the C-terminal of PcrG (Additional file 12) [27]. Though the C-terminal of PcrG is disordered, it is inaccessible to ANS and protected from proteolytic digestion. This may be due to the formation of compact structure.

When PcrG was divided into two fragments one comprising of amino acid 2–40 and the other consisting of amino acid 41–95, only the former fragment interacted with PcrV [13]. Also, PcrG and its homologs like AcrG, LssG and LcrG show maximum identity and conservation between residues 20–35 (Additional file 12) [27]. The COILS/PCOILS server predicted that amino acids 9–31 constitute the first intramolecular coiled-coil region in PcrG (Additional file 13) [34]. The N-terminal intramolecular coiled-coil region is essential in case of LcrG for its interaction with LcrV [29,35]. Based on the above results, we have designed a deletion mutant of PcrG, comprising of amino acids 13–72. This deletion mutant ($\Delta PcrG_{(13-72)}$) interacts with PcrV, undermining the role of first 12 N-terminal residues of PcrG in PcrG-PcrV interaction. Similar to PcrG, $\Delta PcrG_{(13-72)}$ also exists in a dimeric state, as detected by native mass spectrometry. The theoretical mass of $\Delta PcrG_{(13-72)}$ dimer is 17649.6 Da and the observed

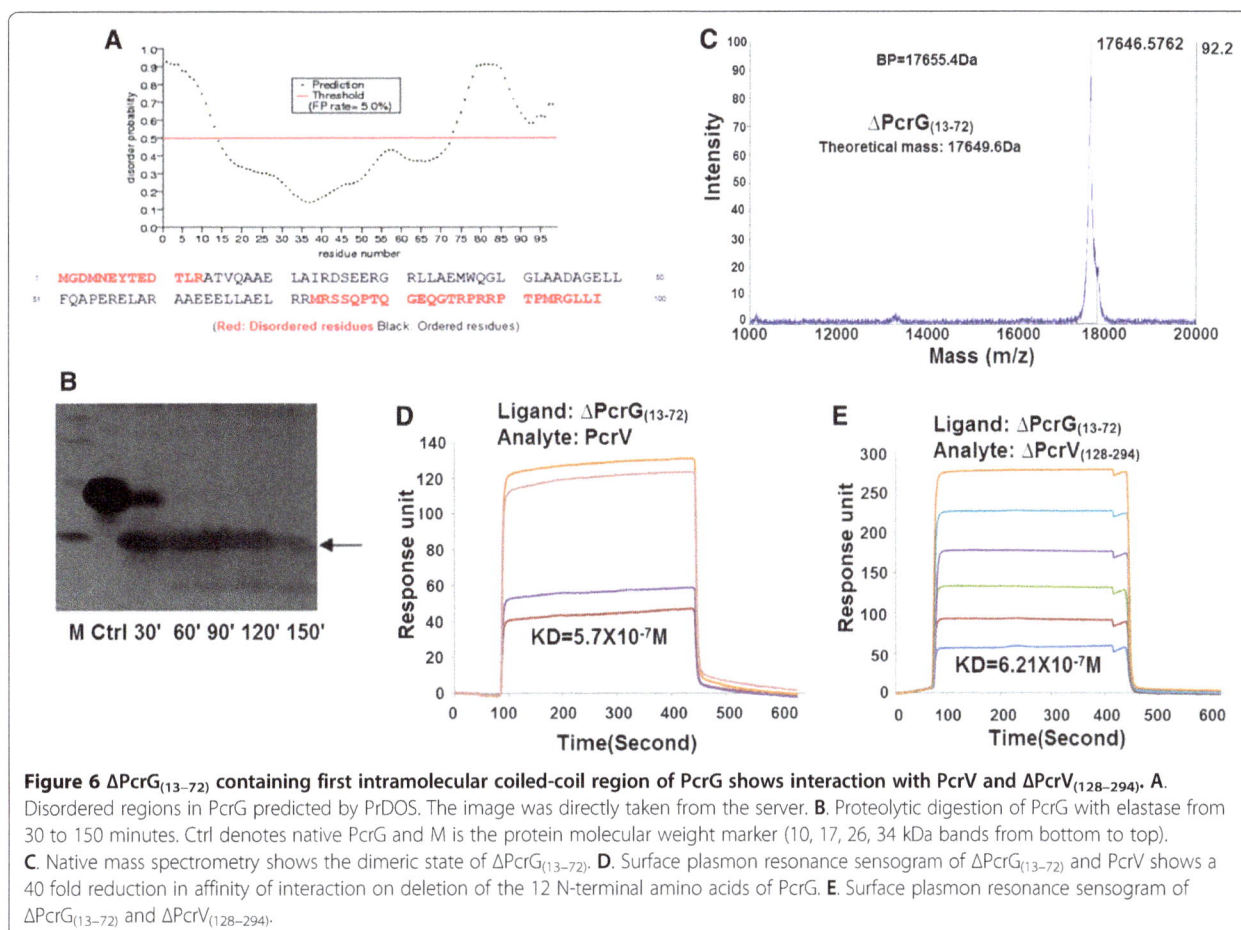

Figure 6 ΔPcrG$_{(13-72)}$ containing first intramolecular coiled-coil region of PcrG shows interaction with PcrV and ΔPcrV$_{(128-294)}$. **A.** Disordered regions in PcrG predicted by PrDOS. The image was directly taken from the server. **B.** Proteolytic digestion of PcrG with elastase from 30 to 150 minutes. Ctrl denotes native PcrG and M is the protein molecular weight marker (10, 17, 26, 34 kDa bands from bottom to top). **C.** Native mass spectrometry shows the dimeric state of ΔPcrG$_{(13-72)}$. **D.** Surface plasmon resonance sensogram of ΔPcrG$_{(13-72)}$ and PcrV shows a 40 fold reduction in affinity of interaction on deletion of the 12 N-terminal amino acids of PcrG. **E.** Surface plasmon resonance sensogram of ΔPcrG$_{(13-72)}$ and ΔPcrV$_{(128-294)}$.

experimental mass is 17646.5762 Da (BP = 17655.4 Da) (Figure 6C). However, due to the deletion of the predicted disordered region, the helicity of ΔPcrG$_{(13-72)}$ increases to ~24% (data to be published).

Deletion of 24 C-terminal residues of PcrG does not have any effect on the affinity of PcrG-PcrV interaction [14]. After deletion of the first 12 N-terminal amino acids and the 26 C-terminal amino acids of PcrG, ΔPcrG$_{(13-72)}$ could still form a high affinity complex with PcrV (K$_D$ 5.7×10^{-7} M). But there are certain alterations in the association and dissociation kinetics (Figure 6D, Table 1) [14]. Since, amino acid 2–40 of PcrG interacts with PcrV [13] and our deletion mutant of PcrG containing residue 13–72, also interacts with PcrV, we predict that residues 13–40 of PcrG form the core region for PcrG-PcrV interaction. Moreover, this region contains almost the entire intramolecular coiled-coil and amino acids A19, S26 and L33. These three amino acids are conserved and are key residues involved in LcrG-LcrV interaction [34,35]. Although the deletion of 12 N-terminal residues abolishes LcrG-LcrV interaction [35], similar deletion of 12 N-terminal residues in PcrG could still allow it to form a high affinity complex with PcrV. However, a 40-fold reduction in binding

affinity was noticed when the K$_D$ value of PcrG-PcrV and ΔPcrG$_{(13-72)}$-PcrV were compared (Figure 6D, Table 1) [14]. ΔPcrG$_{(13-72)}$ also interacts with ΔPcrV$_{(128-294)}$ with a slight reduction in the affinity (K$_D$ 6.21×10^{-7} M), compared to full length PcrV (Figure 6E, Table 1). It may predict an indirect involvement of the N-terminal in the interaction between PcrG and PcrV. The structural domains of PcrG involved in PcrG-PcrV interaction, might differ from those of LcrG in LcrG-LcrV interaction.

During the interaction of ΔPcrG$_{(13-72)}$ with ΔPcrV$_{(1-158)}$, ΔPcrV$_{(1-250)}$, and ΔPcrV$_{(128-250)}$, the K$_D$ reduces to micromolar range. This high reduction in the binding affinity could be attributed to the absence of helix-12 of PcrV and 12 N-terminal residues of PcrG. But ΔPcrG$_{(13-72)}$ and ΔPcrV$_{(159-294)}$ (fragment containing helix-12) still show nanomolar range of interaction, emphasizing the significant role of helix-12 in the PcrG-PcrV interaction (Figure 7). In spite of deletion of various domains of PcrV and PcrG, presence of either of the helices (7 or 12) in PcrV and intramolecular coiled region of PcrG enables the formation of the complex. These observations further established that helix-7 or 12 of PcrV and intramolecular coiled-coil of PcrG contain the sites for PcrG-PcrV interaction.

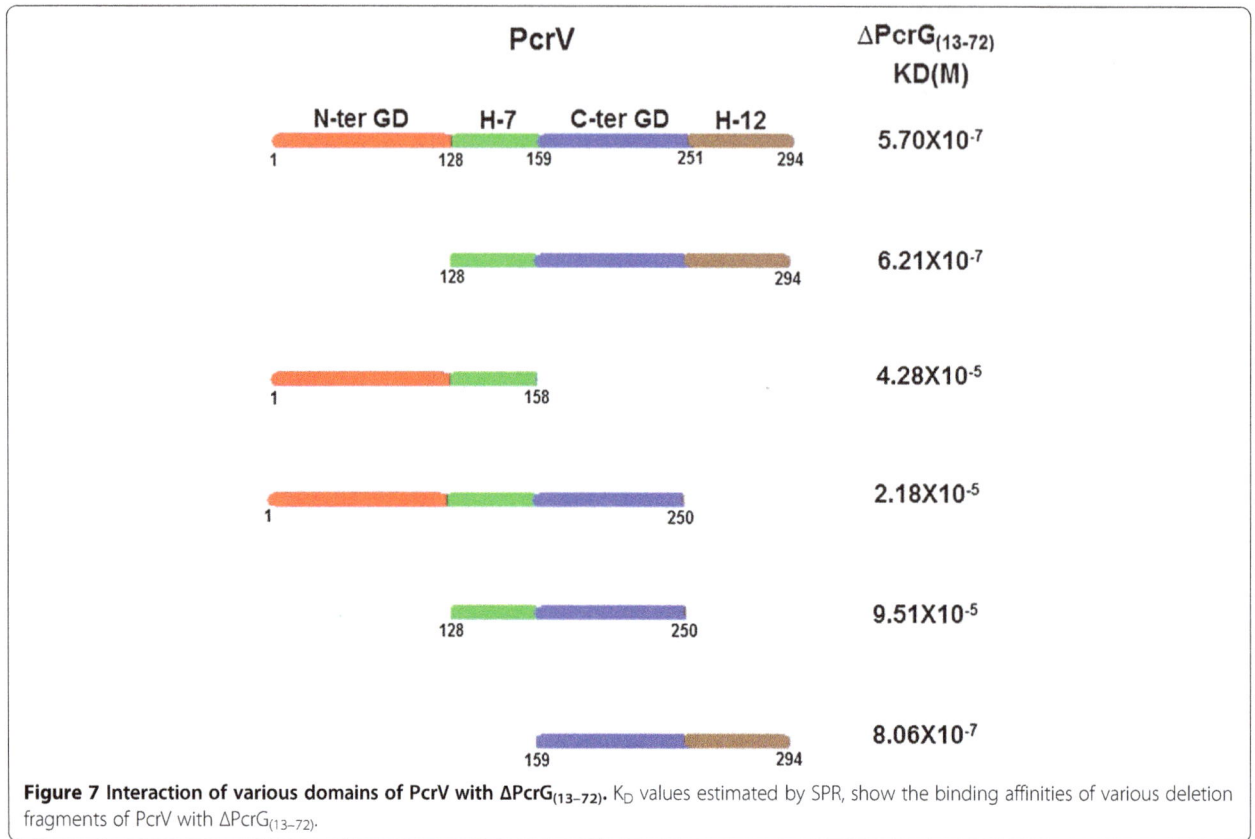

Figure 7 Interaction of various domains of PcrV with ΔPcrG(13–72). K_D values estimated by SPR, show the binding affinities of various deletion fragments of PcrV with ΔPcrG(13–72).

Finally a model of ΔPcrG(13–72) and docking studies put ΔPcrG(13–72) in a groove formed between two globular domains of PcrV

A model of the deletion mutant of PcrG (ΔPcrG(13–72)) was generated using I-Tasser (Additional file 14). Since, experimental 3D structures of orthologs of PcrG were unavailable; I-tasser used a threading algorithm. Deletion of the disordered regions of PcrG led to a significant improvement in the quality of the model. C-score and the TM Score of ΔPcrG(13–72) model are −1.5 and 0.52 ± 0.12, respectively, which satisfied the I-Tasser cut off for a correct model, and confirmed a similar topology between the model and the templates. The model was further validated by PROCHECK (Additional file 15) [22-24]. However, template proteins selected by threading programmes of I-Tasser were not orthologous to the modelled protein, and the process is based on prediction of similar fold between the proteins. If not the actual state of PcrG, the model provides the basic scaffold structure to estimate the binding site of PcrV, which was already verified by experimental techniques. The model of ΔPcrG(13–72) represents four helices interspersed by coiled regions. The predicted residues of PcrG involved in PcrV-binding, exactly map to the first two helices of the model (shown in orange colour) [Figure 4A]. Molecular docking studies were performed using ZDOCK version ZD 3.0.2, where PcrV was designated

as the receptor and ΔPcrG(13–72) as the corresponding ligand. The best model obtained from docking studies was selected and validated by PROCHECK [24,36] (Additional file 16, Additional file 17). The spacefill model of ΔPcrG(13–72)-PcrV complex depicted that the binding pocket of ΔPcrG(13–72) was located within the groove (or grip), in between the two globular terminal domains of PcrV (Figure 8B). The cartoon representation showed that the first two helices of ΔPcrG(13–72), which overlap with the first predicted intramolecular coiled-coil domain, specifically interacts with helix-7 and helix-12 of PcrV (Figure 8C). These observations further confirmed the fact that 13–40 residues of PcrG and two long helices of PcrV are the key mediators for the formation of the complex. Figure 8D depicted the surface representation of ΔPcrG(13–72)-PcrV complex, where the hydrophobic residues were coloured in yellow. This model shows that ΔPcrG(13–72) and PcrV interact through a hydrophobic interface corroborating the concept of hydrophobic residues being involved in interaction of V-antigen with its regulator [13,29,30,35].

Conclusions

PcrG and PcrV form a high affinity complex, which confers stability and maintain the functional physiological states of both the proteins within the bacterial cytoplasm [13,14].

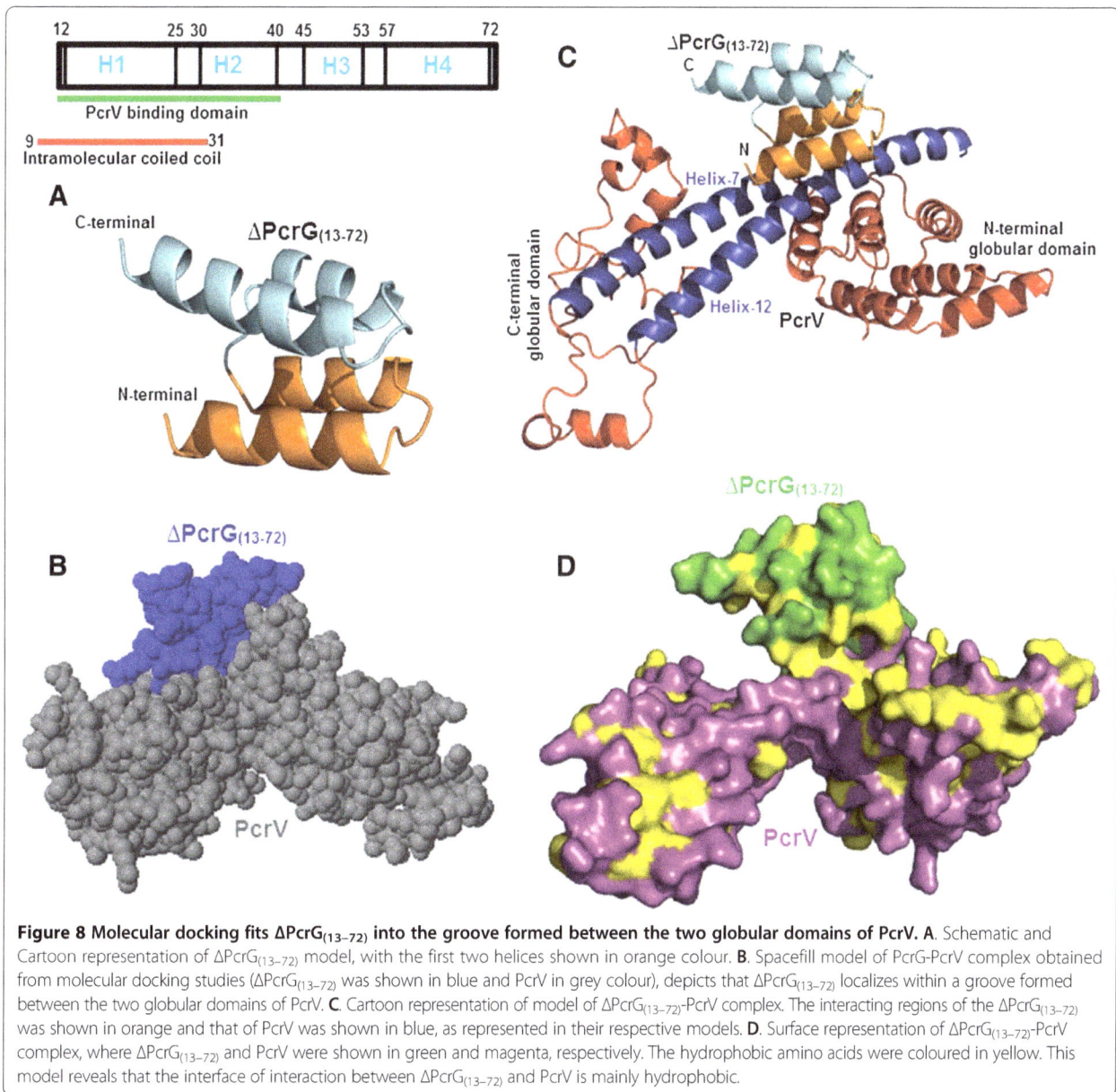

Figure 8 Molecular docking fits ΔPcrG$_{(13-72)}$ into the groove formed between the two globular domains of PcrV. A. Schematic and Cartoon representation of ΔPcrG$_{(13-72)}$ model, with the first two helices shown in orange colour. **B**. Spacefill model of PcrG-PcrV complex obtained from molecular docking studies (ΔPcrG$_{(13-72)}$ was shown in blue and PcrV in grey colour), depicts that ΔPcrG$_{(13-72)}$ localizes within a groove formed between the two globular domains of PcrV. **C**. Cartoon representation of model of ΔPcrG$_{(13-72)}$-PcrV complex. The interacting regions of the ΔPcrG$_{(13-72)}$ was shown in orange and that of PcrV was shown in blue, as represented in their respective models. **D**. Surface representation of ΔPcrG$_{(13-72)}$-PcrV complex, where ΔPcrG$_{(13-72)}$ and PcrV were shown in green and magenta, respectively. The hydrophobic amino acids were coloured in yellow. This model reveals that the interface of interaction between ΔPcrG$_{(13-72)}$ and PcrV is mainly hydrophobic.

Like its homolog LcrV, PcrV exhibits an extended conformation, both in individual and in complex form. PcrG protects a specific region of PcrV extending from helix-7 to helix-12, encompassing the C-terminal globular domain, and devoid of the N-terminal globular domain. However, the deletion of the N-terminal globular domain of PcrV leads to its oligomerization, which could be reverted back by PcrG. Therefore, PcrG can restore the monomeric form of oligomer formed by PcrV. It was seen that helix-12 of PcrV plays the key role in PcrG-PcrV interaction, supported by helix-7. N and C-terminal globular domains have neutral roles in the interaction. However, the N-terminal globular domain of V-antigen maintains a monomeric state. It interacts with needle forming protein

PscF and recruits hydrophobic translocator in the target cell membrane, thereby, indicating towards a chaperoning function [7,19,25].

Interestingly, formation of the functional translocon capable of targeting effectors within the host cell requires oligomerization of PcrV at the needle tip [3-7,11,15,25]. Since, the PcrG interaction domain overlaps with the oligomerization domain of PcrV, PcrG could potentially stop the oligomerization of PcrV by docking into the groove formed by helix-7 and helix-12. The intramolecular coiled-coil region of PcrG contains the PcrV binding domain, but unlike LcrG, the 12 N-terminal residues of PcrG only have an indirect role in PcrG-PcrV interaction. To conclude, PcrG prevents the premature oligomerization of PcrV and maintains its functional state within the bacterial cytoplasm, which is a pre-requisite for the formation of functional translocon.

Methods

Design of expression vectors

The genes of *pcrG*, *pcrV* and *pcrG-pcrV*, Δ*pcrG*$_{(13–72)}$, Δ*pcrVs* (deletion constructs of *pcrV*) were amplified by PCR, using the chromosomal DNA of *P. aeruginosa* strain 2192. *pcrG* was cloned with EcoRI and HindIII, using sense primer TA**GAATTC**TATGGGCGACATGAACGAA and antisense primer AAT**AAGCTT**TCAGATCAACAAGCC ACG in pETDuet-1. Δ*pcrG*$_{(13–72)}$ was cloned in pET-28a (+) with NdeI and XhoI, using sense primer TTAGGAT**CCATATG**CGGGCGACCGTCCAGGCC and antisense primer TTA**CTCGAG**TTAGCGCCGCAGTTCGGCCAG. *pcrV*, Δ*pcrV*$_{(1–127)}$, Δ*pcrV*$_{(1–158)}$, and Δ*pcrV*$_{(1–250)}$ were cloned in pET-22b-Δ50CPD with NdeI and EcoRI, using sense primer AATCCATGGAA**CATATG**GAAGTCAGAAACCT and antisense primers TAA**GAATTC**GGGATCG CGCTGAGAATGTCGC, TAA**GAATTC**GGCTTGCCG TCCTGGGTCTG, TAA**GAATTC**GGCTTGGCCGAC AGCGCGGC, and TAA**GAATTC**GGCGGACGCGAGC GGTCGCT, respectively. Δ*pcrV*$_{(128–294)}$ and Δ*pcrV*$_{(128–250)}$ were cloned in pET-22b-Δ50CPD with NdeI and EcoRI, using sense primer AATCCATGGAA**CATATG**CGCAA GGCGCTGCTCGAC and antisense primers TAA**GAAT TC**GGGATCGCGCTGAGAATGTCGC and TAA**GAA TTC**GGCGGACGCGAGCGGTCGCT, respectively. Δ*pcrV*$_{(159–294)}$ was cloned in pET-28a (+) with NdeI and HindIII, using sense primer AATCCATGGAA**CATATG**CAGG GCATCAGGATCGAC and antisense primer TTA**AAGC TT**CTAGATCGCGCTGAGAAT. *pcrG-pcrV* was cloned in pETDuet-1 with EcoRI and HindIII, using sense primer of *pcrG* TA**GAATTC**TATGGGCGACATGAACGAA and antisense primer of *pcrV* TTA**AAGCTT**CTAGATCGCG CTGAGAAT. The restriction sites were marked in bold letters. *E. coli* Top10 was used as the cloning strain.

Purification of proteins

The proteins were expressed in BL21 DE3 by culturing in Luria Bertani medium and induction was carried with 1 mM IPTG for 4 hours at 37°C. Only ΔPcrG$_{(13–72)}$ was induced overnight at 22°C. 1 mM phenylmethanesulfonyl fluoride was used as protease inhibitor. PcrG, ΔPcrG$_{(13–72)}$ and PcrG-PcrV have N-terminal 6X histidine tag. These proteins were purified using standard nickel-nitrilotriacetic acid (Ni-NTA) affinity chromatography. A step gradient of immidazole was applied in 25 mM Tris–HCl (pH-8) and 150 mM NaCl buffer, for removing the non-specific proteins and eluting the desired protein. PcrV and its deletion mutants (ΔPcrV$_{(1–127)}$, ΔPcrV$_{(1–158)}$, ΔPcrV$_{(1–250)}$, ΔPcrV$_{(128–250)}$, and ΔPcrV$_{(128–294)}$) were fused to a cysteine protease domain (CPD) at their C-terminal. This CPD

possesses a C-terminal histidine tag. The fusion product of the target protein and CPD is engineered in such a manner that inositol hexokisphosphate recognizes a specific leucine residue and cleaves exactly at the junction of the PcrV and the CPD. PcrV-CPD was immobilized in Ni-NTA affinity column and treated with 500 μM inositol hexokisphosphate in 25 mM Tris–HCl (pH-7.5) and 150 mM NaCl buffer. Inositol hexokisphosphate cleaves at the junction of PcrV and CPD to release PcrV (without histidine tag) from the affinity column, and CPD with the C-terminal histidine tag remains attached with the Ni-NTA. The deletion mutants of PcrV were purified in a similar way. So, we obtain PcrV and its various forms without a histidine tag for further downstream experiments. For detailed procedure refer Shen *et al.* [37]. ΔPcrV$_{(159–294)}$ had localized in the inclusion bodies. The inclusion bodies were denatured using 6 M guanidium hydrochloride. ΔPcrV$_{(159–294)}$ was refolded by slow dialysis in 25 mM Tris–HCl (pH-8), 150 mM NaCl and 10% glycerol buffer and subjected to Ni-NTA chromatography. Further the histidine tag of ΔPcrV$_{(159–294)}$ was removed by thrombin digestion which cleaves the leader sequence. All the proteins were checked for their purity and molecular weight by native mass spectrometry. For purification of PcrG-ΔPcrV$_{(128–294)}$, PcrG was immobilized in the Ni-NTA column and incubated with ΔPcrV$_{(128–294)}$ (without histidine tag). The column was washed thoroughly to remove any unbound protein and protein was eluted by a step gradient of immidazole in 25 mM Tris–HCl (pH-8) and 150 mM NaCl buffer. Both the proteins were observed in the elution fraction. Similar incubation of PcrG with ΔPcrV$_{(1–127)}$ and subsequent elution from the affinity column showed the presence of ΔPcrV$_{(1–127)}$ in wash fraction and PcrG in elution fraction, revealing that ΔPcrV$_{(1–127)}$ is not forming a complex with PcrG. Purified proteins were dialyzed to remove immidazole or 500 μM inositol hexokisphosphate, in buffer according to the need of the downstream experiments.

Size exclusion chromatography

Proteins purified by Ni-NTA chromatography were dialyzed in 25 mM Tris–HCl (pH-7.4) and 150 mM NaCl (SEC buffer). HiLoad 16/60 Superdex 200 pg column was equilibrated in the same SEC buffer. Proteins (2–5 mg/ml) were injected into the column and a flow rate of 1 ml/min was maintained throughout the chromatography. Fractions corresponding to the peaks in SEC were collected and observed in SDS PAGE. For calibrations of the SEC column, following gel filtration markers were used (molecular weight and corresponding elution volumes are indicated): Ferritin (440 kDa ~ 51 ml), Aldolase (158 kDa ~ 62 ml), Ovalbumin (43 kDa ~ 76 ml), Carbonic anhydrase (29 kDa ~ 84 ml), Ribonuclease A (13.7 kDa ~ 92 ml).

Near UV CD spectroscopy

Near UV CD spectra were recorded by Jasco J-815 spectrophotometer. Spectra were recorded for 20 μM of proteins from 300 to 250 nm with a scan speed of 10 nm/min. 1 cm path length cuvette was used. Proteins were dialyzed and diluted in 10 mM sodium phosphate (pH-7.4) and 50 mM NaCl buffer. Buffer spectrum was subtracted from the protein CD spectra.

ANS fluorescence spectroscopy

The fluorescence measurements were recorded using Jasco FP-6500 fluorimeter. ANS emission spectra were monitored from 400 to 600 nm with a scan speed of 50 nm/min, using a 1 cm path length cuvette. Both the protein and the ANS were prepared in 10 mM sodium phosphate (pH-7.4) and 50 mM NaCl buffer. 2 μM of protein was used and up to 30 μM ANS was incubated with the protein. The buffer spectrum was subtracted from the ANS and protein fluorescence spectra.

Dynamic light scattering

DLS profile was obtained using a Malvern-Zetasizer nano ZS DLS instrument. The Proteins were concentrated to 2–5 mg/ml and filtered using a 0.22 micron filter to remove any aggregate. Molecular weight corresponding to the experimental hydrodynamic diameter was calculated using the software for the instrument.

Homology modelling, ConSurf and WebLogo analysis, and molecular docking of ΔPcrG$_{(13-72)}$ and PcrV

The spatial model of PcrV was generated by I-Tasser server using the structure of LcrV from *Yersinia pestis* (PDBID: 1R6F) as the template. The sequence of PcrV was loaded in the FASTA format to I-Tasser as the input file. Additional restrained, or templates were not assigned by the user. ΔPcrG$_{(13-72)}$ is also modelled by I-Tasser using a threading approach, where the threading programmes of I-tasser assigned the top 10 threading templates. The best threading templates for ΔPcrG$_{(13-72)}$ have PDBID: 2ZB9, 2Q24, 3SHG, 1GJS, 2B8I, 1LLW, 1ZBP, 2B8I, and 2R78 [22,23]. From the output file the best model was represented in PyMOL Molecular Graphics System [38]. The PDB file of the best model of PcrV generated by I-tasser was submitted as input to the ConSurf server. The default parameters of the server were used for the prediction. To generate the final model the output file- ConSurf modified PDB was loaded in PyMOL, and conservation code information in the script consurf_new.py was ran [26,38]. WebLogo server was used to generate sequence Logos of helix-7 and helix-12 of V-antigens and the sequences of corresponding helices of PcrV were aligned with these sequence Logos. MSA of helix-7 and helix-12 of proteins belonging to LcrV family was loaded as input [28]. To generate a model of ΔPcrG$_{(13-72)}$-PcrV interaction by molecular docking, The PDB file of PcrV was loaded as the receptor and ΔPcrG$_{(13-72)}$ was loaded as the ligand to the Z-Dock server (version ZD 3.0.2) [36]. Finally, best model was selected from top 5 predictions and represented using Jmol and PyMOL [38,39]. All the models were further checked and validated by PROCHECK [24].

Multiple sequence alignment, disorder and coiled-coil prediction

Multiple sequence alignment profile of PcrG and PcrV with their respective homologs, were generated using MultAlin interface [27]. Disordered region in PcrG were predicted by PrDOS, DisEMBL 1.5 and Disopred version 2.0 [31-33]. COILS/PCOILS server from expasy predicted the intramolecular coiled-coil region within PcrG [34].

Proteolytic digestion of PcrV, PcrG-PcrV with α-chymotrypsin and PcrG with elastase

PcrV, PcrG-PcrV and PcrG were dialyzed in 10 mM hepes (pH-7.4) and 150 mM NaCl. Proteases α-chymotrypsin and elastase from Proti-Ace Kit of Hampton research was diluted to 0.02 μg/μl by Proti-Ace dilution buffer from an initial stock of 1 μg/μl of protease in deionised water. 1 μl of protease from the diluted stock (0.02 μg/μl) was used for 10 μg of protein. Proteolytic digestion was carried out at 37°C for different time points and SDS PAGE sample lysis buffer was used to stop the protease activity. Finally the products obtained after digestion, were analyzed by SDS PAGE.

Native mass spectrometry

Native proteins as well as the proteolytic digestion products of PcrV, PcrG-PcrV, and PcrG were diluted to 0.5-1.0 mg/ml concentration using HPLC water and immediately spotted on the MALDI target plate. Proteolytic digestions were stopped by trifluoroacetic acid present in the α-cyano hydroxyl cinnamic acid (CHCA) matrix. Mass was analyzed using an Applied Biosystem 4700 Proteomics Analyser 170.

MS/MS sequence analysis of different fragments of PcrG, PcrV and PcrG-PcrV

After a specific time point, the digestion fragments of PcrG, PcrV and PcrG-PcrV were separated by SDS PAGE. The gel was stained with coomassie R-250 and washed with distilled water. The bands were excised from SDS PAGE and In-Gel tryptic digestion was carried out using trypsin gold from Promega. The "In-Gel tryptic digestion and MS/MS analysis" protocol provided by Promega was followed. Only exceptions to protocol are: the proteins were eluted in 40 mM ammonium bicarbonate, 10% acetonitrile and 0.3% trifluoroacetic acid buffer. During MS/MS analysis CHCA was used as matrix and prominent peaks were identified by MS after trypsinolysis. Peptides corresponding to the prominent peaks were further fragmented by laser and precise peptide fragments were sequenced using MSDB database of MASCOT search engine, using GPS Explorer Software (version 3.6) [40].

Chemical crosslinking

Chemical crosslinking was performed using water soluble crosslinker EGS-sulfonate. The proteins were dialyzed in 10 mM hepes (pH-7.4) and 150 mM NaCl. The cross linking reactions were carried out at room temperature for 30 minutes with 0.5 mM, 1 mM, and 2 mM of EGS-sulfonate, and finally stopped by addition of sample lysis buffer of SDS PAGE. The crosslinked proteins were analyzed in 12% SDS PAGE.

Surface plasmon resonance

Surface plasmon resonance binding analysis was performed by using a BIACORE 3000 systems and NTA sensor chip (GE Healthcare Life Sciences). Initially, both the flow cells (experiment and the reference cell) were thoroughly washed, and equilibrated with running buffer. Nickel chloride solution was charged in the experiment cell for the binding of nickel to the NTA sensor chip. According to the instruction manual of GE Healthcare Life Sciences for Biacore systems, the reference cell was maintained as non activated flow cell and treated with similar concentration of analyte as used in the experiment cell. Histidine tag PcrG and ΔPcrG$_{(13-72)}$ were used as ligands and immobilized on the Ni-NTA surface in experiment cell. PcrV and its deletion mutants were used as analytes. All the proteins were dialyzed in the running buffer (10 mM hepes (pH-7.4), 150 mM NaCl and 50 µM EDTA). Very low concentration of EDTA was recommended in the running buffer to minimize non specific binding. In absence of the ligand, analyte showed negligible interaction with the activated flow cell. 20 µl of 50–100 µg/ml concentration of the ligand was charged for initial coupling to Ni-NTA sensor chip and unbound ligand was thoroughly washed by the running buffer. Binding kinetics were determined by passing PcrV and ΔPcrV (s) (deletion mutants) at different concentration ranging from 10 nM to 400 nM over PcrG and ΔPcrG$_{(13-72)}$. A flow rate of 5 µl/min and temperature of 25°C was maintained throughout the experiment. Binding constants and forward and backward rates of reactions were determined by BIAevaluation software version 4.1, using a Langmuir binding model. In all the cases, the reference cell was subtracted from the experiment cell by the software to eliminate any RU change occurring due to non specific binding.

Availability of supporting data

The data sets supporting the results of this article are included within the article (and its additional files). The mass spectrometry profiles and MS/MS sequence analysis of proteolytically digested protein fragments are given in

Additional files 4, 5, 6, 7, 10, and 11. The PDB files and corresponding PROCHECK analysis of the models of the proteins are given in Additional files 1, 2, 14, 15, 16 and 17.

Additional files

Additional file 1: Homology model of PcrV. PDB file of the homology model of PcrV was provided according to the editorial requirement.

Additional file 2: Ramachandran Plot for homology model of PcrV. For Validation of homology model of PcrV, PROCHECK server was used, which generated the corresponding Ramachandran Plot showing residues in the most favoured, allowed and disallowed region in the model.

Additional file 3: Multiple sequence alignment of PcrV. Identity and similarity in the sequence of PcrV and its homologs (hydrophilic translocators of Ysc family) are shown.

Additional file 4: MS/MS sequence profile of 1st proteolytic digestion fragment of PcrV. Sequence of the approximate region corresponding to the 1st proteolytic digestion fragment of PcrV, as revealed by MS/MS sequence analysis.

Additional file 5: MS/MS sequence profile of 2nd proteolytic digestion fragment of PcrV. Almost entire sequence of the region corresponding to the 2nd proteolytic digestion fragment of PcrV, as revealed by MS/MS sequence analysis.

Additional file 6: Mass spectrometry profile of specifically protected fragment of PcrV (in presence of PcrG) during proteolytic digestion. Molecular weight of the protected fragment of PcrV in presence of PcrG was estimated by mass spectrometry.

Additional file 7: MS/MS sequence profile of specifically protected fragment of PcrV, in presence of PcrG during proteolytic digestion. Almost entire sequence of the region corresponding to the specifically protected fragment of PcrV in presence of PcrG during proteolytic digestion, as revealed by MS/MS sequence analysis.

Additional file 8: Native PAGE showing oligomeric state of ΔPcrV$_{(128-294)}$, and heterodimeric state of PcrG-ΔPcrV$_{(128-294)}$. Both ΔPcrV$_{(128-294)}$ and PcrG-ΔPcrV$_{(128-294)}$ complex were run on the native PAGE. Since, there is no denaturation of the proteins the greater migration of PcrG-ΔPcrV$_{(128-294)}$ compared to ΔPcrV$_{(128-294)}$, shows reversion of the oligomeric state to a lower order species, may be to a heterodimeric form.

Additional file 9: Disordered region of PcrG predicted by DisEMBL 1.5, Disopred version 2.0. DisEMBL 1.5, Disopred version 2.0 disorder prediction servers predicting the disordered regions (regions lacking proper secondary structure) within PcrG, by various algorithms used by these servers.

Additional file 10: Mass spectrometry profile of the proteolytically digested fragment of PcrG. Molecular weight of the proteolytically digested fragment of PcrG was estimated by mass spectrometry.

Additional file 11: MS/MS sequence profile of proteolytically digested fragment of PcrG. Sequence of the approximate region corresponding to the digested fragment of PcrG, as revealed by MS/MS sequence analysis.

Additional file 12: Multiple sequence alignment of PcrG. Identity and similarity in the sequence of PcrG and its homologs, are shown.

Additional file 13: Coiled-coil regions of PcrG predicted by COILS/PCOILS. Probablity of occurrence of intramolecular coiled-coil regions (essential for protein-protein interaction) within PcrG, predicted by COILS/PCOILS server, is shown.

Additional file 14: Model of ΔPcrG$_{(13-72)}$. PDB file of the model of ΔPcrG$_{(13-72)}$ was provided according to the editorial requirement.

Additional file 15: Ramachandran Plot for the model of ΔPcrG$_{(13-72)}$. For Validation of the model of ΔPcrG$_{(13-72)}$, PROCHECK server was used, which generated the corresponding Ramachandran Plot showing residues in the most favoured, allowed and disallowed region in the model.

Additional file 16: Model of ΔPcrG$_{(13-72)}$-PcrV. PDB file of the Model of ΔPcrG$_{(13-72)}$-PcrV was provided according to the editorial requirement.

Additional file 17: Ramachandran Plot for the model of ΔPcrG$_{(13-72)}$-PcrV generated by molecular docking. For Validation of the model of ΔPcrG$_{(13-72)}$-PcrV, PROCHECK server was used, which generated the corresponding Ramachandran Plot showing residues in the most favoured, allowed and disallowed region in the model.

Abbreviations

TTSS: Type three secretion system; ANS- 8: Anilinonapthalene-1-sulfonate; DLS: Dynamic light scattering; MSA: Multiple sequence alignment; SPR: Surface plasmon resonance; SEC: Size exclusion chromatography; EGS: Ethylene glycol bis [succinimidyl succinate]; Ni-NTA: Nickel-nitrilotriacetic acid; CPD: Cysteine protease domain; CHCA: α-cyano hydroxyl cinnamic acid.

Competing interests

The authors declare that they have no competing interests.

Authors' contributions

SDatta conceived the entire idea of the work, designed, suggested the experiments, analyzed the experimental results, and drafted the manuscript. AB framed the idea of the work, designed and carried out all the experiments, and prepared the manuscript. UD and SDey designed and purified the proteins and participated in the downstream experiments with AB. All the authors have read and approved the final manuscript.

Acknowledgements

The facilities and funding of this research were provided by Indian Institute of Chemical Biology (a unit of Council of Scientific and Industrial Research), the Department of Science and Technology (Government of India) and CSIR network project (TREAT and UNSEEN). CSIR provided the fellowships for this research. pET 22b-Δ50CPD vector was a kind gift from Matthew Bogyo, Department of Pathology, Stanford School of Medicine, Stanford, California. We would like to acknowledge Dr. Samir Kumar Roy (senior technical officer, Indian Institute of Chemical Biology, Kolkata) for the SPR experiments and data analysis. Abhishek Basu acknowledges Dr. Madhurima Das (Research Associate, Bose Institute, Kolkata) for the analysis of protein modelling and docking experiments.

References

1. Giamarellou H: Therapeutic guidelines for *Pseudomonas aeruginosa* infections. *Int J Antimicrob Agents* 2000, 16:103–106.
2. Lyczak JB, Cannon CL, Pier GB: Establishment of *Pseudomonas aeruginosa* infection: lessons from a versatile opportunist. *Microbes Infect* 2000, 2:1051–1060.
3. Hauser AR: The type III secretion system of *Pseudomonas aeruginosa*: infection by injection. *Nat Rev* 2009, 7:654–665.
4. Roy-Burman A, Savel RH, Racine S, Swanson BL, Revadigar NS, Fujimoto J, Sawa T, Frank DW, Wiener-Kronish JP: Type III protein secretion is associated with death in lower respiratory and systemic *Pseudomonas aeruginosa* infections. *J Infect Dis* 2001, 183:1767–1774.
5. Galán JE, Wolf-Watz H: Protein delivery into eukaryotic cells by type III secretion machines. *Nature* 2006, 444:567–573.
6. Cornelis GR: The type III secretion injectisome. *Nat Rev Microbiol* 2006, 4:811–825.
7. Cornelis GR: The type III secretion system tip complex and translocon. *Mol Microbiol* 2008, 68:1085–1095.
8. Frithz-Lindsten E, Holmström A, Jacobsson L, Soltani M, Olsson J, Rosqvist R, Forsberg A: Functional conservation of the effector protein translocators PopB/YopB and PopD/YopD of Pseudomonas aeruginosa and Yersinia pseudotuberculosis. *Mol Microbiol* 1998, 29:1155–1165.
9. Håkansson S, Schesser K, Persson C, Galyov EE, Rosqvist R, Homblé F, Wolf-Watz H: The YopB protein of Yersinia pseudotuberculosis is essential for the translocation of Yop effector proteins across the target cell plasma membrane and displays a contact-dependent membrane disrupting activity. *EMBO J* 1996, 15:5812–5823.

10. Neyt C, Cornelis GR: Insertion of a Yop translocation pore into the macrophage plasma membrane by Yersinia enterocolitica: requirement for translocators YopB and YopD, but not LcrG. *Mol Microbiol* 1999, 33:971–981.
11. Goure J, Pastor A, Faudry E, Chabert J, Dessen A, Attree I: The V antigen of *Pseudomonas aeruginosa* is required for assembly of the functional PopB/PopD translocation pore in host cell membranes. *Infect Immun* 2004, 72:4741–4750.
12. Goure J, Broz P, Attree O, Cornelis GR, Attree I: Protective anti-V antibodies inhibit *Pseudomonas* and *Yersinia* translocon assembly within host membranes. *J Infect Dis* 2005, 192:218–225.
13. Lee PC, Stopford CM, Svenson AG, Rietsch A: Control of effector export by the Pseudomonas aeruginosa type III secretion proteins PcrG and PcrV. *Mol Microbiol* 2010, 75:924–941.
14. Nanao M, Ricard-Blum S, Di Guilmi AM, Lemaire D, Lascoux D, Chabert J, Attree I, Dessen A: Type III secretion proteins PcrV and PcrG from *Pseudomonas aeruginosa* form a 1:1 complex through high affinity interactions. *BMC Microbiol* 2003, 3:21.
15. Gébus C, Faudry E, Bohn YS, Elsen S, Attree I: Oligomerization of PcrV and LcrV, protective antigens of *Pseudomonas aeruginosa* and *Yersinia pestis*. *J Biol Chem* 2008, 283:23940–23949.
16. Sato H, Hunt ML, Weiner JJ, Hansen AT, Frank DW: Modified needle-tip PcrV proteins reveal distinct phenotypes relevant to the control of type III secretion and intoxication by *Pseudomonas aeruginosa*. *PLoS One* 2011, 6:e18356.
17. Pallen MJ, Beatson SA, Bailey CM: Bioinformatics, genomics and evolution of non-flagellar type-III secretion systems: a Darwinian perspective. *FEMS Microbiol Rev* 2005, 29:201–229.
18. Troisfontaines P, Cornelis GR: Type III secretion: more systems than you think. *Physiology (Bethesda)* 2005, 20:326–339.
19. Sato H, Frank DW: Multi-Functional Characteristics of the Pseudomonas aeruginosa Type III Needle-Tip Protein. PcrV; Comparison to Orthologs in other Gram-negative Bacteria. *Front Microbiol* 2011, 2:142.
20. Espina M, Ausar SF, Middaugh CR, Baxter MA, Picking WD, Picking WL: Conformational stability and differential structural analysis of LcrV, PcrV, BipD, and SipD from type III secretion systems. *Protein Sci* 2007, 16:704–714.
21. Derewenda U, Mateja A, Devedjiev Y, Routzahn KM, Evdokimov AG, Derewenda ZS, Waugh DS: The structure of yersinia pestis V-antigen, an essential virulence factor and mediator of immunity against plague. *Structure* 2004, 12:301–306.
22. Roy A, Kucukural A, Zhang Y: I-TASSER: a unified platform for automated protein structure and function prediction. *Nat Protocol* 2010, 5:725–738.
23. Zhang Y: I-TASSER server for protein 3D structure prediction. *BMC Bioinformatics* 2008, 9:40.
24. Laskowski RA, MacArthur MW, Moss DS, Thornton JM: PROCHECK - a program to check the stereochemical quality of protein structures. *J App Cryst* 1993, 26:283–291.
25. Broz P, Mueller CA, Müller SA, Philippsen A, Sorg I, Engel A, Cornelis GR: Function and molecular architecture of the Yersinia injectisome tip complex. *Mol Microbiol* 2007, 65:1311–1320.
26. Ashkenazy H, Erez E, Martz E, Pupko T, Ben-Tal N: ConSurf 2010: calculating evolutionary conservation in sequence and structure of proteins and nucleic acids. *Nucleic Acids Res* 2010, 38(Web Server issue):W529–W533.
27. Corpet F: Multiple sequence alignment with hierarchical clustering. *Nucleic Acids Res* 1988, 16:10881–10890.
28. Crooks GE, Hon G, Chandonia JM, Brenner SE: WebLogo: a sequence logo generator. *Genome Res* 2004, 14:1188–1190.
29. Lawton DG, Longstaff C, Wallace BA, Hill J, Leary SEC, Titball RW, Brown KA: Interactions of the type III secretion pathway proteins LcrV and LcrG from *Yersinia pestis* are mediated by coiled-coil domains. *J Biol Chem* 2002, 277:38714–38722.
30. Nilles ML: Dissecting the Structure of LcrV from *Yersinia pestis*, a Truly Unique virulence Protein. *Structure* 2004, 12:357.
31. Ishida T, Kinoshita K: PrDOS: prediction of disordered protein regions from amino acid sequence. *Nucleic Acids Res* 2007, 35(Web Server issue): W460–W464.
32. Linding R, Jensen LJ, Diella F, Bork P, Gibson TJ, Russell RB: Protein disorder prediction: implications for structural proteomics. *Structure* 2003, 11:1453–1459.
33. Ward JJ, Sodhi JS, McGuffin LJ, Buxton BF, Jones DT: Prediction and functional analysis of native disorder in proteins from the three kingdoms of life. *J Mol Biol* 2004, 337:635–645.

34. Lupas A: **Prediction and analysis of coiled-coil structures.** *Methods Enzymol* 1996, **266:**513–525.

35. Matson JS, Nilles ML: **Interaction of the *Yersinia pestis* type III regulatory proteins LcrG and LcrV occurs at a hydrophobic interface.** *BMC Microbiol* 2002, **2:**16.

36. Pierce BG, Hourai Y, Weng Z: **Accelerating Protein Docking in ZDOCK Using an Advanced 3D Convolution Library.** *PLoS One* 2011, **6:**e24657.

37. Shen A, Lupardus PJ, Morell M, Ponder EL, Sadaghiani AM, Garcia KC, Bogyo M: **Simplified, enhanced protein purification using an inducible. Autoprocessing enzyme tag.** *PLoS One* 2009, **4:**e8119.

38. **The PyMOL Molecular Graphics System, Version 1.3 Schrödinger, LLC.** http://www.PyMOL.org.

39. **Jmol: an open-source Java viewer for chemical structures in 3D.** http://www.jmol.org/.

40. **Mascot Server.** www.matrixscience.com.

7

Molecular dynamics simulations of the Nip7 proteins from the marine deep- and shallow-water Pyrococcus species

Kirill E Medvedev[1], Nikolay A Alemasov[1], Yuri N Vorobjev[2], Elena V Boldyreva[3,4], Nikolay A Kolchanov[1,3,5] and Dmitry A Afonnikov[1,3*]

Abstract

Background: The identification of the mechanisms of adaptation of protein structures to extreme environmental conditions is a challenging task of structural biology. We performed molecular dynamics (MD) simulations of the Nip7 protein involved in RNA processing from the shallow-water (*P. furiosus*) and the deep-water (*P. abyssi*) marine hyperthermophylic archaea at different temperatures (300 and 373 K) and pressures (0.1, 50 and 100 MPa). The aim was to disclose similarities and differences between the deep- and shallow-sea protein models at different temperatures and pressures.

Results: The current results demonstrate that the 3D models of the two proteins at all the examined values of pressures and temperatures are compact, stable and similar to the known crystal structure of the *P. abyssi* Nip7. The structural deviations and fluctuations in the polypeptide chain during the MD simulations were the most pronounced in the loop regions, their magnitude being larger for the C-terminal domain in both proteins. A number of highly mobile segments the protein globule presumably involved in protein-protein interactions were identified. Regions of the polypeptide chain with significant difference in conformational dynamics between the deep- and shallow-water proteins were identified.

Conclusions: The results of our analysis demonstrated that in the examined ranges of temperatures and pressures, increase in temperature has a stronger effect on change in the dynamic properties of the protein globule than the increase in pressure. The conformational changes of both the deep- and shallow-sea protein models under increasing temperature and pressure are non-uniform. Our current results indicate that amino acid substitutions between shallow- and deep-water proteins only slightly affect overall stability of two proteins. Rather, they may affect the interactions of the Nip7 protein with its protein or RNA partners.

Keywords: Molecular dynamics simulation, Nip7 protein, High pressure, Adaptation, Salt bridges

Background

High temperatures and pressures cause damage to living cells. For humans and the best studied organisms, conditions with temperature close to 27°C (300 K) and atmospheric pressure around 0.1 MPa are optimal. However, there exist organisms, which colonize habitats extreme, life-incompatible for humans. Such conditions are near deep hot springs colonized by communities of organisms, the extremophiles [1]. Their life is sustained under conditions with temperature as high as 100°C (373 K) and pressure above 20 MPa, exceeding the atmospheric by 200 times. The mechanisms by which cell survival is provided are not well understood. Their elucidation would provide answers to some fundamental questions on the origins of life and the early adaptation of microorganisms [2], also on adaptation to the conditions of diverse ecological niches [3]. A timely challenge is the identification of the molecular mechanisms of the evolutionary adaptation of the genomes and proteomes of the living beings to the conditions of high temperature [4-8] and pressure [9-12]. Extremophiles

* Correspondence: ada@bionet.nsc.ru
[1]Institute of Cytology and Genetics SB RAS, Prospekt Lavrentyeva 10, Novosibirsk 630090, Russia
[3]Novosibirsk State University, Pirogova str. 2, Novosibirsk 630090, Russia
Full list of author information is available at the end of the article

were supposed to provide unprecedented opportunities for biotechnological explorations (single enzyme catalysis) [13-16]. Indeed, based on research on extremophiles, enzymes were developed for biotechnological applications.

To study the possible mechanisms of the influence of high temperature and pressure on the protein dynamics and protein adaptation to altered environmental pressure, we used here computer MD simulations of two homologous Nip7 proteins from hyperthermophilic (optimal growth temperature close to 100°C) archaea, shallow water *P. furiosus* (hydrostatic pressure close to atmospheric) [17] and deep-water, *P. abyssi* (hydrostatic pressure close to ~20 MPa) [18,19]. The goal of our work was to compare the dynamics properties of the models we built at high and low temperatures, also at atmospheric and high pressures to identify their common characteristics, also their differences presumably related to the different depths of the organisms' habitats.

Nip7 was initially identified in yeast as required for processing of the 27S pre-rRNA to form the mature 25S and 5.8S rRNAs [20]. It localizes to the nucleolus but was also found to sediment in the region of free 60S subunits in sucrose density gradients [20], which is consistent with its presence in pre-60S complexes [21]. Experimental evidence suggests that the *P. abyssi* Nip7 may be an exosome regulatory factor. It binds preferentially to U- and AU-rich RNAs and strongly inhibits the exosome due to its association with both the exosome complex and the substrate RNA [22].

The 3D structure of *P. abyssi* Nip7 protein is known (PDB ID 2P38; Figure 1A). The protein polypeptide chain is 166 amino acids long of which 155 residues are represented in the 3D structure. The protein consists of two α-β domains [23], Figure 1. The N-terminal domain (residues 1–90) is composed of five antiparallel β-strands surrounded by three α-helices and one 3_{10} helix. There is an assumption that archaeal Nip7 may interact with exosome via its N-terminal domain, thereby controlling the exosome function [22]. However, the molecular mechanism of this interaction is unknown.

The C-terminal domain is assigned to the PUA class. [24]. It includes amino acid residues 91–155 and is comprised of a mixture of β-sheets, one α-helix, and one short 3_{10} helix [23]. This domain, named after pseudouridine synthases and archaeosine-specific transglycosylases [25], was initially described in tRNA modifying enzymes and in pseudo-uridine synthases from Archaea and eukaryotes [26], and it has been proposed to mediate protein-RNA interactions. Comparative structural analyses revealed that the residues involved in RNA contacts are conservative in the archaeal PUA domains [23]. In the *P. abyssi* Nip7, these are residues K103, L107, D113, P115, E117, R151, R152, K155, L157, K158 (Figure 1), whose interactions with RNA have been confirmed by the results of mutation

experiments [23]. The PUA domain contacts the RNA molecule using a glycine-containing loop, which connects the fifth α-helix and the β_9 strand via residues of the β_{12} strand (Figure 1). The structural alignment suggests that Nip7 may use a mechanism similar to ArcTGT for RNA interaction, which binds to the bottom of the tRNA^{Val} acceptor stem through the major groove. In the yeast and human Nip7, however, some RNA-binding residues are replaced by a glycine and three residues with hydrophobic side chain [23].

The Nip7 from *P. abyssi* and its *P. furiosus* homolog share 70% identity in their sequences (Figure 1B). Comparative analysis identified the excess of radical versus conservative amino acid substitutions fixation rates in Nip7 *P. furiosus* after divergence from the deep-sea ancestor it shared with *P. abyssi* and *P. horikoshii* [27]. It was suggested that Nip7 and some other proteins, which were concerned predominantly with "translation machinery" and "ribosomal function" evolved under positive Darwinian selection, resulted from their adaptation to altered conditions of elusive pressures. However, the molecular mechanisms underlying this mode of gene evolution remained unclear.

A deletion/insertion free alignment is an advantage for simulations of the *P. furiosus* 3D structure allowing to omit the reconstruction of loops. The Nip7 proteins are simulated at atmospheric pressures, also at 50 and 100 MPa. Given the fact that the temperatures optimal for these two organisms are close to 373 K, we consider simulations at room (300 K) and elevated (373 K) temperatures.

The results show that the models of the two proteins at all the examined pressures and temperatures are compact, stable and similar to the crystal Nip7 structure. Analysis of protein structure dynamics at different pressures and temperatures allowed us to disclose similarities and differences between the deep- and shallow-sea protein models at different temperatures and pressures.

Results
Stability of models
Figure 2 shows changes in the root mean – square deviations (the RMSDs) of the protein C_α atoms relative to the starting structures during the MD simulations for the trajectories at 50 MPa. From Figure 2 it follows that the structures achieve equilibrium starting from about 10 ns of the trajectory periods. The deviations from the starting structure are about 2 Å for the NIP7-ABY model, they are about 2.25 Å for the NIP7-FUR model.

We compare the structures during simulations of the two protein models with the crystal structure of the *P. abyssi* Nip7 (2P38:A). Table 1 gives the mean RMSD values and the 95% confidence intervals at different temperatures and pressures calculated using five MD trajectories for each parameter set.

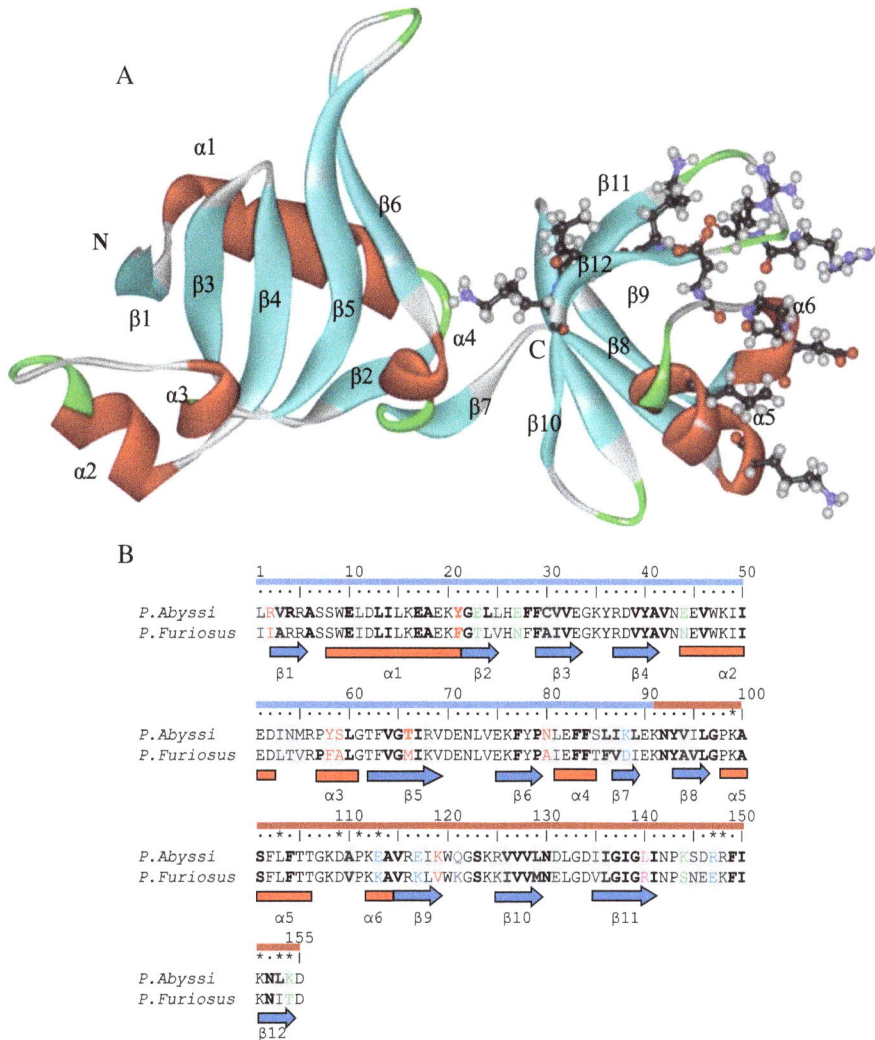

Figure 1 Nip7 protein structure. (A) 3D representation of the *P. abyssi* protein structure (2P38:A) [23]. The secondary structure elements are lettered and colored (helices red, β-strands blue, turns green). The N-terminal domain left, C-terminal domain right. Amino acid residues assumed to bind an RNA molecule [23] shown as ball and stick representation. **(B)** Alignment of the *P. abyssi* and *P. furiosus* sequences. The substitutions in the *P. furiosus* relative to the *P. abyssi* protein are on gray background. Distinguished are the following types of substitutions resulting in replacement of: a polar residue in *P. abyssi* by a nonpolar in *P. furiosus* (red); a charged *P. abyssi* amino acid by an uncharged in *P. furiosus* at retained polarity (green); polar amino acid in *P. furiosus* by nonpolar in *P. abyssi* (pink); *P. furiosus* charged side group by an uncharged (lilac); those resulting in oppositely charged residues (blue). Secondary structure is shown below sequences according to 2P38:A: the helices are indicated in red rectangles, blue arrows indicate β-strands. Residues belonging to the interior of the protein according to the GetArea web-server [27] are shown in bold letters. The symbol *denotes amino acids involved in RNA binding [23]. The N-terminal domain beneath the row of position numbers, blue; the C-terminal domain, red.

As shown in Table 1, the protein deviates from the crystal structure at higher temperatures; however, the mean RMSD is not greater than 2.41 Å for NIP7-FUR and 2.0 Å for NIP7-ABY, thereby supporting the inference that the structures are stable throughout the MD simulations at all the parameters under study. Table 1 also shows that the RMSD for both proteins tends to increase with rising temperature.

The radius of gyration (Rg) is another indicator of stability of a structure during simulation. The average Rgs of the protein structures for trajectories at different pressures and temperatures are compared in Figure 3. As shown in Figure 3, most differences between the Rg values for different trajectories do not exceed the standard errors of means, thereby providing evidence that our models are stable. It should be noted that this is associated with somewhat greater Rg for NIP7-FUR than NIP7-ABY. Two-way ANOVA of the Rg values for NIP7-FUR and NIP7-ABY demonstrated that alterations in pressure and temperature did not result in significant changes in the mean Rg values for both protein models (Additional file 1).

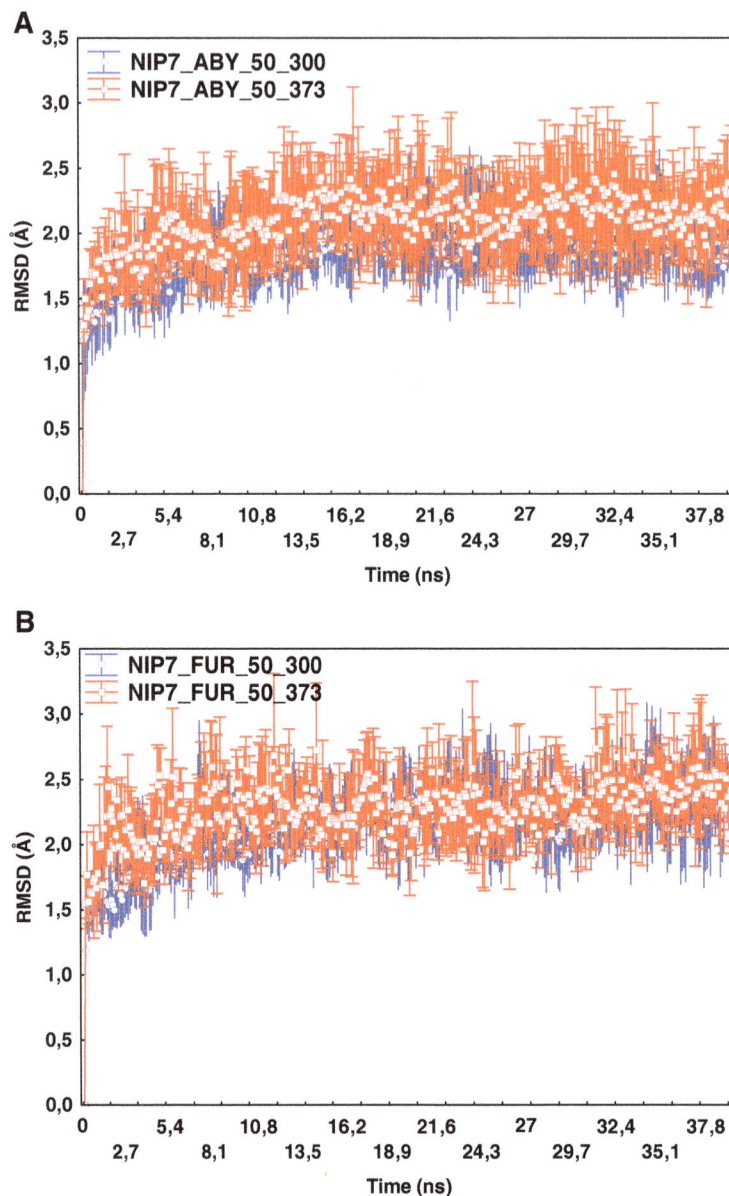

Figure 2 Dependencies of the C$_\alpha$ atoms RMSDs relative to the starting structure on the MD simulation step. Points indicate RMSD values (Y axis) at the simulation time (X axis) averaged over 5 trajectory runs. Whiskers indicate the 95% confidence intervals. **(A)** RMSD for NIP7-ABY model at 50 MPa and 300 and 373 K. **(B)** RMSD for NIP7-FUR model at 50 MPa and 300 and 373 K.

Analysis of solvent accessibility

We compute for each trajectory the average values of solvent accessibility of the residues for the entire protein (SAS$_t$), also separately for the polar (SAS$_p$) and hydrophobic (SAS$_h$) surface fractions. We calculate the mean and standard error of these parameters over five runs for each protein model and pressure-temperature values set. The results are given in Figure 4(A-C).

As seen in Figure 4, the total area is somewhat smaller for NIP7-FUR. Higher values for the solvent accessible polar portion of the residues are observed for the NIP7-ABY than for the NIP7-FUR model (compare the blue/red line values for NIP7-ABY with the brown/green values for NIP7-FUR in Figure 4B). The area of the hydrophobic part of the model is smaller (the same lines in Figure 4C). These observations concerned all the pressure and temperature parameters we studied.

The following trends in the values of the SAS parameters of the NIP7-FUR model are notable: (1) the rise in temperature at the same pressures increases the solvent accessibility of the polar portion of the protein (SAS$_p$ values for NIP7-FUR at 373 K, brown, are higher than the for NIP7-FUR 300 K, green, with the exception of 100 MPa; Figure 4B) and decreases the area of the

Table 1 Comparison of the C_α atom RMSDs in the NIP7-ABY and NIP7-FUR models from their positions in the crystal structure of the *P. abyssi* Nip7 (2P38:A)

Model	NIP7-ABY		NIP7-FUR	
Pressure (MPa)	Temperature 300 K	Temperature 373 K	Temperature 300 K	Temperature 373 K
0.1	1.76 ± 0.04	1.93 ± 0.04	2.08 ± 0.03	2.41 ± 0.02
50	1.81 ± 0.04	1.94 ± 0.06	2.2 ± 0.05	2.2 ± 0.03
100	1.79 ± 0.06	1.78 ± 0.03	2.18 ± 0.02	2.23 ± 0.02

The average values and the 95% confidence intervals at different pressure and temperature values are given in Å.

hydrophobic portion of the surface, this is associated with a small change in the total area (Figure 4C); (2) the increase in pressure systematically reduces the SAS values (brown and green lines, Figure 4B, C). However, most changes are in the range of the mean standard errors.

As for the NIP7-ABY model, the behavior of its SAS parameters was ambiguous. SAS_p at atmospheric pressure and 373 K is greater, than at 300 K, whereas at 100 MPa, on the contrary, is lower (compare the red/blue line Y values, Figure 4B). A significant increase in SAS_h at elevating temperature (red line is above blue one, Figure 4C; 100 MPa was an exception) was other characteristic of this model. At the constant high temperature, increase in pressure resulted in decrease in the SAS values of all the three types (red lines on Figure 4A, B, C). In contrast, at constant T = 300 K, the SAS_p values increased with increasing pressure, although within the limits of standard deviations (blue line, Figure 4B). Changes of SAS_h for the NIP7-ABY model demonstrates no systematic trend at increasing pressure (blue line, Figure 4C); as

for SAS_h, it slightly changes at P = 50 MPa, as compared with P = 0.1 MPa, it rises slightly with pressure increasing to 100 MPa, becoming actually equal to SAS_h at P = 0.1 MPa.

We estimated the influence of temperature and pressure on change in the surface area of the models on the basis of two-way ANOVA (Additional file 2). Taken together, the data agree with the analysis shown in Figure 4. Temperature exerts a significant influence on the hydrophobic part of the NIP7-ABY surface area, thereby demonstrating that the influence of pressure and temperature on change in the total area of protein surface is not additive. This presumably reflects the difference in the SAS values between the high and low temperature NIP7-ABY trajectories. As for the NIP7-FUR model, pressure exerts a significant influence on both the hydrophobic part of the area and SAS_t.

Thus, pressure and temperature differently affect residue solvent accessibility for the two protein models. Pressure results in a decrease in the solvent accessible surface area. This effect is the most conspicuous at high temperatures.

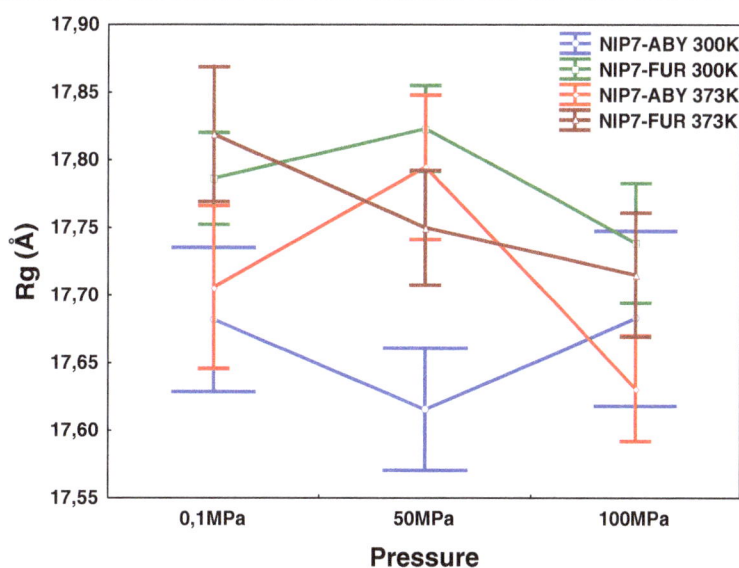

Figure 3 Dependencies of the radius of gyration (Rg, Y axis) on pressure (X axis) at high and low temperatures for the Nip7 protein models NIP7-ABY and NIP7-FUR. Whiskers indicate the 95% confidence intervals. Blue, the NIP7-ABY model, 300 K; green, the NIP7-FUR model, 300 K; red, the NIP7-ABY model, 373 K; brown, the NIP7-FUR model, 373 K.

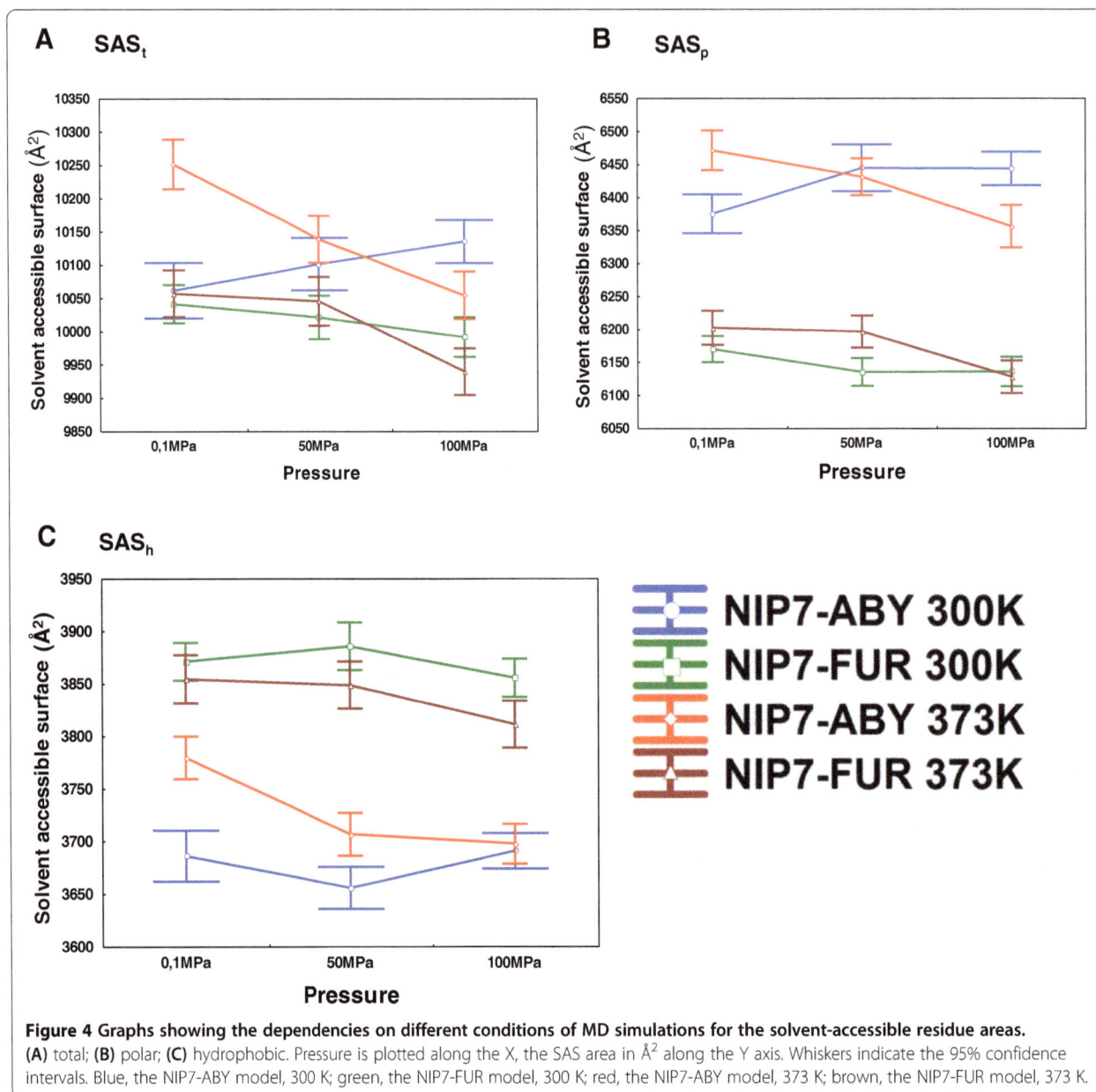

Figure 4 Graphs showing the dependencies on different conditions of MD simulations for the solvent-accessible residue areas.
(A) total; **(B)** polar; **(C)** hydrophobic. Pressure is plotted along the X, the SAS area in Å^2 along the Y axis. Whiskers indicate the 95% confidence intervals. Blue, the NIP7-ABY model, 300 K; green, the NIP7-FUR model, 300 K; red, the NIP7-ABY model, 373 K; brown, the NIP7-FUR model, 373 K.

Local structure of the polypeptide chain

The next step was to define how the conformation of the different regions of the polypeptide chain deviates from the X-ray Nip7 structural model. For different trajectories, we built graphs showing the dependencies of the local structure deviation (RMSDL) parameter (see Methods). These graphs concerned every residue; we estimated also the mean RMSDL error, which characterizes its fluctuation during the simulations. (Additional file 3: Figure S1). The graphs give prominence to the non-uniform local changes in the protein structure during the simulations. There are regions where deviations are quite small (~1 Å), while in the other regions changes in the local chain conformation relative to the crystal structure are considerable

(~4 Å). In such cases, the RMSDLs of the polypeptide chain conformation are greater in those regions, which correspond to the loops connecting the secondary structure elements. Regions with great RMSDL values are, as a rule, located where amino acids substitutions are numerous (see Additional file 3: Figure S1A). As a result, the evolutionary and structurally variable regions are virtually the same (with the exception of positions 85–95 making up the segment between the α_4-α_5 helices, Additional file 3: Figure S1A). For this segment, the RMSDL values are low, whereas the number of differences between the *P. abyssi* and *P. furiosus* Nip7 sequences is high.

Other features concern the differences in the RMSDLs values between the N- and C-terminal domains [23].

There are regions with small RMSDL values and peaks in the loop regions (positions 30–40, which correspond to the β_3-β_4 loop, positions 49–59 and 69–79) for the N-terminal domain. The graphs of structural variations in the C-terminal, the PUA-domain, also contain maxima at positions 105–110, 115–125, 143–148 (Additional file 3: Figure S1) and minima; however, the minima values are substantially higher than those for the N-terminal domain. From comparisons of the deviation graphs for the different trajectories, it is evident that the regions corresponding to the RMSDL maxima are by and large the same for the trajectories that correspond to both high pressures and temperatures (Additional file 3: Figure S2). This means that the local changes in the conformation of the protein chain in our models occur in the same region of the structure despite the different nature of the destabilizing factor (changes in pressure or temperature).

The effects of pressure and temperature on changes in the conformation of the polypeptide chain in the Nip7 models are now considered. The most conspicuous fact is that, at the same pressure, elevated temperature causes an increase in the RMSDLs for some protein residues (compare Additional file 3: Figure S1, panel pairs A-B, C-D, E-F). The changes in the RMSDLs of the polypeptide chain resulting from elevated temperature occur non-uniformly. For example, greater RMSDLs at elevated temperature are observed in the NIP7-ABY model in the region of the α_2-α_3 loop (at positions 49–59). Conversely, for some regions in this model, an increase in temperature results in their decrease, for example residues 109–127 of the C-terminal domain. A similar pattern is observed for this model at low (Additional file 3: Figure S1, panels A-B), moderate (C-D) and high (E-F) pressures. However, for some regions of the protein changes in the RMSDLs at elevated temperature are only slight.

The interesting features brought out by comparing the two proteins are the higher RMSDL values for the C-terminal domain in NIP7-FUR compared to NIP7-ABY at any pressure-temperature parameter (Additional file 3: Figure S1).

To make the local structure changes more prominent with respect to the temperature and pressure, we performed an analysis of the sensitivity of the residue RMSDL parameter to temperature (S_i^t) and pressure (S_i^p) expressed as the ratio of the RMSDL at high and low temperature and high and low pressure (see Methods). We also estimated the dependence of the RMSDL parameter for each residue on the temperature and pressure using two-way ANOVA (see Methods). The influence of temperature and pressure on the RMSDL parameter was significant, if the corresponding F-statistics was above the threshold at $\alpha = 0.05$ (see Additional file 4).

Figure 5A illustrates that temperature sensitivity of the RMSDL parameter is different, being dependent on protein position in the two models. To begin with, there are protein regions for which temperature significantly increases the RMSDL (S_i^t >1) for both models. To such regions belongs, in the first place, the long segment extending from the C-terminal part of the β_4 strand to the N-terminal end of the β_5 strand (positions 41–65). The F-statistics for most of its residues exceeds by many times the 5% critical value for both models (Additional file 4). In the graphs given in Additional file 3: Figure S1, this region corresponds to the peak RMSDL values, which increase when temperature elevates. A small, yet significant, increase in the RMSDL is established in the N-terminal domain for both models at positions 26–27 (the β_3 strand of the N-terminal domain). A significant increase in the RMSDL residues at positions 31–37 (β_3-β_4) under the influence of high temperature is also characteristic of the NIP7-FUR model. High values of the S_i^t parameter are observed for the N-terminal regions of both models, however, the influence of temperature on change in the RMSDL for these residues proved to be insignificant.

An interesting pattern was observed for the C-terminal domain. S_i^t <1 for the great majority of residues, the differences were significant for many of them. This meant that elevation in temperature made the conformation of the C-terminal domain to a great extent similar to that of the crystal structures. These changes were the most prominent for region 109–127 (α_6-β_{10}). This was associated with lower S_i^t values characteristic of the NIP7-ABY model. Another region, which was also subject to conformational change at elevated temperature in the models, joined two protein domains (positions 80–93). A significant decrease in the RMSDL was characteristic of the C-terminal residues in both models, too.

The sensitivity of the RMSDL values to elevated pressure is shown in Figure 5B. We compared S_i^p and S_i^t with respect to the positions at which their proportions are significantly different for both models. We found that their number is considerably smaller for S_i^p than S_i^t. For the NIP7-FUR model, significant changes are observed in the N-terminal domain (positions 30–38). For the NIP7-ABY model, fluctuations in the S_i^p values are more expressed, however, changes at positions 41–44 (a decrease in the RMSDL) and 85–89 (an increase in the RMSDL) are significant.

Analysis of the secondary structure

To describe in more detail the conformational changes, we analyzed the secondary structure of the Nip7 protein models at different MD trajectories. Additional file 3: Figure S2 provides evidence for the stability of the secondary structure during the simulations. For example, although the RMSDL values for the loop 69–79 (the β_5-β_6 region) are high (Additional file 3: Figure S1), the secondary

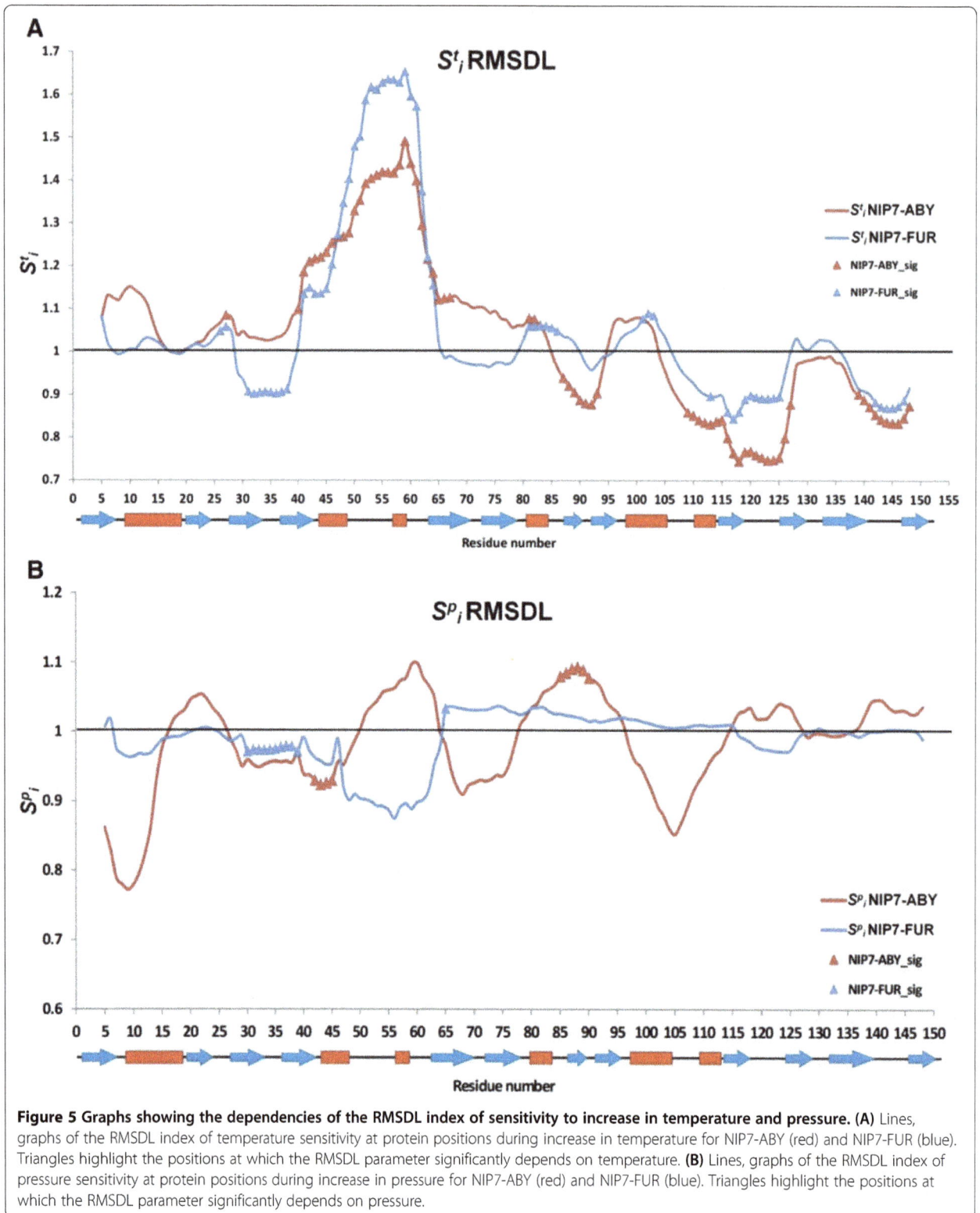

Figure 5 Graphs showing the dependencies of the RMSDL index of sensitivity to increase in temperature and pressure. (A) Lines, graphs of the RMSDL index of temperature sensitivity at protein positions during increase in temperature for NIP7-ABY (red) and NIP7-FUR (blue). Triangles highlight the positions at which the RMSDL parameter significantly depends on temperature. **(B)** Lines, graphs of the RMSDL index of pressure sensitivity at protein positions during increase in pressure for NIP7-ABY (red) and NIP7-FUR (blue). Triangles highlight the positions at which the RMSDL parameter significantly depends on pressure.

structure pattern of the polypeptide chain demonstrates no changes dependent on pressure and temperature. This may suggest that this β-hairpin changes its conformation without breaking hydrogen bonds, just by bending or twisting.

Analysis of the data on the changes in the protein secondary structure (Additional file 3: Figure S2) shows also that a part of its elements is unstable when pressure and temperature change. For example, this is characteristic of the loop formed by residues 48–62 (the α_2-α_3 segment). At different pressures and temperatures, the conformation of these residues varies from the states of the α-helix (H), the 3_{10} –helix (G), the turn (T), and to the bend (B). A local maximum of the deviations of the polypeptide chain from the crystal structure, the RMSDL (Additional file 3: Figure S1) are observed in the segment. Thus, this protein segment proves to be very mobile.

Other regions of smaller size with unstable secondary structures are the first N-terminal residues, the β_{10} terminal residues (transitions from the loop to the extended conformation) and the β_{11}–β_{12} loop (transitions between the turn, the bend and the loop conformations). These regions show characteristic transitions from the loop to the bend, also the maximum RMSDL values (Additional file 3: Figure S1).

The above described conformational changes are supported by the trajectory snapshots providing new dimensions of the credibility of our models (Figure 6; Additional file 3: Figure S3). Thus, residues 69–75, which belong to the β-turn, have a stable secondary structure, but high RMSDL values. These residues bend towards the globular part of the N-terminal domain during the simulations. In so doing, they twist slightly the

β-sheet. The sweep of the loop 69–75 bend depends on pressure and temperature. Similar conformational changes (the bend towards the centre of the N-terminal domain without significant impairment of the secondary structure) are also observed for the loop between the β_3–β_4 strands. It also shows characteristic RMSDL deviations (Additional file 3: Figure S1). As seen in Figure 6B, the range of the fluctuations widens at high temperatures in the NIP7-ABY model, a similar widening is true for the NIP7-FUR model (Figure 6, C, D). The structural fluctuations in the segment 45–55 and those in the loops between the β-strands of the terminal domain are conspicuous, too.

Fluctuations in the polypeptide chain

We estimated the structural flexibility of the two proteins at different temperatures and pressures. The graphs display the dependencies of the Root-Mean-Squared-Fluctuation (the RMSF) on the residue number (Additional file 3: Figure S4).

The peak regions of the RMSF values enclose predominantly loops in the sequence regions at positions 31–41 (the β_3–β_4 loop), 50–60 (the α_3 region), 69–79 (the β_5–β_6 region), 98–110 (the β_8-α_5 region), 120–125 (the

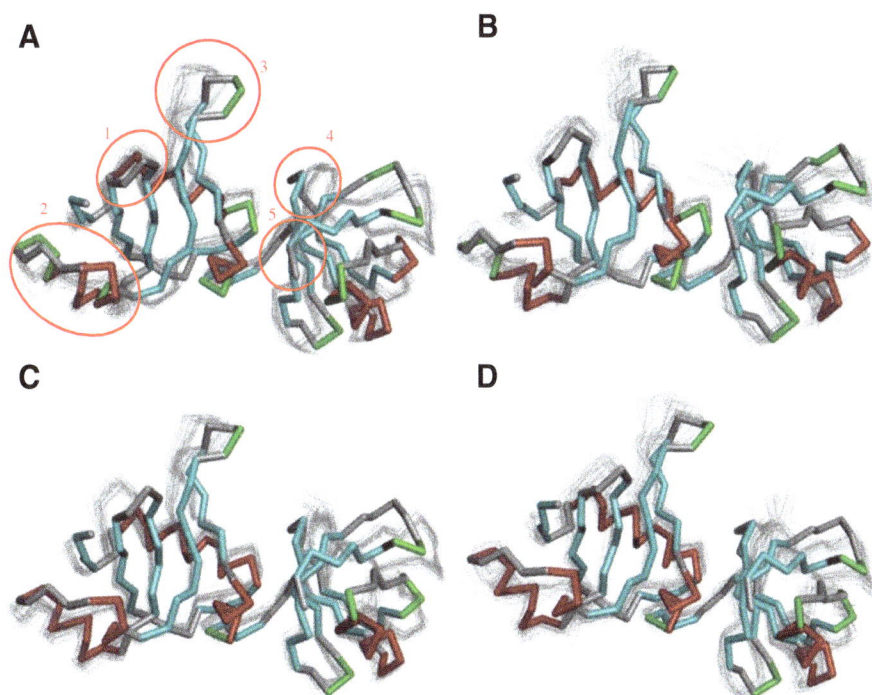

Figure 6 Representative snapshots of the superposition of the trajectories of protein models with the *P. abyssi* Nip7 crystal structure. (A) NIP7-ABY at 300K and 100 MPa; (B) NIP7-ABY at 373K and 100 MPa; (C) NIP7-FUR at 300K and 100 MPa; (D) NIP7-FUR at 373K and 100 MPa. The Nip7 *P. abyssi* crystal structure is in color tube; beta strands, blue; alpha-helices, red; turns, green. Models obtained by the MD simulations are indicated by grey lines. Ovals encircle: (1) the β_3-β_4 loop (residues 29–41), (2) the α_2–α_3 segment (residues 44–61); (3) the β_5-β_6 loop (residues 62–79); (4) the β_9-β_{10} loop (residues 120–125); (5) the C-terminal domain (residues 153–155).

β_9–β_{10} region), 132–135 (the β_{10}–β_{11} region), 140–150 (the β_{11}–β_{12} loop). These regions correspond to those where the RMSFs are the largest. In addition, the highest values for the local root-mean-square deviations of the polypeptide chain (the RMSDLs) are observed in them (compare with Additional file 3: Figure S1).

The other features are the high RMSF values for the residues at the amino (N) and carboxy (C)-termini of the proteins. Notably, the RMSF values are somewhat higher for the trajectories at high temperature than for those at 300 K. These differences were observed for both NIP7-ABY and NIP7-FUR (Additional file 3: Figure S4).

The sensitivity of the RMSF value to increase in temperature and pressure was analyzed for each and every amino acid residue in the same manner as was for the RMSDL parameter. The results are shown in Figure 7 (see Additional file 5 for the two-way ANOVA results). Analysis of the sensitivity profiles of the residues to temperature demonstrated that increase in temperature raised the value of the local polypeptide fluctuations for all the residues at high significance (S_i^t >1). However, the graph displays regions weakly sensitive to increase in temperature, even insensitive to it (these are residues 36, 121, 123 and 131 for the NIP7-ABY model and residue 146 for the NIP7-FUR model). The RMSF values are high in the residues of the loop already at 300 K, while increase in temperature does not cause any considerable increase in their mobility.

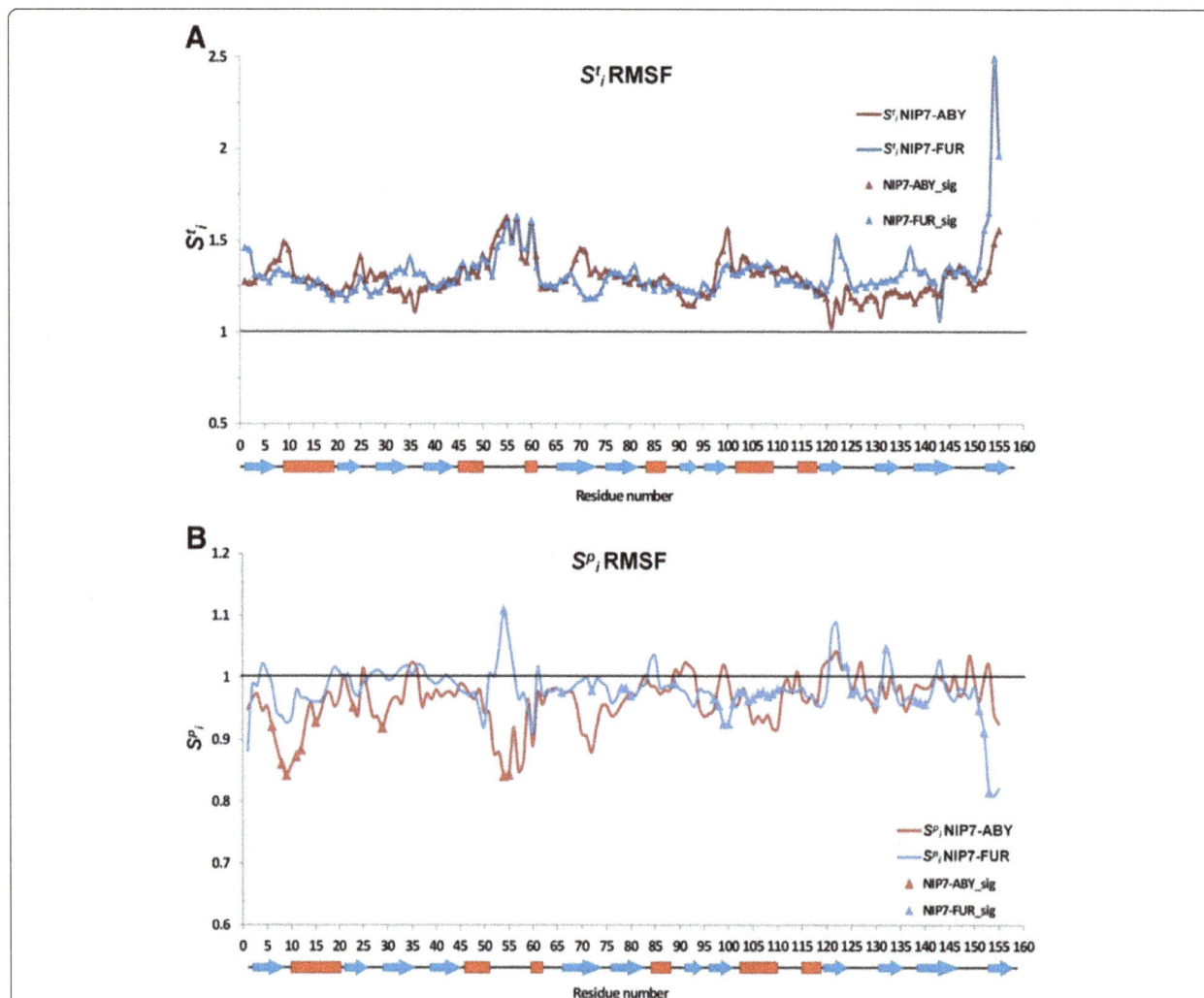

Figure 7 Graphs showing the dependencies of the RMSF index of sensitivity to increase in temperature and pressure. (A) Lines, graphs of the RMSF index of temperature sensitivity at protein positions during increase in temperature for NIP7-ABY (red) and NIP7-FUR (blue). Triangles highlight the positions at which the RMSF parameter significantly depends on temperature. **(B)** Lines, graphs of the RMSF index of pressure sensitivity at protein positions during increase in pressure for NIP7-ABY (red) and NIP7-FUR (blue). Triangles highlight the positions at which the RMSDL parameter significantly depends on pressure.

The results are different for fluctuations in the Nip7 models of the deep- and shallow-sea organisms at increased pressure (Figure 7B). The S_i^p parameter is above unity for a small fraction of residues. For the N-terminal domain, the decrease in fluctuations under the effect of pressure is greater for the NIP7-ABY than the NIP7-FUR model. In this domain, pressure significantly influences residues located at positions 6–15 (the N-terminal part of the α_1 helix), positions 23, 29 (the loop between β_2 and β_3), and 54–55 (the α_2–α_3 loop).

For the NIP7-FUR RMSF, a significant influence of pressure is observed in residues mostly in the C-terminal part of the protein. A wide region of these residues spans positions 97–110 (the α_5 helix), the most sensitive region to increase in pressure and S_i^p <1 (decrease in the RMSF). Several other residues for which RMSF decreased by pressure significantly are located the β_{10} region (positions 125–130), β_{11} (138–140), β_{12} (152–153) and at position 89. Several positions demonstrating significant decrease in the RMSF under pressure are located in the N-terminal domain (66, 72, 78–80). There are, however, residues for which RMSF changes significantly when pressure increases. They are observed for NIP7-FUR only (54, 124, 132). Interestingly, the single position, which demonstrates the significant dependence of its RMSF value on the pressure for both models, 54, is located in the α_2–α_3 loop. However, the influence is inversed for this residue in the NIP7-FUR and NIP7-ABY models.

To analyze the influence of pressure and temperature on fluctuations in the models, we performed two-way ANOVA at different temperatures for both the entire models and separately for each domain. The mean RMSF values for the C_α atoms in the NIP7-ABY and NIP7-FUR models at different trajectories were compared (see Additional file 6). Statistical analysis demonstrates that, for the NIP7-ABY model at 300 K, change in pressure does not affect significantly fluctuations in the N- and C- domains. However, differences in fluctuations between these domains are significant. In general, the C-terminal domain demonstrates larger fluctuations compared with the N-terminal domain. At high temperature, the influence of pressure on the RMSF of two domains becomes significant. The fluctuations in both domains decreased at high pressure. The difference in the fluctuations between two domains at high temperature remains the same as at 300 K. In other words, the fluctuations are greater for the C-terminal domain.

For the NIP7-FUR model, pressure exerts a significant influence on the RMSF values in the two domains both at low and high temperatures (see Additional file 6). The fluctuations in both domains decrease with increase in pressure, like in the NIP7-ABY model. The differences in the RMSF values for the two domains are significant at 300 and 373 K.

Analysis of salt bridges

The crystal structure of *P. abyssi* contains eight pairs of residues forming salt bridges [28]. It should be noted that the positions of the side chain atoms have not been resolved for some residues capable of forming salt bridges probably due to their decreased stability under crystallization conditions. These are GLU51, LYS76, for example. In our simulations it is feasible to reconstruct side chain atoms for these residues and to take them into account in salt-bridge stability analysis. For each and every possible salt bridge, the proportion of structures was estimated where such a bridge was formed (salt bridge persistence; see Methods). This was done for all the trajectories and models under study. Analysis of the NIP7-ABY and NIP7-FUR models revealed that there were stable salt bridges (occurring, on average, in 70% of the structures in the different trajectories of the two proteins), moderately stable (20–70% persistence) and a number of unstable salt bridges occurring in less than 20% of the structures (Additional file 7). The number of stable and moderately stable salt bridges was larger in NIP7-ABY, twenty, than in NIP7-FUR, thirteen. Interestingly, the residues with unresolved side chain atoms in the crystal structure formed moderately stable salt bridges only in the NIP7-FUR and NIP7-ABY models.

Six salt bridges (GLU10-ARG4, GLU75-ARG37, ARG37-GLU10, LYS20-GLU17, ARG148-ASP109, GLU33-ARG4) are the most stable in the NIP7-ABY model (Additional file 7). These form in more than 70% of the structures in both the NIP7-ABY and NIP7-FUR models. The GLU33-ARG4 and GLU131-ARG116 pairs (unresolved in the 2P38:A crystal structure) may be referred to the most conservative in NIP7-FUR (being moderately stable in NIP7-ABY). Four stable salt bridges from the N-terminal domain (GLU10-ARG4, GLU75-ARG37, ARG37-GLU10, GLU33-ARG4) form a network that links the alpha-helix and the beta-strands. The remaining salt bridges may be assigned to the moderately stable. Some stable or moderately stable salt bridges in NIP7-ABY are unstable in the NIP7-FUR (148–109, 147–113). There are also two salt bridges moderately stable in the NIP7-FUR and unstable in the NIP7-ABY models (146–112, 76–17) (Table 2).

Analysis of the functional role of the most stable salt bridges demonstrated (Table 2) that most of them stabilized the packaging of the elements of the protein secondary structure. However, a number of salt bridges in our models formed residues for which interactions with RNA has been demonstrated [23]. There are four such pairs among stable bridges for the NIP7-ABY model and only one for NIP7-FUR model (148–109) is formed by a residue pair with both involved in the interaction with RNA, the other salt bridges are either unstable or missing in the NIP7-FUR model.

Table 2 Sensitivity of the most stable salt bridges in the NIP7-ABY, the NIP7-FUR models to increase in temperatures and pressures

Salt bridge	Marker	Function	NIP7-ABY				NIP7-FUR			
			S^T	$F(T)$	S^P	$F(P)$	S^T	$F(T)$	S^P	$F(P)$
10–4	All	a_1–a_1	0.94	2.74	1.04	0.83	0.99	1	1.01	1
75–37	All	β_6–β_4	1.06	0.52	1.11	0.91	0.99	1.89	1.00	0.36
37–10	All	β_4–a_1	1.04	0.16	1.14	0.99	1.18	2.87	0.88	**4.68**
148–109	All	RNA binding - RNA binding	0.95	0.40	1.09	0.45	u	u	u	u
33–4	All	β_3–β_1	0.92	0.59	0.94	0.69	0.98	0.10	1.00	0.20
20–17	All	a_1–a_1	0.92	2.65	1.02	0.13	0.89	2.96	0.98	0.36
16–12	P.a.&P.f.	a_1–a_1	1.14	3.77	0.91	1.68	1.08	1.60	0.89	2.76
147–113	P.a.&P.f.	RNA binding - RNA binding; a_6	0.98	0.02	0.72	3.18	u	u	u	u
33–2	2p38&P.a.	β_3 - N-terminal	0.98	0.01	0.99	2.07	-	-	-	-
131–116	All	C-terminal - β_9	1.93	**12.23**	0.94	0.49	0.98	0.72	0.99	0.43
51–5	P.a.&P.f.	a_2–β_1	1.25	1.50	1.03	0.03	0.88	1.11	1.18	0.74
76–70	P.a.&P.f.	β_6 - N-terminal	1.20	**6.59**	0.84	**3.42**	0.83	**8.19**	0.95	1.98
147–146	P.a.	RNA binding - C-terminal	1.84	4.12	0.95	0.12	-	-	-	-
48–45	P.a.&P.f.	a_2–a_2	1.06	0.14	0.92	1.42	0.99	0.00	0.88	1.29
88–23	P.a.	β_7–β_2	1.26	**5.91**	0.87	1.72	-	-	-	-
151–109	All	RNA binding; β_{12} - RNA binding	0.93	0.09	0.75	1.49	0.86	2.74	0.92	0.42
119–117	P.a.	β_9–β_9	0.98	0.04	1.26	2.31	-	-	-	-
48–44	2p38&P.a.	a_2–a_2	0.99	0.00	0.98	0.01	-	-	-	-
146–112	All	C-terminal - a_5	u	u	u	u	1.01	0.00	1.11	0.50
76–17	P.a.&P.f.	β_6–a_1	u	u	u	u	0.83	1.07	0.95	0.56

Columns: residue pairs forming salt bridges; occurrence of a salt bridge in the different structures (ALL, the NIP7-ABY, the NIP7-FUR models, and the crystal 2P38:A structure; ABY & FUR for only the NIP7-ABY, NIP7-FUR models and so on); the values of the indices of sensitivity for the NIP7-ABY; for the NIP7-FUR models; the F-value for a concrete parameter. Values expressing significant changes, maintenance of a salt bridge when pressure or temperature changes are in bold. Unstable salt bridges (occurrence less than 20%) are marked with the symbol u.

How high temperatures and pressures may affect the formation of salt bridges in the two different protein models? To answer this question, we introduced for each salt bridge its index of sensitivity to pressure (S^p) and temperature (S^t) in the same manner as we did for the RMSDL and the RMSF parameters (Methods). The higher were the values of these indices, the greater was the persistence of a bridge at high pressure or temperature. To estimate the significance of the influence of pressure and temperature on the stability of a salt bridge, we applied two-way ANOVA and obtained values of the statistics $F(T)$, $F(P)$ (see Methods). We assumed that high temperature or pressure exerted an influence on stability of a salt bridge, if the corresponding $F(T)$ and $F(P)$ statistics exceeded the 0.05 significance threshold. The results for the stable and moderately stable salt bridges common to both models are given in Table 2, Additional files 8 and 9. They demonstrate that for the NIP7-ABY model, of the 20 stable salt bridges 3 (15%) are affected by high temperatures (131–116, 76–70 and 88–23). One salt bridge is significantly subject to the influence of pressure (76–70)

both also become stable at high temperature. This bridge becomes also stable at high temperatures in NIP7-FUR. Both models also contain salt bridges, whose temperature sensitivity considerably, yet insignificantly, deviates from 1 (S^t <0.8 or S^t >1.2). Most become stable when temperature rises (S^t >1). Thus, elevation in temperature for a number of salt bridges in our models changes their persistence mainly in favor of its increase.

A more detailed consideration of the salt bridge stability may reveal certain interesting details. Bridge 131–116, whose stability, in the case of the deep-water protein, is under the significant influence of temperature, and is formed by the side chain groups of residues ASP131 and ARG116 for the NIP7-ABY model. In the case of the shallow-water protein, aspartic acid is substituted by glutamic acid at position 131 and no significant influence of temperature is observed. Bridge 148–109 is formed by the side chain groups of residues ARG148-ASP109 for the NIP7-ABY model. In the case of the shallow-water protein, arginine is substituted by lysine at position 148 and, as a result, this bridge turns out to be unstable for the NIP7-FUR model. Therefore, the structure of the side chain radicals has a strong impact on the stability of these salt bridges.

Influence of protein type on the dynamics of protein structure

To estimate the influence amino acid substitutions exerted on the structural properties of the NIP7 proteins in comparison with factors such as temperature and pressure, we performed three-way ANOVA for each residue. The results are given in Additional file 10.

Analysis of the influence temperature exerts on the RMSDL in the two models shows that the local structure deviates significantly from the crystal 2P38:A for 77 of the 143 positions we analyzed (53%). Pressure exerts a significant influence on the RMSDL at only 9 positions of the protein (6%), while protein type exerts a significant influence on the conformational deviations of the polypeptide chain for 91 positions, 63% (Figure 8).

The most characteristic differences are observed for the C-terminal domain of the protein (Figure 8). Furthermore, significant differences in the RMSDL parameter depending on model type are observed for regions 16–28, 40–45, 67–77 and position 5. As seen, the deviations from the crystal structure are greater for the NIP7-FUR than the NIP7-ABY model.

A similar analysis was carried out for the RMSF parameters of the amino acid residues (Additional file 10). The results showed that a rise in temperature caused a significant increase in the RMSF for all, without exception, amino acid residues of the NIP7 protein. Pressure exerted a significant influence on change in 26 amino acid residues (16%). Type of model exerted an influence on change in fluctuations for 56 (36%) of the amino acid residues.

Significant differences between the RMSF values for the residues of the two models were observed both in the case when fluctuations were higher for the NIP7-ABY model and, conversely, for the NIP7-FUR model (Figure 9). Residues at positions of the N-terminal domain (9, 11, 12, 17, 20, 21, 23, 24–26) corresponding to the α_1 helix and the β_2 strand, also position 34 (the loop between β_3 and β_4), also the C-terminal domain (96–107, 115–118, 154–155) are referred to the first group. Positions 65–67, 78–85, 89, 120–128, 131–133, 138–139 are referred to the second group.

Discussion

Effect of temperature and pressure on the structure and dynamics of Nip7

In our work we compare the dynamics of two Nip7 protein models, NIP7-ABY from *P. abyssi*, a deep water organism, and NIP7-FUR, from *P. furiosus*, a shallow water organism, at different temperatures and pressures.

The hallmark feature of our models is their stability throughout the simulation period (40 ns). It is well maintained at both elevated temperatures and pressures. The RMSDs from the crystal structure (2P38:A) are, on average, smaller for the NIP7-ABY than the NIP7-FUR model. It could be remembered, however, that the NIP7-FUR model is not crystallographic, but it has been rather produced by homology modeling, which could indeed justify a larger RMSD. The secondary structure of the model changes slightly, thereby providing evidence that the structural differences between the two proteins are related to change in loop conformation and in mutual disposition of the secondary structure elements.

Exposure to high temperature for the two proteins results in change in the equilibrated conformations of

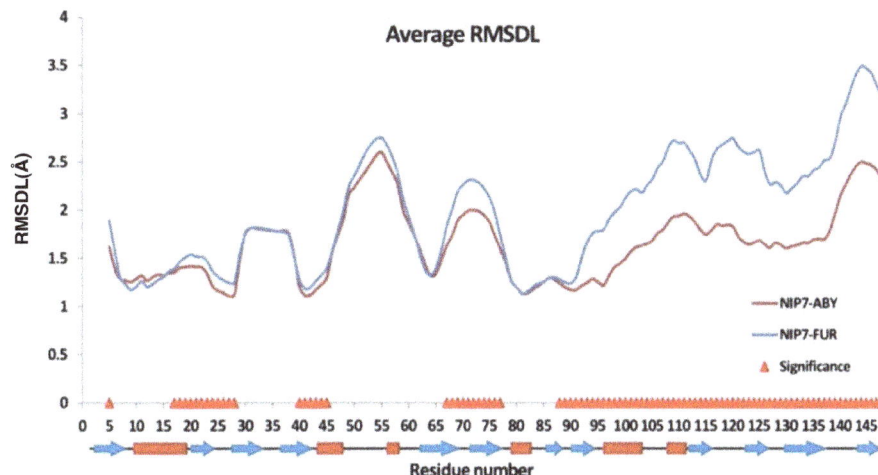

Figure 8 The RMSDL values for the NIP7-ABY and NIP7-FUR proteins averaged for all the trajectories and runs. Red triangles highlight the positions at which the RMSDL parameter significantly depends on model type.

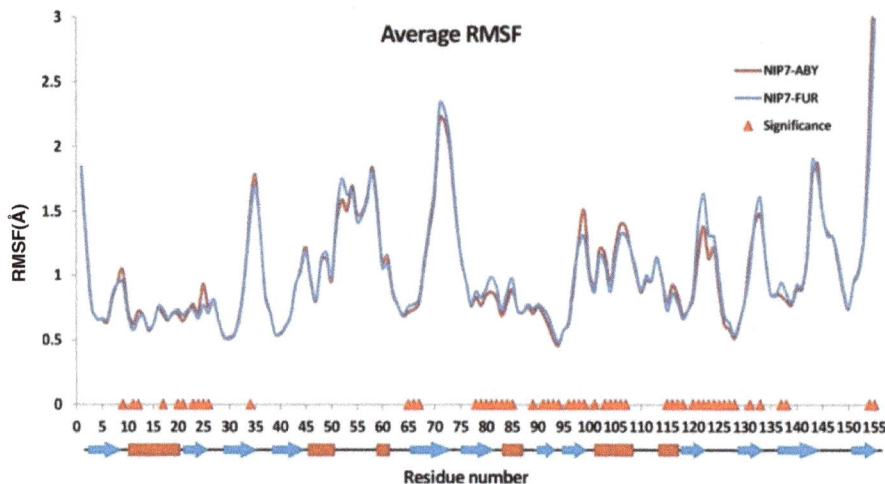

Figure 9 The RMSF values for the NIP7-ABY and NIP7-FUR protein averaged for all trajectories and runs. Red triangles highlight the positions at which the RMSF parameter significantly depends on model type.

the polypeptide chain, on the one hand (seen when comparing the RMSDL profiles in Figure 5), and in increase in its fluctuations, on the other hand.

Conformational changes at increasing temperature are related to decrease in structural deviations from the crystal structure in the C-terminal domain, including the positions that form the interaction with RNA. The deviations increase in the regions of the helices α_2-α_3 in the N-terminal domain. It should be noted that the specific influence of elevated temperature on the conformation of the protein, on its active centers in particular, is one of the possible mechanisms providing optimal protein activity in this region. For example, the important role of temperature-dependent conformational transitions of protein, at its active site particularly, in the enzymatic activity of the NADH oxidase from *Thermus thermophilus* has been well demonstrated at high temperatures [29]. Also it was demonstrated that some mutation may increase and some of them may decrease the thermostability of thioredoxins from *Escherichia coli* and *Bacillus acidocaldarius*; the thermostability of these proteins was revealed to depend on ionic interactions between the thermolabile regions [30]. Such conformational changes may be important to the Nip7 functions as well.

The comparison of the protein models obtained at 373 and 300 K demonstrates that at high temperature the fluctuations in the polypeptide chain increase, although the compactness of the protein structure is preserved (Figure 3). The increase in the fluctuations of the polypeptide chain is non-uniform. The regions subject to the fluctuations are mainly loops and the terminal regions of the protein. Like in the case of the RMSDL, the region α_2-α_3 proved to be the most subject to increase in the RMSF among the internal regions of the chain. It is

pertinent to note that a similar effect of high temperatures on the conformation of the polypeptide chain is characteristic of the high temperature protein models [31-34].

The effect of increased pressure on protein structure is manifold. The structural deviations of the NIP7-ABY from the conformation of the Nip7 model based on the X-ray diffraction structural data at high temperatures decrease as pressure increases. This trend holds true for the NIP7-FUR model (Table 1). The increased pressure results in a decrease in the accessible surface area. This decrease is characteristic of NIP7-FUR for which the influence pressure exerts on the SAS value is significant (Additional file 2). A significant decrease in the SAS area is observed for the NIP7-ABY model, only at high temperature. In this way, increase in pressure makes the Nip7 structure more compact, an effect more expressed at high temperature.

The increase in pressure produces structural rearrangements of the protein. Thus, for the NIP7-ABY model, this increase results in a decrease in the local deviations of protein conformation from the crystal structure. This is in complete agreement with the conclusions that the pressure affects essentially slower motions which imply structural rearrangements of the protein globule [35]. The pressurization effect is more pronounced for the NIP7-ABY model and it affects the conformation non-uniformly, increasing the RMSDL in some regions and decreasing it in others (Figure 5B). As for the NIP7-FUR model most conformational changes occur under high pressure in the N-terminal domain.

A decrease in the fluctuations in the polypeptide is another effect of an increase in pressure we observed both at the domain (Additional file 6) and the residue levels (Figure 7B).

The MD simulations of protein structures subject to increased pressure have yielded abundant evidence of reduced fluctuations in the polypeptide chain due to pressurization [35-38]. They are convergent in demonstrating that, if the protein globule does not denaturate, pressure can stabilize the polypeptide chain. An increase in protein stability resulting from an increase in pressure has been demonstrated experimentally for the exemplary glutamate dehydrogenases from the hyperthermophilic archaea *Pyrococcus furiosus* [39] and *Thermococcus litoralis* [40]. The published data support ours: indeed, an increase in pressure stabilized significantly the protein globule, also the protein complex formed by the enzymes. The authors suggested that such a stabilization may arise through changes in the stability of the native states due to reduced fluctuations in the polypeptide chain at high pressures and temperatures [39]. However, our results demonstrate that the magnitude of the effect can be different for protein parts and for proteins from different organisms.

Functional implications for Nip7

Our current results demonstrate that the C-terminal domain is subject to larger structural displacements and fluctuations than the N-terminal domain during the MD simulations. This domain is plastic possibly because it is small, with just ~60 amino acid residues, and is stabilized predominantly by the hydrophobic core (the few salt bridges in this domain coordinate the positions of the loops and the terminal helical regions). The plasticity of the DNA/RNA-binding domains makes non-specific nucleotide-binding feasible [41-43]. Such a flexible structure, in contrast to the strictly coordinated components, which obey the key-lock rule, provides the possible binding to the poly-U RNA and poly-AU RNA sequences with a weak secondary structure [22,23]. Therefore, the plasticity that we currently observed for the Nip7 PUA-domain may presumably be its significant functional property.

The Nip7 PUA-domain has an interesting feature: the formation of salt bridges by the residues shaping the interaction with RNA. Possibly, this effect allows the conformation of chain side groups of these residues to stabilize so as to facilitate binding to RNA. It is encouraging that a similar effect of the stabilization of the side group amino acids at the expense of the formation of rigidifying salt bridges has been observed in the active centre of acylphosphatase from *Pyrococcus horikoshii* [44]. This is pertinent to our current observations: such a mechanism is to a great extent characteristic of the model for the deep-water NIP7-ABY protein.

The N-terminal domain, on the whole, is subject to smaller fluctuations compared to the C-terminal domain (Additional file 6). In this connection, the structurally unstable segments of this domain (positions 30–40, 49–59,

69–79) are outstanding. Two of these segments are β-hairpins (positions 30–40, 69–79), the deviations from the crystal structure in these segments occur without significant changes in the secondary structure, so that they bend in the direction of the α_1-helix in the two models. These conformational changes may result from interactions in the salt bridges network (GLU10-ARG4, ARG37-GLU10, GLU75-ARG37) bringing close together the β-sheet and the α_1 helix of the N-terminal domain. This interaction network may, in turn, become stabilized through the voluminous hydrophobic nucleus formed by hydrophobic residues in the centre of the β_1-β_5 strands.

In terms of the putative functional role in the N-terminal domain, it appears worthwhile to consider the loop between the second and the third α-helices (positions 49–59). The loop is contiguous with a lengthy nonpolar region at the surface of the N-terminal domain, which extends from the α_2 helix to the groove between the N- and C-terminal domains so that is comes to lie on the side opposite to the polar portion of the Nip7 surface (Figure 1) [23]. The conformational changes in the α_2-α_3 region may be due to the high twisting tension of the main chain in it. It undergoes a sharp bend at the very end of the short α_3-helix, then forms one helix turn, which is converted into the β_5 strand structure (Figure 1A). The positions of the N- and C- terminal residues of this fragment are fixed by the globule of the N-terminal domain. Then, the loop itself comes to lie aside from the globular domain without imposing steric constraints on the conformational changes. Additional destabilizing factors for the conformation of this region are alternating polar and hydrophobic (positions 46, 49, 53, 55) residues, which, as a result of conformational changes in the main chain of the α_2-α_3 segment, can by turns face solvent. It may be assumed that this bend together with the lengthy region of the nonpolar surface may play an important role in the interaction of the Nip7 protein with the exosomal protein partner [20,22]. Interestingly, prediction of the protein-protein interaction sites using 3D structure by the SPPIDER web-server demonstrated that most residues of this region could be involved in protein-protein interactions with high probability both for the NIP7-ABY and NIP7-FUR models (Additional file 11).

The structural lability of proteins may accomplish a role of consequence in molecular recognition. For example, the regions of the polypeptide chain of the monomers devoid of an orderly structure can provide the protein interaction [45] and adopt an orderly conformation in the process of binding [46]. With respect to the Nip7 protein, intrinsic flexibility appears to be more likely [47] because it ensures the presence of a conformational ensemble required for the formation of RNA-protein and protein-protein interactions [48].

Comparison of the protein dynamics between the shallow- and deep-water organisms

The sequences of *P. abyssi* and *P. furiosus* Nip7 in the 3D structure differ by 47 substitutions (~30%; Figure 1B); 19 of them result in radical changes of the physico-chemical properties of amino acids. These substitutions may be assigned to categories I-V.

(I) Substitutions resulting in the replacement of a polar residue in *P. abyssi* by a nonpolar in *P. furiosus* (7 substitutions): 2 (R → I), 21 (Y → F), 58 (Y → F), 59 (S → A), 66 (T → M), 80 (N → A), 119 (K → V).

(II) *P. abyssi → P. furiosus* substitutions, which result in replacement of a charged amino acid by an uncharged one at retained polarity (5 substitutions): 23 (E → T), 27 (E → N), 44 (E → N), 144 (K → S), 154 (K → T).

(III) Substitutions of nonpolar amino acids in *P. abyssi* by polar ones in *P. furiosus* (1 substitution): 140 (L → R).

(IV) *P. abyssi → P. furiosus* substitutions of the uncharged polar side group by the charged side group: 121 (Q → K).

(V) Substitutions *P. abyssi → P. furiosus* resulting in the change of the charge sign to the opposite one (5 substitutions): 88 (K → D), 113 (E → K), 117 (K → Q), 147 (R → E).

As a result of such substitutions, the hydrophobic portion of SAS increased in the shallow-water organism compared with its deep-water counterpart. The question then was: How could this influence the stability, dynamics and functional properties of the proteins? True, both proteins, in all the conditions we examined, proved to be stable and had a similar conformation. However it was perplexing to single out a parameter in the dynamics that could be significantly influenced by these substitutions.

It will be remembered that the majority of substitutions that belong to categories I and II lie in the N-terminal domain of the protein, they are in contact with its surface region, which contains mobile residues (the β_2-strand and the β_2-β_3 loop), also residues at positions α_2-α_3 (Figure 8) and it may presumably be involved in protein-protein interactions (Figure 10; Additional file 11).

The substitutions we observed in the N-terminal domain may, probably, somehow modify the interaction character of this region with its molecular partner. Substitutions in this stretch are associated with increase in hydrophobicity of residues in the shallow-water organisms. Changes of such kind can promote the stabilization of protein interaction of Nip7 and its partner. This is because hydrophobic interactions contribute considerably to the stability of protein-protein interfaces [49]. Another fact will be remembered here. When pressure close to atmospheric and

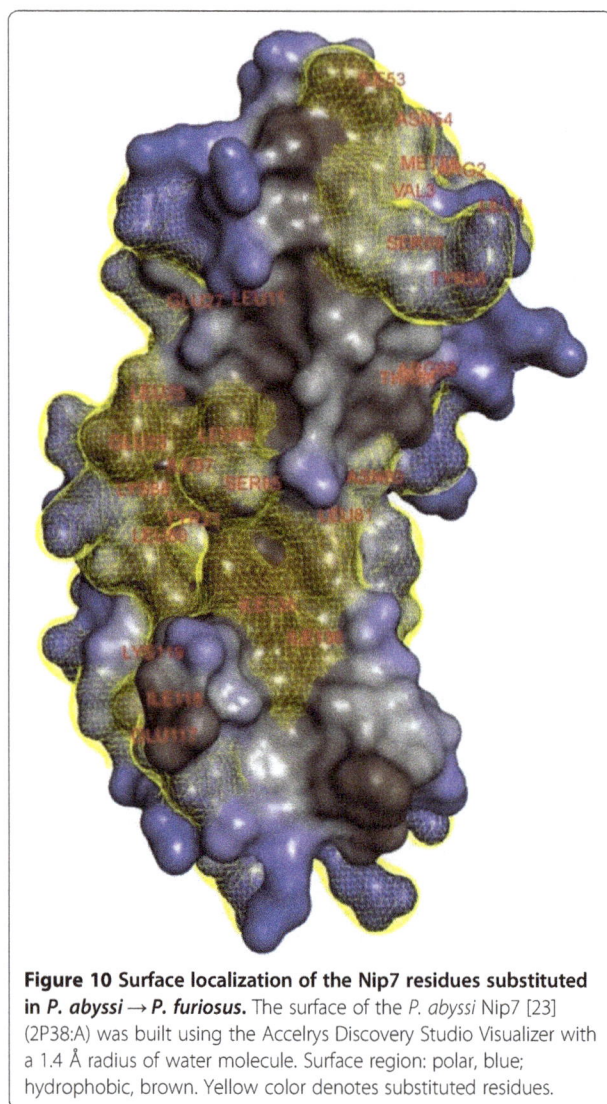

Figure 10 Surface localization of the Nip7 residues substituted in *P. abyssi → P. furiosus*. The surface of the *P. abyssi* Nip7 [23] (2P38:A) was built using the Accelrys Discovery Studio Visualizer with a 1.4 Å radius of water molecule. Surface region: polar, blue; hydrophobic, brown. Yellow color denotes substituted residues.

temperature rises, hydrophobic interactions favor closer interactions of nonpolar protein regions [50], which can additionally stabilize the Nip7-exosome protein complex in *P. furiosus*. If so, the observed substitutions may truly be of adaptive character. We have previously analyzed [51] the indices of asymmetry for amino acid substitutions in hundreds of *Pyrococcus* proteins starting from its common deep-sea ancestor *P. furiosus*, *P. horikoshii* and *P. abyssi* to the extant shallow-sea organism *P. furiosus*. According to the results, the values of the indices are positively correlated with the scales of amino acid hydrophobicity and negatively with the polarity scales indicating that polar residues were predominantly substituted by nonpolar when the shallow-sea habitats replaced the deep-sea ones. However, the issue as to whether this trend is a reflection of the stabilization of protein-protein interactions or of the entire structure requires further study.

Furthermore, the character of changes in fluctuations for the chain stretch in the α_2-α_3 region at increasing pressure is drastically different for the two proteins: they significantly increase in the shallow-water protein and decrease, conversely, in its deep-water counterpart (Figure 7A). A general decrease in fluctuations throughout the entire N-terminal domain at increasing pressure is also characteristic of the deep-sea protein. The magnitude of fluctuations for efficient binding must be within definite limits and their general decrease at reduction in pressure is also additive in the deep-water protein.

Substitutions exert a significant influence on the conformational parameters in proteins connecting two protein domains (the RMSDL and the RMSF parameters) in the β_6-α_4 region. This is the stretch that joins together two protein domains. Probably, mutations in it favor change in the mutual orientation of the domains and result in an increase in the structural deviations in the C-terminal domain of the protein in the NIP7-FUR as compared with the NIP7-ABY model (Figure 8).

As for the RNA-binding domain, radical changes that could alter residue charge are characteristic of the shallow-water protein. As a consequence, the formation of certain salt bridges, those between residues involved in RNA recognition in particular, is impaired. It is also worth to note, that the number of residues associated with significant differences in RMSDL/RMSF parameters between the shallow- and deep-water protein models are higher in the RNA-binding domain. The differences we observed in the course of MD simulations may reflect the different ways and means RNA interacts with the Nip7 proteins at high and low pressures.

Conclusions

The MD simulations we performed currently for the Nip7 protein from the shallow-water (*P. furiosus*) and deep-water (*P. abyssi*) marine hyperthermophylic archaea demonstrated that their structures are stable at different temperatures (300 and 373 K) and pressures (0.1–100 MPa). Increase in temperatures and pressures caused conformational changes in about the same protein regions, in the loops first of all. In the examined ranges of temperatures and pressures, increase in temperature had a more pronounced effect on changes in the dynamic properties of the protein globule than increase in pressure. A number of highly mobile segments of protein globule presumably involved in protein-protein interactions were identified. Substitutions of the polar residues in the deep-water organisms by hydrophobic in the shallow-water ones were observed in some of the regions. These replacements may be evidence of change in the interaction pattern of the Nip7 protein and the proteins of the exosome complex in the course of the divergence of the shallow-water organisms from the deep-water ones. Taken together, the results of such a complex comparative analysis of the MD of proteins from organisms belonging to different ecological niches may be useful in refining ideas of their functioning and adaptation to extreme environmental conditions.

Methods
Protein models

The model of the *P. abyssi* Nip7 protein was retrieved from the PDB [52], identifier 2P38 [23]. The atomic model of the *P. abyssi* Nip7 contains two independent monomers and 252 water molecules. The monomers are symmetric with respect to each other, with a head-to-tail orientation, they form a dimer without showing structural differences (the RMSD of the C_α atoms between two monomers is 0.11 Å). The contact surface between the monomers is about 700 Å^2 per molecule, or 8% of its total surface. These estimates suggest that the dimeric structure might have resulted from crystallization and, if so, it may be biologically irrelevant [23]. Hence, 2P38:A is the only monomer included in analysis. The first four and the last seven amino acid residues not located in the experiment were discarded. As a result, the total length of the *P. abyssi* Nip7 protein model is 155 amino acid residues. This model is referred to as NIP7-ABY, the amino acid residues are numbered according to their order numbers in the structure (1VAL-155ASP).

The *P. furiosus* Nip7 model (NIP7-FUR) was built using the method of reconstruction based on homology. It was retrieved from the ModBase databank [53], ID Q8TZP7. The *P. abyssi* and *P. furiosus* Nip7 alignments contain neither deletions nor insertions and, hence, NIP7-FUR reliability is ensured (Figure 1B). The number of residues of the NIP7-FUR model is 155, their numbering corresponds to the one for NIP7-ABY (1ILE-155ASP). Based on the C_α atoms, the RMSD of the initial structures of NIP7-FUR and NIP7-ABY is 0.1 Å.

Here we examine the MD simulations of the Nip7 protein models at various pressures (0.1, 50, and 100 MPa) and at temperatures 300 and 373 K.

Preparation of models for MD simulation

The MD of the protein 3D structure is simulated using the program package GROMACS [54] 4.5.3. The algorithm for preparation of the models is as follows. To remove undesirable tensions, the structure energy is minimized in vacuum. Then, the molecule is placed in a cubic cell (edge 200 Å) containing water molecules in the SPCE model. Then, ion molecules are added to the cell to neutralize the charge. The NIP7-ABY model consists of 2 585 protein atoms, 32 900 water molecules, and 1 ion. The NIP7-FUR model is composed of 2 544 protein atoms, 32 900 water molecules, and 2 ions. Thereafter, the protein solvent system is minimized for 2 000 steps. There are two different techniques for further preparation of the models. In the case of simulation at 300 K, there are two preparatory steps of simulation at the given pressure-temperature values, 200 ps each; in that of simulation at 373 K, the number of preparatory simulations is nine, with a gradual increase in temperature by 8 K at each simulation. Pressure in the system is set at simulation onset.

MD simulations

MD simulations are performed using the LINCS [55] algorithm in the amber99sb-ildn force field [56]. Electrostatic interactions are computed using the Partial Mesh Ewald (PME) algorithm. Pressure coupling was done using Parrinello-Rahman method [57]. Temperature coupling was done using v-rescale method [58].

The simulation time was 40 ns. To take into account slower dynamics of proteins at high pressure due to increase in viscosity of water for each set of parameters we perform 5 independent trajectory runs. As a result, 30 experiments are considered for each protein model.

Data were saved every 80 ps. We collected statistics for protein conformation parameters for model structures after equilibration from 20 to 40 ns.

Structure similarity analysis

We characterized changes in the structures by the C_α atoms RMSDs and the solvent accessible surface area of the residues (the SAS); the secondary structure parameters based on the DSSP algorithm [59] were calculated using the GROMACS package with the help of the g_rms, g_sas and do_dssp programs, respectively.

To define the regions of the protein structure most susceptible to pressure/temperature impact, in a sliding window of the size $q + 1$, the local deviation of the structure for each residue, the RMSDL after superposition of two protein structures using all C_α atoms was calculated as:

$$RMSDL_i(\mathbf{v}, \mathbf{w}, q)$$
$$= \sqrt{\left(\frac{1}{q} \sum_{j=i-q/2}^{j=i+q/2} \left(v_{jx} - w_{jx}\right)^2 + \left(v_{jy} - w_{jy}\right)^2 + \left(v_{jz} - w_{jz}\right)^2\right)},$$

where $\mathbf{v}$, $\mathbf{w}$ are two structures superposed to minimize their C_α RMSD; i is the index of residue for which RMSDL is calculated; v_{jk}, w_{jk} ($k = x,y,z$) are x, y, z coordinates of the C_α atoms of the aligned residues j in each of the $\mathbf{v}$, $\mathbf{w}$ structures; the q parameter was set to 10 yielding a window of 11 amino acids to calculate the RMSDL.

Secondary structure analysis

To analyze the stability of the secondary structure during the simulations, we analyze the 3D model at each step using the DSSP program [59]. A hundred 3D structures obtained at the last steps of the MD simulation were used

as input for the DSSP program. The occurrence frequencies of a secondary structure of a particular type at a concrete position in the secondary structure were determined. The total secondary structure for a given trajectory included the type of the secondary structure with the highest occurrence frequency. The procedure was performed on all the 24 trajectories.

To estimate the variability in the secondary structure types in the 12 trajectories for each of two models we used the Protein Variability Server [60].

Analysis of the fluctuation in the polypeptide chain

Fluctuations in the polypeptide chain were characterized by the C_α atom RMSFs computed using the GROMACS package program g_rmsf. The average C_α atoms RMSF value in Angstroms was computed for all the residues in the Nip7 model and N- and C- terminal domains. To determine the significance of the differences between the average C_α atoms RMSFs of the different trajectories or models, we used two-way ANOVA (see below).

Dependence of the models' conformational parameters on temperature, pressure and protein sequence

To determine the influence of temperature, pressure and amino acid substitutions on the protein parameters, we used ANOVA of two types [61].

First, to estimate the influence of pressure and temperature on the protein structural characteristics (for example, Rg, SAS, salt bridge persistence or RMSF of the i-th residue) we used two-way ANOVA. In this analysis the influence of two factors was examined: temperature (T = 300, 373 K) and pressure (P = 0.1, 50, 100 MPa). In each set of modeling parameters, we estimated the means of this characteristic on the basis of the results of analysis of 5 runs. In this way, for the structural characteristic we analyzed the ANOVA table with $r = 2$ (temperature), $c = 3$ (pressure), the same number of observations for each parameter set, whose sum total was $n = 30$ ($2 \cdot 3 \cdot 5$). We tested the hypothesis that the means of the characteristic are independent of temperature and pressure. The critical F-values were set at $\alpha = 0.05$ for F (P) $= 3.4$ ($df = 2$), temperature F (T) $= 4.26$ ($df = 1$), and for interaction of these factors F (PxT) $= 3.4$ ($df = 24$).

Second, to estimate the effect of increased temperature on the conformational parameters of the protein, their index of sensitivity (S^t) was computed. To this end, we calculated the average value of a model's conformational parameter, X (for example, the RMSF of the i-th residue), for the trajectories obtained at high temperatures ($X_{T=373}$) and for those obtained at low temperatures ($X_{T=300}$). The index of sensitivity was expressed as $S^t = \frac{X_{T=373}}{X_{T=300}}$. The S^t value close to unity meant that increased temperature had no considerable effect on the value of the conformational parameter compared to its value at low temperature. The S^t value considerably larger than unity implied that increased temperature increased the value of the structural parameter compared to its value at low temperature. The S^t value considerably smaller than unity imply that increase in temperature reduces the value of the structural parameter. A number of conformational parameters, such as the $RMSDL_i$, the local structural deviations of the polypeptide chain for the i-th residue, the $RMSF_i$, the root-mean-square fluctuations in the i-th residue, and salt bridge persistence (see below) were considered.

Similarly, the sensitivity of the structural parameters to increased pressure S^p was estimated. We determined the value of the parameters for the trajectories at high pressures (50 and 100 MPa), X_{P-high}, atmospheric and moderate pressure (0.1 MPa), X_{P-low}. The sensitivity to pressure was expressed as $S^p = \frac{X_{P-high}}{X_{P-low}}$.

Third, in analysis of the structural characteristics of the amino acid residues the RMSDL, the RMSF, we estimated the influence of protein type (NIP7-ABY, NIP7-FUR); in such a way, we treated the three-way ANOVA model that considered temperature, pressure and protein type ($r = 2$, $c = 3$, $l = 2$; $n = 60$). The additional factor (protein type) reflected the influence amino acid substitutions exerted on the RMSDL and RMSF parameters for each of the residues. It should be stipulated, however, that such an analysis did not provide information about the influence of specific amino acid substitutions on changes in a conformational parameter of a residue. The critical F-values in this analysis were set at $\alpha = 0.05$ for pressure F (P) = 3.4 ($df = 2$), for temperature F (T) = 4.26 ($df = 1$), and for interaction of these factors F (PxT) = 3.4 ($df = 24$).

Analysis of salt bridges

To identify salt bridges, we used the criterion according to which the distance between two oppositely changed atoms should not exceed 4 Å [28]. Salt bridge persistence was expressed as the proportion of structures in the equilibrium region of trajectories (20–40 ns) for which the salt bridge was identified.

The sensitivity of the salt bridge to temperature (S^t) and pressure (S^p) was defined as the ration of the salt bridge persistence at high and low temperature and high and low pressure. The S^t (S^p) value close to unity meant that increased temperature (pressure) had no considerable effect on the persistence of the salt bridge. The S^t (S^p) value considerably larger than unity implied that increased temperature (pressure) increased the persistence of the salt bridge compared to its value at low temperature. The S^t (S^p) value considerably smaller than unity imply that increase in temperature (pressure) reduces the persistence of the salt bridge.

The dependence of salt bridge persistence on temperature and pressure was estimated using two-way ANOVA (see above).

Prediction of the protein-protein interactions

To predict the sites of protein-protein interactions, the SPPIDER server (http://sppider.cchmc.org/) was used [62].

Structure visualization

The Accelrys Discovery Studio Visualizer was applied to visualize the structure of the models (http://accelrys.com/products/discovery-studio/).

Statistical analysis

Statistical analysis of the data was performed using the Excel and Statistica 6.0 package.

Additional files

Additional file 1: 2-way ANOVA of change in the radius of gyration, the NIP7-ABY and NIP7-FUR models.

Additional file 2: 2-way ANOVA of change in the SAS parameter, the NIP7-ABY and NIP7-FUR models.

Additional file 3: Analysis of the RMSDL, the NIP7-ABY and NIP7-FUR models; Analysis of the secondary structure, the NIP7-ABY and NIP7-FUR models; Snapshots of the NIP7-ABY and NIP7-FUR models at different temperatures and pressures; Analysis of the C_α RMSF, the NIP7-ABY and NIP7-FUR models.

Additional file 4: 2-way ANOVA of RMSDL values of each residue, the NIP7-ABY and NIP7-FUR models.

Additional file 5: 2-way ANOVA of RMSF values of each residue, the NIP7-ABY and NIP7-FUR models.

Additional file 6: 2-way ANOVA of the fluctuations in the C- and N-terminal domains of the NIP7-ABY and NIP7-FUR models.

Additional file 7: Analysis of salt bridge persistence during molecular dynamics simulation at different pressures and temperatures.

Additional file 8: 2-way ANOVA of salt bridges persistence, the NIP7-ABY model.

Additional file 9: 2-way ANOVA of salt bridges persistence, the NIP7-FUR model.

Additional file 10: 3-way ANOVA of RMSDL and RMSF fluctuations, the NIP7-ABY and NIP7-FUR models.

Additional file 11: Prediction of the protein-protein interaction sites for the NIP7-ABY and NIP7-FUR models by the SPPIDER web-server.

Abbreviations

MD: Molecular dynamics; Rg: Radius of gyration; RMSD: Root mean squared deviation; RMSF: Root mean squared fluctuation; RMSDL: Local root mean squared deviation; SAS: Solvent accessible surface; NIP7-ABY: Model of the Nip7 protein of *Pyrococcus abyssi*; NIP7-FUR: Model of the Nip7 protein of *Pyrococcus furiosus*.

Competing interests

The authors declare that they have no competing interests.

Authors' contributions
KEM and NAA performed the MD simulations of the Nip7 proteins. KEM, DAA and YNV performed the analysis of the MD results. EVB, NAK and DAA initiated the comparative molecular dynamics study of the Nip7 proteins and participated in its coordination. All authors read and approved the final manuscript.

Acknowledgements
The SB RAS Siberian Supercomputing Center and the Novosibirsk State University High-Performance Computing Center are gratefully acknowledged for providing computer facilities. This work was in part supported by the Ministry of Education and Science of the Russian Federation, RFBR (projects 11-04-01771 and 12-04-00135); Integration Projects № 130 and 39 from the SB RAS; Program № 6.6 and "Biosphere Origin and Evolution" from the RAS; EVB acknowledges the support from a grant of RSF 14-13-00834.
The authors are grateful to the referees for their helpful comments, to K.V. Gunbin for helpful discussions, to A.N. Fadeeva for translating the manuscript from Russian into English and to N.V. Kuchin for technical support.

Author details
[1]Institute of Cytology and Genetics SB RAS, Prospekt Lavrentyeva 10, Novosibirsk 630090, Russia. [2]Institute of Chemical Biology and Fundamental Medicine SB RAS, Prospekt Lavrentyeva 8, Novosibirsk 630090, Russia. [3]Novosibirsk State University, Pirogova str. 2, Novosibirsk 630090, Russia. [4]Institute of Solid Chemistry and Mechanochemistry, SB RAS, Novosibirsk 630090, Russia. [5]NRC Kurchatov Institute, 1, Akademika Kurchatova pl., Moscow 123182, Russia.

References
1. Rothschild LJ, Mancinelli RL: Life in extreme environments. *Nature* 2001, **409**(6823):1092–1101.
2. Daniel I, Oger P, Winter R: Origins of life and biochemistry under high-pressure conditions. *Chem Society Rev* 2006, **35**(10):858–875.
3. Brooks AN, Turkarslan S, Beer KD, Yin Lo F, Baliga NS: Adaptation of cells to new environments. *Wiley Interdiscip Rev Syst Biol Med* 2011, **3**(5):544–561.
4. Sterner R, Liebl W: Thermophilic adaptation of protein. *Crit Rev Biochem Mol Biol* 2001, **36**:39–106.
5. Makarova KS, Wolf YI, Koonin EV: Potential genomic determinants of hyperthermophily. *Trends Genet* 2003, **19**(4):172–176.
6. Berezovsky IN, Shakhnovich EI: Physics and evolution of thermophilic adaptation. *Proc Natl Acad Sci U S A* 2005, **102**(36):12742–12747.
7. McDonald JH: Temperature adaptation at homologous sites in proteins from nine thermophile–mesophile species pairs. *Genome Biol Evol* 2010, **2**:267.
8. Jollivet D, Mary J, Gagniere N, Tanguy A, Fontanillas E, Boutet I, Hourdez S, Segurens B, Weissenbach J, Poch O, Lecompte O: Proteome adaptation to high temperatures in the ectothermic hydrothermal vent Pompeii worm. *PLoS One* 2012, **7**(2):e31150.
9. Di Giulio M: A comparison of proteins from *Pyrococcus furiosus* and *Pyrococcus abyssi*: barophily in the physicochemical properties of amino acids and in the genetic code. *Gene* 2005, **346**:1–6.
10. Simonato F, Campanaro S, Lauro FM, Vezzi A, D'Angelo M, Vitulo N, Bartlett DH: Piezophilic adaptation: a genomic point of view. *J Biotechnol* 2006, **126**(1):11–25.
11. Campanaro S, Treu L, Valle G: Protein evolution in deep sea bacteria: an analysis of amino acids substitution rates. *BMC Evol Biol* 2008, **8**(1):313.
12. Oger PM, Jebbar M: The many ways of coping with pressure. *Res Microbiol* 2010, **161**(10):799–809.
13. Podar M, Reysenbach AL: New opportunities revealed by biotechnological explorations of extremophiles. *Curr Opin Biotechnol* 2006, **17**(3):250–255.
14. Frock AD, Kelly RM: Extreme thermophiles: moving beyond single-enzyme biocatalysis. *Curr Opin Chem Eng* 2012, **1**(4):363–372.
15. Liszka MJ, Clark ME, Schneider E, Clark DS: Nature versus nurture: developing enzymes that function under extreme conditions. *Annu Rev Chem Biomol Eng* 2012, **3**:77–102.
16. Singh OV: *Extremophiles: Sustainable Resources and Biotechnological Implications.* New York: Wiley-Verlag; 2012.
17. Fiala G, Stetter KO: Pyrococcus furiosus sp. nov. represents a novel genus of marine heterotrophic archaebacteria growing optimally at 100°C. *Arch Microbiol* 1986, **145**:56–61.
18. Erauso G, Reysenbach A-L, Godfroy A, Meunier J-R, Crump B, Partensky F, Baross JA, Marteinsson V, Barbier G, Pace NR, Prieur D: Pyrococcus abyssi sp. nov., a new hyperthermophilic archaeon isolated from a deep-sea hydrothermal vent. *Arch Microbiol* 1993, **160**:338–349.
19. González JM, Masuchi Y, Robb FT, Ammerman JW, Maeder DL, Yanagibayashi M, Tamaoka J, Kato C: Pyrococcus horikoshii sp. nov., a hyperthermophilic archaeon isolated from a hydrothermal vent at the Okinawa Trough. *Extremophiles* 1998, **2**:123–130.
20. Zanchin NIT, Goldfarb DS: Nip7p interacts with Nop8p, an essential nucleolar protein required for 60S ribosome biogenesis, and the exosome subunit Rrp43p. *Mol Cell Biol* 1999, **19**:1518–1525.
21. Bassler J, Grandi P, Gadal O, Lessmann T, Petfalski E, Tollervey D, Lechner J, Hurt E: Identification of a 60S preribosomal particle that is closely linked to nuclear export. *Mol Cell Biol* 2001, **8**:517–529.
22. Luz JS, Ramos CR, Santos MC, Coltri PP, Palhano FL, Foguel D, Oliveira CC: Identification of archaeal proteins that affect the exosome function in vitro. *BMC Biochem* 2010, **11**(1):22.
23. Coltri PP, Guimaraez BG, Granato DC, Luz JS, Teixeira EC, Oliveira CC, Zanchin NIT: Structural insights into the interaction of the Nip7 PUA domain with polyuridine RNA. *Biochemistry* 2007, **46**:14177–14187.
24. Pérez-Arellano I, Gallego J, Cervera J: The PUA domain– a structural and functional overview. *FEBS J* 2007, **274**(19):4972–4984.
25. Aravind L, Koonin E: Novel predicted RNA-binding domains associated with the translation machinery. *J Mol Evol* 1999, **48**:291–302.
26. Hallberg BM, Ericsson UB, Johnson KA, Andersen NM, Douthwaite S, Nordlund P, Beuscher AE, Erlandsen H: The structure of the RNA m5C methyltransferase YebU from Escherichia coli reveals a C-terminal RNArecruiting PUA domain. *J Mol Biol* 2006, **360**:774–787.
27. Gunbin KV, Afonnikov DA, Kolchanov NA: Molecular evolution of the hyperthermophilic archaea of the Pyrococcus genus: analysis of adaptation to different environmental conditions. *BMC Genomics* 2009, **10**(1):639.
28. Barlow DJ, Thornton JM: Ion-pairs in proteins. *J Mol Biol* 1983, **168**(4):867–885.
29. Merkley ED, Daggett V, Parson WW: A temperature-dependent conformational change of NADH oxidase from Thermus thermophilus HB8. *Proteins* 2012, **80**(2):546–555.
30. Polyansky AA, Kosinsky YA, Efremov RG: Correlation of local changes in the temperature-dependent conformational flexibility of thioredoxins with their thermostability. *Russ J Bioorg Chem* 2004, **30**:421–430.
31. Martinez R, Schwaneberg U, Roccatano D: Temperature effects on structure and dynamics of the psychrophilic protease subtilisin S41 and its thermostable mutants in solution. *Protein Eng Des Sel* 2011, **24**(7):533–544.
32. Lee KJ: Molecular dynamics simulations of a hyperthermophilic and a mesophilic protein L30e. *J Chem Inf Model* 2011, **52**(1):7–15.
33. Priyakumar UD, Ramakrishna S, Nagarjuna KR, Reddy SK: Structural and energetic determinants of thermal stability and hierarchical unfolding pathways of hyperthermophilic proteins, Sac7d and Sso7d. *J Phys Chem B* 2010, **114**(4):1707–1718.
34. Tiberti M, Papaleo E: Dynamic properties of extremophilic subtilisin-like serine-proteases. *J Struct Biol* 2011, **174**(1):69–83.
35. Calandrini V, Kneller GR: Influence of pressure on the slow and fast fractional relaxation dynamics in lysozyme: a simulation study. *J Chem Phys* 2008, **128**:065102.
36. Capece L, Marti MA, Bidon-Chanal A, Nadra A, Luque FJ, Estrin DA: High pressure reveals structural determinants for globin hexacoordination: neuroglobin and myoglobin cases. *Proteins* 2009, **75**(4):885–894.
37. McCarthy AN, Grigera JR: Effect of pressure on the conformation of proteins. A molecular dynamics simulation of lysozyme. *J Mol Graph Model* 2006, **24**(4):254–261.
38. Laurent AD, Mironov VA, Chapagain PP, Nemukhin AV, Krylov AI: Exploring structural and optical properties of fluorescent proteins by squeezing: modeling high-pressure effects on the mStrawberry and mCherry red fluorescent proteins. *J Phys Chem B* 2012, **116**(41):12426–12440.
39. Sun MM, Tolliday N, Vertiani C, Robb FT, Clark DS: Pressure-induced thermostabilization of glutamate dehydrogenase from the hyperthermophile Pyrococcus furiosus. *Protein Sci* 1999, **8**(5):1056–1063.
40. Sun MM, Caillot R, Mark G, Robb FT, Clark DS: Mechanism of pressure-induced thermostabilization of proteins: studies of glutamate

dehydrogenases from the hyperthermophile Thermococcus litoralis. *Protein Sci* 2001, **10**(9):1750–1757.

41. Kalodimos CG, Biris N, Bonvin AMJJ, Levandoski MM, Guennuegues M, Boelens R, Kaptein R: **Structure and flexibility adaptation in nonspecific and specific protein-DNA complexes.** *Science* 2004, **305**(5682):386–389.

42. Brown C, Campos-León K, Strickland M, Williams C, Fairweather V, Brady RL, Cramp MP, Gaston K: **Protein flexibility directs DNA recognition by the papillomavirus E2 proteins.** *Nucleic Acids Res* 2011, **39**(7):2969–2980.

43. Boehr DD: **Promiscuity in protein-RNA interactions: conformational ensembles facilitate molecular recognition in the spliceosome.** *Bioessays* 2012, **34**(3):174–180.

44. Lam SY, Yeung RC, Yu TH, Sze KH, Wong KB: **A rigidifying salt-bridge favors the activity of thermophilic enzyme at high temperatures at the expense of low-temperature activity.** *PLoS Biol* 2011, **9**(3):e1001027.

45. Fong JH, Shoemaker BA, Garbuzynskiy SO, Lobanov MY, Galzitskaya OV, Panchenko AR: **Intrinsic disorder in protein interactions: insights from a comprehensive structural analysis.** *PLoS Comput Biol* 2009, **5**(3):e1000316.

46. Uversky VN, Dunker AK: **The case for intrinsically disordered proteins playing contributory roles in molecular recognition without a stable 3D structure.** *F1000 Biol Rep* 2013, **5**:1.

47. Janin J, Sternberg MJ: **Protein flexibility, not disorder, is intrinsic to molecular recognition.** *F1000 Biol Rep* 2013, **5**:2.

48. Boehr DD, Nussinov R, Wright PE: **The role of dynamic conformational ensembles in biomolecular recognition.** *Nat Chem Biol* 2009, **5**(11):789–796.

49. Jones S, Thornton JM: **Principles of protein-protein interactions.** *Proc Natl Acad Sci U S A* 1996, **93**(1):13–20.

50. Ferrara CG, Ghara O, Grigera JR: **Aggregation of nonpolar solutes in water at different pressures and temperatures: the role of hydrophobic interaction.** *J Chem Phys* 2012, **137**(13):135104.

51. Afonnikov DA, Medvedev KE, Gunbin KV, Kolchanov NA: **Important role of hydrophobic interactions in high-pressure adaptation of proteins.** *Dokl Biochem and Biophys* 2011, **438**(1):113–116.

52. Berman HM, Westbrook J, Feng Z, Gilliland G, Bhat TN, Weissig H, Shindyalov IN, Bourne PE: **The protein data bank.** *Nucleic Acids Res* 2000, **28**:235–242.

53. Pieper U, Eswar N, Davis FP, Braberg H, Madhusudhan MS, Rossi A, Marti-Renom M, Karchin R, Webb BM, Eramian D, Shen MY, Kelly L, Melo F, Sali A: **ModBase, a database of annotated comparative protein structure models, and associated resources.** *Nucleic Acids Res* 2006, **34**:D291–D295.

54. Hess B, Kutzner C, Van Der Spoel D, Lindahl E: **GROMACS 4: algorithms for highly efficient, load-balanced, and scalable molecular simulation.** *J Chem Theory Comput* 2008, **4**(3):435–447.

55. Hess B, Bekker H, Berendsen HJC, Fraaije JGEM: **LINCS: a linear constraint solver for molecular simulations.** *J Comp Chem* 1997, **18**(12):1463–1472.

56. Lindorff-Larsen K, Piana S, Palmo K, Maragakis P, Klepeis JL, Dror RO, Shaw DE: **Improved side-chain torsion potentials for the Amber ff99SB protein force field.** *Proteins* 2010, **78**(8):1950–1958.

57. Parrinello M, Rahman A: **Polymorphic transitions in single crystals: a new molecular dynamics method.** *J Appl Phys* 1981, **52**(12):7182–7190.

58. Bussi G, Donadio D, Parrinello M: **Canonical sampling through velocity rescaling.** *J Chem Phys* 2006, **126**(1):014101–014107.

59. Kabsch W, Sander C: **Dictionary of protein secondary structure: pattern recognition of hydrogen-bonded and geometrical features.** *Biopolymers* 1983, **22**(12):2577–2637.

60. Garcia-Boronat M, Diez-Rivero CM, Reinherz EL, Reche PA: **PVS: a web server for protein sequence variability analysis tuned to facilitate conserved epitope discovery.** *Nucleic Acids Res* 2008, **36**:W35–W41.

61. Kendall MG, Stuart A: *The Advanced Theory of Statistics. Volume 3.* London: Charles Griffin & Co; 1968.

62. Porollo A, Meller J: **Prediction-based fingerprints of protein-protein interactions.** *Proteins* 2007, **66**:630–645.

8

Sequence analysis on the information of folding initiation segments in ferredoxin-like fold proteins

Masanari Matsuoka[1,2] and Takeshi Kikuchi[1*]

Abstract

Background: While some studies have shown that the 3D protein structures are more conservative than their amino acid sequences, other experimental studies have shown that even if two proteins share the same topology, they may have different folding pathways. There are many studies investigating this issue with molecular dynamics or Go-like model simulations, however, one should be able to obtain the same information by analyzing the proteins' amino acid sequences, if the sequences contain all the information about the 3D structures. In this study, we use information about protein sequences to predict the location of their folding segments. We focus on proteins with a ferredoxin-like fold, which has a characteristic topology. Some of these proteins have different folding segments.

Results: Despite the simplicity of our methods, we are able to correctly determine the experimentally identified folding segments by predicting the location of the compact regions considered to play an important role in structural formation. We also apply our sequence analyses to some homologues of each protein and confirm that there are highly conserved folding segments despite the homologues' sequence diversity. These homologues have similar folding segments even though the homology of two proteins' sequences is not so high.

Conclusion: Our analyses have proven useful for investigating the common or different folding features of the proteins studied.

Keywords: Folding initiation segment prediction, Sequence analysis, Inter-residue average distance statistics, Evolutionarily conserved folding, Ribosomal protein S6, Procarboxypeptidase A2, U1A Spliceosomal protein, mt-Acylphosphatase

Background

Clarifying how a protein folds into its unique 3D structure is a very significant yet unsolved problem in molecular biophysics and bioinformatics [1]. In particular, some recent experimental studies have revealed that proteins sharing the same topology can take different folding pathways [2-10].

Ferredoxin-like fold proteins are well-known proteins that fold via different folding pathways. Their topology is composed of 2 α helices and 4 β strands, and the secondary structure arrangement seems similar to the $\beta/\alpha/\beta$ triad motif in flavodoxin [11,12] or TIM-barrel proteins [13]. However, the connectivity of the secondary structures differs. While flavodoxin or TIM-barrel proteins have a parallel β sheet, ferredoxin-like proteins have an anti-

parallel β sheet. Therefore, it is hard to explain the ferredoxin-like proteins' folding behavior only with the formation of $\beta/\alpha/\beta$ triads and the interaction among subdomains as in the case of flavodoxin, even if it is true that most of the hydrophobic contacts are composed of Ile, Leu and Val residues as reported in the literature [14].

Ferredoxin-like fold proteins are relatively small as shown in Figure 1, but they have interesting features in the structural transformation from denatured states to transit or native states. For example, one ferredoxin-like protein called ribosomal protein S6 contains two overlapping foldons, which fold cooperatively, located at different termini, and the overlapping makes this protein fold in a two-state manner as reported by Haglund et al. [15]. However, other proteins such as U1A spliceosomal protein or procarboxypeptidase A2 fold via the N- or C-terminal foldon as reported by Ternström et al. [16] or Villegas et al. [17], respectively.

---
* Correspondence: tkikuchi@sk.ritsumei.ac.jp
[1]Department of Bioinformatics, College of Life Sciences, Ritsumeikan University, 1-1-1 Nojihigashi, Kusatsu, Shiga 525-8577, Japan
Full list of author information is available at the end of the article

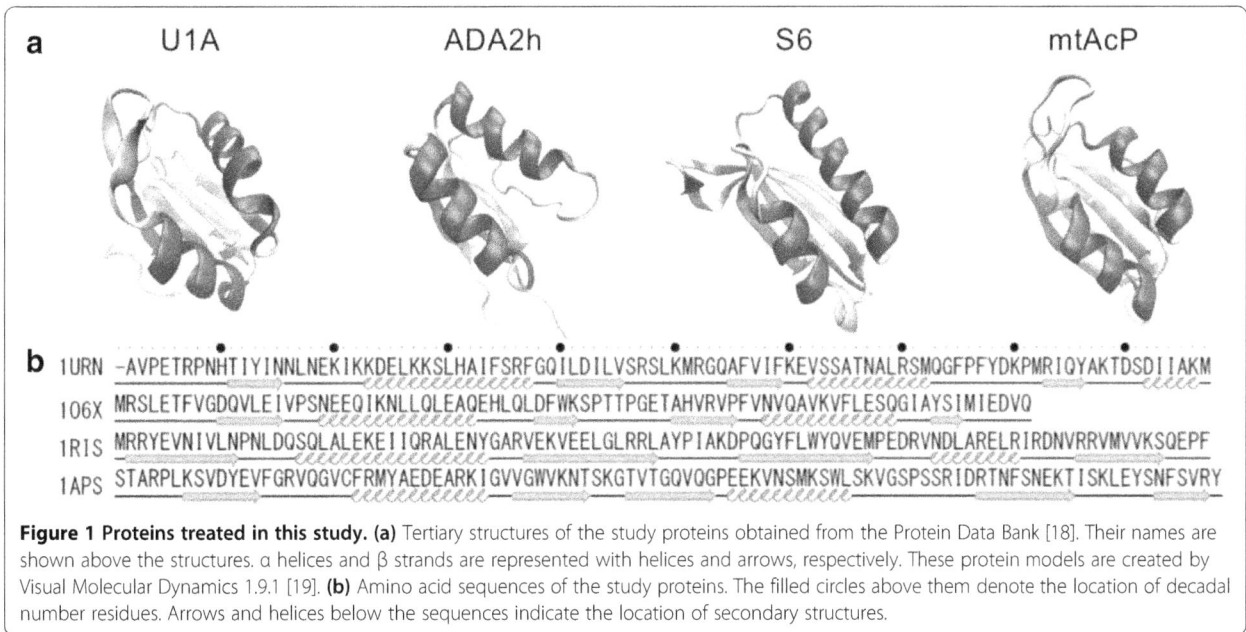

Figure 1 Proteins treated in this study. (a) Tertiary structures of the study proteins obtained from the Protein Data Bank [18]. Their names are shown above the structures. α helices and β strands are represented with helices and arrows, respectively. These protein models are created by Visual Molecular Dynamics 1.9.1 [19]. **(b)** Amino acid sequences of the study proteins. The filled circles above them denote the location of decadal number residues. Arrows and helices below the sequences indicate the location of secondary structures.

If all the information related to the formation of 3D structures is encoded in the amino acid sequences, we should be able to decode these sequence differences to obtain their folding features. Still, this is a challenging task. How the folding mechanism of a protein is coded in the amino acid sequence information remains an important issue to be clarified.

There are some bioinformatics approaches for predicting some aspects of folding mechanisms, like folding rate, from the amino acid sequence [20-27]. Nevertheless, it is still difficult to extract more detailed information on the folding mechanism, that is, how each protein folds.

The fact that the topologically equivalent proteins do not always fold via the same folding pathway leads us to the question of whether evolutionarily related proteins really have common folding properties. Evolutionarily related proteins have been observed to be possible to fold via different folding pathways. For example, in spite of the fact that PDZ2 and PDZ3, both of which contain mainly β sheets, are evolutionarily related and have more than 30% sequence identity, they do not share the same folding mechanism, at least in the early stage of folding [28]. On the other hand, fibronectin and titin, which are evolutionarily unrelated proteins but have the same topology, share the folding mechanism involving four key residues and their peripheral residues [29]. There are also some other studies focusing on the differences in the folding of topologically equivalent proteins [3]; yet, these experiments were performed only for several proteins of each topology and were not applied to a whole family.

In this study, we aim to decode such information for well-studied ferredoxin-like fold [30] proteins by analyzing their amino acid sequences, not only with the previously

mentioned bioinformatics approaches but also with our own analyses. The methods we apply here are homologous sequence search, phylogenetic analysis and sequence-based analyses by means of inter-residue average distance statistics.

The methods, which are based on the inter-residue average distance statistics [31,32] using the amino acid sequences as input, have so far provided valuable information on the initial folding segments that play crucial roles in the structural formation in the cases of lysozyme, leghemoglobin, fatty acid-binding protein, azurin, and two ancient TIM-barrel proteins [33-36]. We also apply our methods to some evolutionarily related proteins of four ferredoxin-like fold proteins to examine whether evolutionarily related proteins have common folding properties.

Methods

Proteins treated in this study

The proteins treated in this study are U1A spliceosomal protein (U1A) [PDB: 1URN] [37], procarboxypeptidase A2 (ADA2h) [PDB: 1O6X] [38], ribosomal protein S6 (S6) [PDB: 1RIS] [39], and muscle-type acylphosphatase (mtAcP) [PDB: 1APS] [40] as shown in Figure 1. These were selected through the Protein Folding Database 2.0, [41] which provides experimental folding data on proteins [15-17,42]. We call these proteins our "study proteins". The amino acid sequences of these proteins were obtained from the structured region in their PDB files. Their sequence identities are quite low, ranging from 11 to 23%. Many experimental studies on these proteins have been performed with respect to the ferredoxin-like fold, and some of these studies suggest the existence of different folding segments [15-17,42].

Inter-residue average distance statistics

To prepare the statistics for our methods, we calculate the average distance and its standard deviation for each inter-residue pair in 42 various proteins, considering the amino acid types and the sequence separation. For the sequence separation, we simplify the sequence separations k: $1 \sim 8$, $9 \sim 20$, $21 \sim 30$, $31 \sim 40$... in terms of the ranges M: 1, 2, 3, 4..., respectively. The 42 representative proteins were carefully chosen so as not to be biased towards some specific structures and have been confirmed to extract the regions corresponding to the structural domains. Because our analysis results are strongly affected by the particular protein datasets used, we chose not to alter the datasets based on the analysis results and to instead use the same datasets as in the first paper on ADM (Ref. [31]) to allow for comparability. We present the 42 proteins (Additional file 1: Table S3).

Average Distance Map analysis

The regions predicted by Average Distance Map (ADM) analysis correspond to the regions that tend to be compact in their 3D structure. We believe that these compact regions might be structured in the early stage of folding.

The ADM analysis itself is a method for predicting the location of possible structural units in a protein by analyzing a predicted contact map [31] based on inter-residue average distance statistics. This map is referred to as an ADM and is used to extract standard structural units, such as structural domains or compact regions. In the ADM, any pair of residues with smaller average distance is considered to be in contact more than the other pairs, so the segment with many such pairs is considered to form a structural unit like folding segments by mechanisms such as hydrophobic collapse. These segments are automatically extracted by analyzing the ADM (as explained in the following text).

In the current study, we use this method to extract compact regions (not structural domains). For each compact region, the strength as a predicted folding segment is expressed as an η value. The η value tends to be higher if the corresponding compact regions have many contacts within the region or with the other regions including non-local areas (for more details, see Refs. [31,43] or [Additional file 1]). Contact density has been reported to correctly identify the nucleating subdomains in T4 lysozyme and interleukin-1β [12]. Studies on flavodoxin-like proteins also suggest a relationship betrween contact density and the folding rate of the corresponding area by showing that low contact density leads to structuring late in folding [44]. Thus, it would be reasonable to consider the regions with many contacts (a high η value) as the region structured in the early stage of folding.

Finally, all high-η-value units which do not overlap with other high-η-value units are designated as predicted folding segments, except for units that cover the whole sequence. When a predicted folding segment covers 70 to 100% of the whole sequence length, we conduct an additional search for folding segments overlapping in this unit. Because we could find only two folding segments for each protein, in this study, we call the segment with the higher η value the primary segment and the other, the auxiliary segment.

In Additional file 1: Figure S5, we compare the ADMs of the ferredoxin-like proteins with the actual contact map constructed based on the contacts defined later.

Comparison of the regions predicted by two ADMs

Suppose that a multiple alignment of homologous sequences in a ferredoxin-like protein is obtained. Since it is convenient to define the similarity of location predicted by any two ADMs in the multiple alignment, we define the similarity as follows: First, two sequences are chosen from the aligned sequences, as shown in Figure 2. Second, all sites with a gap in either one or both sequences are removed. Here, "site" refers to the common sequential number in the multiple alignment. Finally, the number of sites that are commonly included in or excluded from the regions predicted by the ADMs for two given sequences is calculated. The ratio of this number to the number of all the non-gapped sites is defined as the similarity of location predicted by the two ADMs.

It should be noted that the similarity calculated by this procedure does not take η values and gapped sites into account: the present method is therefore not suitable in cases where the sequences in an alignment show large gaps. Having said that, the multiple alignments using the sequences obtained in this study contain few small gaps. For this reason, we can apply this definition of the similarity to the present results.

F-value analysis

We performed additional analyses to determine the location where initial folding events, such as hydrophobic collapse, happen [32]. Using other statistical potentials like Miyazawa-Jernigan [45] and Skolnick [46] do may return similar information, but it makes difficult to interpret the results with ADM. Because there is not only the average distance but also the standard deviation of distance in inter-residue average distance statistics, we expect that the potential used in our F-value analysis to better reproduce the dynamics of the denatured state ensemble compared to the potential based on the contact energy. In F-value analysis, we use the Cα bead model to represent a protein's structure, as well as the Metropolis Monte Carlo method with the potential energy $\varepsilon_{i,j}$ derived from average distance $\bar{r}_{i,j}$ and its standard deviation $\sigma_{i,j}$.

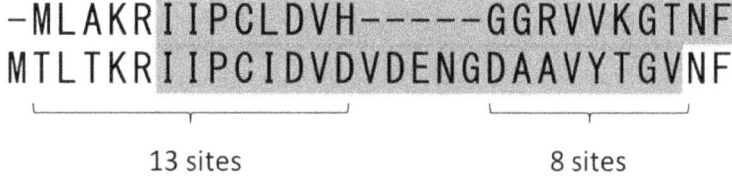

Figure 2 Example of two sample aligned sequences from multiple alignment. A region in gray corresponds to a compact region predicted by the ADM. Because there are 23 sites that have no gap in either sequence and there are 21 sites that are commonly included in or excluded from the predicted compact regions, 21/23 ~ 91.3% is the ADM similarity for this example.

The bond and dihedral angles of the initial conformation are randomly selected. The movement in the simulation is done as follows: The bond angle between the residue i and $i+1$ is bent and rotated randomly from -10 to $10°$ followed by the Metropolis judgment to decide if the new conformation is acceptable or not. Within a step, $i = 1...N-1$ is performed, that is, all the bond angles are altered and judged.

The probability density with the potential energy between two residues, $P(\varepsilon_{i,j})$, is hypothesized as being equivalent to the probability density based on the standard Gaussian distribution calculated with its average distance and standard deviation, $\rho(\bar{r}_{i,j}, \sigma_{i,j})$, as follows:

$$P(\varepsilon_{i,j}) = \rho(\bar{r}_{i,j}, \sigma_{i,j}) \tag{1}$$

Where this equation can be expressed by equation (2);

$$\frac{\exp\left(-\frac{\varepsilon_{i,j}}{kT}\right)}{Z} = \frac{1}{\sqrt{2\pi}\sigma_{i,j}} \exp\left\{-\frac{\left(r_{i,j}-\bar{r}_{i,j}\right)^2}{2\sigma_{i,j}^2}\right\} \tag{2}$$

Finally we obtain equations (3) and (4);

$$-\frac{\varepsilon_{i,j}}{kT} - \ln Z = -\ln\left(\sqrt{2\pi}\sigma_{i,j}\right) - \frac{\left(r_{i,j}-\bar{r}_{i,j}\right)^2}{2\sigma_{i,j}^2} \tag{3}$$

$$\frac{\varepsilon_{i,j}}{kT} = \frac{\left(r_{i,j}-\bar{r}_{i,j}\right)^2}{2\sigma_{i,j}^2} - \ln\frac{Z}{\sqrt{2\pi}\sigma_{i,j}} \tag{4}$$

where kT is set so that the acceptance ratio is 0.5. This potential is designed to sample the ensembles which can reproduce the inter-residue average distance statistics. From the simulation, the contact frequency, g(i,j), for each pair of residues is calculated with sampled structures generated using the potential energy function. Then we normalize the residue contact frequencies, g(i,j), in the same range M as follows:

$$D(M) = \sqrt{\sum_{|\mu-\nu|=m} \frac{\left(\tilde{g}_{|\mu-\nu|\in M} - g(\mu,\nu)_{|\mu-\nu|\in M}\right)^2}{\Sigma_{|\mu-\nu|\in M}}} \tag{5}$$

$$Q(i,j) = \frac{g(i,j)_{|i-j|\in M} - \frac{\Sigma_{|\mu-\nu|\in M} \; g(\mu,\nu)}{\Sigma_{|\mu-\nu|\in M}}}{D(M)} \tag{6}$$

where μ or ν is the residue number. Finally, the relative contact frequency, F_i, is obtained by summing the normalized contact frequencies, Q(i,j), from $j = 1$ to N for each residue i, where N is the protein sequence length:

$$F_i = \sum_j Q(i,j) \tag{7}$$

The peaks of the plots of the F_i- or F-value peaks are thought to be located in the center of many inter-residue contacts, such as a hydrophobic cluster. Therefore, the regions around the peaks are assumed to be important for folding, especially for its initial state. F-value analysis therefore allows us to estimate the location where a folding initiation occurs, except for the termini with their expected extreme flexibility in the simulation: due to the flexibility, the F_i values at the terminal residues become unrealistically high, and this value is then considered not to be true [47]. We performed this simulation with 60000 steps 100 times, calculating the average F_i value for residue i.

Analysis of evolutionarily conserved residues
Evolutionarily conserved residues maintain a protein's function, contribute to its stability, or relate to its structural formation [48-53]. Therefore, conserved residues that have many contacts with other conserved residues in the native structure are thought to be significant indicators of potential folding segments. Based on this idea, we gather the homologous sequences for each study protein with the Basic Local Alignment Search Tool (BLAST) [54] (DB: Uniref100, Threshold: 0.01, Gapped: No) and aligned them with the MUltiple Sequence Comparison by Log-Expectation tool (MUSCLE) [55]. We applied the neighbor-joining method [56] to construct the study protein's phylogenetic tree for inputting into the Phylogenetic Analysis by Maximum Likelihood software package (PAML) [57]. With PAML, for each site without any gap, we can count the number of residue substitutions by using JTT matrices [58] for the substitution matrix and a Poisson distribution for the substitution model. This

procedure allows us to estimate the conservation or substitution of a specific residue during evolution based on branch lengths and bifurcations in a phylogenetic tree. Because only the conservation of hydrophobic residues is taken into account in this study, the hydrophobic residues with more than 99% conservation are regarded as conserved residues, that is, we still regard a residue as conserved, when one of the hydrophobic residues A, M, W, L, F, V, I, and Y has mutated to another one.

We employed the BLAST to identify potential homologous sequences. Only sequences that cover the whole sequence of a study protein were selected as homologous sequences based on the BLAST results. The BLAST search identified at least 100 homologous sequences for each study protein were obtained (see Additional file 1: Table S1).

Definition of inter-residue contacts
The Shrake-Rupley algorithm [59] was used to define a contact by the decrease in the Solvent Accessible Surface Area (SASA) upon folding. The reduction in the surface area is calculated by the difference between a sidechain's SASA in the presence of contacts with other residues and that in the absence of contacts. In this study, only heavy atoms are considered, and when the decrease in the SASA reaches 27 Å^2, the corresponding hydrophobic residue pair is defined as being in contact. This threshold was determined from the decrease in the SASA when two carbon atoms form a contact, namely, 27.27 Å^2.

Summary of the experimental results from the literature
Figure 3 provides a summary of the results reported in the literature from various Φ-value studies which provide information on structured sites in the transition state [60]. We compare the regions predicted by ADM with the location of secondary structures with high Φ values. Averaging the Φ values for each secondary structure is a good way to understand the differences in folding mechanisms among a set of proteins with the same topology and this method has been performed by many researchers (for instance, TNfn3/CD2d1 [61], mt/sso-AcP, [62] mt-AcP/ADA2h, [17,42] wt-S6/permutants [15], and so on). We validated our predictions by comparing them to experimental results interpreted in the same manner. (We need to note that there are a few residues with high Φ values which should be excluded. For example, P54A in mtAcP has a high Φ value of 0.98. However, according to the 3D structure, its side chain seems exposed; thus, its high Φ value seems to be derived from the unique dihedral angles of the proline residue, and we did not treat it as a member of the folding segment).

In this study, a few secondary structures with relatively high Φ values are defined as an experimental folding segment. Even though the resolution is somewhat lower because the folding segment is defined by average Φ values, this approach is still similar in concept to the "folding nucleus" first introduced by Shaknovich and his colleagues [63] as a set of contacts in denatured states that are considered sufficient and necessary for transitioning or

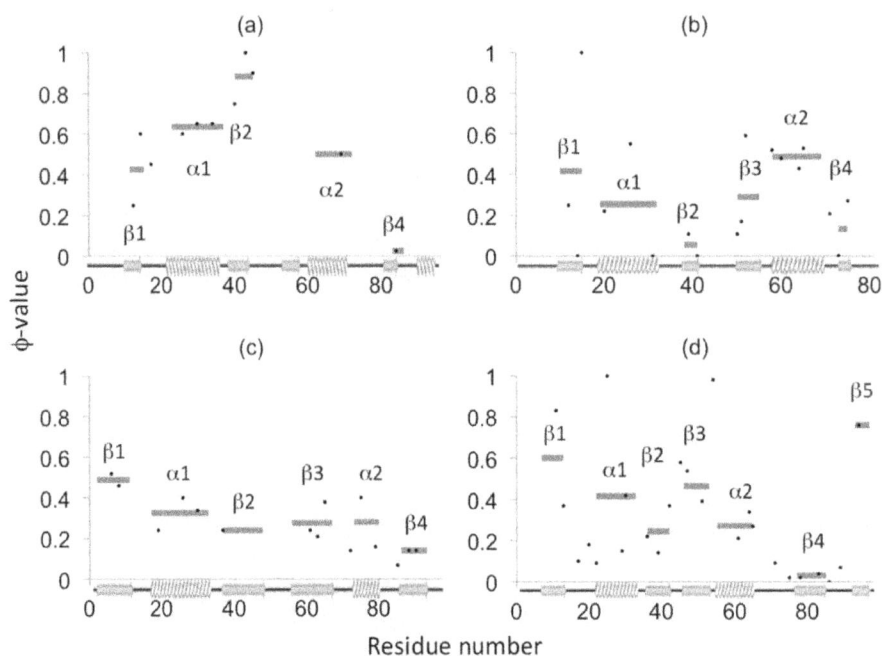

Figure 3 Experimental Φ values and average values for each secondary structure. (a) U1A, **(b)** ADA2h, **(c)** S6, and **(d)** mtAcP. Dots denote the experimental Φ values. Gray bars indicate the average Φ value for each secondary structure. Because no Φ values in the 3rd β strand of U1A have been reported, its average value is not shown.

molten globule states to occur as observed by a computational technique. The formation of these contacts is rate-limiting step and should be done by its transition state.

Some later studies support the idea of a folding nucleus by means of Φ-value analysis or combining experimental Φ values with computational techniques [64,65]. In other words, the folding segment is thought to be relatively structured compared to other regions from the denatured state to transition state. (This is because high-Φ-value sites correspond to the sites which are energetically stable in the transition state, and we expect that such an energetically stable region forms even in the early stage of folding.) For example, for the wild-type ribosomal protein S6 from *Thermus Thermophilus*, one of the ferredoxin-like folds studied by Haglund et al., [15] one folding nucleus is reported to consist of β1, α1, and β3 (despite the α1 and α2 having very similar average Φ values). However, for the circular permutants prepared by connecting N- and C-termini and disconnecting other loops between neighboring secondary structures, sometimes the folding nucleus noticeably shifts to β1, α2, and β4 [15]. Therefore Haglund et al. [15] found that in S6, there are two competing and overlapping folding nuclei, and the relative magnitude of significance for folding could be perturbed by circular permutation. (In Additional file 1: Figures S3 and S4, we also summarized the results of the ADM and F-value analyses for the circular permutant of S6 whose X-ray crystallographic structure is available. This means there is a guarantee that the native structure is not disordered or structured in other conformations, thereby making its Φ values seem reliable. We obtained similar results as in previous analyses). Ternström et al. [16] also performed an experimental study on U1A spliceosomal protein, which also folds through the α1 formation and its surrounding secondary structures in an early stage. However, Villegas et al. [17] reported that procarboxypeptidase A2, which has the same topology as S6 and U1A, folds through the α2 formation and part of β1/β3. Its folding segments seem to be in the C-terminus (β1, α2, β3) like the S6 permutants (β1, α2, β4 in the S6 wild type), while there are some differences around the β strands. Specifically, there seems to be two tendencies with respect to the location of the folding segments: the N-terminus with α1 and the C-terminus with α2, but it remains difficult to say which β strands contribute to folding. Therefore in this study, we simply chose the β strands that have higher average Φ values compared to the values of the α helix with a lower average Φ value than the other α helix.

Results

Folding segments predicted by ADM analysis

The predicted contact maps and the location of the compact regions, namely, the predicted folding segments are shown in Figure 4 and summarized in Table 1.

According to these data, all four proteins have two compact regions, and each region contains one α helix with a couple of β strands.

For U1A spliceosomal protein ([PDB: 1URN]) and muscle-type acylphosphatase 2 [PDB: 1APS], the N-terminal compact region has a higher η value than that of the C-terminus, which suggests that the N-terminal region is stable compared to the C-terminal region during the early stage of folding, while procarboxypeptidase A2 [PDB: 1O6X] shows the opposite trend. As for ribosomal protein S6 [PDB: 1RIS], the two compact regions have similar η values, which is interpreted as meaning that both of these regions play equally important roles in structural formation. It is also notable that except for the case of mtAcP, β3 is always included in the primary folding segments (the predicted region with a higher η value) for each protein.

We are interested in comparing the predicted folding segments with the secondary structures whose average Φ values are high. Figure 5a-d show the results. According to these figures, the secondary structures within the predicted folding segments often correspond to the secondary structures with high average Φ values.

In Figure 5, the positions of the secondary structures with higher average Φ values than those of the α helix (taking the lower Φ value of the two α helices) are colored red in the right panel, while in the left panel, the positions of the predicted primary folding segment at the N- or C-terminus are colored yellow or green, respectively. For S6, however, we color both segments in red or yellow/green, because according to the average Φ values and η values of S6 (see Figure 3 and Table 1, respectively), both N-terminal and C-terminal folding segments are equally important in the formation of its 3D structure. Figures 3, 4, and 5 indicate that almost all the important secondary structures for folding, as defined by experimental Φ value results, are included in the folding segments predicted by the ADMs, although β3 in mtAcP, which shows a relatively high Φ value, is not included in the region predicted by the ADM.

Evolutionary conservation analysis with F-value analysis - Location of predicted hydrophobic clusters

The solid line in Figure 6 indicates the F-value result, while the broken line shows the smoothed plots of the number of contacts with other conserved hydrophobic residues. (The conserved hydrophobic sites are shown in Additional file 1: Figure S8.) Smoothing was performed with a Gaussian kernel [66]. A conserved hydrophobic contact means that a pair of conserved hydrophobic residues form a contact. The locations of the peaks are indicated by single or double daggers for F-value or conserved hydrophobic contacts, respectively. A number near a dagger indicates the corresponding residue number in a

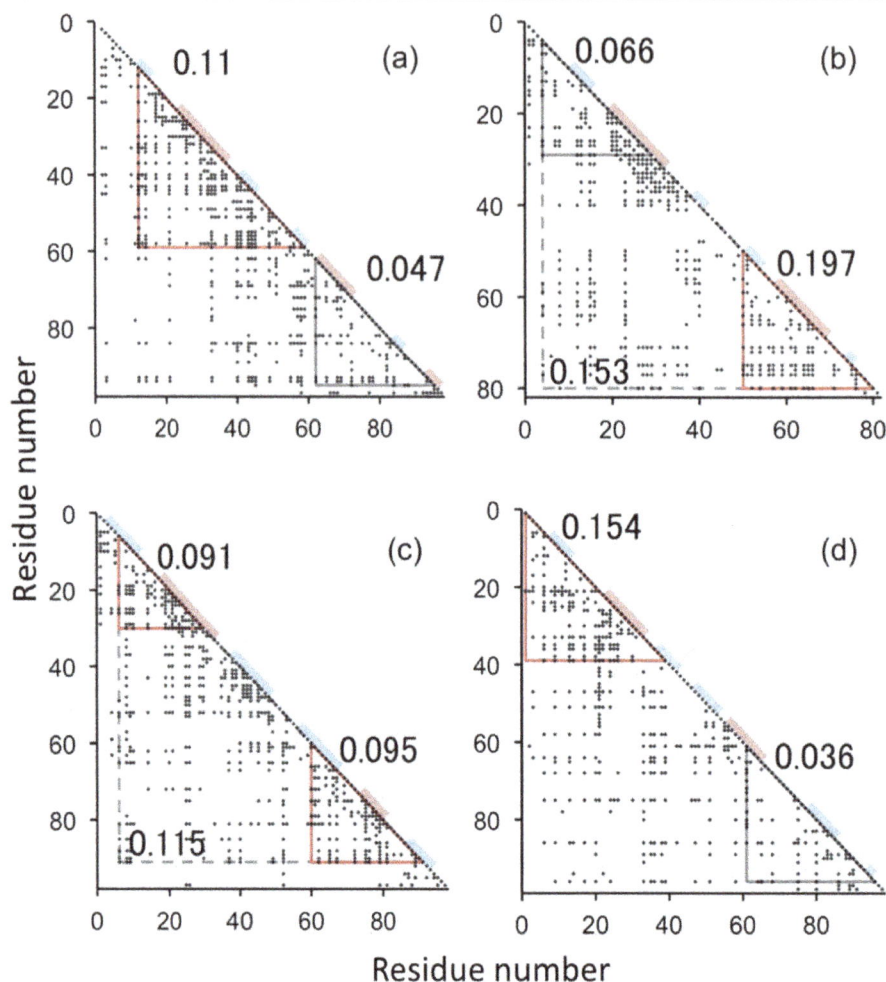

Figure 4 Results of ADM analyses. (a) U1A, **(b)** ADA2h, **(c)** S6, and **(d)** mtAcP. The color bars on the diagonal of a predicted contact map indicate the location of secondary structures. The abscissa and ordinate denote residue numbers, and triangles with a solid line in red or black indicate the location of primary or auxiliary compact regions, respectively. A large triangle with a broken line means it is ignored because it covers more than 70% of the entire sequence. η values are shown beside the triangles.

protein. We follow the PDB system concerning the residue number in a protein. The secondary structures and conserved hydrophobic residues are shown below the plot.

Except for several sites, most of the conserved hydrophobic residues are distributed somewhat sparsely but uniformly, which implies that it is hard to extract folding segments from only their amino acid sequences and conservation analyses. According to Figure 6, most of the F-value peaks are close to those of the smoothed line within ± 3 residues. This indicates that F-value peaks, which can be mainly regarded as hydrophobic clusters in the initial nucleation stage, also correspond to the region with many conserved hydrophobic contacts, which are important for the formation of a native structure.

Direct comparisons of these regions with high-Φ-value sites are shown in Additional file 1: Figure S9. For high-Φ-value sites, only the sites with a Φ value higher than

the average Φ value of the protein are shown along with each site's residue type and number. The peaks of smoothed conserved hydrophobic contacts are marked by double daggers as in Figure 6. High-Φ-value sites are found to exist near the peaks of the conserved hydrophobic contacts (within ± 3 residues), suggesting that some of these contacts are responsible for structural formation. These high-Φ-value sites are also found to exist near the F-value peaks within ± 3 residues, as shown in Additional file 1: Figure S10.

Folding segments predicted by ADMs in the homologous proteins of the study proteins

To confirm whether the folding segments are conserved among evolutionarily related proteins, we performed our sequence analysis on the homologues of the four study proteins. The results of applying ADM analyses to these homologues are shown in Figure 7.

Table 1 Summary of the Average Distance Map (ADM) Analyses

PDB entry	N-termini	C-termini	η	Dominance
	12	95	0.112^b	
1URN	12	59	0.110	N
	62	95	0.047	
	4	29	0.066	
1O6X	50	80	0.197	C
	6	91	0.115^b	
	6	30	0.091	N
1RIS	52	91	0.087^a	
	60	91	0.095	C
	1	39	0.154	N
1APS	1	47	0.132^a	
	61	96	0.036	

N or C denote the N- or C-terminal borders of compact regions. The primary compact regions are shown with N or C in the Dominance column. In the η column, compact regions extended by the 85% rule [31] are identified with a superscript a; compact regions ignored due to covering more than 70% of the entire sequence are identified with a superscript b.

This figure denotes the respective multiple alignments of the homologues. The location of the predicted folding segments are colored dark gray: the brighter the color, the higher the region's η value. It can be visually confirmed in Figure 7a-c that there are several bands indicating that for most of the homologues the same regions are predicted.

In Figure 7, we ordered the sequences based on the similarity of the location of the regions predicted by the ADMs. In the right column, the phylogenetic tree based on an ADM similarity matrix and the neighbor-joining method is shown. Another result based on the sequence identity is shown in Additional file 1: Figure S7. It is difficult to determine the relationship between the location of the folding segments and their evolutionary distance (as specified by the calculated sequence identity). However, we can conclude that for U1A, ADA2h, and S6, the folding segments themselves are conserved among their homologues, while those of mtAcP are not.

To represent these common folding segments, we calculate the percentage of residues that are members of the predicted folding segments for each site. The results are also shown as a histogram colored black in Figure 7.

In the case of U1A, four secondary structures $\beta\alpha\beta\beta$ at the N-terminus form one strong folding segment for most of the homologues, while the other C-terminal region comprises a weak folding segment. For ADA2h, the C-terminal folding segment $\beta\alpha\beta$ is conservative and strong, and the N-terminal folding segment is conservative but weak.

As for S6, there are many homologues, and they share almost the same folding segments. One segment consists

of $\beta\alpha\beta$ at the C-terminus and the other one consists of $\beta\alpha$ at the N-terminus. The dominance of these two folding segments at the termini often differs among the homologues. It is also notable that for some homologous proteins, the region from $\beta2$ to the hairpin-loop comprises the weakest folding segment, which forms a β-hairpin with $\beta3$ in the C-terminal folding segment.

Finally, for mtAcP, the folding segments are not conservative among its homologous proteins. However, the locations of the folding segments appear similar to those of S6, ADA2h, and U1A.

Discussion

The ADM analyses of the four proteins predict two compact regions including one α helix and a couple of β strands for each protein. These regions contain the secondary structures with high average Φ values (Figure 3). Therefore, we consider the predicted compact regions to correspond well to the folding initiation segments as was the case for the other proteins we treated in previous studies, including lysozyme, leghemoglobin, fatty acid binding protein, azurin, and two ancient TIM-barrel proteins [33-36]. According to the η values, mtAcP and U1A have the primary predicted folding initiation segment at the N-terminus, whereas ADA2h has one at the C-terminus. On the other hand, the two folding segments of S6 have similar η values (see Figure 4). Figure 5 shows good agreement between the ADM predictions and the experimental results; however, the resolution of this analysis is too low to predict the folding mechanisms.

By means of F-value analysis, we increased the resolution of the prediction made by ADM, allowing us to identify the regions that would form some hydrophobic clusters. According to Figure 6, almost all the peaks are located on the secondary structures or their edges, and the highest peak is located in the primary folding segment for each protein. For example, U1A has the primary folding segment at the N terminus from $\beta1$ to $\beta3$, and its highest F-value peak is located in $\beta3$. In addition, a peak of F values is located at a peak of the smoothed line of the conserved hydrophobic contacts within ± 3 residues, except for the broad peak observed at the C-terminus of ADA2h, which contains several minor peaks. These conserved contacts are thought to play important roles in the structural formation or stabilization of U1A's native structure [48-53]. Taking these facts into account, the conserved hydrophobic residues near the F-value peaks are considered to be significant for the folding initiation. The basis for predicting the folding mechanisms from only sequence information is the fact that the regions predicted by ADM analysis contain the high-Φ-value residues measured by experiments [15-17,42] and that the F-value analysis reflects the conserved hydrophobic contacts. Let us now make

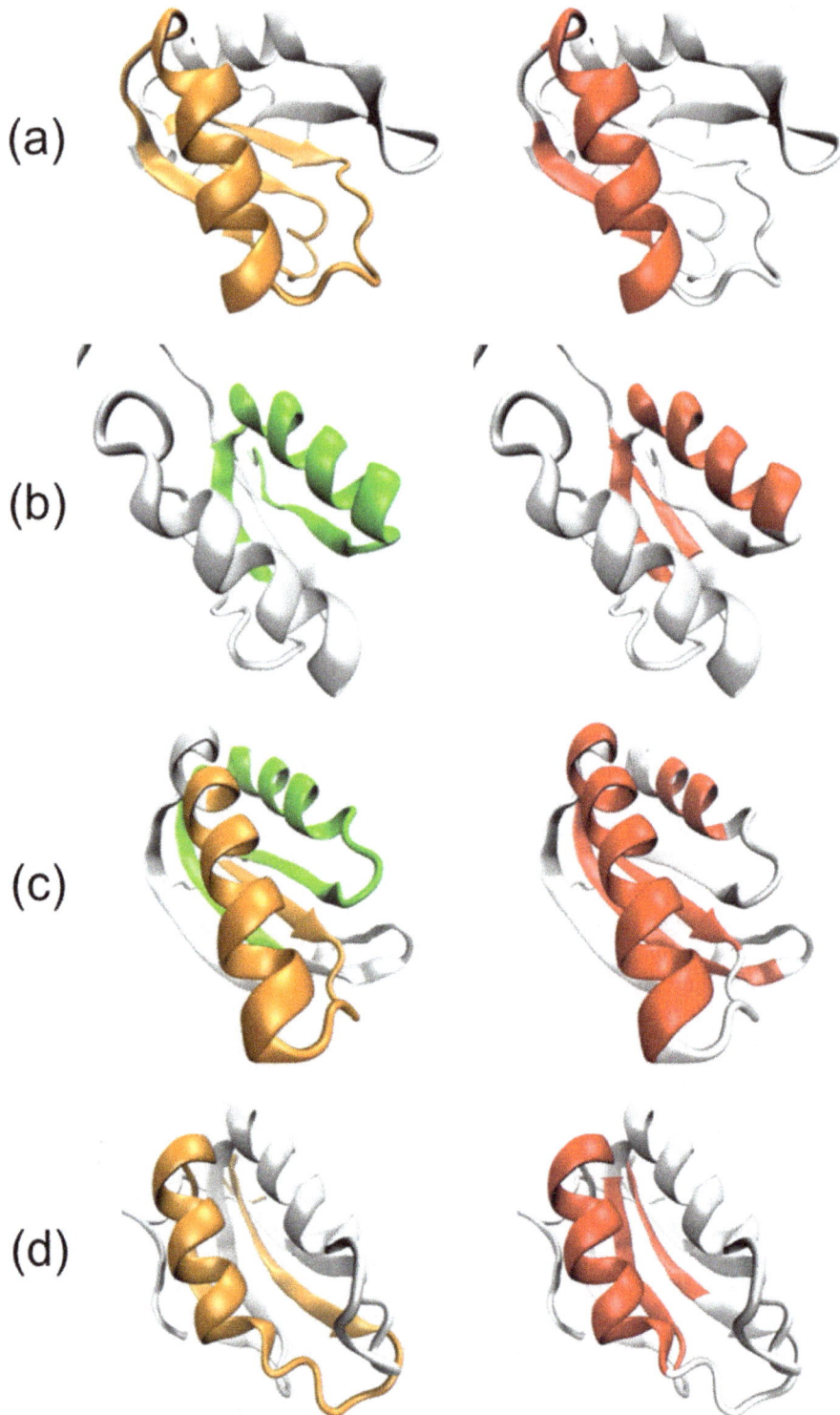

Figure 5 Comparison of predicted folding segments and experimental folding segments. (a) U1A, **(b)** ADA2h, **(c)** S6, and **(d)** mtAcP. In the left column, the predicted primary folding segments located at the N- or C-termini are colored orange or green, respectively. In the right column, all the secondary structures with an average Φ value higher than that of the α helix with the lower average Φ value are colored red. However, for S6, the β-strand 3 and α-helix 1 are also colored in red, because their average Φ values are not significantly lower than the averge Φ value of the α helix with the higher value, unlike the case in other proteins.

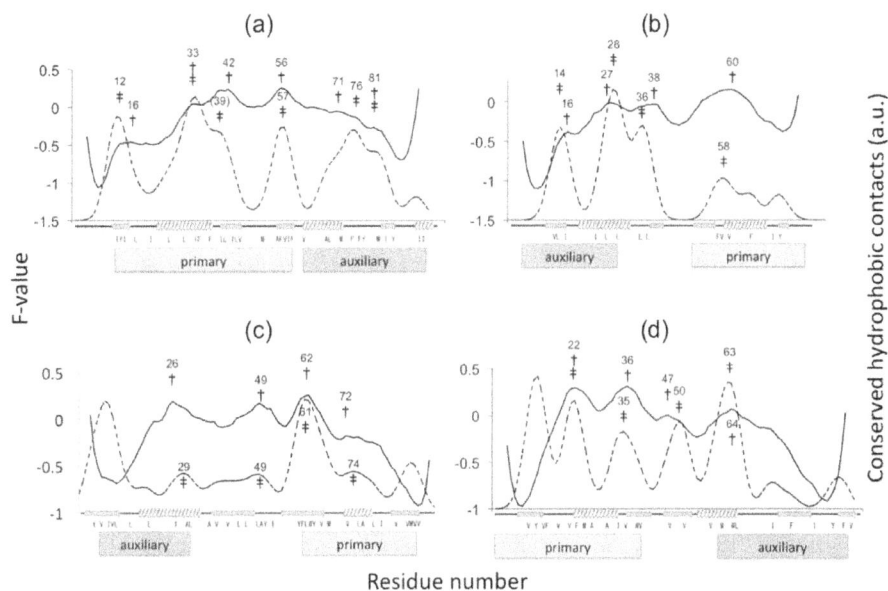

Figure 6 Results of F-value analyses and the distribution of conserved hydrophobic contacts. **(a)** U1A, **(b)** ADA2h, **(c)** S6, and **(d)** mtAcP. F values or the smoothed number of conserved hydrophobic contacts are shown as a solid or broken line, respectively. The ordinate denotes the F value or the number of conserved hydrophobic contacts and the patterns along the abscissa show the location of secondary structures. The conserved amino acid residues and the location of predicted folding segments are also shown below the plot. The F-value peaks that were the focus of this study are marked with single daggers (†), and the number above each dagger denotes the residue number of the respective peak. The smoothed number of conserved hydrophobic contacts is in arbitrary units, and the peak location is shown with a double dagger (‡) like the F-value peaks. Only for U1A, the shoulder is indicated with parentheses.

inferences regarding the folding mechanisms for proteins based on the results of the ADMs and F-value analyses.

U1A spliceosomal protein; U1A

As shown in Figure 6a, the primary compact region of U1A covers β1, α1, β2, and β3, and each region of α1, β2, and β3 contains just one F-value peak. The auxiliary compact region of U1A ranges from α2 to α3. Because the auxiliary compact region has a lower η value, it is thought to participate in the structural formation after the primary compact region has been formed. Since Ternström et al. [16] suggest that the region from β1 to β3 is more structured compared to α2 and β4 [16,67], we find that our results agree well with their experimental Φ-value analysis. Figure 8a presents the packing formed by conserved hydrophobic residues near the F-value peaks. The residues that contribute to the hydrophobic packing are represented in the CPK model in this figure. The regions colored yellow or green correspond to the predicted N- or C-terminal compact regions, respectively.

Procarboxypeptidase A2; ADA2h

The auxiliary compact region of ADA2h at the N-terminus has two secondary structures, β1 and α1, whereas the primary compact region at the C-terminus has three secondary structures, β3, α2, and β4. The former has a

high F-value peak around α1, indicating that α1 is the center of folding within the auxiliary compact region. On the other hand, the primary compact region has the highest and broadest peak from β3 to α2. Therefore, we can predict that after β3 and α2 form a folding segment, β1 and α1, pack with this segment and stabilize it.

This prediction can also be validated by Figure 3b. Villegas et al. [17] state that the folding segment in this protein consists of α2 and its surrounding β strands. This agrees with our result. However, we could not confirm any packing between the conserved hydrophobic residues near the F-value peaks in Figure 8b within the primary compact region in the native structure as observed in U1A. This is because the region with the largest broad F-value peak in the C-terminal region seems to have only a few conserved hydrophobic residues as indicated by the smoothed plot of the conserved contacts which shows several minor peaks here. In this case, the resolution of the F-value line is too low to detect the residues important for folding.

Ribosomal protein S6; S6

The relative auxiliary folding segment of S6 at the N-terminus contains β1 and α1, while the primary folding segment at the C-terminus contains β3, α2, and β4. The η values are quite similar, so we cannot say which region folds more dominantly. S6 has two significant F-value

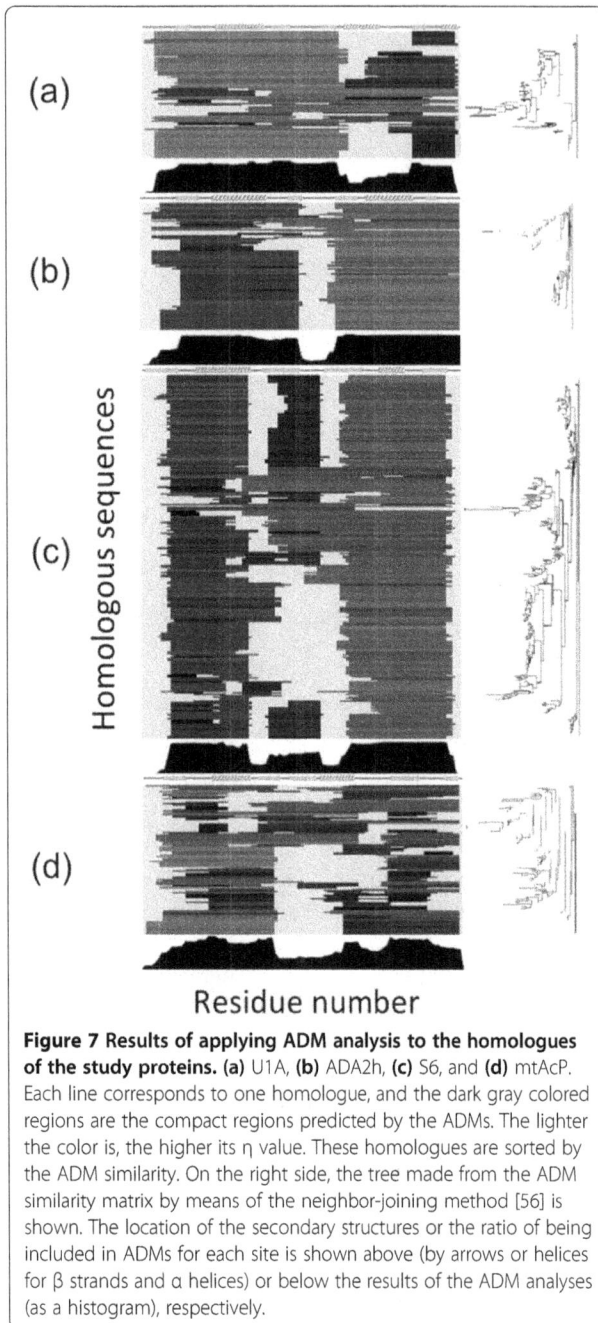

Figure 7 Results of applying ADM analysis to the homologues of the study proteins. (a) U1A, **(b)** ADA2h, **(c)** S6, and **(d)** mtAcP. Each line corresponds to one homologue, and the dark gray colored regions are the compact regions predicted by the ADMs. The lighter the color is, the higher its η value. These homologues are sorted by the ADM similarity. On the right side, the tree made from the ADM similarity matrix by means of the neighbor-joining method [56] is shown. The location of the secondary structures or the ratio of being included in ADMs for each site is shown above (by arrows or helices for β strands and α helices) or below the results of the ADM analyses (as a histogram), respectively.

peaks within the predicted folding segments: one is around α1 at the N-terminus, and the other is around β3 at the C-terminus. Lindberg et al. [67] suggest that the primary folding segment consists of β1, α1, and β3 in the early folding stage, and our results reflect this. Figure 8c shows the residue packing near the F-value peaks inside the predicted folding segments.

It is also notable that in the case of S6, there is a highly frustrated region between the C-terminal unstructured coil and the β sheet based on the structure of S6 [68]. However, the corresponding C-terminal region does not

have any specific contact with other regions in the ADM. This is confirmed by its NMR structure ([PDB: 2KJV]). At least as far as concerning the ADM result and the NMR structure, the frustration between the C-terminal unstructured coil and the β sheet does not seem strong.

Muscle-type acylphosphatase 2; mtAcP

The primary folding segment of this protein, which has a significantly higher η value than the other folding segment, is located at the N-terminus. This result is the same as the result from Selvaraj et al. [69] who suggested the existence of a hydrophobic cluster surrounding α1 based on the distribution of contacts. The primary segment contains β1, α1 and β2, whereas the auxiliary segment contains α2, β4, and β5. There are four F-value peaks near α1, β2, β3, and α2; three of them are located in the predicted folding segments, but the peak near β3 is not. (In fact, β3 belongs to the primary folding segment and seems to play a critical role in the folding in other proteins.) Therefore we propose that after α1 and β2 play a role in early structural formation, α2 participates in the last structural formation, followed by contributions from β3. This scenario does not seem to fit the results of the average Φ-value analysis [42], which indicate that β3 plays a more important role than β2 (see Figure 2d).

According to Parrini et al. [70], when β3 is forced to join the folding process by a disulfide bond between β1 and β3, the folding rate is improved dramatically. This result suggests that the participation of β3 in the folding process is rate limiting and may reflect the present findings. For this reason, we do not consider the two analyses to conflict with each other. The inter-segment packing between the conserved residues is represented in Figure 8d.

Meanwhile, it should be noted that in their analysis, Chiti et al. [42] ignore the highest Φ value of the 23rd residue located in α1 and then compare the result with that of ADA2h. In this case, the average Φ value of α2 is higher than that of α1, indicating that the more structured secondary structure in its transition state is α2, the same as for ADA2h. There are also several studies that suggest that the α2 in mtAcP is more important than α1 [71-74]. Interestingly, one of the studies refers to the large effect on α1 induced by point-mutation. Taddei et al. [71] consider the Φ values of α1 to be unreliable because, according to their experiment, inducing a point-mutation on α1 makes mtAcP form amyloid fibrils. Thus, interpreting the folding segment of mtAcP is difficult.

Our previous simplified Go-model simulations reveal that the interactions between the folding segments in the present definition are significant in the formation of transition state ensembles [75].

According to the discussion above, the conserved hydrophobic residues among homologues are distributed near F-value peaks (Figure 6), and they seem to be

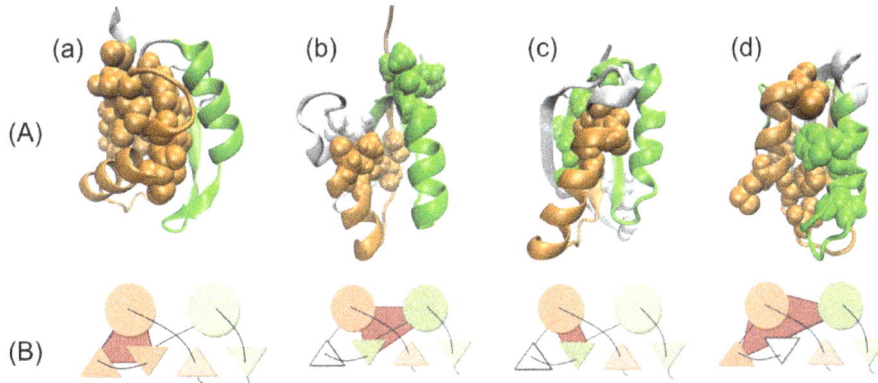

Figure 8 Hydrophobic interactions observed in the ferredoxin-like fold proteins. (A) Representation of internal conserved hydrophobic contacts. N- and C-terminal compact regions are colored yellow and green, respectively. The conserved hydrophobic residues near the F-value peaks have a space-filling representation. **(B)** Illustrations of important interactions among the secondary structures discussed in the current study. α helices are shown as circles, and the β strands are shown as triangles. The N- or C-terminal compact regions are colored in yellow or green, respectively, as in **(A)**. The important interactions are indicated in red.

involved in folding. Yet, as we mentioned in the introduction section, there are some studies that have shown that several sequences that fold into the same 3D structure have the same folding process, while other studies have shown that two evolutionarily related sequences with the same 3D structure could have different folding processes. In this sense, whether all the homologues of our study proteins really have the same folding segments or not would be an interesting question.

In the present study, we aim to analyze the conservation of the predicted folding segments in the homologues of a protein by applying ADM analysis to them. The predicted folding segments are highly conserved among the highly homologous proteins (with roughly 50% sequence identity on average) except for mtAcP and its homologous proteins, as shown in Figures 7 and 9. In Figure 7, the conservation of the predicted folding segments is summarized as a histogram below each result. As seen in this figure, the histogram of mtAcP is uneven compared to the other histograms. Figure 9 depicts this property from another aspect.

The abscissa in Figure 9 represents the lower limit of the sequence identity for calculating the average ADM similarity, that is, an average ADM similarity value is calculated using homologues with a sequence identity of more than a lower limit, while the ordinate denotes the average ADM similarity. The doubled line, solid line, thick line, and dotted line correspond to U1A, ADA2h, S6, and mtAcP, respectively. While the other three proteins maintain their folding segment similarity of more than 75% even when the sequence identity decreases; only mtAcP loses similarity down to 62%. This result suggests a diversity of folding processes in mtAcP compared to those in the other proteins, especially when sequence identity is low. On the other hand, as we mentioned in the results

section, the relationship between ADM similarity and sequence identity is not parallel (see Additional file 1: Figure S7). This is an unexpected result: we expected that the more similar the sequence identity is, the more similar the protein folding segments are. Yet the present results suggest that a property related to folding segments is conserved more than sequence identity.

Summarizing the discussion above, there are mainly two situations. One of them clearly comprises a main large folding segment around α1, like in U1A. The other situation comprises complex folding segments in which,

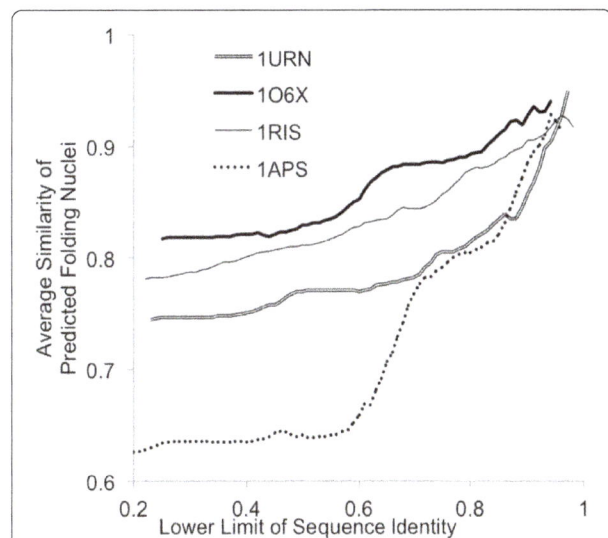

Figure 9 The relationship between the similarity of the predicted folding segments and sequence identity. The abscissa denotes the lower limit of sequence identity and the ordinate denotes the average similarity of the predicted folding segments. The double line, solid line, thick line, and dotted line correspond to U1A, ADA2h, S6, and mtAcP, respectively.

one of the α helices and its surrounding β strands play a key role at first, immediately followed by the other helix and its surrounding strands, as in ADA2h or S6. The homologous proteins of mtAcP have either property: some of them have folding segments similar to those of U1A proteins, and some others have folding segments similar to those of S6 or ADA2h proteins. This implies that mtAcP and its homologous proteins do not seem to have any common or rigid folding segments.

Conclusions

The secondary structures that are thought to play important roles in folding as revealed by their average Φ values correspond to the folding segments predicted by ADM analyses at least for the proteins treated in this study, as was the case in our previous studies [33-36,47,76]. There are two predicted folding segments at the termini of each protein; however, which segment is primary is completely determined on a case-by-case basis. This tendency was also in good agreement with the experimental results for the present four study proteins. Some of the conserved hydrophobic contacts considered to play important roles in structural formation [49,53] are located near the F-value peaks. Therefore, we can predict the folding mechanisms by extracting the conserved hydrophobic residues near them. For the four proteins we studied above, we conclude that we succeeded in predicting their folding mechanisms correctly from only their sequences.

According to the ADM results of the homologues, their folding segments seem to be conserved, especially when the sequence identity is above 80%. Below this level, only mtAcP represents a diversity of folding segments, whereas the other three proteins show high conservations.

Our findings suggest that it should be possible to predict the folding mechanisms or properties of many other kinds of proteins from only the amino acid sequences by means of our ADM analysis and F-value analysis.

Abbreviations

ADM: Average Distance Map; BLAST: Basic Local Alignment Search Tool; SASA: Solvent Accessible Surface Area; PDB: Protein Data Bank; MUSCLE: MUltiple Sequence Comparison by Log Expectation; PAML: Phylogenetic Analysis with Maximum Likelihood; JTT: Jones Taylor Thornton.

Competing interests

The authors declare that they have no competing interests.

Authors' contributions

TK conceived and designed the basis of this study. MM performed all the calculations and data analysis. TK and MM wrote the manuscript. Both authors read and approved the final manuscript.

Acknowledgments

The authors wish to acknowledge Asst. Prof. Yosuke Kawai. This work is supported by JSPS KAKENHI to MM, Grant-in-Aid for JSPS Fellows, Grant Number 259198. One of the authors (TK) expresses his gratitude to the Ministry of Education, Culture, Sports, Science, and Technology for the support of the present work through a program for strategic research foundations at private universities, 2010–2014. (Grant Number S1001042).

Author details

¹Department of Bioinformatics, College of Life Sciences, Ritsumeikan University, 1-1-1 Nojihigashi, Kusatsu, Shiga 525-8577, Japan. ²Japan Society for the Promotion of Science (JSPS), Tokyo, Japan.

References

1. Dill KA, MacCallum JL: The protein-folding problem, 50 years on. Science 2012, 338:1042–1046.
2. Arai M, Ito K, Inobe T, Nakao M, Maki K, Kamagata K, Kihara H, Amemiya Y, Kuwajima K: Fast compaction of α-Lactalbumin during folding studied by stopped-flow X-ray scattering. J Mol Biol 2002, 321:121–132.
3. Nickson AA, Clarke J: What lessons can be learned from studying the folding of homologous proteins? Methods 2010, 52:38–50.
4. Nishimura C, Prytulla S, Dyson HJ, Wright PE: Conservation of folding pathways in evolutionarily distant globin sequences. Nat Struct Biol 2000, 7:679–686.
5. Cavagnero S, Dyson HJ, Wright PE: Effect of H helix destabilizing mutations on the kinetic and equilibrium folding of apomyoglobin. J Mol Biol 1999, 285:269–282.
6. Jennings PA, Wright PE: Formation of a molten globule intermediate early in the kinetic folding pathway of apomyoglobin. Science 1993, 262:892–896.
7. Burns LL, Dalessio PM, Ropson IJ: Folding mechanism of three structurally similar beta-sheet proteins. Proteins 1998, 33:107–118.
8. Kim DE, Fisher C, Baker D: A breakdown of symmetry in the folding transition state of protein L. J Mol Biol 2000, 298:971–984.
9. Park SH, Shastry MCR, Roder H: Folding dynamics of the B1 domain of protein G explored by ultrarapid mixing. Nat Struct Biol 1999, 6:943–947.
10. Radford SE, Dobson CM, Evans PA: The folding of hen lysozyme involves partially structured intermediates and multiple pathways. Nature 1992, 358:302–307.
11. Nakamura T, Makabe K, Tomoyori K, Maki K, Mukaiyama A, Kuwajima K: Different folding pathways taken by highly homologous proteins, goat α-Lactalbumin and Canine milk lysozyme. J Mol Biol 2010, 396:1361–1378.
12. Hills RD Jr, Brooks CL III: Subdomain competition, cooperativity, and topological frustration in the folding of CheY. J Mol Biol 2008, 382:485–495.
13. Gangadhara BN, Laine JM, Kathuria SV, Massi F, Matthews CR: Clusters of branched aliphatic side chains serve as cores of stability in the native state of the HisF TIM barrel protein. J Mol Biol 2013, 425:1065–1081.
14. Hills RD Jr, Kathuria SV, Wallace LA, Day IJ, Brooks CL III, Matthews CR: Topological frustration in βα-repeat proteins: sequence diversity modulates the conserved folding mechanisms of α/β/α sandwich proteins. J Mol Biol 2010, 398:332–350.
15. Haglund E, Lindberg MO, Oliveberg M: Changes of protein folding pathways by circular permutation. Overlapping nuclei promote global cooperativity. J Biol Chem 2008, 283:27904–27915.
16. Ternström T, Mayor U, Akke M, Oliveberg M: From snapshot to movie: phi analysis of protein folding transition states taken one step further. Proc Natl Acad Sci U S A 1999, 96:14854–14859.
17. Villegas V, Martínez JCJ, Avilés FXF, Serrano LL: Structure of the transition state in the folding process of human procarboxypeptidase A2 activation domain. J Mol Biol 1998, 283:1027–1036.
18. Bernstein FC, Koetzle TF, Williams GJ, Meyer EF, Brice MD, Rodgers JR, Kennard O, Shimanouchi T, Tasumi M: The Protein Data Bank: a computer-based archival file for macromolecular structures. J Mol Biol 1977, 112:535–542.
19. Humphrey W, Dalke A, Schulten K: VMD: Visual molecular dynamics. J Mol Graph 1996, 14:33–38.
20. Xi L, Li S, Liu H, Li J, Lei B, Yao X: Global and local prediction of protein folding rates based on sequence autocorrelation information. J Theor Biol 2010, 264:1159–1168.

21. Lin G, Wang Z, Xu D, Cheng J: **SeqRate: sequence-based protein folding type classification and rates prediction.** *BMC Bioinf* 2010, **11**:S1.

22. Guo J-X, Rao N: **Predicting protein folding rate from amino acid sequence.** *J Bioinform Comput Biol* 2011, **09**:1–13.

23. Gao J, Zhang T, Zhang H, Shen S, Ruan J, Kurgan L: **Accurate prediction of protein folding rates from sequence and sequence-derived residue flexibility and solvent accessibility.** *Proteins* 2010, **78**:2114–2130.

24. Ma B-G, Guo J-X, Zhang H-Y: **Direct correlation between proteins' folding rates and their amino acid compositions: an ab initio folding rate prediction.** *Proteins* 2006, **65**:362–372.

25. Chang L, Wang J, Wang W: **Composition-based effective chain length for prediction of protein folding rates.** *Phys Rev E* 2010, **82**:051930.

26. Huang J-T, Tian J: **Amino acid sequence predicts folding rate for middle-size two-state proteins.** *Proteins* 2006, **63**:551–554.

27. Huang JT, Xing DJ, Huang W: **Relationship between protein folding kinetics and amino acid properties.** *Amino Acids* 2012, **43**:567–572.

28. Calosci N, Chi CN, Richter B, Camilloni C, Engström A, Eklund L, Travaglini-Allocatelli C, Gianni S, Vendruscolo M, Jemth P: **Comparison of successive transition states for folding reveals alternative early folding pathways of two homologous proteins.** *Proc Natl Acad Sci U S A* 2008, **105**:19241–19246.

29. Fowler SB, Clarke J: **Mapping the folding pathway of an immunoglobulin domain: structural detail from Phi value analysis and movement of the transition state.** *Structure* 2001, **9**:355–366.

30. Murzin AG, Brenner SE, Hubbard T, Chothia C: **SCOP: a structural classification of proteins database for the investigation of sequences and structures.** *J Mol Biol* 1995, **247**:536–540.

31. Kikuchi T, Némethy G, Scheraga HA: **Prediction of the location of structural domains in globular proteins.** *J Protein Chem* 1988, **7**:427–471.

32. Kikuchi T: **Analysis of 3D structural differences in the IgG-binding domains based on the interresidue average-distance statistics.** *Amino Acids* 2008, **35**:541–549.

33. Nakajima S, Kikuchi T: **Analysis of the differences in the folding mechanisms of c-type lysozymes based on contact maps constructed with interresidue average distances.** *J Mol Model* 2007, **13**:587–594.

34. Ichimaru T, Kikuchi T: **Analysis of the differences in the folding kinetics of structurally homologous proteins based on predictions of the gross features of residue contacts.** *Proteins* 2003, **51**:515–530.

35. Kawai Y, Matsuoka M, Kikuchi T: **Analyses of protein sequences using inter-residue average distance statistics to study folding processes and the significance of their partial sequences.** *Protein Pept Lett* 2011, **18**:979–990.

36. Matsuoka M, Kabata M, Kawai Y, Kikuchi T: **Analyses of Sequences of (β/α) Barrel Proteins Based on the Inter-Residue Average Distance Statistics to Elucidate Folding Processes.** In *Chemical Biology*. Rijeka, Croatia: InTech; 2012:83–98.

37. Oubridge C, Ito N, Evans PR, Teo CH, Nagai K: **Crystal structure at 1.92 Å resolution of the RNA-binding domain of the U1A spliceosomal protein complexed with an RNA hairpin.** *Nature* 1994, **372**:432–438.

38. Jiménez MA, Villegas V, Santoro J, Serrano L, Vendrell J, Avilés FX, Rico M: **NMR solution structure of the activation domain of human procarboxypeptidase A2.** *Protein Sci* 2003, **12**:296–305.

39. Lindahl M, Svensson LA, Liljas A, Sedelnikova SE, Eliseikina IA, Fomenkova NP, Nevskaya N, Nikonov SV, Garber MB, Muranova TA: **Crystal structure of the ribosomal protein S6 from Thermus thermophilus.** *EMBO J* 1994, **13**:1249–1254.

40. Pastore A, Saudek V, Ramponi G, Williams RJ: **Three-dimensional structure of acylphosphatase. Refinement and structure analysis.** *J Mol Biol* 1992, **224**:427–440.

41. Fulton KF, Bate MA, Faux NG, Mahmood K, Betts C, Buckle AM: **Protein Folding Database (PFD 2.0): an online environment for the International Foldeomics Consortium.** *Nucleic Acids Res* 2007, **35**:D304–D307.

42. Chiti F, Taddei N, White PM, Bucciantini M, Magherini F, Stefani M, Dobson CM: **Mutational analysis of acylphosphatase suggests the importance of topology and contact order in protein folding.** *Nat Struct Biol* 1999, **6**:1005–1009.

43. Kikuchi T: **Decoding amino acid sequences of proteins using inter-residue average distance statistics to extract information on protein folding mechanisms.** In *Protein Folding*. Edited by Walters EC. New York, USA: Nova Science Publishers Inc; 2011:465–487.

44. Hills RD, Brooks CL: **Coevolution of function and the folding landscape: correlation with density of native contacts.** *Biophys J* 2008, **95**:L57–L59.

45. Miyazawa S, Jernigan RL: **Residue-residue potentials with a favorable contact pair term and an unfavorable high packing density term, for simulation and threading.** *J Mol Biol* 1996, **256**:623–644.

46. Skolnick J: **In quest of an empirical potential for protein structure prediction.** *Curr Opin Struct Biol* 2006, **16**:166–171.

47. Ishizuka Y, Kikuchi T: **Analysis of the local sequences of folding sites in sandwich proteins with inter-residue average distance statistics.** *Open Bioinform J* 2011, **5**:59–68.

48. Dasmeh P, Serohijos AWR, Kepp KP, Shakhnovich EI: **Positively selected sites in Cetacean myoglobins contribute to protein stability.** *PLoS Comput Biol* 2013, **9**:e1002929.

49. Mirny L, Shakhnovich E: **Evolutionary conservation of the folding nucleus.** *J Mol Biol* 2001, **308**:123–129.

50. Rorick MM, Wagner GP: **Protein structural modularity and robustness are associated with evolvability.** *Genome Biol Evol* 2011, **3**:456–475.

51. Liao H, Yeh W, Chiang D, Jernigan RL, Lustig B: **Protein sequence entropy is closely related to packing density and hydrophobicity.** *Protein Eng Des Sel* 2005, **18**:59–64.

52. Ting K-LH, Jernigan RL: **Identifying a folding nucleus for the lysozyme/alpha-lactalbumin family from sequence conservation clusters.** *J Mol Evol* 2002, **54**:425–436.

53. Ptitsyn OB, Ting KL: **Non-functional conserved residues in globins and their possible role as a folding nucleus.** *J Mol Biol* 1999, **291**:671–682.

54. Altschul SF, Gish W, Miller W, Myers EW, Lipman DJ: **Basic local alignment search tool.** *J Mol Biol* 1990, **215**:403–410.

55. Edgar RC: **MUSCLE: multiple sequence alignment with high accuracy and high throughput.** *Nucleic Acids Res* 2004, **32**:1792–1797.

56. Saitou N, Nei M: **The neighbor-joining method: a new method for reconstructing phylogenetic trees.** *Mol Biol Evol* 1987, **4**:406–425.

57. Yang Z: **PAML 4: phylogenetic analysis by maximum likelihood.** *Mol Biol Evol* 2007, **24**:1586–1591.

58. Jones DT, Taylor WR, Thornton JM: **The rapid generation of mutation data matrices from protein sequences.** *Comput Appl Biosci* 1992, **8**:275–282.

59. Shrake A, Rupley JA: **Environment and exposure to solvent of protein atoms. Lysozyme and insulin.** *J Mol Biol* 1973, **79**:351–371.

60. Matouschek A, Kellis JT Jr, Serrano L, Fersht AR: **Mapping the transition state and pathway of protein folding by protein engineering.** *Nature* 1989, **340**:122–126.

61. Hamill SJ, Steward A, Clarke J: **The folding of an immunoglobulin-like Greek key protein is defined by a common-core nucleus and regions constrained by topology.** *J Mol Biol* 2000, **297**:165–178.

62. Bemporad F: *Folding and aggregation studies in the acylphosphatase-like family.* Firenze, Italy: Firenze University Press; 2009.

63. Abkevich VI, Gutin AM, Shakhnovich EI: **Specific nucleus as the transition state for protein folding: evidence from the lattice model.** *Biochemistry* 1994, **33**:10026–10036.

64. Faísca PFN, Travasso RDM, Ball RC, Shakhnovich EI: **Identifying critical residues in protein folding: Insights from φ-value and P[sub fold] analysis.** *J Chem Phys* 2008, **129**:095108.

65. Faísca PFN: **The nucleation mechanism of protein folding: a survey of computer simulation studies.** *J Phys Condens Matter* 2009, **21**:373102.

66. Parzen E: **On estimation of a probability density function and mode.** *Ann Math Stat* 1962, **33**:1065–1076.

67. Lindberg MO, Oliveberg M: **Malleability of protein folding pathways: a simple reason for complex behaviour.** *Curr Opin Struct Biol* 2007, **17**:21–29.

68. Truong HH, Kim BL, Schafer NP, Wolynes PG: **Funneling and frustration in the energy landscapes of some designed and simplified proteins.** *J Chem Phys* 2013, **139**:121908.

69. Selvaraj S, Gromiha MM: **Importance of hydrophobic cluster formation through long-range contacts in the folding transition state of two-state proteins.** *Proteins* 2004, **55**:1023–1035.

70. Parrini C, Bemporad F, Baroncelli A, Gianni S, Travaglini-Allocatelli C, Kohn JE, Ramazzotti M, Chiti F, Taddei N: **The folding process of Acylphosphatase from Escherichia coli is remarkably accelerated by the presence of a disulfide bond.** *J Mol Biol* 2008, **379**:1107–1118.

71. Taddei N, Capanni C, Chiti F, Stefani M, Dobson CM, Ramponi G: **Folding and aggregation are selectively influenced by the conformational preferences of the alpha-helices of muscle acylphosphatase.** *J Biol Chem* 2001, **276**:37149–37154.

72. Taddei N, Chiti F, Fiaschi T, Bucciantini M, Capanni C, Stefani M, Serrano L, Dobson CM, Ramponi G: **Stabilisation of alpha-helices by site-directed**

mutagenesis reveals the importance of secondary structure in the transition state for acylphosphatase folding. *J Mol Biol* 2000, **300:**633–647.

73. Paci E, Vendruscolo M, Dobson CM, Karplus M: **Determination of a transition state at atomic resolution from protein engineering data.** *J Mol Biol* 2002, **324:**151–163.

74. Arad-Haase G, Chuartzman SG, Dagan S, Nevo R, Kouza M, Mai BK, Nguyen HT, Li MS, Reich Z: **Mechanical unfolding of acylphosphatase studied by single-molecule force spectroscopy and MD simulations.** *Biophys J* 2010, **99:**238–247.

75. Sugita M, Kikuchi T: **Analyses of the folding properties of ferredoxin-like fold proteins by means of a coarse-grained Gō model: relationship between the free energy profiles and folding cores.** *Proteins Struct Funct Bioinform* 2014, **82:**954–965.

76. Nakajima S, Alvarez-Salgado E, Kikuchi T, Arredondo-Peter R: **Prediction of folding pathway and kinetics among plant hemoglobins using an average distance map method.** *Proteins* 2005, **61:**500–506.

9

Crystal structure of the DNA polymerase III β subunit (β-clamp) from the extremophile *Deinococcus radiodurans*

Laila Niiranen[1], Kjersti Lian[1], Kenneth A Johnson[1] and Elin Moe[1,2*]

Abstract

Background: *Deinococcus radiodurans* is an extremely radiation and desiccation resistant bacterium which can tolerate radiation doses up to 5,000 Grays without losing viability. We are studying the role of DNA repair and replication proteins for this unusual phenotype by a structural biology approach. The DNA polymerase III β subunit (β-clamp) acts as a sliding clamp on DNA, promoting the binding and processivity of many DNA-acting proteins, and here we report the crystal structure of *D. radiodurans* β-clamp (*Dr*β-clamp) at 2.0 Å resolution.

Results: The sequence verification process revealed that at the time of the study the gene encoding *Dr*β-clamp was wrongly annotated in the genome database, encoding a protein of 393 instead of 362 amino acids. The short protein was successfully expressed, purified and used for crystallisation purposes in complex with Cy5-labeled DNA. The structure, which was obtained from blue crystals, shows a typical ring-shaped bacterial β-clamp formed of two monomers, each with three domains of identical topology, but with no visible DNA in electron density. A visualisation of the electrostatic surface potential reveals a highly negatively charged outer surface while the inner surface and the dimer forming interface have a more even charge distribution.

Conclusions: The structure of *Dr*β-clamp was determined to 2.0 Å resolution and shows an evenly distributed electrostatic surface charge on the DNA interacting side. We hypothesise that this charge distribution may facilitate efficient movement on encircled DNA and help ensure efficient DNA metabolism in *D. radiodurans* upon exposure to high doses of ionizing irradiation or desiccation.

Keywords: DNA polymerase III β subunit, *Deinococcus radiodurans*, Radiation resistance

Background

The bacterial DNA polymerase III β subunit (β-clamp), and the corresponding eukaryotic and archaeal proliferating cell nuclear antigen (PCNA) are ring-shaped proteins that encircle double-stranded DNA. They act as a processivity factor, or a sliding clamp, for a wide variety of proteins that act on DNA including DNA polymerases, DNA ligase, endonucleases and glycosylases (reviewed in [1]). For *Escherichia coli* DNA polymerase catalytic core the replication speed is increased from approximately 20 nt/s with frequent dissociation [2] to approximately 750 nt/s with a processivity of >50 kb in the presence of

the β-clamp [3]. To be loaded onto DNA, the sliding clamps need the help of ATP-dependent clamp loaders. Clamp loaders are multi-subunit complexes where ATP hydrolysis is coupled to conformational changes that enable the clamp loader to open the sliding clamp and place it on DNA [4]. Once loaded, the sliding clamp allows the binding of other polymerase subunits.

The crystal structure of a β-clamp was first determined for *E. coli* in 1992 [5], and after that for five other bacteria so far [6-8]. The structures show that the bacterial sliding clamp is a head-to-tail dimer [5], where one of the interfaces is opened by the clamp loader to allow DNA to enter the ring interior [9]. In eukaryotes, the PCNA clamp is also ring shaped but consist of a homotrimer [10] and in archaea such as *Sulfolobus solfataricus* a heterodimer [11]. There are also available two structures of a sliding clamp bound to DNA (*E. coli*, PDB code 3BEP [12] and

* Correspondence: elin.moe@uit.no
[1]The Norwegian Structural Biology Center (NorStruct), Department of Chemistry, UIT – the Arctic University of Norway, N-9037 Tromsø, Norway
[2]The Macromolecular Crystallography Unit, Instituto de Tecnologia Química e Biológica (ITQB), Universidade Nova de Lisboa, Oeiras 2780-157, Portugal

S. cerevisiae, PDB code 3K4X [13]). In spite of the different quaternary structures and a low sequence identity of the different clamp types [14], their overall shape and internal architecture with six similarly folded domains are strikingly similar also compared to bacteriophage clamps. Due to its central role in many DNA related cellular functions the β-clamp is an active target for inhibitor drug design in the development of new antibiotics to combat drug resistant strains [15-17].

In this paper, we describe the crystallographic structure of the DNA polymerase III β-clamp from the extremely radiation resistant bacterium *Deinococcus radiodurans* (*Dr*β-clamp). *D. radiodurans* exhibits an outstanding resistance to ionising radiation and desiccation and tolerates radiation doses up to 5,000 Gray (Gy) without loss of viability whereas most other organisms cannot survive doses above 50 Gy [18]. Such a massive radiation dose is estimated to induce several hundred double-strand breaks (DSB), thousands of single-strand gaps and about one thousand sites of DNA base damage per chromosome (reviewed by [19]). The overall structure of *Dr*β-clamp is similar to *E.coli* β-clamp (*Ec*β-clamp) and consists of a dimer which forms a ring lined by 12 α-helices, with an opening big enough to accommodate dsDNA. Each monomer consists of three domains (A, B, and C) with identical topology. *Dr*β-clamp displays a strong negative electrostatic surface potential on the outside of the ring, but a more even charge distribution inside the ring on the DNA binding surface and in the dimer forming interface. We hypothesise that the evenly distributed surface charge inside the ring helps ensure efficient clamp loading and DNA processivity which are needed to tackle the substantial amount of DNA metabolic processes that are activated upon exposure to high doses of irradiation and desiccation.

Results and discussion

Sequence analysis

During the initial part of this work we discovered that the *D. radiodurans* gene sequence *DR_0001* deposited in the GeneBank (Q9RYE8) was incorrectly annotated, encoding a protein of 393 instead of 362 amino acids. This was confirmed by sequence analysis and expression tests. The mistake was most likely caused by the automated gene recognition program used in the annotation of the sequenced genome. These programs can fail to recognise frame shifts caused by insertions or deletions (as demonstrated by [20]). Our discovery is in line with the findings of Baudet et al. [21] who showed that the original annotation of over a hundred *D. radiodurans* R1 genes is wrong and needs to be corrected. In 2014 the *D. radiodurans* R1 genome was re-annotated by the NCBI Ref Seq project, and the new version of *DR_0001* gene product (accession number WP_027480259.1 (GI:653293780),

published June 12th 2014) corresponds to our short version of the protein (except for the first Val). The reannotation confirms that we have been working with the biologically relevant version of the protein.

The short *Dr*β-clamp protein sequence shares over 70% identity with other *Deinococcus* β-clamp sequences, and 40 – 70% identity to sequences from other members of the phylum *Deinococcus-Thermus*. Interestingly, the sequence identity to other Gram-negative species is as low as to Gram-positive species, below 32%.

Overall structure

The crystal structure of *Dr*β-clamp was determined in space group P3$_2$21 to 2.0 Å resolution using molecular replacement. Despite our efforts, the DNA oligomer the *Dr*β-clamp was co-crystallised with was not clearly visible in the electron density, and could not be modelled into the structure. The asymmetric unit contains one *Dr*β-clamp dimer (residues 1 to 361 of chains A and B), with 292 solvent atoms. A schematic representation of the structural model is presented in Figure 1. The majority (97.7%) of the main-chain torsion angles were in the favoured regions of the Ramachandran plot, with the remaining 2.3% in the allowed regions. The final model had R_{work} and R_{free} values of 19.8% and 23.5%, respectively. The details of the data collection and refinement statistics are given in Table 1.

Like other bacterial β-clamps, the *Dr*β-clamp forms a head to tail dimer of two monomers, each monomer composed of three domains with identical α/β topology (Figure 1). The domains are slightly shifted in position when comparing the two dimers, so that in addition to NCS restrains between chains A and B we also used TLS refinement where each domain was defined as a separate group. The final RMSD between the chains is 0.9 Å and between the domains on average 1.8 Å. Compared to the *Ec*β-clamp the overall RMSD is 3.3 Å.

The shape of the *Dr*β-clamp dimer is slightly more elliptic than that of the circular *Ec*β-clamp, with an internal diameter of 32 to 41 Å. A comparison of *Dr*β-clamp core structure and surface contour (Figure 1, left panel) suggests that the elliptic shape of *Dr*β-clamp cavity is enhanced by the conformation of certain long side chains and loops. The positively charged residues on the ring inner surface (Figure 1, right panel) appear flexible and side chains not involved in inter- or intramolecular interactions display poor electron density (Figure 2). In *Ec*β-clamp many of the basic side chains inside the clamp ring have been found to display flexibility and become ordered first upon contact with DNA [12].

Surface potential

A comparison of the electrostatic surface potential of β-clamps from *D. radiodurans*, *E. coli*, *Mycobacterium*

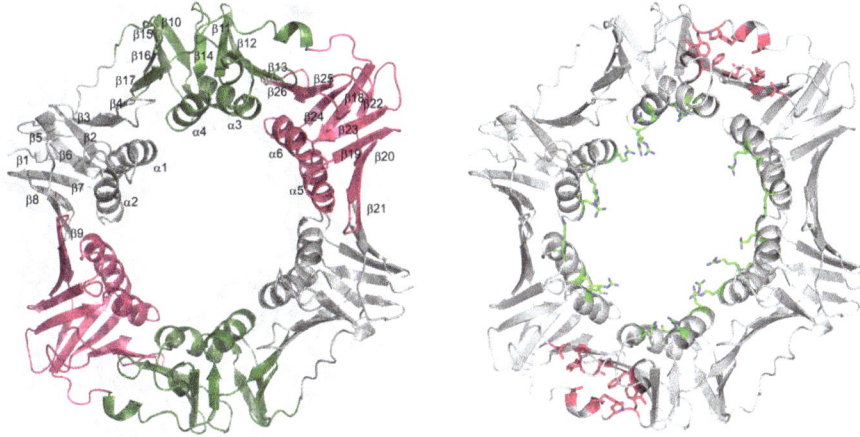

Figure 1 A schematic model of the *Dr*β-clamp dimer crystal structure. In the left panel the secondary structure elements are labelled in chain A, and the molecular surface is shown. Each monomer has three domains (A, B and C) with identical topology, here coloured grey, green and pink, respectively. The right panel shows stick models of the positively charged residues on the ring interior (green), and the residues of the hydrophobic pocket located between domains B and C (pink).

tuberculosis and *Thermotoga maritima* (Figure 3) reveals some interesting differences. All molecules have a more or less uniform negative charge on the outside of the ring, with this effect being strongest in *T. maritima* and weakest in *E. coli*. On the inside of the ring all molecules have positively charged residues for interacting with DNA. In *E. coli*, *M. tuberculosis* and *T. maritima* clamps the positive charge forms a relatively continuous band pattern across the surface whereas in *Dr*β-clamp the charge is more spread forming small positive patches separated by negatively charged areas. The surface charge distribution of *Dr*β-clamp may suggest that DNA binding is less tight or less specific compared to the other clamps. An even distribution of both positive and negative charge may facilitate efficient clamp sliding on DNA by hindering the formation of strong local interactions that might slow down the sliding movement.

The *Dr*β-clamp dimer interfaces have a spread, even charge distribution compared to the very strong positive (N-terminal domain) and negative charge (C-terminal domain) of the *Ec*β-clamp interfaces. The *M. tuberculosis* and *T. maritima* interfaces fall in between these two opposites, with *T. maritima* resembling more *Dr*β-clamp. In *T. maritima* β-clamp and *Dr*β-clamp electrostatic interactions may be less important in dimer formation and stability. Analysis of *Dr*- and *Ec*β-clamp interfaces with the Protein Interfaces, Surfaces and Assemblies service (PISA; [22]) shows that the *Ec*β-clamp interface is larger and has more hydrogen bonds and ionic interactions (24 and 7, respectively) compared to *Dr*β-clamp (15 and 2 interactions, respectively). The effect of dimer interface electrostatic interactions on clamp loading and dimer stability is currently not known.

The hydrophobicity pocket

A groove lined with hydrophobic residues called the hydrophobicity pocket or protein interaction pocket is located between domains B and C in *Ec*β-clamp. This pocket has been shown to serve as a ssDNA interaction site during clamp loading [12] in addition to being important for protein-protein interaction [23]. An analysis of the amino acid content (Figure 4) and structure (Figure 1, right panel) between domains B and C in *Dr*β-clamp suggests that this hydrophobicity pocket is present also in *Dr*β-clamp which thus may serve the same function as in *Ec*β-clamp.

DNA interacting residues

The positions of positively charged residues on the β-clamp inner surface appear to be only moderately conserved (Figure 4), indicating that DNA backbone positioning is not critical. The DNA complex structure of *Ec*β-clamp identified two residues inside the β-clamp ring, Arg24 and Gln149, which are important for dsDNA interaction and necessary for clamp loading [12]. In *Dr*β-clamp these correspond to Arg25 and Glu147 (Figure 2). The consequence of having a negatively charged residue in *Dr*β-clamp in the same position as Glu147 in *Ec*β-clamp is not known, since no protein-DNA interaction data is available at the moment. However, this residue is completely conserved among the sequenced *Deinococcus* species, and is found also in *M. tuberculosis* and *T. maritima* β-clamps, and probably serves a similar function in these organisms as Glu147 in *Ec*β-clamp.

Structural explanation for lack of DNA binding

An analysis of the *Ec*β-clamp-DNA complex suggests that the DNA in the structure is oriented in the opposite

Table 1 X-ray data collection and crystallographic refinement statistics for *Dr*β-clamp

Data collection	
X-ray source	ESRF ID29
Space group	$P3_221$
Unit cell (Å)	$a = b = 84.41$, $c = 198.74$
Resolution (Å)	30 – 2.00 (2.05 – 2.00)
Wavelength (Å)	0.9724
No. unique reflections	55 287
Multiplicity	3.3 (3.3)
Completeness (%)	98.0 (98.2)
Mean ($<I> / <\sigma_I>$)	17.4 (2.6)
R-sym (%)[a]	4.5 (52.8)
Wilson B-factor (Å^2)	40.2
Refinement	
PDB entry	4TRT
R-factor (all reflections) (%)	20.0
R-free (%)[b]	23.5
Number of atoms	5758
Number of water molecules	292
RMSD bond lengths (Å)	0.008
RMSD bond angles (°)	1.255
Average B-factor (Å^2)	
All atoms	46.4
Protein	46.2
Water	49.2
Ramachandran plot	
Favoured regions (%)	713 (97.7)
Allowed regions (%)	17 (2.3)
Outliers (%)	0 (0)

Values in parenthesis are for the highest resolution shell.
[a]$R - sym = (\Sigma_h\Sigma_i|I_i(h) - <I(h)>|) / (\Sigma_h\Sigma_i I(h))$, where $I_i(h)$ is the ith measurement of reflection h and $<I(h)>$ is the weighted mean of all measurements of h.
[b]5 % of the reflections were used in the R-free calculations.

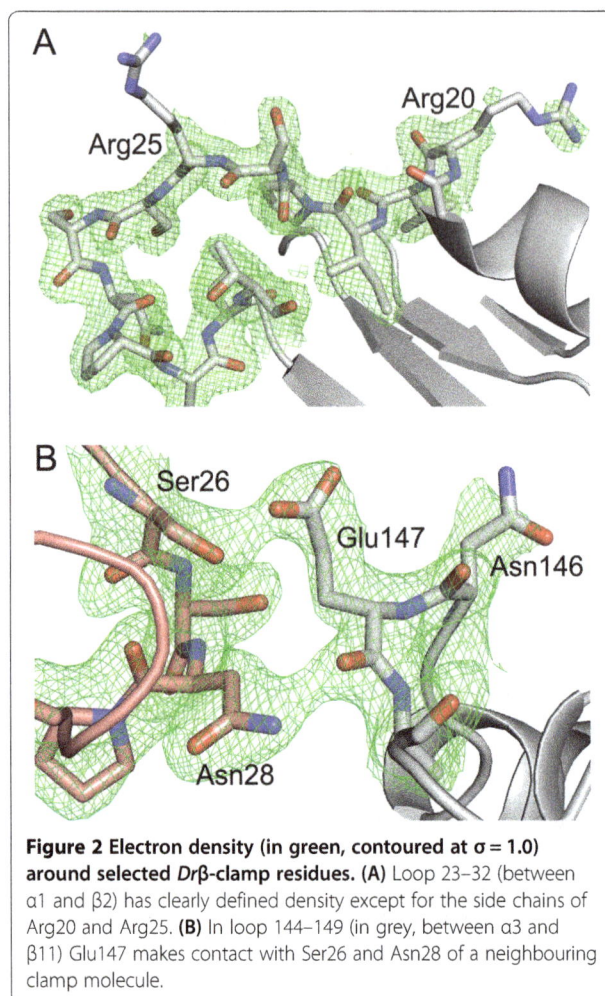

Figure 2 Electron density (in green, contoured at σ = 1.0) around selected *Dr*β-clamp residues. (A) Loop 23–32 (between α1 and β2) has clearly defined density except for the side chains of Arg20 and Arg25. **(B)** In loop 144–149 (in grey, between α3 and β11) Glu147 makes contact with Ser26 and Asn28 of a neighbouring clamp molecule.

direction from what could be expected. This is caused by an interaction between the 5′ end of the DNA (4 T) and the hydrophobicity pocket of a neighbouring clamp molecule. Although this interaction is biologically relevant, it should normally occur on the same molecule which the DNA is inside. The strong interaction with the hydrophobicity pocket is caused by stacking interactions of the third thymidine and the first adenine base of the oligomer with two tyrosines (Tyr153 and Tyr154) in the pocket [12]. In *Dr*β-clamp these tyrosines are substituted by Ala151 and Val152, which may explain why the DNA is bound in a too disorganised fashion to be clearly visible in the electron density. The electron density for DNA is weak also in the eukaryotic complex structure [13], which may indicate that the true DNA binding mode is too unspecific and flexible to be well recorded by crystallographic methods. Co-crystallization with different length and type oligonucleotides should be tested to screen for more specific binding enabling DNA visualization.

Conclusion

We have determined the crystal structure of *Dr*β-clamp to 2.0 Å resolution. The protein is a ring shaped dimer with a head-to-tail orientation and is similar to previously determined structures of bacterial β-clamps. Based on the observation of an even charge distribution inside the ring we hypothesise that the protein is optimised for efficient function during high turnover of DNA metabolic processes.

Methods

Cloning, expression and purification

The gene encoding *Dr*β-clamp (*DR_0001*) was cloned from genomic DNA of *D. radiodurans* strain R1 into expression vector pDEST14 (Invitrogen). All primers used in cloning are listed in Table 2. We used a two-step

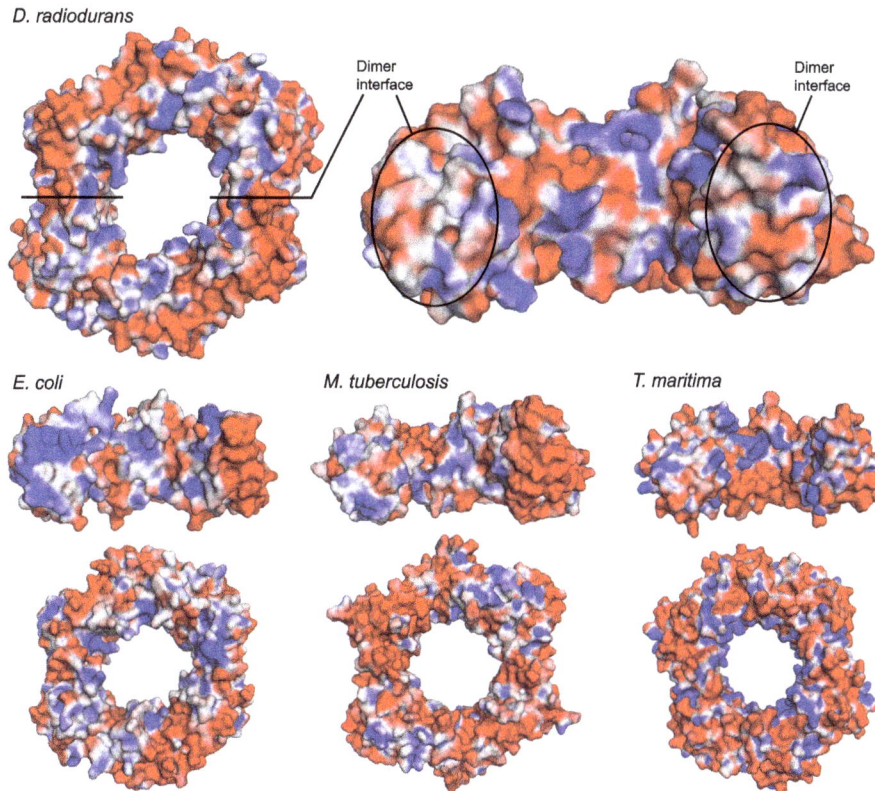

Figure 3 Electrostatic surface potential of β-clamps from *D. radiodurans* (this work), *E. coli* (PDB code 1POL), *M. tuberculosis* (PDB code 3P16) and *T. maritima* (PDB code 1VPK). The dimer molecules are depicted with the C-terminal protein-interacting side facing up (dimer), and the monomers showing the inside of the half-ring and the dimer interfaces. The surface is coloured according to the electrostatic potential at 298 K (−8 to 8 kT/e) with negative potential in red and positive potential in blue.

Gateway method with gene specific primers Fw1 and Rev1 which introduced a TEV-cleavable His$_7$-tag to the C-terminus of the protein, and extension primers *att*B1 and *att*B2 that contained the rest of the *att*-sites. During sequence verification it was discovered that the cloned gene contained a deletion of cytosine 1039 compared to the published genomic sequence of *D. radiodurans* strain R1 [25]. Bioinformatic analysis of the gene and protein sequence of *Dr*β-clamp showed that the absence of cytosine 1039 leads to a frameshift and an earlier stop codon producing a shorter version of the gene (1089 nt compared to the 1182 nt GeneBank version) which encodes a 362 aa protein more similar both in length and in sequence to other known β-clamps (Figure 4) than the longer version. We cloned both the short and the long version of the gene to see if they could both be expressed. We recreated the long version by reintroducing the missing cytosine 1039 to the already cloned, deletion containing gene by site-directed mutagenesis using the QuikChange kit from Stratagene (primers Ins-fw and Ins-rev, Table 2). The short version was cloned from genomic DNA using the same procedure

as for the long version, except that the primers were Fw1 and Rev2.

Only the short version of *Dr*β-clamp could be successfully expressed and was verified by MS peptide fingerprinting. The *Dr*β-clamp was expressed in *E. coli* BL21 (DE3)Star pLysS pRARE (Invitrogen) with 0.5 mM IPTG induction overnight at 293 K. The cells were suspended in 50 mM Tris, 150 mM NaCl, pH 7.5, and disrupted by sonication followed by centrifugation at 20 000xg for 25 min, 277 K. The protein was purified with affinity and ion exchange chromatography (HisTrap HP and HiTrap Q columns from GE Healthcare), and the His$_7$-tag was cleaved off by incubating the protein with 1/50 w/w TEV protease with 1 mM DTT and 0.5 mM EDTA added. Unprocessed protein was removed by using a HisTrap HP column in flow through mode, and an Amicon Ultra Centrifugal Filter (10 000 MWCO, Millipore) was used for concentration of the protein and buffer exchange to the original conditions. For size and tertiary structure analysis *Dr*β-clamp was run on Superdex 75 10/30 (GE Healthcare) where it behaved as a dimer of approximately 80 kDa.

Figure 4 sequence alignment (row labels, top to bottom): Dr-clampShort, Mt-clamp, Spn-clamp, Spy-clamp, Tm-clamp, Ec-clamp

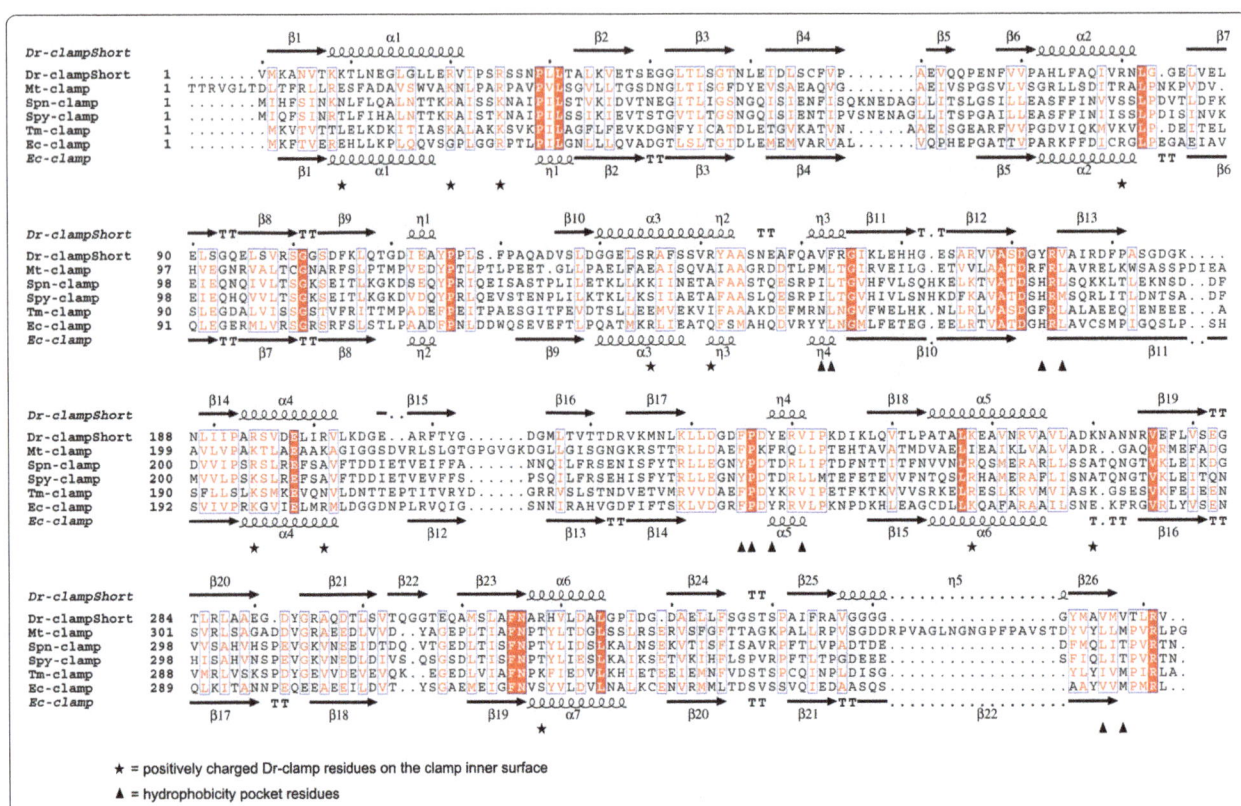

Figure 4 Sequence alignment of the bacterial β-clamps with determined crystal structures. *Dr*β-clamp short, *Mycobacterium tuberculosis* (*Mt*β-clamp; Gui et al., [7]), *Streptococcus pneumoniae* clamp (*Spn*β-clamp; unpublished, PDB code 2AWA), *Streptococcus pyogenes* clamp (*Spy*β-clamp; Argiriadi et al., [6]), *Thermotoga maritima* clamp (*Tm*β-clamp; unpublished, PDB code 1VPK), and *Ec*β-clamp (Kong et al., [5]). Stars denote the positively charged *Dr*β-clamp residues facing the inside of the ring, and triangles indicate hydrophobicity pocket residues. The figure was prepared with ESPript 3.0 [24].

★ = positively charged Dr-clamp residues on the clamp inner surface

▲ = hydrophobicity pocket residues

Oligonucleotide annealing and purification

We obtained from Sigma-Genosys two unlabelled DNA oligos (5′-TTTT ATACGATGGG, 5′-TTTTTT ATACG ATGGG) and one Cy5-labeled oligo (5′-Cy5-CCCAT CGTAT) in order to create two different Cy-5 labelled double-stranded oligos with 10 nt double-stranded region and either 4 or 6 nt long single-stranded thymidine overhang (4 T and 6 T). This method was previously used successfully by both [12] and [13]. The oligos were dissolved in 50 mM HEPES, 50 mM NaCl and 0.5 mM EDTA, pH 8.0, the labelled oligos mixed with a slight excess of unlabelled oligos (1:1.1), and annealed by placing in a 343 K heat-block and allowed to cool to room temperature overnight. The annealed oligos were purified with a GE Healthcare Mono Q HR 5/5 column, concentrated with an Amicon 4 ml 3000 MWCO concentrator (Millipore) and dialysed overnight at 277 K in Pierce Slide-A-Lyzer MINI dialysis tubes (MWCO 2 000) against 50 mM HEPES pH 8.0. Oligomer concentration was measured with a NanoDrop 2000c spectrophotometer (NanoDrop Technologies) at 650 nm using the extinction coefficient for Cy5 (2.5×10^5 M^{-1} cm^{-1}).

Table 2 Cloning primers used in this work

Name	Sequence
Fw1	5′-AAAAAGCAGGCTTCGAAGGAGATAGAACC ATG**GTGATGAAAGCCAAT GTCACC**
Rev1	5′-AGAAAGCTGGGTCTTAGTGATGGTGATGGT GATGTCCCTGGAAATACA GGTTTTC**CGCGAA CTCTGGCCTCGGTTC**
Rev2	5′-AGAAAGCTGGGTCTTAGTGATGGTGATGGT GATGTCCCTGGAAATACA GGTTTTC**AACGCG CAGCGTGACCATGACC**
attB1 adapter	5′-GGGGACAAGTTTGTACAAAAAAGCAGGCT
attB2 adapter	5′-GGGGACCACTTTGTACAAGAAAGCTGGGT
Ins-fw	5′-CATTTTCCGCGCC**C**GTAGGTGGGGGA
Ins-rev	5′-TCCCCCACCTAC**G**GGCGCGGAAAATG

Gene specific parts are in bold, except in mutagenesis primers where bold indicates the inserted base.

Crystallisation and data collection

For crystallisation, pure *Dr*β-clamp was mixed with purified Cy5-labeled 4 T/6 T oligos to give a solution of 5 mg/ml protein with 1:1.1 protein:DNA ratio. Crystallisation was done in sitting-drop format using a Crystal Phoenix liquid dispenser (Art Robbins Instruments) with

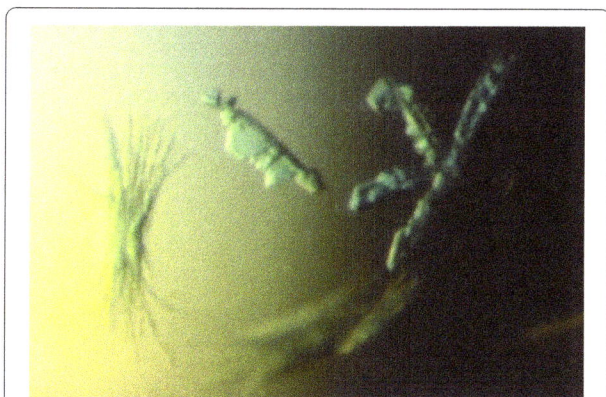

Figure 5 An image of the blue *Drβ*-clamp crystals obtained with 4 T-DNA oligo. Pure *Drβ*-clamp was mixed with purified Cy5-labeled 4 T oligos in a 1:1.1 protein:DNA ratio. Crystallization in 21% PEG 5000 MME, 0.12 M Tris, 3.6% hexanediol, pH 8.0, using the 4 T oligo containing protein solution yielded blue needles and clusters of small blue crystals.

MRC-2 plates (Molecular Dimensions). Home-made stochastic screens were used for initial screening. Clusters of small blue crystals (Figure 5) were obtained in conditions with 21% PEG 5000 MME, 0.12 M Tris, 3.6% hexanediol, pH 8.0, using the 4 T oligo containing protein solution. A strong blue colour indicated that the crystals contained DNA. Single crystals were separated from the clusters, transferred into a cryo solution containing 23% PEG 5000 MME, 0.15 M Tris, 4% hexanediol, and 19% glycerol, pH 8.0, and immediately flash-frozen in liquid nitrogen. Diffraction data were collected on a single crystal at the European Synchrotron Radiation Facility (ESRF), Grenoble, France, at beamline ID-29 equipped with a Pilatus 6 M detector [26] at 100 K to 2.0 Å.

Structure determination and refinement

The data were indexed, integrated, scaled and converted to structure factors using the XDS program package [27]. The space group P3$_2$21 was chosen after analysis with the program Pointless [28]. The unit cell dimensions are a = b = 84.66 Å and c = 199.08 Å. We found two monomers (one β-clamp dimer) in the asymmetric unit, giving a solvent content of 53.8% and a Matthews coefficient of 2.66 Å^2/Da. The structure was solved by molecular replacement using MOLREP [29] in CCP4 [30] with the *Ecβ*-clamp (PDB code 2POL, 29% identity) as a search model. Because of the relatively low identity between the model and *Drβ*-clamp, ARP/wARP tracing [31] was used to find the correct backbone position. The structure was refined in REFMAC5 [32] using TLS refinement where each of the three domains of a monomer was defined as a group, and automatically determined NCS restraints (chain A to B). Inspection and manual building of the model between refinement runs was done using

Coot [33], and water molecules were added using Coot findwaters. During the refinement it became obvious that although the crystals contained DNA, the observed electron density for it (in the central cavity of the β-clamp ring) was too weak to support the placement of any nucleotides in the model.

Structure analysis

The structural model quality, geometry and fit to electron density were evaluated using Coot tools, and finally validated with the program MolProbity [34]. Structural images were drawn with PyMol (http://www.pymol.org/) and the APBS plugin [35] was used for calculation of the electrostatic surface potentials.

Abbreviations

β-clamp: DNA polymerase III β subunit; *Drβ*-clamp: *D. radiodurans* β-clamp; PCNA: Proliferating cell nuclear antigen; *Ecβ*-clamp: *E.coli* β-clamp; ESRF: European Synchrotron Radiation Facility; RMSD: Root mean square deviation.

Competing interests

The authors declare that they have no competing interests.

Authors' contributions

LN cloned the constructs described in this study, expressed, purified, crystallised and participated in data processing, structure determination and refinement of the structure, and drafted the manuscript. KL performed the initial cloning, sequencing and expression tests of *Drβ*-clamp, and participated in the crystallisation experiments. KAJ participated in the crystallisation experiments, processed the data, determined the structure and participated in refinement of the structure. EM conceived and designed the study, coordinated and helped to draft the manuscript. All authors read and approved the final manuscript.

Acknowledgements

Provision of beamtime at the European Synchrotron Radiation Source ESRF beamline ID29 is gratefully acknowledged. This work was supported by the Research Council of Norway, the National Functional Genomics Program (FUGE).

References

1. Warbrick E. The puzzle of PCNA's many partners. Bioessays. 2000;22(11):997–1006.
2. Maki H, Kornberg A. The polymerase subunit of DNA polymerase III of Escherichia coli. II. Purification of the alpha subunit, devoid of nuclease activities. J Biol Chem. 1985;260(24):12987–92.
3. O'Donnell ME, Kornberg A. Dynamics of DNA polymerase III holoenzyme of Escherichia coli in replication of a multiprimed template. J Biol Chem. 1985;260(23):12875–83.
4. Indiani C, O'Donnell M. The replication clamp-loading machine at work in the three domains of life. Nat Rev Mol Cell Biol. 2006;7(10):751–61.
5. Kong XP, Onrust R, O'Donnell M, Kuriyan J. Three-dimensional structure of the beta subunit of E. coli DNA polymerase III holoenzyme: a sliding DNA clamp. Cell. 1992;69(3):425–37.
6. Argiriadi MA, Goedken ER, Bruck I, O'Donnell M, Kuriyan J. Crystal structure of a DNA polymerase sliding clamp from a Gram-positive bacterium. BMC Struct Biol. 2006;6:2.
7. Gui WJ, Lin SQ, Chen YY, Zhang XE, Bi LJ, Jiang T. Crystal structure of DNA polymerase III beta sliding clamp from Mycobacterium tuberculosis. Biochem Biophys Res Commun. 2011;405(2):272–7.
8. Wolff P, Amal I, Olieric V, Chaloin O, Gygli G, Ennifar E, et al. Differential modes of peptide binding onto replicative sliding clamps from various bacterial origins. J Med Chem. 2014;57(18):7565–76.

9. Stewart J, Hingorani MM, Kelman Z, O'Donnell M. Mechanism of beta clamp opening by the delta subunit of Escherichia coli DNA polymerase III holoenzyme. J Biol Chem. 2001;276(22):19182–9.

10. Krishna TS, Fenyo D, Kong XP, Gary S, Chait BT, Burgers P, et al. Crystallization of proliferating cell nuclear antigen (PCNA) from Saccharomyces cerevisiae. J Mol Biol. 1994;241(2):265–8.

11. Williams GJ, Johnson K, Rudolf J, McMahon SA, Carter L, Oke M, et al. Structure of the heterotrimeric PCNA from Sulfolobus solfataricus. Acta Crystallogr Sect F: Struct Biol Cryst Commun. 2006;62(Pt 10):944–8.

12. Georgescu RE, Kim SS, Yurieva O, Kuriyan J, Kong XP, O'Donnell M. Structure of a sliding clamp on DNA. Cell. 2008;132(1):43–54.

13. McNally R, Bowman GD, Goedken ER, O'Donnell M, Kuriyan J. Analysis of the role of PCNA-DNA contacts during clamp loading. BMC Struct Biol. 2010;10:3.

14. Neuwald AF. Evolutionary clues to DNA polymerase III beta clamp structural mechanisms. Nucleic Acids Res. 2003;31(15):4503–16.

15. Wijffels G, Johnson WM, Oakley AJ, Turner K, Epa VC, Briscoe SJ, et al. Binding inhibitors of the bacterial sliding clamp by design. J Med Chem. 2011;54(13):4831–8.

16. Wolff P, Olieric V, Briand JP, Chaloin O, Dejaegere A, Dumas P, et al. Structure-based design of short peptide ligands binding onto the E. coli processivity ring. J Med Chem. 2011;54(13):4627–37.

17. Yin Z, Kelso MJ, Beck JL, Oakley AJ. Structural and thermodynamic dissection of linear motif recognition by the E. coli sliding clamp. J Med Chem. 2013;56(21):8665–73.

18. Mattimore V, Battista JR. Radioresistance of Deinococcus radiodurans: functions necessary to survive ionizing radiation are also necessary to survive prolonged desiccation. J Bacteriol. 1996;178(3):633–7.

19. Krisko A, Radman M. Biology of extreme radiation resistance: the way of Deinococcus radiodurans. Cold Spring Harb Perspect Biol. 2013;5(7).

20. Baytaluk MV, Gelfand MS, Mironov AA. Exact mapping of prokaryotic gene starts. Brief Bioinform. 2002;3(2):181–94.

21. Baudet M, Ortet P, Gaillard JC, Fernandez B, Guerin P, Enjalbal C, et al. Proteomics-based refinement of Deinococcus deserti genome annotation reveals an unwonted use of non-canonical translation initiation codons. Mol Cell Proteomics. 2010;9(2):415–26.

22. Krissinel E, Henrick K. Inference of macromolecular assemblies from crystalline state. J Mol Biol. 2007;372(3):774–97.

23. Burnouf DY, Olieric V, Wagner J, Fujii S, Reinbolt J, Fuchs RP, et al. Structural and biochemical analysis of sliding clamp/ligand interactions suggest a competition between replicative and translesion DNA polymerases. J Mol Biol. 2004;335(5):1187–97.

24. Robert X, Gouet P. Deciphering key features in protein structures with the new ENDscript server. Nucleic Acids Res. 2014;42(Web Server issue):W320–4.

25. White O, Eisen JA, Heidelberg JF, Hickey EK, Peterson JD, Dodson RJ, et al. Genome sequence of the radioresistant bacterium Deinococcus radiodurans R1. Science. 1999;286(5444):1571–7.

26. de Sanctis D, Beteva A, Caserotto H, Dobias F, Gabadinho J, Giraud T, et al. ID29: a high-intensity highly automated ESRF beamline for macromolecular crystallography experiments exploiting anomalous scattering. J Synchrotron Radiat. 2012;19(Pt 3):455–61.

27. Kabsch W. Xds. Acta Crystallogr D Biol Crystallogr. 2010;66(Pt 2):125–32.

28. Evans P. Scaling and assessment of data quality. Acta Crystallogr D Biol Crystallogr. 2006;62(Pt 1):72–82.

29. Vagin A, Teplyakov A. MOLREP: an automated program for molecular replacement. J Appl Crystallogr. 1997;30:1022–5.

30. Winn MD, Ballard CC, Cowtan KD, Dodson EJ, Emsley P, Evans PR, et al. Overview of the CCP4 suite and current developments. Acta Crystallogr D Biol Crystallogr. 2011;67(Pt 4):235–42.

31. Langer G, Cohen SX, Lamzin VS, Perrakis A. Automated macromolecular model building for X-ray crystallography using ARP/wARP version 7. Nat Protoc. 2008;3(7):1171–9.

32. Murshudov GN, Vagin AA, Dodson EJ. Refinement of macromolecular structures by the maximum-likelihood method. Acta Crystallogr D Biol Crystallogr. 1997;53(Pt 3):240–55.

33. Emsley P, Cowtan K. Coot: model-building tools for molecular graphics. Acta Crystallogr D Biol Crystallogr. 2004;60(Pt 12 Pt 1):2126–32.

34. Chen VB, Arendall 3rd WB, Headd JJ, Keedy DA, Immormino RM, Kapral GJ, et al. MolProbity: all-atom structure validation for macromolecular crystallography. Acta Crystallogr D Biol Crystallogr. 2010;66(Pt 1):12–21.

35. Baker NA, Sept D, Joseph S, Holst MJ, McCammon JA. Electrostatics of nanosystems: application to microtubules and the ribosome. Proc Natl Acad Sci U S A. 2001;98(18):10037–41.

10

A Sco protein among the hypothetical proteins of *Bacillus lehensis* G1: Its 3D macromolecular structure and association with Cytochrome C Oxidase

Soo Huei Tan[1], Yahaya M Normi[1*], Adam Thean Chor Leow[1], Abu Bakar Salleh[1], Roghayeh Abedi Karjiban[1,2], Abdul Munir Abdul Murad[3], Nor Muhammad Mahadi[4] and Mohd Basyaruddin Abdul Rahman[1,2,4]

Abstract

Background: At least a quarter of any complete genome encodes for hypothetical proteins (HPs) which are largely non-similar to other known, well-characterized proteins. Predicting and solving their structures and functions is imperative to aid understanding of any given organism as a complete biological system. The present study highlights the primary effort to classify and cluster 1202 HPs of *Bacillus lehensis* G1 alkaliphile to serve as a platform to mine and select specific HP(s) to be studied further in greater detail.

Results: All HPs of *B. lehensis* G1 were grouped according to their predicted functions based on the presence of functional domains in their sequences. From the metal-binding group of HPs of the cluster, an HP termed Bleg1_2507 was discovered to contain a thioredoxin (Trx) domain and highly-conserved metal-binding ligands represented by Cys69, Cys73 and His159, similar to all prokaryotic and eukaryotic Sco proteins. The built 3D structure of Bleg1_2507 showed that it shared the $\beta\alpha\beta\alpha\beta\beta$ core structure of Trx-like proteins as well as three flanking β-sheets, a 3_{10} –helix at the N-terminus and a hairpin structure unique to Sco proteins. Docking simulations provided an interesting view of Bleg1_2507 in association with its putative cytochrome c oxidase subunit II (COXII) redox partner, Bleg1_2337, where the latter can be seen to hold its partner in an embrace, facilitated by hydrophobic and ionic interactions between the proteins. Although Bleg1_2507 shares relatively low sequence identity (47%) to BsSco, interestingly, the predicted metal-binding residues of Bleg1_2507 i.e. Cys-69, Cys-73 and His-159 were located at flexible active loops similar to other Sco proteins across biological taxa. This highlights structural conservation of Sco despite their various functions in prokaryotes and eukaryotes.

Conclusions: We propose that HP Bleg1_2507 is a Sco protein which is able to interact with COXII, its redox partner and therefore, may possess metallochaperone and redox functions similar to other documented bacterial Sco proteins. It is hoped that this scientific effort will help to spur the search for other physiologically relevant proteins among the so-called "orphan" proteins of any given organism.

Keywords: Hypothetical proteins, Bleg1_2507, Sco, Thioredoxin, Copper binding, Redox reaction, Cytochrome c oxidase

* Correspondence: normi_yahaya@upm.edu.my
[1]Center for Enzyme and Microbial Biotechnology (EMTECH), Faculty of Biotechnology and Biomolecular Sciences, Universiti Putra Malaysia, Serdang, Selangor 43400, Malaysia
Full list of author information is available at the end of the article

Background

Hypothetical proteins (HPs) are generally known as proteins of unknown structures and functions [1]. They constitute approximately 25% of any sequenced genome even in simple, model organisms such as *Escherichia coli*, *Bacillus subtilis* and *Saccharomyces cerevisiae* [2]. Annotating their biological function has remained a challenge due to their low sequence and structural homology to other known proteins [3]. This in turn creates difficulties in attempts to understand a biological system as a whole due to the information gaps posed by these "orphan" proteins. Therefore, efforts in trying to unravel their structures and functions are crucial in filling these information gaps for any particular system and hence, should be encouraged. To deepen our understanding on these proteins, *in silico* methods of analysis are the most time and cost effective in generating a wealth of reliable and useful information on HPs ranging from their localization, properties to their possible structures and functions. It also provides a quick means to preliminary screen and chooses potential HPs for subsequent downstream *in vitro* and *in vivo* experiments, particularly to initiate structural genomic projects in the future [4].

With the wealth of information which can be generated using the *in silico* approach, this present study highlights the utilization of various *in silico* tools and methods to functionally predict and cluster all HPs present in the complete genome of a locally isolated alkaliphile, *Bacillus lehensis* G1, with the purpose of establishing a platform to mine and later select specific HPs to be studied further in greater detail. Whilst the genome of this alkaliphile has been completely sequenced [5], the structural and functional omics of this extremophile is not well characterised, particularly its HPs. Hence, our attempt at firstly predicting their functions and clustering them accordingly will serve as a platform in mining and selecting specific HPs to be studied further in greater detail. Stemming from this effort, our attention was led to a particular HP, Bleg1_2507, in the metalloprotein category of the cluster. This HP contained a Sco1 domain of the Thioredoxin (Trx) superfamily and showed up to 50% of sequence identity to other bacterial HPs and 47% to bacterial Sco proteins.

It is important to note that Sco proteins are present in both prokaryotic and eukaryotic organisms and are required for the proper assembly of cytochrome c oxidase (COX), a terminal enzyme in the respiratory chain [6]. Sco was suggested to be involved in the delivery of copper ion to COX complex [7–9]. To date, the only bacterial Sco1 protein which was structurally studied and analysed at length is from *Bacillus subtilis* [6] while its eukaryotic counterpart is from yeast [10], human [11] and plant [12]. Sco proteins have garnered importance in recent years due to their roles in the correct assembly of the copper center (CuA) of COX subunit II (COX II).

Improper assembly of COX has been reported to cause fatal infantile encephalopathy due to mutations in *Sco1* and *Sco2* genes in humans [11,13].

Due to the physiological relevance of Sco proteins, we embarked on the task of building the structure of HP Bleg1_2507 of *B. lehensis* G1 via homology modelling to investigate the structural similarities and differences between this protein with Sco1 protein from *B. subtilis* (BsSco), yeast and human. As Sco1 was suggested to be involved in the delivery of copper ion to COX complex, specifically COXII [7–9], we subsequently mined the genome of *B. lehensis* G1 for the sequence encoding COXII by performing a BLASTP search. As a result, Bleg1_2337 was retrieved and its structure was built via homology modelling. Docking of both HP Bleg1_2507 and Bleg1_2337 models was performed to investigate their possible interaction. Based on the results obtained, the possible structure, function and mechanism of HP Bleg1_2507 from *B. lehensis* G1 are duly discussed. Lastly, the structure of HP Bleg1_2507 of *B. lehensis* G1 was compared to other Sco proteins to highlight its similarity and distinctness.

Methods

Domain and sequence similarity analysis of HPs of *B. lehensis* G1: Development of an HP cluster

All sequences encoding HPs in the genome of *B. lehensis* G1 were firstly subjected to INTERPROSCAN [14] and Conserved Domain Search (NCBI-CDD) [15] analyses to detect possible functional domains within the sequences. Sequence similarity of all the HPs with other proteins in the database was investigated also using BLASTP [16]. Based on the predicted functional domains and sequence similarity, an HP cluster was built to categorize the proteins accordingly. A scan on HPs which showed acceptable similarity to other proteins with low e-value was carried out and the potential candidate was chosen for further analyses and structure prediction.

Sequence analysis of selected HP (Bleg1_2507) and putative Cytochrome c Oxidase subunit II, COXII (Bleg1_2337)

From the above scan, a particular HP in the metal-binding group of proteins encoded by *bleg1_2507* gene was chosen for further analyses based on the presence of the conserved copper chaperone Sco1 domain related to Trx-like superfamily. Firstly, the possible presence of consensus amino acids pattern in Bleg1_2507 sequence was identified using ScanProsite [17]. Important and metal-binding residues were subsequently identified using Consurf [18] and MetalDetector v 2.0 [19] respectively. The physicochemical aspect of the protein such as its theoretical pI value was investigated using ProtParam [20]. SigCleave from EMBOSS [21] was used to investigate the possible presence of signal peptide in Bleg1_2507, while the evolutionary relatedness of the protein with other known proteins was investigated using PHYLIP [22].

Subsequently, a genome-wide scan for Bleg1_2507 redox partner i.e. cytochrome c oxidase subunit II (COX II), was performed. As a result, Bleg1_2337, a putative COXII protein was identified from the genome sequence of *B. lehensis* G1. Further analyses similar to the ones mentioned above were performed on Bleg1_2337.

Homology modeling of HP Bleg1_2507 and Bleg1_2337
PSI-BLAST [16] search against Protein Data Bank (PDB) was performed to retrieve potential templates for Bleg1_2507 and Bleg1_2337 model construction. Subsequently, Multiple Sequence Alignment (MSA) of the templates with the query sequences was performed using ClustalW [23] to determine the degree of similarity and conservation of specific motifs and amino acids in the sequences. 3D models of both proteins were developed via homology modeling using MODELLER 9v10 [24]. The best built models for Bleg1_2507 and Bleg1_2337 were chosen based on their lowest Discrete Optimized Protein Energy (DOPE) values and GA 341 score of one, which signify that the models resemble the native structure and hence, reliable. The generated models were visualized by PYMOL [25]. To aid assessment and ease discussion on the models, a color gradient scheme representing different levels of hydrophobicity of the amino acid side chains [26] were utilized. Further confirmation on the validity of the models was made based on their Root Mean Square Deviation (RMSD) calculated by superimposing them with their respective templates using Chimera USCF 1.6.1 [27].

Model refinement and validation
Refinement of the built models was performed using FoldX [28] (RepairPDB) to remove Van Der Waals clashes and bad contacts of amino acid side chains. Subsequently, structural evaluation and stereochemical analyses of the refined models were performed using ERRAT which sets a 95% confidence limit being the cut-off value to evaluate any incorrect residues present in the protein structure based on the average of six atomic interactions in the protein i.e. f(CC), f(CN), f(CO), f(NN), f(NO) and f(OO) [29]. Further supporting analysis to evaluate the models was performed using PROCHECK Ramachandran plot [30].

Docking of Bleg1_2507 and Bleg1_2337 models and energy minimization of protein complex
Both of the models were docked by using Cluspro v 2.0 [31] to predict protein interaction. Subsequently, the docked protein complex was refined using AMBER force in YASARA [32].

Results and discussion
Domain and sequence similarity analyses and clustering of HPs of *B. lehensis* G1
Based on the BleG1DB v1.0 complete genome of the locally isolated alkaliphilic *B. lehensis* G1 (released Aug 24th, 2012), the 3.5 Mbp genome consisted of 4021 predicted genes, in which 2819 of them encode proteins with putative functions while 1202 of them encode uncharacterized, hypothetical proteins [http://27.126.156.144/]. INTERPROSCAN analysis on all the HPs revealed that 51.5% of them were predicted with functions, while 48.5% were of unknown functions and hence categorized as unknown proteins (Table 1). In the category of proteins with predicted functions, majority of the HPs (286 in total) were signal peptides, in which 221 of them were transmembrane peptides. Other than these signal peptides, there were other 47 transmembrane proteins as well. The second most abundant type of HPs (totaling up to 115) was predicted to have enzymatic functions. The remaining numbers of HPs were predicted to be regulatory proteins (52), metal-binding proteins (43) and antibiotic-related (12) (Table 1).

Bleg1_2507 and Bleg1_2337 (putative COX II) Sequence Analyses
Within the pool of metal-binding proteins, HP Bleg1_2507 which consists of 198 amino acids was discovered to possess a copper chaperone Sco1 domain related to Trx-like superfamily from both INTERPROSCAN and NCBI-CDD analyses (data not shown).

Sequence similarity analysis of the HP using BLASTP indicated that Bleg1_2507 is similar to HPs of other *Bacillus* species with 40-50% identities (Table 2). Additionally in the top 10 hits of the BLASTP analysis, results

Table 1 Number of hypothetical proteins based on predicted functions

Predicted functions	No. of HPs	Percentage %
Unknown protein functions	583	48.5
Signal peptide, transmembrane	221	18.3
Enzymes	115	9.5
Signal peptides	65	5.4
Regulatory proteins	52	4.3
Transmembranes	47	3.9
Metal-binding proteins	43	3.6
Lipoproteins	27	2.2
Transporters	20	1.7
Antibiotic-related proteins	12	1.0
Spore proteins	9	0.7
Others	8	0.7
Total	1202	100.0

Table 2 Top 10 hits of BLASTP similarity search against NR database for Bleg1_2507

Accession No.	Description	Identity %	E-value
NP 244201.1	Hypothetical Protein BH 3335 (*Bacillus halodurans* C-125)	50	2e-63
YP 176895.1	Hypothetical Protein ABC 3401 (*Bacillus clausii* KSM-K16)	47	1e-43
YP 003425399.1	Unnamed protein product (*Bacillus pseudofirmus* OF4)	45	3e-61
YP 173730.1	Hypothetical Protein ABC 0226 (*Bacillus clausii* KSM-K16)	46	3e-46
NP 243770.1	Hypothetical Protein BH 2904 (*Bacillus halodurans* C-125)	40	9e-46
EIM 07312.1	Electron transport protein SCO1/SenC (*Planococcus antarcticus* DSM 14505)	47	1e-42
ZP 08680789.1	Sco1 family electron transport protein (*Sporosarcina newyorkensis* 2681)	42	8e-44
EIJ 83190.1	Electron transport protein SCO1/SenC (*Bacillus methanolicus* MGA3)	38	2e-35
ZP 10043231.1	Sco1/SenC (*Bacillus* sp.5B6)	35	2e-36
YP 003920672.1	Assembly factor BSco of the CuA site of cytochrome c oxidase (*Bacillus amyloliquefaciens* DSM7)	35	4e-38

also indicated that Bleg1_2507 shared similarity to electron transport protein Sco1 of various microorganisms. Sco1/SenC protein from *Planococcus antarcticus* DSM 14505 has the highest identity of 47% while the model Sco1 protein from *B. subtilis* (BsSco) surprisingly shared only 35% identity with Bleg1_2507 (Table 2).

Further analysis on Bleg1_2507 using ScanProsite revealed that amino acids 31–196 matched a thioredoxin_2 consensus pattern with total score of 10. Apart from this, analysis on the degree of amino acids conservation in Bleg1_2507 using Consurf revealed that Cys-69, Cys-73 and His-159 were highly conserved and similar to the ones in the top 10 hits from the UniRef 90 database of the Consurf program. The hits include the assembly factor of copper II (CuA) site of COX in *B. atrophaeus* 1942, Sco1 protein homolog from *B. subtilis* (BsSco) and putative uncharacterized protein from *B. clausii* KSM-16. In addition to this, these residues were also identified as potential metal-binding ligands via MetalDetector v2.0 analysis. These metal-binding residues resemble those of Sco1 proteins which similarly contain two Cys and a His residues at their metal-binding sites [33]. Hence, there is a possibility that the conserved Cys-69, Cys-73 and His-159 residues of Bleg1_2507 form the metal binding site of Bleg1_2507. Further analysis on the sequence of Bleg1_2507 revealed that the protein also possessed CXXXC and DXXXD motifs, which are very well preserved in all eukaryotic and prokaryotic Sco proteins [34,35]. Both of the conserved Cys in the CXXXC motif of Bleg1_2507 are made up of Cys-69 and Cys-73 respectively while the conserved Asp in the DXXXD motif are made up of Asp-102 and Asp-106 respectively (Figure 1). The conserved Asp residues of the motif have been implicated in copper ion binding during CuA assembly [35].

Since the results above highlight the possibility of HP Bleg1_2507 to be a Sco1 protein, a genome search for the sequence encoding the well-documented redox partner of

Sco1, COXII [35] was performed. According to [35], 82% of regular prokaryotic genomes possessed the same number of *sco* and *coxII* genes or the absence of both. Based on BLASTP result of Bleg1_2507, there were 5 HPs which exhibited 40-50% of sequence similarity to Bleg1_2507, originating from *B. halodurans*, *B. clausii* KSM-K16 and *B. pseudofirmus* OF4 (Table 2). Search for COXII from the genomes of these microorganisms revealed that COXII was present in these microorganisms with the accession number of NP_243481.1, YP_175889.1 and YP_003425144.1 respectively. This suggested a correlation between these genes and highlights the possible presence of the *coxII* gene in the genome of *B. lehensis* G1. The genome search led to the retrieval of Bleg1_2337 sequence whereby BLASTP analysis on this 344 amino acids protein showed 80% identity to COXII of *B. clausii* KSM-16. ScanProsite analysis showed the presence of Cox2_CuA (PS 50857) domain in Bleg1_2337 spanning from amino acid number 125–236. Three conserved amino acid residues namely Cys-261, Cys-264 and His-265 were identified to be metal-binding ligands. In addition to these residues, Metal Detector v2.0 program predicted Cys-207, Cys-211 and His-215 to be metal-binding ligands as well. ClustalW alignment of Bleg1_2337 showed that its predicted metal-binding residues, Cys-207, Cys-211 and His-215 are very well aligned with those of COXII sequences from other organisms such as *Thermus thermophiles*, *Bos taurus* and *Paracoccus denitrificans* (Figure 2). This result strengthens the prediction that Bleg1_2337 is indeed a COXII protein and that the sets of conserved residues are indeed metal-binding ligands, similar with reported metal-binding ligands of other COXII proteins [36].

Further probing on Bleg1_2337 using Consurf indicated that several hydrophobic residues such as Val-170, Ser-173, Phe-174, Trp-175 and Pro-177 were found to be highly conserved as well (Figure 2). Such hydrophobic residues have been highlighted to play an important role in the hydrophobic interaction with Sco protein [37].

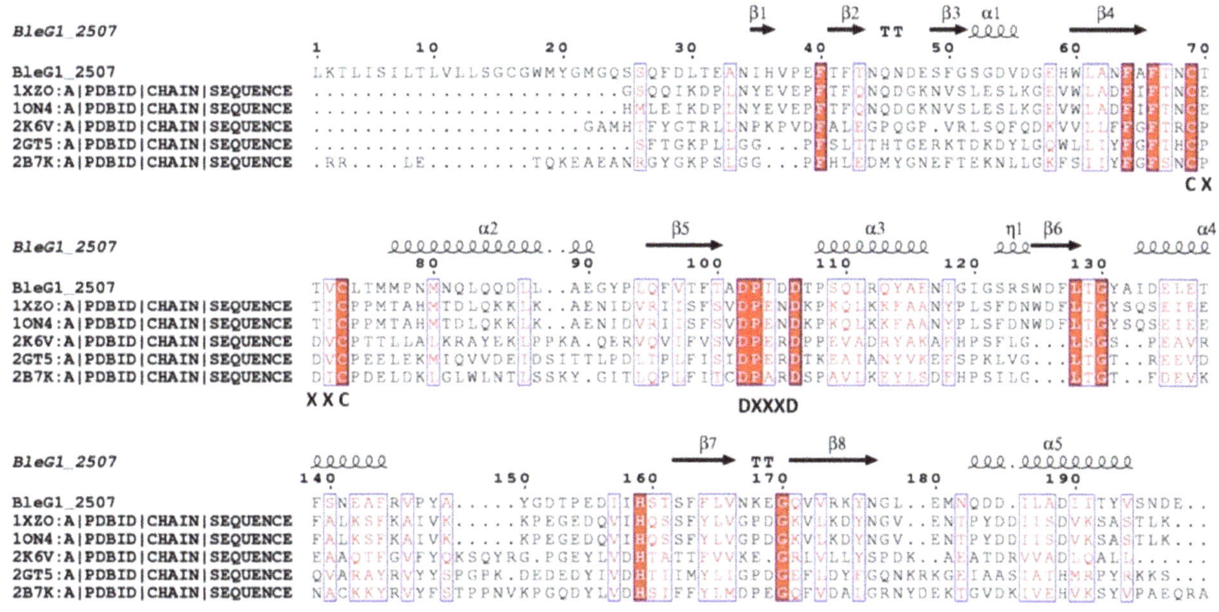

BleG1_2507

β1　β2　TT　β3　α1　β4

```
BleG1_2507
                                      1        10        20        30        40        50        60        70
BleG1_2507                            LKTLISILTLVLLSGCGWMYGMGQSSQFDLTEANIHVPEFTFINQNDESFGSGDVDGEHWLANEAFFNCT
1XZO:A|PDBID|CHAIN|SEQUENCE           ...................GSQQIKDPLNYEVEPFTFQNQDGKNVSLESLKGEVWLADFIFTNCE
1ON4:A|PDBID|CHAIN|SEQUENCE           ..................HMLEIKDPLNYEVEPFTFQNQDGKNVSLESLKGEVWLADTIFINCE
2K6V:A|PDBID|CHAIN|SEQUENCE           ...............GAMHTFYGTRLLNPKPVDPALEGPQGP.VRLSQFQDKVVLLFFGFTRCP
2GT5:A|PDBID|CHAIN|SEQUENCE           ...............SFTGKPLLGG...PFSLTHTGERKTDKDYLGQWLLIYFGFTHCP
2B7K:A|PDBID|CHAIN|SEQUENCE           .RR....LE......TQKEAEANRGYGKPSLGG...PFHLRDMYGNEFTEKNLLGKFSILYFGGSNCP
                                                                                                      C X
```

BleG1_2507

α2　β5　α3　η1　β6　α4

```
BleG1_2507
                                            80        90        100       110       120       130
BleG1_2507                            TVGLTMMPNMNQLQQDLIL..AEGYPLQFVTFTADPIDDTPSQIRQYAFNIGIGSRSWDFLTGYAIDELET
1XZO:A|PDBID|CHAIN|SEQUENCE           TTCPPMTAHMTDLQKKLK..AENIDVRIISFSVDPENDKPKQLKKFAANYPLSFDNWDFLTGYSQSEIEE
1ON4:A|PDBID|CHAIN|SEQUENCE           TTCPPMTAHMTDLQKKLK..AENIDVRIISFSVDPENDKPKQLKKFAANYPLSFDNWDFLTGYSQSEIEE
2K6V:A|PDBID|CHAIN|SEQUENCE           DVCPTTLLALKRAYEKLPPKA.QERVQVIFVSVDPERDPPEVADRYAKAFHPSFLG...LSGS..PEAVR
2GT5:A|PDBID|CHAIN|SEQUENCE           DVGPEELEKMIQVVDELDSITTLPDLTPLFISIDPERDTKEAIANYVKEFSPKLVG...LTGT..REEVD
2B7K:A|PDBID|CHAIN|SEQUENCE           DLCPDELDKLGLWLNTLSSKY.GITLQPFITCDPARDSPAVLKEYLSDFHPSILG...LTGT..FDEVK
                                      X X C                                 DXXXD
```

BleG1_2507

β7　TT　β8　α5

```
BleG1_2507
                                            140       150       160       170       180       190
BleG1_2507                            FSNEAFRVPYA.....YGDTPEDIIHSTSFFLVNKEGQVVRKYNGL..EMNQDD.ILADIITYVSNDE..
1XZO:A|PDBID|CHAIN|SEQUENCE           FALKSFKAIVK.....KPEGEDQVIHQSSFYLVGPDGKVLKDYNGV..ENTPYDDIISDVKSASTLK...
1ON4:A|PDBID|CHAIN|SEQUENCE           FALKSFKAIVK.....KPEGEDQVIHQSSFYLVGPDGKVLKDYNGV..ENTPYDDIISDVKSASTLK...
2K6V:A|PDBID|CHAIN|SEQUENCE           EAAQTFGVFYQKSQYRG.PGEYLVDHTATTFVVKE.GRLMLLYSPDK..AEATDRVVADLQALI......
2GT5:A|PDBID|CHAIN|SEQUENCE           QVARAYRVYYSGPK.DEDEDYIMDHTIIMYLIGPDGEFYFGQNKRKGEIAASIATHMRPYKKS...
2B7K:A|PDBID|CHAIN|SEQUENCE           NACKKYRVYFSTPPNVKPGQDYLVDHSIIFFYIMDPEGQFVDALGRNYDEKTGVDKLVEHVKSYVPAEQRA
```

Figure 1 Multiple sequence alignment of Bleg1_2507 with prokaryotic and eukaryotic Sco templates. The red boxes indicate conserved amino acids residues. The secondary structures of Bleg1_2337 are shown at the top of the alignment.

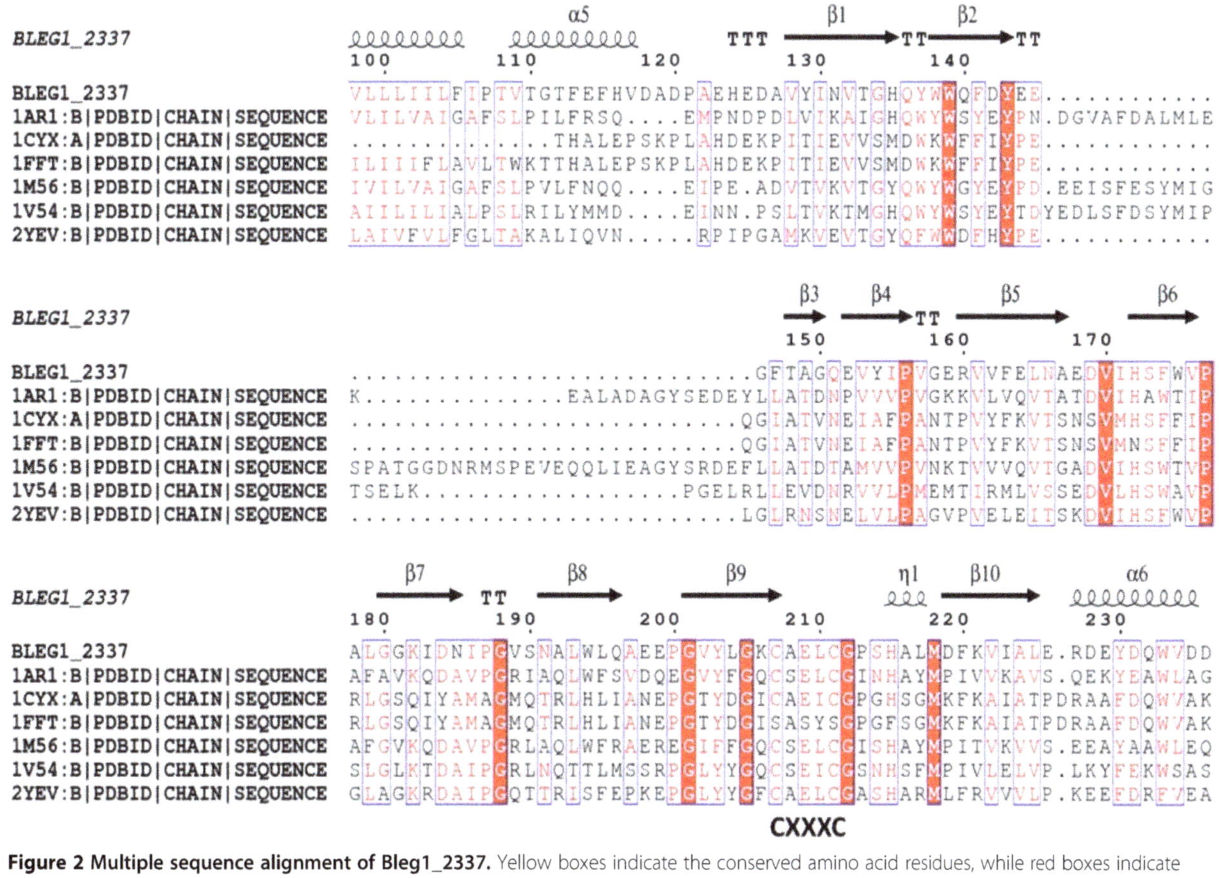

BLEG1_2337

α5　TTT　β1　TT　β2　TT

```
BLEG1_2337
                                            100       110       120       130       140
BLEG1_2337                            VLLLIILFIPTVTGTFEFHVDADPAEHEDAVYINVTGHQYVWQFDYEE...............
1AR1:B|PDBID|CHAIN|SEQUENCE           VLILVAIGAFSLPILFRSQ....EMPNDPDLVIKAIGHQWYWSYEYPN.DGVAFDALMLE
1CYX:A|PDBID|CHAIN|SEQUENCE           .............THALEPSKPLAHDEKPITIEVVSMDWKWFFIYPE...............
1FFT:B|PDBID|CHAIN|SEQUENCE           ILIIIFLAVLTWKTTHALEPSKPLAHDEKPITIEVVSMDWKWFFIYPE...............
1M56:B|PDBID|CHAIN|SEQUENCE           IVILVAIGAFSLPVLFNQQ....EIPE.ADVTVKVTGYQWYWGYEYPD.EEISFESYMIG
1V54:B|PDBID|CHAIN|SEQUENCE           AIILILIALPSLRILYMMD....EINN.PSLTVKTMGHQWYWSYEYTDYEDLSFDSYMIP
2YEV:B|PDBID|CHAIN|SEQUENCE           LAIVFVLFGLTAKALIQVN.....RPIPGAMKVEVTGYQFWWDFHYPE...............
```

β3　β4　TT　β5　β6

```
BLEG1_2337
                                            150       160       170
BLEG1_2337                            ...............GFTAGQEVYIPVGERVVFELNAEDVIHSFWVP
1AR1:B|PDBID|CHAIN|SEQUENCE           K...............EALADAGYSEDEYLLATDNPVVVPVGKKVLVQVTATDVIHAWTIP
1CYX:A|PDBID|CHAIN|SEQUENCE           ...............QGIATVNEIAFPANTPVYFKVTSNSVMHSFFIP
1FFT:B|PDBID|CHAIN|SEQUENCE           ...............QGIATVNEIAFPANTPVYFKVTSNSVMNSFFIP
1M56:B|PDBID|CHAIN|SEQUENCE           SPATGGDNRMSPEVEQQLIEAGYSRDEFLLATDTAMVVPVNKTVVVQVTGADVIHSWTVP
1V54:B|PDBID|CHAIN|SEQUENCE           TSELK...............PGELRLLEVDNRVVLPMEMTIRMLVSSEDVLHSWAVP
2YEV:B|PDBID|CHAIN|SEQUENCE           ...............LGLRNSNELVLPAGVPNELEITSKDVIHSFWVP
```

β7　TT　β8　β9　η1　β10　α6

```
BLEG1_2337
                                            180       190       200       210       220       230
BLEG1_2337                            ALGGKIDNIPGVSNALWLQAEEPGVYLGKCAELCGPSHALMDFKVIALE.RDEYDQWVDD
1AR1:B|PDBID|CHAIN|SEQUENCE           AFAVKQDAVPGRIAQLWFSVDQEGVYFGQCSELCGINHAYMPIVVKAVS.QEKYEAWLAG
1CYX:A|PDBID|CHAIN|SEQUENCE           RLGSQIYAMAGMQTRLHLIANEPGTYDGICAEICGPGHSGMKFKAIATPDRAAFDQWVAK
1FFT:B|PDBID|CHAIN|SEQUENCE           RLGSQIYAMAGMQTRLHLIANEPGTYDGISASYSGPGFSGMKFKAIATPDRAAFDQWVAK
1M56:B|PDBID|CHAIN|SEQUENCE           AFGVKQDAVPGRLAQLWFRAEREGIFFGQCSELCGISHAYMPITVKVVS.EEAYAAWLEQ
1V54:B|PDBID|CHAIN|SEQUENCE           SLGLKTDAIPGRLNQTTLMSSRPGLYYGQCSEICGSNHSFMPIVLELVP.LKYFEKWSAS
2YEV:B|PDBID|CHAIN|SEQUENCE           GLAGKRDAIPGQTTRISFEPKEPGLYYGFCAELCGASHARMLFRVVVLP.KEEFDRFVEA
                                                                     CXXXC
```

Figure 2 Multiple sequence alignment of Bleg1_2337. Yellow boxes indicate the conserved amino acid residues, while red boxes indicate metal-binding residues.

Homology modeling of Bleg1_2507 and Bleg1_2337

Potential templates for homology modeling of both Bleg1_2507 and Bleg1_2337 were retrieved using PSI-BLAST [16] search against PDB. Four templates were obtained for HP Bleg1_2507. Alignment of these sequences with Bleg1_2507 showed that Cys-69, Cys-73 of the CXXXC motif, His-159, Asp-102 and Asp-106 of the DXXXD motif of the HP aligned perfectly with all of the potential templates (Figure 1). 1XZO, chain A, which is a crystal structure of a disulfide switch in *B. subtilis* Sco (BsSco) and a well-studied member of the Sco family of COX assembly proteins [37] was selected as the template for homology modeling as it possessed the highest sequence similarity to Bleg1_2507 with 39% identity and lower e-value of 5e-31 compared to other templates (data not shown). In addition to this, parsimony phylogenetic analysis from PHYLIP also revealed that 1XZO is closely related to Bleg1_2507 (Figure 3(A)), lending further credence for it to be used as the template for Bleg1_2507 model generation.

As for Bleg1_2337, six templates were obtained and alignment of their sequences with Bleg1_2337 showed that highly the conserved residues, Cys-207, 211 and His-215 as well as the hydrophobic residues mentioned above were very well aligned. 2yev chain B, which is a crystal structure of caa3-type cytochrome oxidase *Thermus thermophilus* HB8 [36] was selected as the template for homology modeling of Bleg1_2337 as it has the highest sequence identity of 37% and low e-value of 4e-50 when compared to other templates. Phylogenetic analysis from PHYLIP also revealed that 2YEV chain B is closely related to Bleg1_2337 (Figure 3(B)).

The best built models for Bleg1_2507 and Bleg1_2337 (Figure 4(A) & (C)) have the lowest DOPE values and GA 341 scores of one, suggesting that the developed models resemble the native structure and hence reliable. Further supporting results from the superimposition of all Cα atoms of both developed models and templates gave forth low RMSD of less than 0.5 Å, suggesting that the built models are indeed reliable.

Bleg1_2507 and Bleg1_2337 model refinement and validation

Refinement of the built models using FoldX in YASARA resulted in minimization of the free energy of Bleg1_2507 from 296.2 kcal/mol to 62.2 kcal/mol and from 657.5 kcal/mol to 247.1 kcal/mol for Bleg1_2337.

Evaluations on the quality of the refined protein structures using ERRAT [29] indicated that 84.2% of the amino acids of the refined Bleg1_2507 model were located in the acceptable region within 95% confidence limit, as opposed to only 70.5% before refinement. As for refined Bleg1_2337 model, 85.4% of its amino acids were located in the acceptable region within the 95% confidence limit, as opposed to only 75.4% before refinement. These indicated that the refined protein models for Bleg1_2507 and Bleg1_2337 (Figure 4(A) & (C)) were precise and reliable since the frequencies of atom randomizations were low. Further supporting results from PROCHECK Ramachandran analysis on refined Bleg1_2507 model revealed that 85.7% of the residues were located in favored region, 12.3% in additional allowed region, 1.3% and 0.6% in generously allowed and disallowed regions respectively. As for Bleg1_2337 refined model, 87.7% of the residues were located in favored region, 10.3% in additional allowed region and 2% were located in generously allowed region. Overall, both ERRAT and PROCHECK evaluations further reaffirmed that both of the predicted protein models were reliable.

Probing the structures of Bleg1_2507 and Bleg1_2337

Overall, the built model for HP Bleg1_2507 showed that it adopted a global topology unique to Thioredoxin (Trx)-like superfamily of proteins including Sco proteins whereby four β-sheets (β4, β5, β7 and β8) are flanked by three α-helices (α1, α3 and α4) at the core of the structure (Figure 4(A)). This is similar to Trx-like superfamily members which have a characteristic, common core βαβαββ secondary structural pattern with different insertions of secondary structural elements to distinguish the various structural families such as thioltransferases,

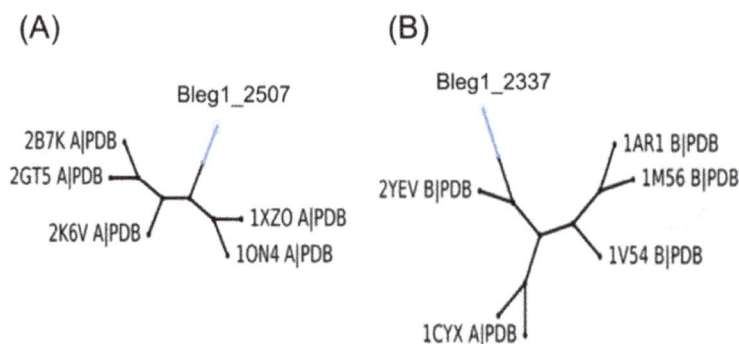

Figure 3 Phylogenetic analysis of (A) Bleg1_2507 and (B) Bleg1_2337 with their possible structural templates.

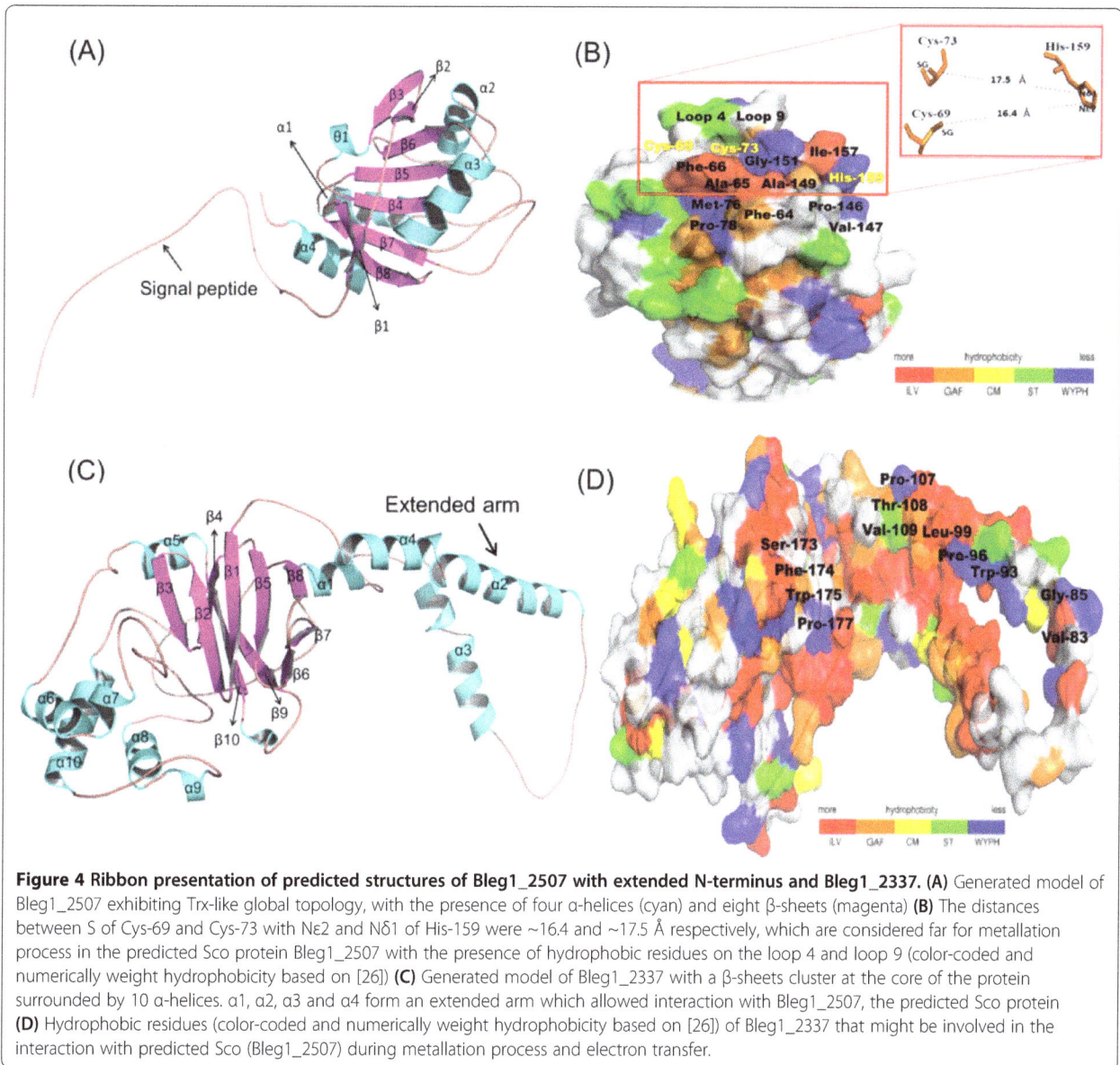

Figure 4 Ribbon presentation of predicted structures of Bleg1_2507 with extended N-terminus and Bleg1_2337. **(A)** Generated model of Bleg1_2507 exhibiting Trx-like global topology, with the presence of four α-helices (cyan) and eight β-sheets (magenta) **(B)** The distances between S of Cys-69 and Cys-73 with Nε2 and Nδ1 of His-159 were ~16.4 and ~17.5 Å respectively, which are considered far for metallation process in the predicted Sco protein Bleg1_2507 with the presence of hydrophobic residues on the loop 4 and loop 9 (color-coded and numerically weight hydrophobicity based on [26]) **(C)** Generated model of Bleg1_2337 with a β-sheets cluster at the core of the protein surrounded by 10 α-helices. α1, α2, α3 and α4 form an extended arm which allowed interaction with Bleg1_2507, the predicted Sco protein **(D)** Hydrophobic residues (color-coded and numerically weight hydrophobicity based on [26]) of Bleg1_2337 that might be involved in the interaction with predicted Sco (Bleg1_2507) during metallation process and electron transfer.

glutaredoxins, bacterial arsenate reductases and disulfide bond isomerases. They are in general involved in cellular thiol-redox reactions to maintain the reduction and oxidation of proteins or certain small molecules [38]. Other than this core structure, Bleg1_2507 also consisted three flanking β-sheets (β1, β2 and β3), a 3_{10} –helix at the N-terminus and a hairpin formed by β2 and β3, which are the additional secondary structure elements only present in Sco proteins [39].

Other similarities between include the presence of a transmembrane region in the N-terminal of Bleg1_2507 in which an external loop comprising of 25 amino acids at the N-terminus could be observed (Figure 4(A)). Sig-Cleave analysis from EMBOSS [21] revealed that the amino acids from Leu-4 to Cys-16 in this loop were predicted to be a signal peptide, which can be subjected to

cleavage to give forth the mature protein. Previous study has shown that in the Sco protein of *B. subtilis*, (BsSco) the first 20 amino acids are located in the transmembrane and Cys-16 is the signal peptidase II recognition site. To this cleaved site a diacyl glycerol moiety will be attached to the protein and the post-translationally modified protein will subsequently be able to anchor to the membrane through the attached lipid by covalent bond [39]. This in turn highlights that Cys-16 in HP Bleg1_2507 is most probably a cleavage recognition site similar to the one found in BsSco for processing of the native Sco protein before it can be partially integrated into the membrane.

Unique to Bleg1_2507 however, there is an α-helix (α2) near loop 4 and a β-strand (β6) that are parallel with β5 and β4 (Figure 4(A)), which are not present in

Trx. In addition to this, the location of the metal-binding ligands in Bleg1_2507 is different from the solution structure of BsSco. For instance, in the predicted model of Bleg1_2507, both metal-binding Cys-69 and Cys-73 residues were located in loop 4 (situated in between α1 and β4) while His-159 was positioned in loop 9 (situated in between α3 and β7) (Figure 4(B)). In BsSco on the other hand, its metal-binding residues, namely Cys-45 and Cys-49, were located in loop 3 and the conserved His-135 was located in loop 8 [6].

Looking closely at the predicted metal-binding cavity of Bleg1_2507, it featured the highly conserved Cys-69, Cys-73 (of the CXXXC motif) and His-159 residues. These residues which are strictly preserved in all Sco proteins have been implicated to be responsible in copper-binding and redox reaction [39]. In proximity to the CXXXC motif, a hydrophobic groove which is formed by Phe-64, 66, Ala-65, Val-72, Met-76, 77 and Pro-78 was observed (Figure 4(B)). Besides this, another hydrophobic groove formed by the His-ligand loop consisting of Val-146, Pro-147, 154, Ala-149, Gly-151, Ile-157, 158, and Phe-163 was also observed (Figure 4(B)). Both of the hydrophobic grooves formed a hydrophobic pocket which may allow proteins and small molecules such as copper ion to be accommodated [6].

According to Balatri et al., 2003 [6], the interaction of hydrophobic residues within the protein are essential in the metallated state of Sco, because it is able to stabilize the metal-ligand region of the protein, which are located far apart from each other within the two hydrophobic grooves, by coordinating the metal-ligand geometry. Based on the Sco1 protein crystal structures from B. subtilis [6], yeast [10] and human [40] the distances between the N atoms of His residues and the S atom of Cysteine pair are in the range of 10–19 Å. In the case of Bleg1_2507, the distance between Nε2 of His-159 and S of Cys-69 was ~16.4 Å, while the distance between Nδ1 of His-159 and S of Cys-73 was ~17.5 Å (Figure 4(B)), similar to the range noted above for the respective Sco1 proteins from B. subtilis, yeast and human. These distances are considered far for metal-ligand interaction in the apo-form state of the predicted Sco protein. It has been proposed by Balatri et al., 2003 [6] that the hydrophobic residues in the hydrophobic grooves of Bleg1_2507 might be responsible in regulating the appropriate geometric coordination for metal-ligand interaction to occur. According to Andruzzi et al. [41], the oxidized BsSco was coordinated to Cu(II) by a distorted tetragonal square planar, in which the Cys-45, Cys-49 and His-135 are ligated equatorially and water was ligated axially to the Cu(II) respectively. Hence, Cys-69, Cys-73 and His-159 of the predicted Sco of Bleg1_2507 was proposed to coordinate the copper through square planar geometry. Based on these observations, we propose that this particular HP is potentially a Sco protein with an ability to bind and chaperone copper to the CuA site in COXII.

Interaction between Bleg1_2507 and putative COXII redox protein, Bleg1_2337 and their possible mechanism of action

To test the above hypothesis, HP Bleg1_2507 was docked to its predicted redox partner, Bleg1_2337, a putative COXII protein to investigate possible interaction between these two proteins. Docking results showed that Bleg1_2507 was indeed able to interact with Bleg1_2337 with lowest energy of –1163.8 KJ/mol. From the results, Bleg1_2337 could be seen embracing Bleg1_2507 with an extended arm (Figure 5(A)). Refinement of the docked protein complex successfully decreased the potential energy from –257189.2 KJ/mol to –295640.7 KJ/mol signifying favorable protein conformation.

Inspecting the interface of both interacting proteins, Bleg1_2507 possessed a set of acidic residues such as Asp-47, 55, 56 and 126, Glu-48, 141 and 169 on its surface (Figure 5(B)). This observation lends support to ProtParam analysis which revealed that Bleg1_2507 has an acidic pI of 3.95, indicating this HP has abundant number of acidic residues. Its putative COX partner, Bleg1_2337, on the other hand, possessed a set of positively charged, basic residues on its surface such as Lys-3, 34, 69,73, 182 and 206, Arg-8, 71 and 72 and His-215 (Figure 5(C)), which gave the protein a basic pI value of 4.21 according to ProtParam analysis. The presence of these oppositely charged residues on the respective proteins undoubtedly facilitated ionic interaction between the two macromolecules. Such ionic interaction between Sco and COX proteins has been similarly observed and was proposed to generate a charged-mediated Sco-apo-CuA located in the COX protein which will allow copper exchange to occur [39].

Based on the mechanism proposed for BsSco protein, for copper to be bound and later transferred from Sco to COX, both of the conserved Cysteine residues in the CXXXC motif of BsSco will be firstly reduced to di-thiol state via formation of SH-SH bond. This will result in a reduced BsSco protein that is able to strongly interact with Cu(II) ion to form BsSco-Cu(II) complex. From the BsSco-Cu(II) complex, the metal ion will then be transferred to its intended target i.e. apo-subunit II of COX via association of both of the proteins (BsSco and COXII). In this process, the Histidine residue was postulated to play an important role in this second phase of copper coordination by regulating the structural dynamism of bacterial Sco [6]. During association of both of the proteins, apart from metal ion transfer, an oxidation-reduction (redox) process also takes place via electron transfer [39]. During interaction of the proteins, BsSco will undergo oxidation by releasing electrons which will

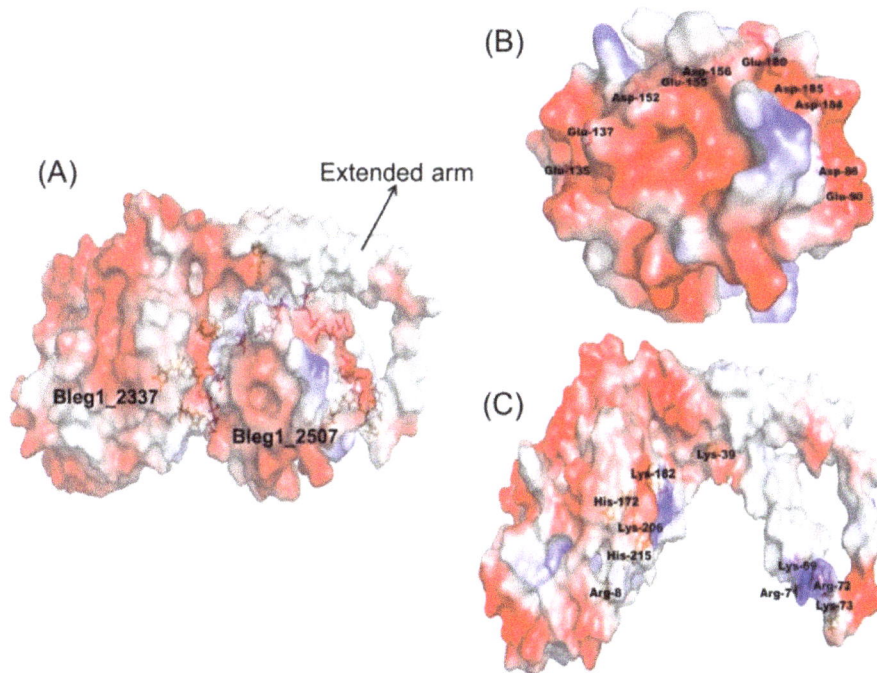

Figure 5 Docking between the generated models of Bleg1_2507 and Bleg1_2337. (A) Interaction of predicted models of Bleg1_2507 and Bleg1_2337 with the assistance of acidic residues of Bleg1_2507 **(B)** and basic residues of Bleg1_2337 **(C)** in addition to the hydrophobic residues.

be subsequently delivered to the binuclear CuA site of COXII. Consequently, the received electrons in the CuA site of COXII which made up of Cys-217, 221 and His-215 will reduce Fe(III) to Fe(II) or Cu(II) to Cu(I) [39]. The di-thiol state of BsSco is vital in this electron transfer process and maintaining the redox state of the partner.

In the case of Bleg1_2507, it is therefore highly possible that it may possess redox properties by potentially acting as a thiol disulfide oxidoreductase. Similar to BsSco, both the conserved Cys-69 and Cys-73 in Bleg1_2507 may also form the SH-SH bond and give forth the di-thiol state of the protein. From this point onwards, Bleg1_2507 HP may be able to undergo oxidation to release electrons which are subsequently delivered to the binuclear CuA site of reduced COXII of Bleg1_2337 which were made up of Cys-207, 211 and His-215 and Cys-261, 264 and His-265 (highly similar to the other documented COXII proteins), to reduce the respective metal ion at the site. This postulation is further supported by the presence of hydrophobic grooves in HP Bleg1_2507 which form an extensive uncharged surface surrounding the active disulfide bond formed by Cys-69, Cys-73 and His-159 encompassed in a solvent exposed environment which might facilitate the electron transfer process, similar to the features observed in BsSco [35].

It is worthy to note that other than binding metal ions, the conserved Cys pair and His-ligand loops of Sco proteins are important for Sco to interact with other proteins as well [33]. This feature was clearly observed in our docking results where these residues of HP Bleg1_2507 are shown to interact with putative COXII (Bleg1_2337) via hydrophobic interaction, stabilized by uncharged surface features surrounding the hydrophobic pocket between loop 4 and loop 9 (Figure 4(B)). In the Cysteine pair loop (loop 4), it possessed uncharged polar residues such as Thr-67, 70, 71 and 75 and Asn-68. While, Thr-153, 161 and Ser-160 were located in the His-ligand loop (loop 9). Similarly in its redox partner, the putative COXII (Bleg1_2337), crucial hydrophobic residues such as Val-82 and 109, Gly-84, Trp-92 and 175, Ile-95, Pro-96 and 107, Leu-99, Phe-106 and 174, Thr-108, Ser-173 and Pro177 may have contributed to the interaction of both proteins (Figure 4(D)), in accordance to other similarly reported study [39]. All of these residues are crucial in maintaining the intactness of Sco-CuA protein complex for electron transfer [39].

Structural and functional comparison of Sco proteins

Comparing the sequences of the predicted Sco Bleg1_2507, BsSco (1XZO, chain A), human Sco1 (2GT5, chain A) and yeast Sco1 (2BTK, chain A), these proteins share many conserved residues regardless of their taxonomic and sequence differences (Figure 1). Superimposition of predicted protein model of Bleg1_2507 with the 3D structure of BsSco showed that they are highly similar with RMSD value of 0.286 Å (Figure 6(A)). Interestingly, superimposition of

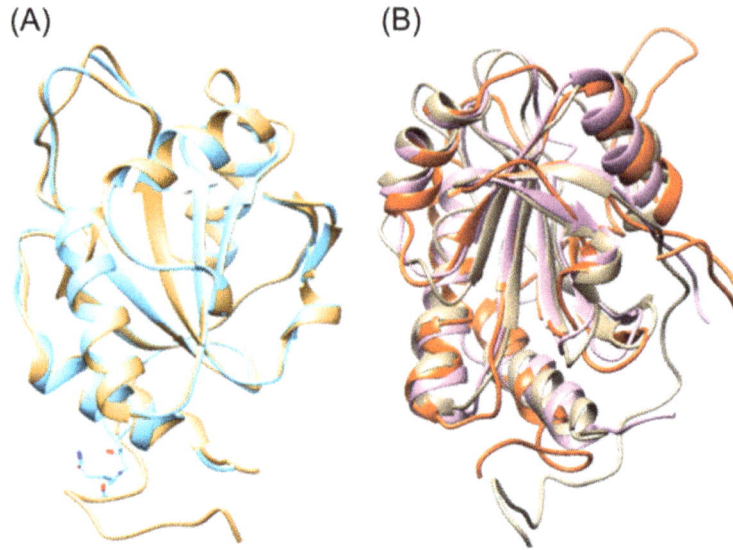

Figure 6 Structural alignment of Bleg1_2507 protein model with BsSco, human and yeast Sco1 proteins. (A) Superimposition of Cα atoms between Bleg1_2507 (brown) and BsSco [1XZO, Chain A] (blue), **(B)** human Sco1 [2GT5, Chain A] (orange) and yeast Sco1 [2BTK, Chain A] (magenta) proteins exhibited structural resemblance across biological taxa.

predicted protein model of Bleg1_2507 with the 3D structures of human and yeast Sco1 showed that their structures are comparatively similar as well even at RMSD values of 1.215 Å and 1.060 Å respectively (Figure 6(B)).

It is important to note that Bleg1_2507, BsSco and Sco1 of yeast contain the same number of β-sheets and α-helices i.e. eight and four respectively. The 3D structure of human Sco1 (2GT5, chain A), however, contained only six β-sheets. The two additional β sheets in Bleg1_2507 are β2 and β3, which forms the hairpin structure located before the 3_{10}- helix at the N-terminus.

In addition, although Bleg1_2507 shares relatively low sequence identity to BsSco (47%), human Sco1 (2GT5, chain A) (27%) and yeast Sco1 (27%), the predicted metal-binding residues of Bleg1_2507 i.e. Cys-69, Cys-73 and His-159, were located at flexible active loops (loop 4 and 9) similar to other Sco proteins across biological taxa.

It is indeed interesting to note that despite their remarkable structural similarity, Sco proteins are rather diverse in function. In eukaryotes, Sco proteins are involved in the assembly of the CuA cofactor of mitochondrial cytochrome c oxidase, redox signaling and regulation of copper homeostasis [13]. In prokaryotes, Sco proteins are more promiscuous in functions. While some are required for COX biosynthesis, others have been implicated in different processes such as copper delivery to other enzymes and protection against oxidative stress [33]. More investigations as to the physiological function and mechanism of Sco proteins will be useful in determining its exact role and importance in the system biology of an organism.

Conclusions

In this study, genome mining of hypothetical protein sequences of *B. lehensis* G1 alkaliphile led to the discovery of Bleg1_2507 which showed the presence of a Trx-like domain linked to Sco protein. Showing only 35% of similarity to BsSco, the prokaryotic model of Sco, the 3-D model of Bleg1_2507 was built by homology modeling. The built protein model showed good preservation of the βαβαββ core structure (characteristic of many Trx-like redox proteins), three flanking β-sheets, a 3_{10} – helix at the N-terminus and a hairpin (characteristic of Sco proteins). Unique to Bleg1_2507, there is an α-helix (α2) near loop 4 and a β-strand (β6) that are parallel with β5 and β4, which are not present in Trx. Another distinct feature of Bleg1_2507 compared to BsSco is the difference in the location of the metal-binding ligands in Bleg1_2507. Docking simulations interestingly provided a view of Bleg1_2507 in association with its putative COXII redox partner, Bleg1_2337, where the latter can be seen to hold its partner in an embrace, facilitated via hydrophobic and ionic interaction between the two proteins. Although Bleg1_2507 shares relatively low sequence homology to BsSco, interestingly, it is structurally similar to the protein as well as other Sco proteins across biological taxa. This observation highlights that despite its varying functions in prokaryotes and eukaryotes, Sco proteins are structurally well preserved.

Based on all the results obtained, we hereby propose that HP Bleg1_2507 of *B. lehensis* G1 is a Sco protein which is able to interact with COXII, its redox partner. Hence, it is highly possible that HP Bleg1_2507 may possess metallochaperone and disulfide redox functions similar to other documented bacterial Sco proteins. It is hoped that the predicted structure and function of Bleg1_2507 from the hypothetical protein dataset of *B. lehensis* G1 will help to spur the search for other physiologically relevant proteins among the so-called "orphan" proteins of any given organism as well as direct future biological studies in improving our understanding of COX complex assembly process.

Competing interests

The authors declare that they have no competing interests.

Authors' contributions

TSH carried out bioinformatics analyses and clustering of all the hypothetical proteins, homology modelling and protein docking in this study. TSH and YMN drafted the manuscript. ABS, MBAR, RAK, ALTC, AMAM, NMM revised and proofread the manuscript. YMN conceived the study and participated in its design and coordination together with ABS, MBAR. RAK, ALTC, AMAM gave technical advice to the study. All authors read and approved the final manuscript.

Acknowledgement

We would like to thank Mr. Yusuf Muhammad Noor of Malaysia Genome Institute (MGI) in providing the initial genome information, Ministry of Science, Technology and Innovation for Sciencefund, MOSTI-MGI for Sciencefund Special Allocation (Project code: 02-05-20-SF11112) and facility, Universiti Putra Malaysia (UPM) for Graduate Research Fellowship awarded to Ms. Tan Soo Huei.

Author details

[1]Center for Enzyme and Microbial Biotechnology (EMTECH), Faculty of Biotechnology and Biomolecular Sciences, Universiti Putra Malaysia, Serdang, Selangor 43400, Malaysia. [2]Department of Chemistry, Faculty of Science, Universiti Putra Malaysia, Serdang, Selangor 43400, Malaysia. [3]School of Biosciences and Biotechnology, Faculty of Science and Technology, Universiti Kebangsaan Malaysia, 43600 UKM, Bangi, Selangor, Malaysia. [4]Malaysia Genome Institute, Ministry of Science, Technology and Innovation, Jalan Bangi, Kajang, Selangor 43000, Malaysia.

References

1. Bork P: Powers and pitfalls in sequence analysis: the 70% hurdle. *Genome Res* 2000, **10**(4):398–400.
2. Galperin MY, Koonin EV: 'Conserved hypothetical' proteins: prioritization of targets for experimental study. *Nucl Acid Res* 2004, **32**(18):5452–5463.
3. Roberts RJ: Identifying Protein Function- A Call for Community Action. *PLoS Biol* 2004, **2**(3):e42.
4. Desler C, Suravajhala P, Sanderhoff M, Rasmussen M, Rasmussen LJ: *In Silico* screening for functional candidates amongst hypothetical proteins. *BMC Bioinforma* 2009, **10**(289):1471–2105.
5. Samsulrizal NH, Abdul Murad AM, Noor YM, Jema'on NA, Najimudin N, Illias RM, Abu Bakar FD, Mahadi NM: *Genome Sequencing and Annotation of Alkaliphilic Bacterium Bacillus lehensis G1*. India: The Eighth Asia Pacific Bioinformatics Conference, Bangalore (APBC 2010); 2010:18–21.
6. Balatri E, Banci L, Bertini I, Cantini F, Baffoni-Ciofi S: Solution Structure of Sco1: A Thioredoxin-like Protein Invovled in Cytochrome c Oxidase Assembly. *Structure* 2003, **11**(11):1431–1443.
7. Glerum DM, Shtanko A, Tzagoloff A: Characterization of COX17, a yeast gene involved in copper metabolism and assembly of cytochrome oxidase. *J Biol Chem* 2003, **271**(24):14504–14509.
8. Krummeck G, Rödel G: Yeast Sco1 protein is required for a post-translational step in the accumulation of mitochondrial cytochrome c oxidase subunits I and II. *Curr Genet* 1990, **18**(1):13–15.
9. Schulze M, Rödel G: SCO1, a yeast nuclear gene essential for accumulation of mitochondrial cytochrome c oxidase subunit II. *Mol Gen Genet* 1988, **211**(3):492–498.
10. Abajian C, Rosenzweig CA: Crystal structure of yeast Sco1. *J Biol Inorg Chem* 2006, **11**(4):459–466.
11. Papadopoulou LC, Sue CM, Davidson MM, Tanji K, Nishino I, Sadlock JE, Krishna S, Walker W, Selby J, Glerum DM, Coster RV, Lyon G, Scalais E, Lebel R, Kaplan P, Shanske S, De Vivo DC, Bonilla E, Hirano M, DiMauro S, Schon EA: Fatal infantile cardioencephalomyopathy with COX deficiency and mutations in SCO2, a COX assembly gene. *Nat Genet* 1999, **23**(3):333–337.
12. Attallah CV, Welchen E, Martin AP, Spinelli SV, Bonnard G, Palatnik JF, Gonzalez DH: Plants contain two SCO proteins that are differentially involved in cytochrome c oxidase function and copper and redox homeostasis. *J Exp Bol* 2011, **62**(12):4281–4294.
13. Valnot I, Osmond S, Gigarel N, Mehaye B, Amiel J, Cormier-Daire V, Munnich A, Bonnefont JP, Rustin P, Rötig A: Mutations of the SCO1 gene in mitochondrial cytochrome c oxidase deficiency with neonatal-onset hepatic failure and encephalopathy. *Am J Hum Genet* 2000, **67**(5):1104–1109.
14. Quevillon E, Silventoinen V, Pillai NS, Harte N, Mulder N, Apweile R, Lopez R: InterProScan: protein domains identifier. *Nucl Acids Res* 2005, **33**(Web Server Issue):W116–W120.
15. Marchler-Bauer A, Anderson JB, Derbyshire MK, DeWeese SC, Gonzales NR, Gwadz M, Hao L, He S, Hurwitz DI, Jackson JD, Ke Z, Krylov D, Lanczycki CJ, Liebert CA, Liu C, Lu F, Lu S, Marchler GH, Mullokandov M, Song JS, Thanki N, Yamashita RA, Yin JJ, Zhang D, Bryant: CDD: a conserved domains and protein three-dimensional structure. *Nucl Acids Res* 2013, **41**(D1):D348–D352.
16. Altschul SF, Madden TL, Schäffer AA, Zhang J, Zhang Z, Miller W, Lipman DJ: Gapped-BLAST and PSI-BLAST: a new generation of protein database search programs. *Nucl Acids Res* 1997, **25**(17):3389–3402.
17. De Castro E, Sigrist CJA, Gattiker A, Bulliard V, Langendijk-Genevaux PS, Gasteiger E, Bairoch A, Hulo N: ScanProsite: detection of PROSITE signature matches and ProRule-associated functional and structural residues in proteins. *Nucl Acids Res* 2006, **34**(Web Server Issue):W362–W365.
18. Ashkenazy H, Erez E, Martz E, Pupko T, Ben TN: ConSurf 2010: calculating evolutionary conservation in sequence and structure of proteins and nucleic acids. *Nucl Acids Res* 2010, **38**(Suppl 2):W529–W533.
19. Passerini A, Lippi M, Frasconi P: MetalDetector v2.0: predicting the geometry of metal binding sites from protein sequence. *Nucl Acids Res* 2011, **39**(Suppl 2):W288–W292.
20. Gasteiger E, Hoogland C, Gattiker A, Duvaud S, Wilkins MR, Appel RD Bairoch A: Protein Identification and Analysis Tools on the ExPASy Server. In *The Proteomics Protocols Handbook*. Edited by John MW. Totowa, New Jersey, USA: Humana Press; 2005:571–607.
21. Rice P, Longen T, Bleasby A: The European Molecular Biology Open Software Suite. *Trends Genet* 2000, **16**(6):276–277.
22. Felsenstein J: PHYLIP – Phylogeny Inference Package (Version 3.2). *Cladistics* 1989, **5**:164–166.
23. Larkin MA, Blackshields G, Brown NP, Chenna R, McGettigan PA, McWilliam H, Valentin F, Wallance IM, Wilm A, Lopez R, Thompson JD, Gibson TJ, Higgins DG: Clustal W and Clustal X version 2.0. *Bioinformatics* 2007, **23**(21):2947–2948.
24. Šali A, Blundell TL: Comparative protein modeling by satisfaction of spatial restraints. *J Mol Biol* 1993, **234**(3):779–815.
25. Delano WL: *The PyMOL Molecular Graphics System*. San Carlos, CA, USA: Delano Scientific; 2002.
26. Kojetin DJ, Thompson RJ, Cavanagh J: Hypothesis: Sub-classification of response regulators using the surface characteristics of their receiver domains. *FEBS Lett* 2004, **5601**(1–3):227–228.
27. Pettersen EF, Goddard TD, Huang CC, Couch GS, Greenblatt DM, Meng EC, Ferrin TE: UCSF Chimera—a visualization system for exploratory research and analysis. *J Comput Chem* 2004, **25**(13):1605–1612.
28. Schymkowitz JW, Rousseau F, Martins IC, Ferkinghoff-Berg J, Stricher F, Serrano L: Prediction of water and metal-binding sites and their affinities by using the Fold-X force field. *Proc Natl Acad Sci USA* 2005, **102**(29):10147–10152.
29. Colovos C, Yeates OT: Verification of protein structures: Patterns of nonbonded atomic interactions. *Prot Sci* 1993, **2**(9):1511–1519.
30. Laskoswki RA, MacArthur MW, Moss DS, Thornton JM: PROCHECK: a program to check the stereochemical quality of protein structures. *J Appl Cryst* 1993, **26**(2):283–291.
31. Kozakov D, Hall DR, Beglov D, Brenke R, Comeau SR, Shen Y, Keyong L, Jiefu Z, Vakili P, Paschalidis CI, Vajda S: Achieving reliability and high accuracy in automated protein docking: ClusPro, PIPER, SDU, and stability analysis in CAPRI rounds 13–19. *Proteins* 2004, **78**:3124–3130.

32. Krieger E, Joo K, Lee J, Raman S, Thompson J, Tyka M, Baker D, Karplus L: **Improving physical realism, stereochemistry, and side-chain accuracy in homology modeling: Four approaches that performed well in CASP8.** *Proteins* 2009, **77**(Suppl 9):114–122.

33. Banci L, Bertini I, Cavallaro G, Baffoni-Ciofi S: **Seeking the determinants of the elusive functions of Sco proteins.** *FEBS J* 2011, **278**(13):2244–2262.

34. Blundell KL, Wilson MT, Svistunenko DA, Vijgenboom E, Worrall JA: **Morphological development and cytochrome c oxidase activity in Streptomyces lividans are dependent on the action of a copper bound Sco protein.** *Open Biol* 2013, **3**(1):120163.

35. Banci L, Bertini I, Cavallaro G, Rosato A: **The Functions of Sco Proteins from Genome-Based Analysis.** *J Proteome Res* 2007, **6**(4):1568–1579.

36. Lyons JA, Aragao D, Slattery O, Pisliakov AV, Soulimane T, Caffrey M: **Structural insights into electron transfer in caa3-type cytochrome oxidase.** *Nature* 2012, **487**(7408):514–518.

37. Ye Q, Iveta SI, Hill BC, Chao JZ: **Identification of a Disulfide Switch in BsSco, a Member of the Sco Family of Cytochrome c Oxidase Assembly Proteins.** *Biochemistry* 2005, **44**(14):5552–5560.

38. Kinch LN, Baker D, Grishin NV: **Deciphering a Novel Thioredoxin-Like Fold Family.** *Proteins* 2003, **52**(3):323–331.

39. Banci L, Bertini I, Baffoni-Ciofi S, Kozyreva T, Mori M, Wang S: **Sco proteins are involved in electron transfer processes.** *J Biol Inorg Chem* 2011, **16**(3):391–403.

40. Banci L, Bertini I, Calderone V, Ciofi-Baffoni S, Mangani S, Martinelli M, Palumaa P, Wang S: **A hint for the function of human Sco1 from different structures.** *Proc Natl Acad Sci USA* 2006, **103**:8595–8600.

41. Andruzzi L, Nakano M, Nilges Mark J, Blackburn Ninian J: **Spectroscopic Studies of Metal-Binding and Metal Selectivity in Bacillus subtilis BSco, a Homologue of the Yeast Mitochondrial Protein Sco1p.** *J Am Chem Soc* 2005, **127**(47):16548–16558.

11

Molecular details of ligand selectivity determinants in a promiscuous β-glucan periplasmic binding protein

Parthapratim Munshi[1,2,3], Christopher B Stanley[1], Sudipa Ghimire-Rijal[1], Xun Lu[1], Dean A Myles[1] and Matthew J Cuneo[1*]

Abstract

Background: Members of the periplasmic binding protein (PBP) superfamily utilize a highly conserved inter-domain ligand binding site that adapts to specifically bind a chemically diverse range of ligands. This paradigm of PBP ligand binding specificity was recently altered when the structure of the *Thermotoga maritima* cellobiose-binding protein (tmCBP) was solved. The tmCBP binding site is bipartite, comprising a canonical solvent-excluded region (subsite one), adjacent to a solvent-filled cavity (subsite two) where specific and semi-specific ligand recognition occur, respectively.

Results: A molecular level understanding of binding pocket adaptation mechanisms that simultaneously allow both ligand specificity at subsite one and promiscuity at subsite two has potentially important implications in ligand binding and drug design studies. We sought to investigate the determinants of ligand binding selectivity in tmCBP through biophysical characterization of tmCBP in the presence of varying β-glucan oligosaccharides. Crystal structures show that whilst the amino acids that comprise both the tmCBP subsite one and subsite two binding sites remain fixed in conformation regardless of which ligands are present, the rich hydrogen bonding potential of water molecules may facilitate the ordering and the plasticity of this unique PBP binding site.

Conclusions: The identification of the roles these water molecules play in ligand recognition suggests potential mechanisms that can be utilized to adapt a single ligand binding site to recognize multiple distinct ligands.

Keywords: Periplasmic binding protein, Carbohydrate recognition, Laminarin, ABC transport, Ligand specificity

Background

The periplasmic binding proteins (PBP) are a protein superfamily that serve as primary receptors for a diverse group of metabolic solutes in signaling [1], chemotaxis [2] and metabolite transport systems in bacteria [3], eukaryotes and archaea. PBP mediated transmembrane transport of ligands are coupled to either ATP hydrolysis (ABC transport) [4] or H^+/M^+ motive force (TRAP transport or tripartite tricarboxylate transport) [5]. In addition, the PBP module is also found in enzymes [6], transcriptional control elements [7] and eukaryotic neurotransmission systems [8]. PBPs bind multiple ligands that range in size from a few Daltons to as large as 1 kDa, including ions [9], amino acids [10], peptides [11], monosaccharides [12], oligosaccharides [13], polyamines [14], oxidized inorganics [15]. Many other ligands continue to be discovered through current genome sequencing technology [16].

Despite the wide variation in PBP cognate ligand size and chemical functionality, the three-dimensional structure is highly conserved across all PBPs. PBPs are comprised of two α/β domains connected by a flexible linker region that serves as a pivot point for the ligand induced hinge-bending motion that this protein superfamily is known for [17-20]. PBPs were initially classified into three distinct sub-groups based upon the topology of β-strands in each domain [21]. Recently, the PBP super-family was

---

* Correspondence: cuneomj@ornl.gov
[1]Neutron Sciences Directorate, Oak Ridge National Laboratory, Oak Ridge, TN 37831, USA
Full list of author information is available at the end of the article

re-categorized into six distinct clusters by combining known ligand specificities with the wealth of structural information available in the Protein Data Bank [22].

PBPs typically bind cognate ligands with exquisite specificity, discriminating among anomeric/epimeric carbohydrates or different ions [23,24]. This remarkable ability to specifically bind their cognate ligands from pools of similarly related molecules has been attributed to the localization of the PBP ligand binding site at the interdomain interface [25]. In the apo form, ligand binds to a highly adaptable solvent exposed surface which upon complexation with ligand, and the other PBP domain, produces an environment similar to a less adaptable solvent excluded protein core [26,27]. In most cases PBP function relies on differential recognition of the apo and ligand bound forms of the protein by transmembrane-bound proteins [28]. Ligand binding at the interdomain interface stimulates a conformational change which is characterized as a rigid body hinge-bending/twisting motion about the interdomain linker region. The relative orientation of the two domains changes by as much as 60-70° [29,30], although the magnitude of the hinge-bending motion is variable and can be rather small in some cases [31]. The conformational coupling of ligand binding and function is also conserved when the PBP module is found in larger multidomain proteins, such as the eukaryotic glutamate receptor [8] and the LacI family of transcriptional regulators [7].

This PBP ligand binding paradigm was recently altered when the crystal structure of the *Thermotoga maritima* cellobiose binding protein (tmCBP) was solved [13]. Unlike other PBPs, the tmCBP binding site is bipartite, being composed of a typical PBP solvent excluded disaccharide binding site (subsite one) that is adjacent to an atypical large solvent filled cavity (subsite two) where three additional saccharide rings could be placed (Figure 1). The structure of tmCBP was solved in the presence of β(1,4) linked sugars, however the size of the tmCBP binding cavity and molecular modeling suggested additional glucan sugar linkages could be accommodated in both the disaccharide binding site and the solvent filled cavity (Figure 1). tmCBP is found in an operon that consists of an ABC transport system and an endoglucanase, the natural substrate of which has been predicted to be the algae-based storage polysaccharide laminarin [32-34]. Using a series of laminarin-based β(1,3) linked carbohydrates we sought to further identify the molecular mechanisms underlying simultaneous encoding of specificity and promiscuity in tmCBP subsites. These studies suggests ways that the bound hydrogen bonding rich water molecules bound in subsite two can potentially be used to adapt and expand ligand binding sites beyond the functionality encoded by the fixed protein scaffold.

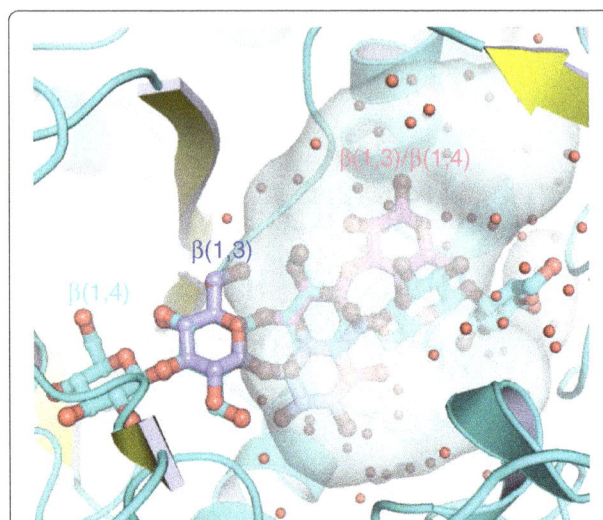

Figure 1 Carbohydrates accommodated in the tmCBP binding site. Previous molecular modeling studies of tmCBP identified that in addition to β(1,4) glucosaccharides (cyan), β(1,3) (blue) and mixed β(1,3)/β(1,4) (magenta) could be accommodated in the water-filled (red spheres) non-specific ligand binding subsite. Adapted from [13].

Results and discussion
Thermal stability and ligand binding specificity of tmCBP
Previous molecular modeling of the tmCBP binding [13] site suggested that in addition to β(1,4) oligosaccharides, additional linkages such as β(1,3) carbohydrates or mixed β(1,4)/β(1,3) could be accommodated in the bipartite binding site (Figure 1). Thermal denaturation of tmCBP, monitored by the change in circular dichroism (CD) signal, was used to assess the binding of the xylan-based β(1,4) linked xylose pentasaccharide, xylopentaose, and the laminarin-based β(1,3) linked glucose disaccharide, laminaribiose (LR2), and pentasaccharide, laminaripentaose (LR5). To bring the thermal melting point (T_m) into a measurable range, experiments were carried out in the presence of the chemical denaturant guanidine hydrochloride at a concentration of 2 M [17,35]. Addition of LR2 and LR5 shifted the T_m of the protein from 94.8°C to 99.2°C and 105.2°C respectively (Figure 2). These studies indicate that the tmCBP binding site accommodates β(1,3) glucosaccharides ranging in size from two to five sugar rings, which is consistent with earlier CD binding studies for the β(1,4) glucosaccharides cellobiose and cellopentaose [13]. Unexpectedly, the β(1,4) linked xylose-based sugar, xylopentaose, did not induce a change in the T_m indicating a lack of a stabilizing or binding interaction with tmCBP (Figure 2). In the previous molecular modeling of the tmCBP binding site, the first sugar ring of either the β(1,3) or β(1,4) glucosaccharides, and in-turn the C6 hydroxyl, is coincident among the two ligand bound forms (Figure 1). It is likely that the hydrogen bonding interactions of the carbohydrate ring C6 hydroxyl with the protein are important for the discrimination of xylo-

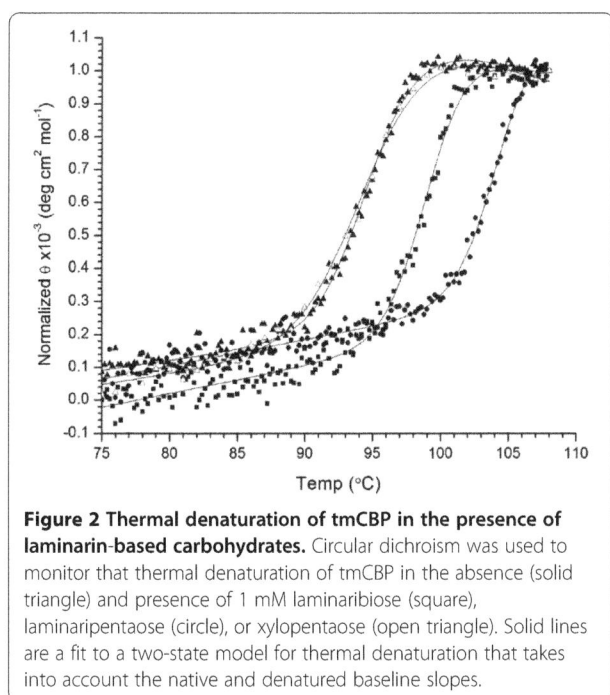

Figure 2 Thermal denaturation of tmCBP in the presence of laminarin-based carbohydrates. Circular dichroism was used to monitor that thermal denaturation of tmCBP in the absence (solid triangle) and presence of 1 mM laminaribiose (square), laminaripentaose (circle), or xylopentaose (open triangle). Solid lines are a fit to a two-state model for thermal denaturation that takes into account the native and denatured baseline slopes.

and glucosaccharides, as the xylosaccharides lack the C6 carbon and hydroxyl atoms.

Solution structure of apo and ligand-bound tmCBP

In order to characterize the conformational changes induced upon addition of ligand, small-angle neutron scattering (SANS) data was used to characterize both the apo and ligand bound forms of tmCBP. Cellobiose was used for these studies, which based on the previous crystal structures produces a closed state essentially identical to the laminarin-based carbohydrates. Comparison of the apo and ligand-bound curves show significant differences in both the high q and low q data, indicative of a ligand induced conformational change (Figure 3a). These raw data can be transformed into a Kratky plot where geometrical differences, such as compactness and flexibility, in the scattering particles can be highlighted. In the case of a multi-domain protein connected by flexible linkers, the Kratky plot would show a broad peak at lower q-values, with an upturn at the higher q-values. The Krakty plots for the apo and cellobiose-bound protein are similar in shape and indicative of a globular protein rather than two domains connected by a flexible linker (Figure 3b). This suggests that the tmCBP hinge does not allow for significant conformational flexibility in the absence of ligand which is consistent with the previous small-angle scattering studies of the group II maltose binding protein [36]. It is interesting to speculate as to whether the flexibility of the PBP hinge may be inherent to a particular PBP group as structures of group I PBPs in the absence of ligand have been shown to adopt

a series of domain closure angles, perhaps suggesting flexibility in the absence of ligand [29].

Upon addition of ligand the protein undergoes large scale conformational changes as evidenced in the decrease in the radius of gyration and D_{max} (Table 1). The large differences in these biophysical parameters of the scattering particles are of a greater magnitude than one would expect based upon previous small-angle scattering studies of other group II PBPs [36]. Molecular weight determination, based upon the intensity at zero scattering angle suggests that the apo protein forms inter-protein associations that are alleviated upon addition of cellobiose (Table 1). Inter-protein associations have previously been reported for this protein superfamily [37,38]. No molecular modeling of the apo protein was carried out. The SANS data of the cellobiose-bound protein is well accounted for by the previously determined cellobiose-bound tmCBP crystal structure (Figure 3a). This is also observed in comparison of the cellobiose-bound crystal structure and the *ab-initio* model generated from the SANS data in the presence of 5 mM cellobiose (Figure 3c).

Crystal structure of laminaribiose complex

The crystal structure of tmCBP complexed with LR2 was solved to a resolution of 2.05 Å by molecular replacement using the previously determined tmCBP structure [13] (Figure 4a). The structure was refined to R_{work} and R_{free} values of 18.4% and 20.3%, respectively. The final model consists of 582 amino acids, a larminaribiose molecule and 293 water molecules. The overall fold and conformation of the protein is similar to the structure of the cellobiose complexed protein (all atom RMSD= 0.4 Å). Data collection, stereochemistry and refinement statistics are summarized in Table 2.

An extensive network of polar and non-polar amino acids and water molecules bind the LR2 ligand (Figure 4b). A LigPlot+ representation of the LR2 binding pocket is shown in Additional file 1: Figure S1a [39]. As in other periplasmic carbohydrate binding proteins, a network of aromatic amino acids envelops the ligand between the N- and C-terminal domains. A total of six tryptophan residues (Trp15, Trp380, Trp383, Trp426, Trp510 and Trp535) form van der Waals interactions with the two sugar rings. In total, ten hydrogen bonds can be formed with the first sugar ring. Two hydrogen bonds are formed with the C3 and C4 hydroxyl each, whereas the C2 and C6 hydroxyls form three hydrogen bonds each. All but one hydrogen bond are directly formed with the protein, with the C6 hydroxyl being ligated by a single specifically bound water molecule, W20. For second sugar ring of the laminaribiose, only the C2 hydroxyl forms a direct hydrogen bond with the main chain carbonyl of Gly12, which also forms a hydrogen bond with the C2 hydroxyl of the first ring. The other six hydrogen bonds with the second ring are with

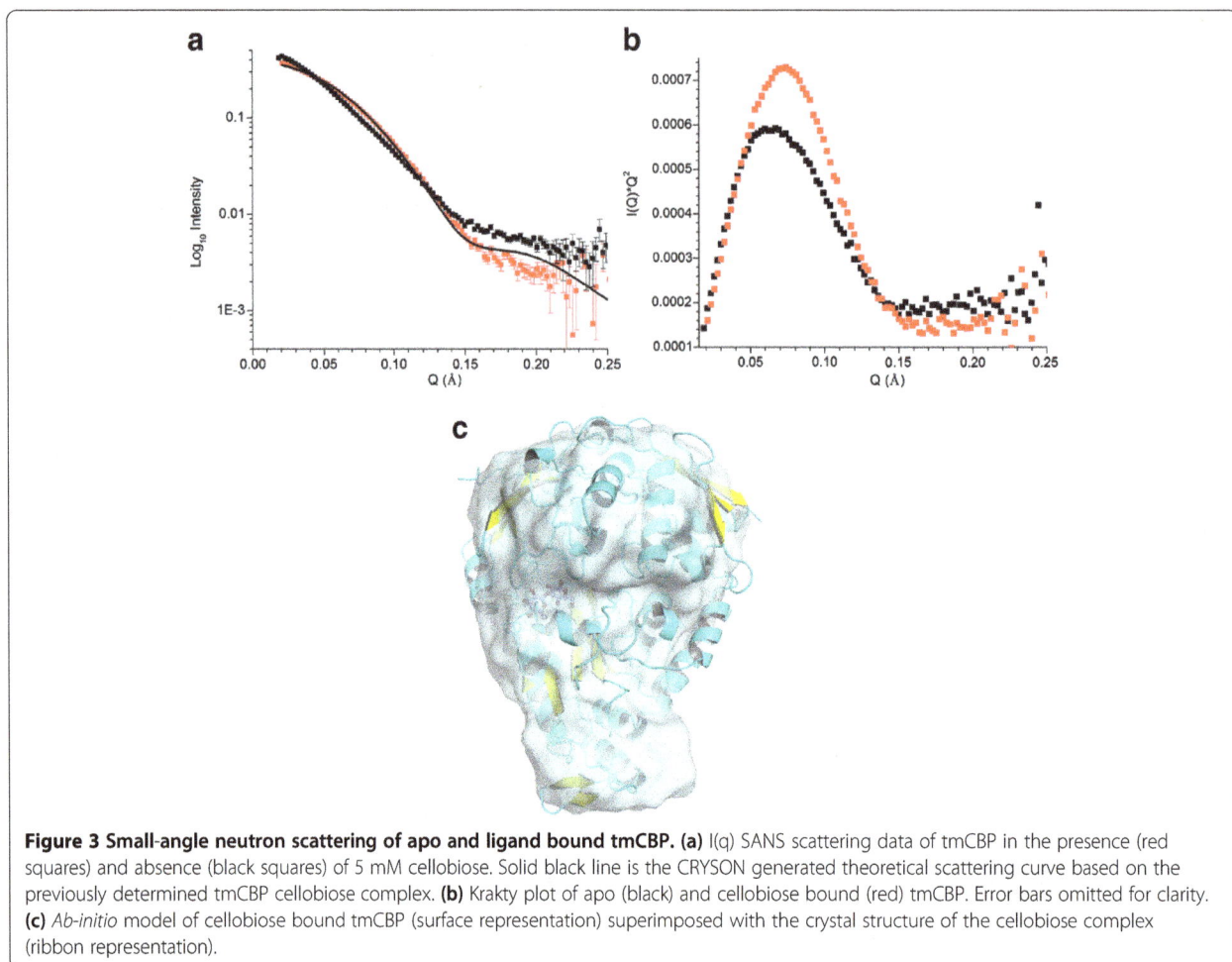

Figure 3 Small-angle neutron scattering of apo and ligand bound tmCBP. (a) I(q) SANS scattering data of tmCBP in the presence (red squares) and absence (black squares) of 5 mM cellobiose. Solid black line is the CRYSON generated theoretical scattering curve based on the previously determined tmCBP cellobiose complex. **(b)** Krakty plot of apo (black) and cellobiose bound (red) tmCBP. Error bars omitted for clarity. **(c)** *Ab-initio* model of cellobiose bound tmCBP (surface representation) superimposed with the crystal structure of the cellobiose complex (ribbon representation).

water molecules. The C1 hydroxyl of second ring forms hydrogen bonds with specifically bound water molecules, W47 and W150, whereas the C4 and C6 hydroxyls form hydrogen bonds with W39 and W6 water molecules. C2 hydroxyl also forms hydrogen bonds with W47, which forms another hydrogen bond with the main chain carbonyl of Ala13. The O5 hemiacetal oxygen of second ring

forms a hydrogen bond with another water molecule, W179 (Figure 4b).

Unlike the other tmCBP structures that have been solved thus far, both the LR2 complex and the LR5 complex contain a pentagonal bipyrimidally-bound calcium ion (Figures 4a and 5). This calcium ion was bound from the crystallization precipitant solution that contained

Table 1 SANS data analysis

Sample	Concentration (mg/mL)	Io	Io/C	MW (kDa) Calculated	MW (kDa) Expected	$R_{g(exp)}$ (Å)	$R_{g(calc)}$ (Å)	D_{max} (Å)
Apo	4.2	0.47± 0.001	0.1	73.9	68.7	30.6± 0.04	*	88
	8.4	1.0± 0.001	0.1	73.9	68.7	31.4± 0.03	*	90
	12.5	1.5± 0.002	0.1	73.9	68.7	31.5± 0.03	*	90
	16.7	1.9± 0.002	0.1	73.9	68.7	31.5± 0.015	*	90
Sat (5 mM Cellobiose)	4.2	0.39±0.001	0.09	66.5	68.7	24.1± 0.09	23.4	68
	8.4	0.72±0.0007	0.09	66.5	68.7	24.1± 0.02	23.4	67
	12.5	1.05±0.001	0.08	59.1	68.7	24.1± 0.01	23.4	67
	16.7	1.34±0.001	0.08	59.1	68.7	24.0± 0.01	23.4	67

*No model available for apo form.

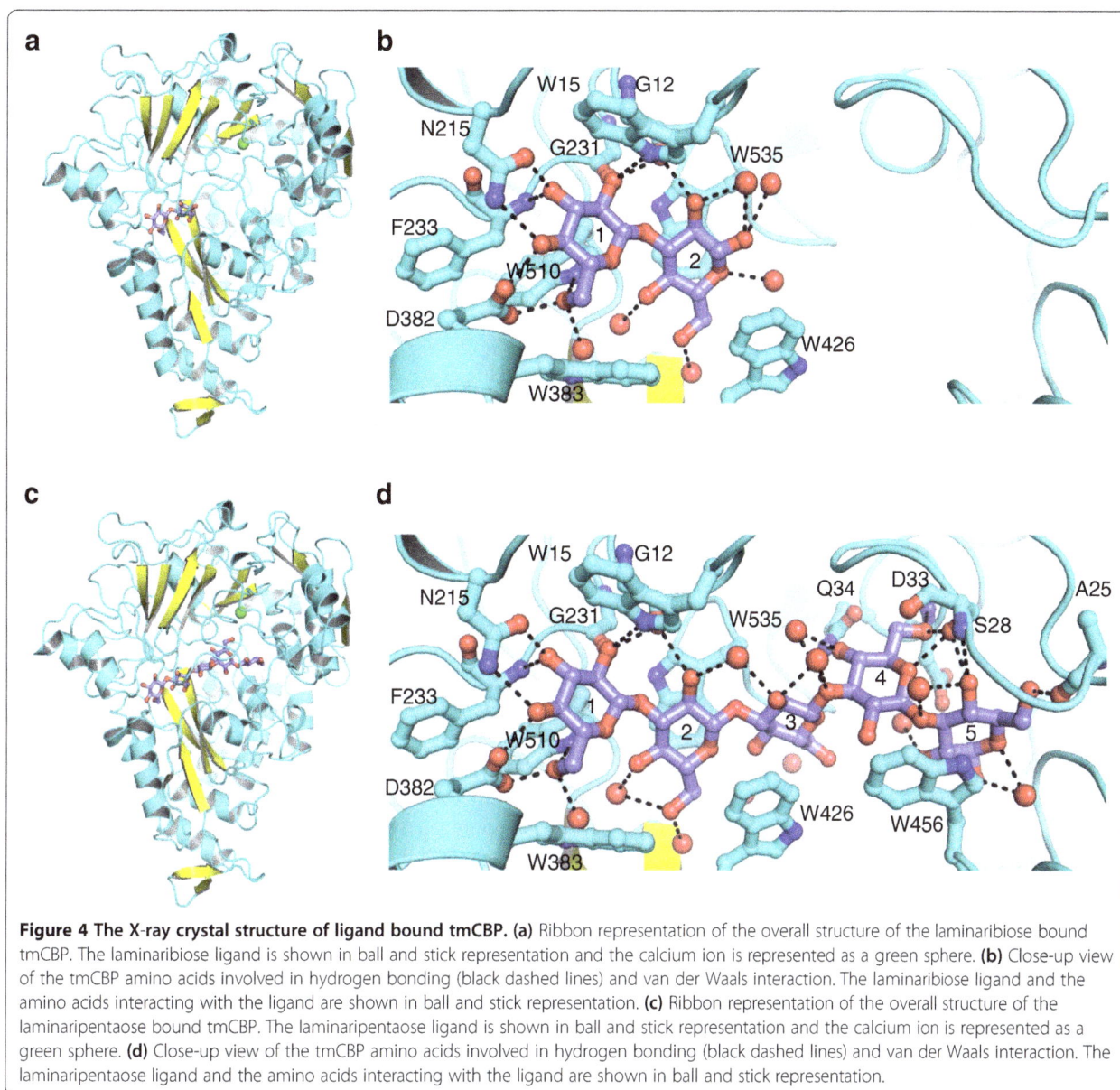

Figure 4 The X-ray crystal structure of ligand bound tmCBP. (a) Ribbon representation of the overall structure of the laminaribiose bound tmCBP. The laminaribiose ligand is shown in ball and stick representation and the calcium ion is represented as a green sphere. **(b)** Close-up view of the tmCBP amino acids involved in hydrogen bonding (black dashed lines) and van der Waals interaction. The laminaribiose ligand and the amino acids interacting with the ligand are shown in ball and stick representation. **(c)** Ribbon representation of the overall structure of the laminaripentaose bound tmCBP. The laminaripentaose ligand is shown in ball and stick representation and the calcium ion is represented as a green sphere. **(d)** Close-up view of the tmCBP amino acids involved in hydrogen bonding (black dashed lines) and van der Waals interaction. The laminaripentaose ligand and the amino acids interacting with the ligand are shown in ball and stick representation.

calcium acetate. The carbonyl of Tyr37, the side-chain Gln142, and the main carbonyl and side chain carboxylate of Asp33 fill four of the seven coordination sites, while the remainders are filled by a network of specifically bound water molecules (Figure 5).

Crystal structure of laminaripentaose complex

The crystal structure of tmCBP complexed with laminaripentaose (LR5) was solved to 2.07 Å resolution by molecular replacement using the previously determined tmCBP structure [13] (Figure 4c). This structure was refined to R_{work} and R_{free} values of 18.7% and 22.1%, respectively. The final model consists of 582 amino acids, a laminaripentaose molecule, and 237 water molecules. A calcium ion is also bound in an identical manner as the

LR2 complex. The overall fold of the protein is similar to the structure of the cellopentaose complexed protein (all atom RMSD= 0.2 Å). Data collection, stereochemistry and refinement statistics are summarized in Table 2.

Like the LR2 complex, an extensive network of polar and non-polar amino acids and water molecules also bind the LR5 ligand (Figure 4d). A LigPlot+ representation of the LR5 binding pocket is shown in Additional file 1: Figure S1a [39]. The first and second LR5 sugar rings are bound in an identical manner as the LR2 sugar rings. However, The O5 hemiacetal oxygen of the second LR5 sugar ring forms an intra-molecular interaction with the third sugar ring C4 hydroxyl, which replaces the water molecule W179 of LR2. All the hydroxyls of ring 3 of LR5 have only water mediated hydrogen bonds. The C2 hydroxyl and the

Table 2 Data collection and refinement statistics

	LR2	LR5
Data collection		
Space group	$P2_12_12_1$	$P2_12_12_1$
Cell parameters (Å)	a=56.6 b=89.6 c=108.2	a=56.9 b=89.8 c=108.3
Resolution range (Å)	50.0-2.05	50.0-2.07
Unique reflections	34613	33022
Redundancy[a]	3.4 (3.2)	4.3 (4.2)
Mean I/σ[a]	15.4 (2.5)	16.6 (3.5)
Completeness (%)[a]	98.1 (98.3)	95.9 (98.5)
R_{merge} (%)[a]	6.2 (48.8)	5.2 (49.4)
Refinement		
Num. of reflections (working /test set)	32549/2000	30944/1996
R_{work}/R_{free} (%)	18.4/20.3	18.8/22.1
Non-hydrogen atoms in refinement		
Protein	4831	4833
Water	293	237
Metal Ion	1	1
Carbohydrate	23	56
r.m.s.d.[b] from ideal		
Bond lengths (Å)	0.007	0.005
Bond angles (°)	1.7	1.1
B-factors (Å^2)		
Protein	33.5	38.7
Ligand	26.2	40.9
Metal Ion	28.6	29.0
Solvent	35.8	40.7
Ramachandran		
Ramachandran favored (%)	96.0	95.9
Ramachandran allowed (%)	99.5	99.3

[a]Number in parentheses represent values in the highest resolution shell.
[b]r.m.s.d. indicates root mean square deviation.

LR5 ring 3/4 hemiacetal each form two hydrogen bonds with three water molecules, while C4 and C6 hydroxyls form single hydrogen bond with W73 and W15, respectively. Unlike the ring three of LR5, ring four forms direct hydrogen bonds with the protein. The C6 hydroxyl forms two hydrogen bonds, while the C4 hydroxyl and O5 oxygen each form a single hydrogen bond with the protein. An additional water molecule, W129, hydrogen bonds with the C4 hydroxyl of the LR5 ring 4. The C4 hydroxyl of the fifth LR5 ring forms two hydrogen bonds, while the C6 hydroxyl forms a single hydrogen bond with the protein. Three specifically bound water molecules hydrogen bond with the C1, C2 and C4 hydroxyls; two of these waters also separately hydrogen bond with the ring 4/5 O3 hemiacetal oxygen and the O5 oxygen of ring 5. Ring 4 is

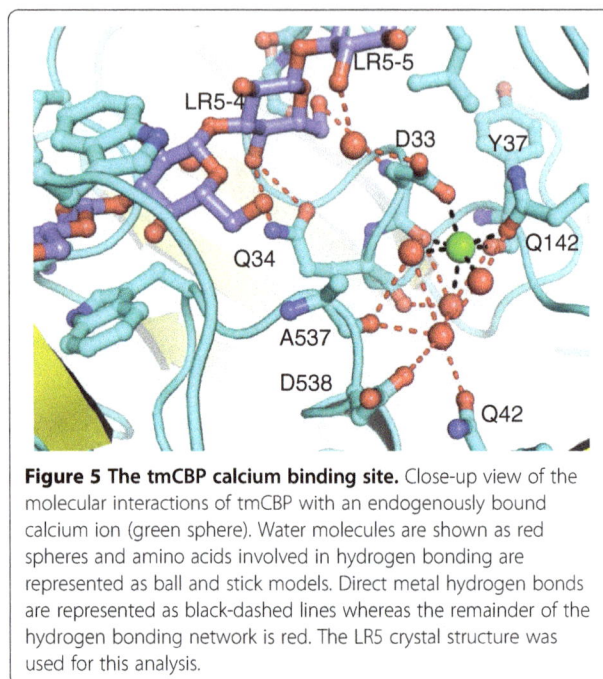

Figure 5 The tmCBP calcium binding site. Close-up view of the molecular interactions of tmCBP with an endogenously bound calcium ion (green sphere). Water molecules are shown as red spheres and amino acids involved in hydrogen bonding are represented as ball and stick models. Direct metal hydrogen bonds are represented as black-dashed lines whereas the remainder of the hydrogen bonding network is red. The LR5 crystal structure was used for this analysis.

essentially occupying the positions of W49, W87 and W171 of LR2 complex while ring 5 replaces the W119 water of the LR2 complex.

The calcium ion of the LR5 complex (Figure 4c), with slight variation in coordination distances, has identical coordination geometry as the calcium ion present in the LR2 structure. It is interesting to note that the Asp33, which primarily coordinates the calcium ion also forms hydrogen bonds to C6 hydroxyl of LR5 ring 4 and a water molecule that hydrogen bonds to the C2 hydroxyl of the fifth sugar ring (Figure 5). Moreover, the side chain of Gln34, of which the main chain carbonyl interacts with a water molecule that coordinates with the calcium ion, also forms two hydrogen bonds with the C4 hydroxyl of the LR5 ring four. The localization of this calcium ion suggests it may potentially have a functional role in ligand binding, rather than a structural role as found in other PBPs [22].

Comparison of laminarin and cellodextrin bound structures

Comparison of the laminarin and cellodextrin bound tmCBP structures allow for the identification of the molecular details of ligand selectivity in this semi-specific periplasmic binding protein. Superposition of the laminaripentaose (LR5) and cellopentaose (CP5) complexes demonstrates that the conformation of almost every amino acid that forms van der Waals interactions or hydrogen bonds with the LR5 or CP5 in both subsites is in an identical conformation (Figure 6). A single amino acid in subsite two, Gln142, adopts an alternate rotamer between the two structures. As the ligands occupy

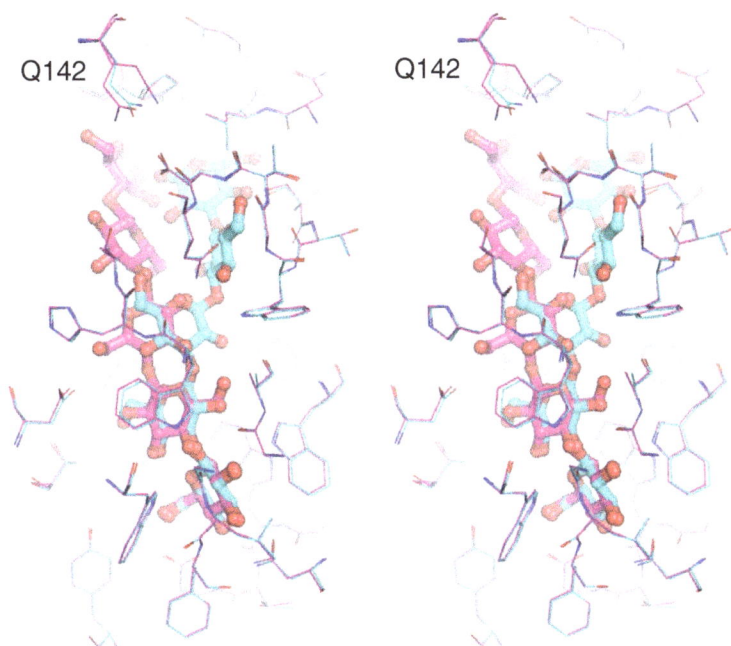

Figure 6 The tmCBP cellopentaose and laminaripentaose binding site. Stereo view of the superposition of tmCBP bound with cellopentaose (magenta) and laminaripentaose (cyan). All amino acids that interact with either ligand are shown in line representations.

distinct regions of the binding pocket, which remain fixed in position regardless of which ligand is bound, we postulate that the identical conformation of the hydrogen bonding and van der Waals ligand interaction network suggests the protein binding pocket is pre-ordered for binding of either ligand.

Role of water molecules in organizing the tmCBP binding site

The tmCBP binding site accommodates both β(1,3) and β(1,4) carbohydrates by utilizing the non-specific, subsite two, binding site [13]. Although the ligands occupy different regions of the non-specific subsite, there is significant overlap of the amino acids involved in the recognition of either ligand. We sought to understand how this occurs when the protein amino acid network remains in a fixed conformation. Superposition of the LR5 and CP5 bound structures suggest the bound water network may play an important role in the plasticity of the tmCBP binding site (Figure 7). Five distinct classes of water molecules that are specifically bound to either the N or C-terminal domain are found potentially mediating and modulating ligand selectivity in the tmCBP ligand binding site.

1. *Waters in identical positions coordinating identical atoms on different ligands* (Figure 7, red water molecule): The first LR5 and CP5 sugar rings are in identical positions and the water molecule coordinating the C6 hydroxyl is also found in an identical position. This water simultaneously forms

hydrogen bonds with both the N and C-terminal domains of the protein, the C6 hydroxyl of the ligand, and another bound water molecule that interacts with the second sugar ring. As five membered β-xylan carbohydrates do not bind to tmCBP it is possible that the positioning and the bonding network of this water molecule make it important in transducing to the protein that a six membered ring is bound in subsite one. The combination of the lack of this water molecule, and the lacking of the molecular interactions with the C6 carbon and hydroxyl group, potentially impede formation of the closed ligand bound state of the protein in the presence of xylans. These types of bridging, ligand-binding-induced interdomain contacts have been postulated to be important in stimulating the PBP conformational change [27].

2. *Waters in identical positions coordinating different ligand atoms* (Figure 7, green water molecules): Beyond the first sugar ring, the conformation or localization of the sugars in the binding pocket significantly differ. Although coordinating distinct ligand atoms which are located in different regions of subsite two, several water molecules are found conserved in the LR5 and CP5 bound structures. Water molecules that hydrogen bond with ring 2, 3, or 5 of the LR5 ligand are also found in the CP5 bound structure. The conservation of water molecules is also observed in the water molecules interacting with ring 2, 3, or 5 of the CP5 ligand. The rich hydrogen bonding potential of water molecules allows for conservation of water position in

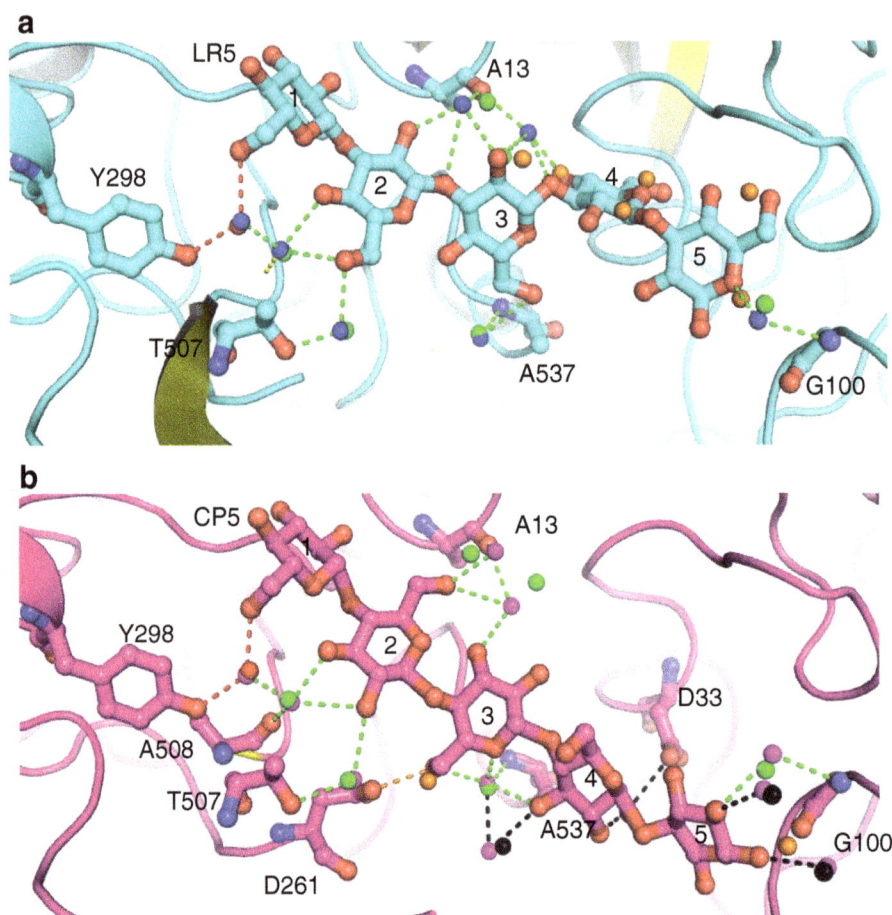

Figure 7 The network of water molecules modulating the ligand selectivity of tmCBP. (a) View of the LR5 water network. Water molecules found interacting with the LR5 are shown as blue spheres. The different classes of conserved water molecules also found in the CP5 binding site are colored as follows: red, waters in identical positions coordinating identical atoms; green, waters identical positions coordinating different ligand atoms; black, waters in identical positions forming ligand contacts in one form and not the other; orange, waters mimicking hydroxyl atoms of ligand. **(b)** View of the CP5 water network. Water molecules found interacting with the CP5 are shown as magenta spheres. The different classes of conserved water molecules also found in the LR5 binding site are colored as in **(a)**.

both the CP5 and LR5 forms and thereby potentially permits the preordering of the hydrogen bonding potential of subsite one for either type of ligand.

3. *Waters in identical positions forming ligand contacts in one form and not the other* (Figure 7b, black water molecules): Beyond the first two sugar rings found in subsite one, the localization of the LR5 and the CP5 in the tmCBP subsite two differ significantly. However, several water molecules that are involved in forming hydrogen bonds with LR5 are still present when CP5 is bound in subsite two. The same is also true for subsite two waters that hydrogen bond with CP5 and not LR5. This class of water molecule is involved in pre-ordering the rotameric state of subsite two hydrogen bonding residues. Although crystal structure of apo form of tmCBP protein could not be obtained, it is possible

the same water-mediated hydrogen bonding network pre-orders subsite 2 for binding of $\beta(1,3)$ or $\beta(1,4)$ ligands in the apo state.

4. *Waters that mimic hydroxyls/hemiacetals of other ligand* (Figure 7, orange water molecules): Several water molecules are present in either the LR5 or CP5 bound structure that mimic the localization in subsite two of ligand hydroxyl groups of the other ligand. Similar to the class 3 water molecules, this class of water molecules preforms both the water and protein hydrogen bonding interactions for either ligand, the role of which is potentially important for promiscuous ligand recognition of the apo protein.

5. *Secondary shell waters involved in coordinating primary shell waters (and/or involved coordinating preordering of binding pocket)*: Beyond the primary shell water molecules that directly interact with the

ligands, a conserved network of at least twelve water molecules is found ordering either the primary shell waters or the amino acids that are involved in interacting with the ligands.

Conclusions

Depending on biological function, PBPs ligand selectivity is modulated through a combination of binding pocket adaptations that mediate ligand positioning, alter ligand size selection, or alter the free energy of ligand binding in such a manner that excludes incorrect ligands [35,40]. The novel bipartite tmCBP binding site represents an additional, interesting alteration of PBP ligand recognition, exemplifying how both specificity and promiscuity are encoded in a single binding site. Comparison of the tmCBP structures bound to laminarin-based and cellodextrin-based carbohydrates allows for identification of the novel binding selectivity determinants found in this binding site.

The structural changes accompanying ligand binding in PBPs typically involves a re-organization of side-chain rotamers or the protein backbone. Although in some cases one of the two binding half sites in each domain undergoes conformational changes upon ligand binding and this has been suggested as a mechanism of ordering ligand binding [19]. Although lacking the apo crystal structure, analysis of the tmCBP ligand bound state suggests the binding site may be pre-ordered for binding of either laminarins or cellodextrins. Essentially no structural differences are observed among the amino acids in the ligand recognition sphere of either class of ligands. It should be pointed out that no side-chain movement is observed even when the same residue forms direct or indirect interactions with either ligand.

The conservation and positioning of water molecules trapped in the tmCBP binding pocket suggest they potentially play a role in tuning tmCBP ligand selectivity. Several classes of water molecules, playing distinct functional roles, are found in the tmCBP binding site. This network of water molecules preforms the ligand hydrogen bonding network and side chain conformations of ligand interacting amino acids, and thereby reduces the entropic penalty of ligand binding. Additionally, these water molecules are rich in hydrogen bonding potential, allowing for conservation of water placement while facilitating the plasticity of this bipartite binding site.

The mode of ligand binding found in tmCBP represents an interesting adaptation mechanism not previously observed in other PBPs. The downstream carbohydrate transport systems have a narrow, predefined limit to the size and type of carbohydrate that can be processed. The tmCBP binding cavity pre-filters this pool of ligands, potentially optimizing the transport process and in-turn eliminating the energetic penalty of presentation of incorrect carbohydrates to the transport machinery. In E. coli maltose binding protein ligands that do not fit within the binding site are still

bound and presented to the transport machinery [4]. However, with tmCBP a single protein is used to promiscuously select a molecular class of ligands while at the same time sterically restricting the number of rings that can be placed within the constraints of the tmCBP binding cavity. Larger ligands are likely not bound as they would impede the hinge bending motion and in-turn specific ligand recognition.

Seven additional oligosaccharide binding proteins with varying substrate specificities are found in the T. maritima genome [32]. It remains to be observed whether the mode of ligand recognition in tmCBP is unique or found across this subset of periplasmic carbohydrate binding proteins. The adaptation mechanisms observed in tmCBP allow for expansion of binding site selectivity while maintaining specificity for a molecular class of ligands. These types of adaptation mechanisms could potentially be recapitulated in drug design studies where the rich hydrogen bonding potential of water molecules can be utilized to expand binding sites, or enable multiple drugs to bind to a single target site.

Methods
Protein expression and purification
The tmCBP plasmid was transformed into BL21-RIL cells for heterologous expression in either terrific broth media, M9 minimal media or Enfor's minimal media supplemented with carbennicillin and chloramphenicol. Growth on terrific broth or M9 minimal media produced protein that was bound with a disaccharide ligand as determined by circular dichroism (CD) and X-ray crystallography (data not shown). Ligand-free protein was produced by growth on the glycerol based medium, Enfor's minimal media. In all cases, tmCBP was purified as previously described [13,32], with slight modifications. Cell pellets were lysed by sonication. The resulting lysate was clarified by centrifugation (34,000 × g) for 20 minutes. Following nickel chelation chromatography purification of the lysate, the protein was loaded on to a Superdex S75 26/60 (Amersham) gel filtration column that was equilibrated with 20 mM Tris, pH 8.0, 150 mM NaCl. This purified material was used for all other experiments.

Circular dichroism
CD experiments were performed on a Jasco CD spectrophotometer. Thermal denaturations were determined by measuring the CD signal at 225 nm as a function of temperature using 0.5 μM protein in 10mM Tris–HCl pH 8.0 and 40 mM NaCl. In the absence of guanidinium chloride, tmCBP is too stable to exhibit temperature-induced denaturation and all measurements were carried out in the presence of 2 M guanidine hydrochloride and 1.0 mM ligand. CD measurements were fit to a two-state model that takes into account the slope of the native and denatured baselines [41].

Small angle neutron scattering data collection and analysis

Small-angle neutron scattering (SANS) experiments were performed on the extended Q-range small-angle neutron scattering (EQ-SANS, BL-6) beam line at the Spallation Neutron Source (SNS) located at Oak Ridge National Laboratory (ORNL) [42]. Protein was concentrated to 16.7 mg/mL and dialyzed in to 20 mM Tris pH 8.0, 40 mM NaCl in 100% D_2O for SANS measurements that were performed at 20°C. Cellobiose was added to the apo protein at a concentration of 5 mM for measurements of the ligand bound form. Data reduction followed standard procedures using MantidPlot (http://www.mantidproject.org/) [43].

Upon verifying a Guinier regime [44] in the SANS profiles, the pair distance distribution function, $P(r)$, was calculated from the scattering intensity using the indirect Fourier transform method implemented in the GNOM program [45] (Table 1). The real-space radius of gyration, R_g, and scattering intensity at zero angle, $I(0)$, were determined from the $P(r)$ solution to the scattering data. The molecular mass, M, was calculated by $I(0) = M (\Delta\rho)^2 \bar{v}^2 / N_A$, where $\Delta\rho$ = contrast in scattering length density between protein and D_2O buffer solution (= $\rho_{prot} - \rho_{buf}$), $\bar{v}$ = protein partial specific volume (= 0.73 ml/g), and N_A = Avogadro's number. The GASBOR program [46] was used to generate *ab initio* shape reconstructions.

Crystallization and X-ray data collection

tmCBP was concentrated to 20 mg/mL and dialyzed into 10 mM Tris, 40 mM NaCl 0.5 mM TCEP for crystallization. Laminaribiose (LR2) or laminaripentaose (LR5) was added to a final concentration of 1 mM prior to crystallization trials. Crystals were grown by hanging drop vapor diffusion in drops containing 2 μL of the protein solution mixed with 2 μL of 0.2–0.3 M magnesium acetate or calcium acetate, 20–30% (wt/vol) PEG 3350 equilibrated against 900 μL of the same solution. Crystals were transferred to 35% (wt/vol) PEG 3350 for cryoprotection, mounted in a nylon loop, and flash frozen in liquid nitrogen. All X-ray diffraction data were collected at 100 K on a Rigaku 007HFmicromax X-ray generator with a Raxis IV++ detector. The diffraction data were scaled and indexed using HKL3000 [47]. The data collection statistics are listed in Table 2.

Structure determination, model building and refinement

The LR2 bound and LR5 bound tmCBP structures were solved by molecular replacement using the Phaser program [48]. The crystal structure of the previously determined tmCBP was used as the initial model for fitting the X-ray data ([13], PDB code 3I5O). Manual model building was carried out in COOT [49] and refined using REFMAC5 [48] and PHENIX [50]. The models exhibit

good stereochemistry as determined by MolProbity [51]; final refinement statistics are listed in Table 2.

Accession numbers

Atomic coordinates and structure factors have been deposited in the Protein Data Bank [52] under the accession codes 4JSD and 4JSO for the LR2 complex and LR5 complex, respectively.

Additional file

> **Additional file 1: Figure S1.** Interaction network of tmCBP laminarin ligands. The polar and non-polar contacts of the LR2 (a) and LR5 (b) ligands as generated by LigPlot [39]. The interaction network that is coincident among the LR2 and LR5 structures are highlighted in red, hydrogen bonding interactions are represented as black dashed lines and water molecules as red spheres.

Abbreviations

tmCBP: *Thermotoga maritima* cellobiose binding protein; CD: Circular dichroism; SANS: Small-angle neutron scattering; LR2: Laminaribiose; LR5: Laminaripentaose.

Competing interests

The authors declare that they have no competing interests.

Authors' contributions

MJC, PM and DAM designed the research and drafted the manuscript. MJC and CBC performed small-angle scattering experiments. MJC, PM, SG, and XL performed CD and X-ray crystallography experiments. All authors read and approved the final manuscript.

Acknowledgments

A portion of this research was performed at Oak Ridge National Laboratory's Spallation Neutron Source, sponsored by the U.S. Department of Energy, Office of Basic Energy Sciences. PM was funded in part through a research grant from the National Science Foundation (Award 0922719).

Author details

[1]Neutron Sciences Directorate, Oak Ridge National Laboratory, Oak Ridge, TN 37831, USA. [2]Department of Chemistry, Middle Tennessee State University, Murfreesboro, TN 37132, USA. [3]Department of Chemistry & Center for Informatics, Shiv Nadar University, Dadri, Uttar Pradesh 203207, India.

References

1. Neiditch MB, Federle MJ, Miller ST, Bassler BL, Hughson FM: **Regulation of LuxPQ receptor activity by the quorum-sensing signal autoinducer-2.** *Mol Cell* 2005, **18**(5):507–518.
2. Aksamit RR, Koshland DE Jr: **Identification of the ribose binding protein as the receptor for ribose chemotaxis in Salmonella typhimurium.** *Biochemistry* 1974, **13**(22):4473–4478.
3. Iida A, Harayama S, Iino T, Hazelbauer GL: **Molecular cloning and characterization of genes required for ribose transport and utilization in Escherichia coli K-12.** *J Bacteriol* 1984, **158**(2):674–682.
4. Shuman HA: **Active transport of maltose in Escherichia coli K12. Role of the periplasmic maltose-binding protein and evidence for a substrate recognition site in the cytoplasmic membrane.** *J Biol Chem* 1982, **257**(10):5455–5461.
5. Kelly DJ, Thomas GH: **The tripartite ATP-independent periplasmic (TRAP) transporters of bacteria and archaea.** *FEMS Microbiol Rev* 2001, **25**(4):405–424.
6. Campobasso N, Costello CA, Kinsland C, Begley TP, Ealick SE: **Crystal structure of thiaminase-I from Bacillus thiaminolyticus at 2.0 A resolution.** *Biochemistry* 1998, **37**(45):15981–15989.

7. Friedman AM, Fischmann TO, Steitz TA: Crystal structure of lac repressor core tetramer and its implications for DNA looping. *Science* 1995, **268**(5218):1721–1727.

8. Frandsen A, Pickering DS, Vestergaard B, Kasper C, Nielsen BB, Greenwood JR, Campiani G, Fattorusso C, Gajhede M, Schousboe A, *et al*: Tyr702 is an important determinant of agonist binding and domain closure of the ligand-binding core of GluR2. *Mol Pharmacol* 2005, **67**(3):703–713.

9. Bruns CM, Nowalk AJ, Arvai AS, McTigue MA, Vaughan KG, Mietzner TA, McRee DE: Structure of Haemophilus influenzae Fe(+3)-binding protein reveals convergent evolution within a superfamily. *Nat Struct Biol* 1997, **4**(11):919–924.

10. Sun YJ, Rose J, Wang BC, Hsiao CD: The structure of glutamine-binding protein complexed with glutamine at 1.94 A resolution: comparisons with other amino acid binding proteins. *J Mol Biol* 1998, **278**(1):219–229.

11. Berntsson RP, Thunnissen AM, Poolman B, Slotboom DJ: Importance of a hydrophobic pocket for peptide binding in lactococcal OppA. *J Bacteriol* 2011, **193**(16):4254–4256.

12. Chaudhuri BN, Ko J, Park C, Jones TA, Mowbray SL: Structure of D-allose binding protein from Escherichia coli bound to D-allose at 1.8 A resolution. *J Mol Biol* 1999, **286**(5):1519–1531.

13. Cuneo MJ, Beese LS, Hellinga HW: Structural analysis of semi-specific oligosaccharide recognition by a cellulose-binding protein of thermotoga maritima reveals adaptations for functional diversification of the oligopeptide periplasmic binding protein fold. *J Biol Chem* 2009, **284**(48):33217–33223.

14. Sugiyama S, Vassylyev DG, Matsushima M, Kashiwagi K, Igarashi K, Morikawa K: Crystal structure of PotD, the primary receptor of the polyamine transport system in Escherichia coli. *J Biol Chem* 1996, **271**(16):9519–9525.

15. Tirado-Lee L, Lee A, Rees DC, Pinkett HW: Classification of a Haemophilus influenzae ABC transporter HI1470/71 through its cognate molybdate periplasmic binding protein, MolA. *Structure* 2011, **19**(11):1701–1710.

16. Mauchline TH, Fowler JE, East AK, Sartor AL, Zaheer R, Hosie AH, Poole PS, Finan TM: Mapping the Sinorhizobium meliloti 1021 solute-binding protein-dependent transportome. *Proc Natl Acad Sci USA* 2006, **103**(47):17933–17938.

17. Cuneo MJ, Beese LS, Hellinga HW: Ligand-induced conformational changes in a thermophilic ribose-binding protein. *BMC Struct Biol* 2008, **8**:50.

18. Magnusson U, Chaudhuri BN, Ko J, Park C, Jones TA, Mowbray SL: Hinge-bending motion of D-allose-binding protein from Escherichia coli: three open conformations. *J Biol Chem* 2002, **277**(16):14077–14084.

19. Magnusson U, Salopek-Sondi B, Luck LA, Mowbray SL: X-ray structures of the leucine-binding protein illustrate conformational changes and the basis of ligand specificity. *J Biol Chem* 2004, **279**(10):8747–8752.

20. Shilton BH, Flocco MM, Nilsson M, Mowbray SL: Conformational changes of three periplasmic receptors for bacterial chemotaxis and transport: the maltose-, glucose/galactose- and ribose-binding proteins. *J Mol Biol* 1996, **264**(2):350–363.

21. Fukami-Kobayashi K, Tateno Y, Nishikawa K: Domain dislocation: a change of core structure in periplasmic binding proteins in their evolutionary history. *J Mol Biol* 1999, **286**(1):279–290.

22. Berntsson RP, Smits SH, Schmitt L, Slotboom DJ, Poolman B: A structural classification of substrate-binding proteins. *FEBS Lett* 2010, **584**(12):2606–2617.

23. Borrok MJ, Kiessling LL, Forest KT: Conformational changes of glucose/galactose-binding protein illuminated by open, unliganded, and ultra-high-resolution ligand-bound structures. *Protein Sci* 2007, **16**(6):1032–1041.

24. Bagaria A, Kumaran D, Burley SK, Swaminathan S: Structural basis for a ribofuranosyl binding protein: insights into the furanose specific transport. *Proteins* 2011, **79**(4):1352–1357.

25. Marvin JS, Hellinga HW: Manipulation of ligand binding affinity by exploitation of conformational coupling. *Nat Struct Biol* 2001, **8**(9):795–798.

26. Toth-Petroczy A, Tawfik DS: Slow protein evolutionary rates are dictated by surface-core association. *Proc Natl Acad Sci USA* 2011, **108**(27):11151–11156.

27. Dwyer MA, Hellinga HW: Periplasmic binding proteins: a versatile superfamily for protein engineering. *Curr Opin Struct Biol* 2004, **14**(4):495–504.

28. Gould AD, Telmer PG, Shilton BH: Stimulation of the maltose transporter ATPase by unliganded maltose binding protein. *Biochemistry* 2009, **48**(33):8051–8061.

29. Bjorkman AJ, Mowbray SL: Multiple open forms of ribose-binding protein trace the path of its conformational change. *J Mol Biol* 1998, **279**(3):651–664.

30. Alicea I, Marvin JS, Miklos AE, Ellington AD, Looger LL, Schreiter ER: Structure of the Escherichia coli phosphonate binding protein PhnD and

31. Karpowich NK, Huang HH, Smith PC, Hunt JF: Crystal structures of the BtuF periplasmic-binding protein for vitamin B12 suggest a functionally important reduction in protein mobility upon ligand binding. *J Biol Chem* 2003, **278**(10):8429–8434.

32. Nanavati DM, Thirangoon K, Noll KM: Several archaeal homologs of putative oligopeptide-binding proteins encoded by Thermotoga maritima bind sugars. *Appl Environ Microbiol* 2006, **72**(2):1336–1345.

33. Zverlov VV, Volkov IY, Velikodvorskaya TV, Schwarz WH: Highly thermostable endo-1,3-beta-glucanase (laminarinase) LamA from Thermotoga neapolitana: nucleotide sequence of the gene and characterization of the recombinant gene product. *Microbiology* 1997, **143**(Pt 5):1701–1708.

34. Conners SB, Montero CI, Comfort DA, Shockley KR, Johnson MR, Chhabra SR, Kelly RM: An expression-driven approach to the prediction of carbohydrate transport and utilization regulons in the hyperthermophilic bacterium Thermotoga maritima. *J Bacteriol* 2005, **187**(21):7267–7282.

35. Cuneo MJ, Changela A, Beese LS, Hellinga HW: Structural adaptations that modulate monosaccharide, disaccharide, and trisaccharide specificities in periplasmic maltose-binding proteins. *J Mol Biol* 2009, **389**(1):157–166.

36. Telmer PG, Shilton BH: Insights into the conformational equilibria of maltose-binding protein by analysis of high affinity mutants. *J Biol Chem* 2003, **278**(36):34555–34567.

37. Cuneo MJ, Changela A, Miklos AE, Beese LS, Krueger JK, Hellinga HW: Structural analysis of a periplasmic binding protein in the tripartite ATP-independent transporter family reveals a tetrameric assembly that may have a role in ligand transport. *J Biol Chem* 2008, **283**(47):32812–32820.

38. Gonin S, Arnoux P, Pierru B, Lavergne J, Alonso B, Sabaty M, Pignol D: Crystal structures of an Extracytoplasmic Solute Receptor from a TRAP transporter in its open and closed forms reveal a helix-swapped dimer requiring a cation for alpha-keto acid binding. *BMC Struct Biol* 2007, **7**:11.

39. Wallace AC, Laskowski RA, Thornton JM: LIGPLOT: a program to generate schematic diagrams of protein-ligand interactions. *Protein Eng* 1995, **8**(2):127–134.

40. Millet O, Hudson RP, Kay LE: The energetic cost of domain reorientation in maltose-binding protein as studied by NMR and fluorescence spectroscopy. *Proc Natl Acad Sci USA* 2003, **100**(22):12700–12705.

41. Cohen DS, Pielak GJ: Stability of yeast iso-1-ferricytochrome c as a function of pH and temperature. *Protein Sci* 1994, **3**(8):1253–1260.

42. Zhao JK, Gao CY, Liu D: The extended Q-range small-angle neutron scattering diffractometer at the SNS. *J Appl Cryst* 2010, **43**:1068–1077.

43. Wignall GD, Bates FS: Absolute calibration of small-angle neutron scattering data. *J Appl Cryst* 1987, **20**:28–40.

44. Guinier A, Fournet G: *Small-angle scattering of X-rays.* New York: Wiley; 1955.

45. Svergun DI: Determination of the regularization parameter in indirect-transform methods using perceptual criteria. *J Appl Cryst* 1992, **25**:495–503.

46. Svergun DI, Petoukhov MV, Koch MH: Determination of domain structure of proteins from X-ray solution scattering. *Biophys J* 2001, **80**(6):2946–2953.

47. Otwinowski ZaM W: Processing of X-ray diffraction data collected in oscillation mode. *Methods Enzymol* 1997, **276A**:307–326.

48. Collaborative Computational Project N: The CCP4 suite: programs for protein crystallography. *Acta Crystallogr D Biol Crystallogr* 1994, **50**(Pt 5):760–763.

49. Emsley P, Cowtan K: Coot: model-building tools for molecular graphics. *Acta Crystallogr D Biol Crystallogr* 2004, **60**(Pt 12 Pt 1):2126–2132.

50. Adams PD, Afonine PV, Bunkoczi G, Chen VB, Davis IW, Echols N, Headd JJ, Hung LW, Kapral GJ, Grosse-Kunstleve RW, *et al*: PHENIX: a comprehensive Python-based system for macromolecular structure solution. *Acta Crystallogr D Biol Crystallogr* 2010, **66**(Pt 2):213–221.

51. Davis IW, Murray LW, Richardson JS, Richardson DC: MOLPROBITY: structure validation and all-atom contact analysis for nucleic acids and their complexes. *Nucleic Acids Res* 2004, **32**:W615–619. Web Server issue.

52. Berman HM, Westbrook J, Feng Z, Gilliland G, Bhat TN, Weissig H, Shindyalov IN, Bourne PE: The Protein Data Bank. *Nucleic Acids Res* 2000, **28**(1):235–242.

12

The crystal structure of JNK from *Drosophila melanogaster* reveals an evolutionarily conserved topology with that of mammalian JNK proteins

Sarin Chimnaronk[1], Jatuporn Sitthiroongruang[1], Kanokporn Srisucharitpanit[2], Monrudee Srisaisup[1], Albert J. Ketterman[1] and Panadda Boonserm[1*]

Abstract

Background: The c-Jun N-terminal kinases (JNKs), members of the mitogen-activated protein kinase (MAPK) family, engage in diverse cellular responses to signals produced under normal development and stress conditions. In *Drosophila*, only one JNK member is present, whereas ten isoforms from three JNK genes (JNK1, 2, and 3) are present in mammalian cells. To date, several mammalian JNK structures have been determined, however, there has been no report of any insect JNK structure.

Results: We report the first structure of JNK from *Drosophila melanogaster* (DJNK). The crystal structure of the unphosphorylated form of DJNK complexed with adenylyl imidodiphosphate (AMP-PNP) has been solved at 1.79 Å resolution. The fold and topology of DJNK are similar to those of mammalian JNK isoforms, demonstrating their evolutionarily conserved structures and functions. Structural comparisons of DJNK and the closely related mammalian JNKs also allow identification of putative catalytic residues, substrate-binding sites and conformational alterations upon docking interaction with *Drosophila* scaffold proteins.

Conclusions: The DJNK structure reveals common features with those of the mammalian JNK isoforms, thereby allowing the mapping of putative catalytic and substrate binding sites. Additionally, structural changes upon peptide binding could be predicted based on the comparison with the closely-related JNK3 structure in complex with pepJIP1. This is the first structure of insect JNK reported to date, and will provide a platform for future mutational studies in *Drosophila* to ascertain the functional role of insect JNK.

Background

The c-Jun N-terminal kinases (JNKs), the extracellular signal-regulated kinases (ERK1/2) and the 38 kDa MAP kinases (p38 MAPKs) are members of the mitogen-activated protein kinase (MAPK) family [1]. Of the MAPK pathways, the JNK signaling pathway is of particular importance for various biological processes such as cell proliferation, differentiation, apoptosis, cellular responses to stress as well as embryonic development and morphogenesis [1]. Activation of JNKs requires a dual phosphorylation, by upstream JNK kinases (MKK4 and MKK7), on two conserved threonine and tyrosine residues within the Thr-X-Tyr motif in their activation

loops [2]. The phosphorylated JNKs subsequently activate their substrates such as transcriptional factors, resulting in mediating the gene expression regulation in response to stress stimuli.

In mammalian cells, at least 10 different splicing isoforms of three JNK genes (JNK1, 2, and 3) have been identified [2]. JNK1 and JNK2 are widely expressed in a variety of tissues, whereas JNK3 is selectively expressed in the brain, heart and testis [2]. Similar to mammalian cells, the JNK signaling pathway is also conserved in *Drosophila*. The *Drosophila* JNK pathway consists of *Drosophila* JNK or basket (DJNK) and JNK kinase Hep, which are homologs of JNK and MKK7 in mammals, respectively [3]. The important role of *Drosophila* JNK pathway in embryogenesis has been emphasized by the findings that mutations in the *Drosophila* JNK pathway

* Correspondence: panadda.boo@mahidol.ac.th
[1]Institute of Molecular Biosciences, Mahidol University, Salaya, Phuttamonthon, Nakhon Pathom 73170, Thailand
Full list of author information is available at the end of the article

components resulted in the disruption of the morphogenetic process of dorsal closure during embryogenesis [4]. Although there are ten JNK isoforms in mammals, only one JNK isoform (DJNK) exists in *Drosophila* [4]. Due to a lower genetic complexity and redundancy of the JNK signaling components in comparison to their mammalian counterparts, the *Drosophila* JNK pathway thus becomes a simpler model system for a study of JNK regulation under normal development and stress conditions.

Several efforts have been made to determine the structures of different mammalian JNK isoforms either in the presence or absence of the JNK pathway scaffolding proteins. In contrast, the structural information of DJNK and other related *Drosophila* JNK signaling components is still unavailable. In this study, we have solved the crystal structure of unphosphorylated DJNK in complex with adenylyl imidodiphosphate (AMP-PNP) and magnesium at 1.79 Å resolution. This is the first insect JNK structure to be solved and, together with the mammalian JNK structures, provides crucial insights into the evolutionary conservation of structures and catalytic regulation across insect and mammalian JNK proteins.

Methods

Protein expression and purification

The *Drosophila* JNK gene (*DJNK* or *basket* [GenBank: AAB97094.1]) was cloned in-frame with the hexahistidine tag at the N-terminus as previously described [5]. *E. coli* BL21 (DE3)pLysS cells containing the recombinant plasmid were grown in Luria-Bertani (LB) broth containing 100 µg/ml kanamycin and 34 µg/ml chloramphenicol at 37 °C until an OD_{600} reached 0.6. The protein expression was induced by the addition of IPTG (isopropyl β-D thiogalactopyranoside) at a final concentration of 0.2 mM. The cultured cells were further grown for 15 h at 25 °C, and harvested by centrifugation at $6000 \times g$ for 10 min. Cell pellet was resuspended in 20 ml of buffer A (20 mM sodium phosphate buffer pH 7.4, 0.5 M NaCl) containing 4 mg/ml lysozyme and 1 mM β-mercaptoethanol. Cells were disrupted by the French press at 1000 psi, and the lysate was clarified by centrifugation at $10,000 \times g$ for 30 min, followed by being filtered through a 0.22 µm membrane. The supernatant was loaded onto a 5-ml HiTrap™ Chelating HP column packed with precharged Ni^{2+} resin (GE Healthcare Life Sciences). The bound (His)$_6$-tagged JNK protein was eluted by buffer A containing 200 mM imidazole. The imidazole was further removed using a HiTrap™ desalting column (GE Healthcare Life Sciences) pre-equilibrated with 50 mM Tris–HCl, pH 8.0 and 10 mM DTT. The eluted fractions were concentrated to 10 mg/ml by using a Microsep-10 (PALL). The proteins collected at every step were analyzed by 12 % sodium dodecyl sulfate–polyacrylamide electrophoresis (SDS-PAGE). The concentration of protein was determined by Bradford's assay using BSA as standard protein.

Protein crystallization

Crystallization trials were conducted using hanging- and sitting-drop vapour-diffusion methods in 24 and 96-well plates at 295 K (Molecular Dimensions, UK and QIAGEN, Germany). Before setting up the crystallization, the purified DJNK was mixed with a substrate, AMP-PNP (Adenosine 5′-(β,γ-imido) triphosphate lithium salt hydrate) and $MgCl_2$ at the concentrations of 1 mM and 2 mM, respectively. Initial screening was performed using Hampton Research Crystal Screen kits (Hampton Research, USA) and positive hits were then optimized. Drops were prepared by mixing 1.2 µl of 10 mg/ml protein with an equivalent volume of reservoir solution and were equilibrated against 500 µl of reservoir solution. After optimizing the conditions, prism crystals were obtained from a condition consisting of 25 % (w/v) PEG 4000, 0.1 M Tris–HCl pH 8.5 and 0.2 M $MgCl_2$. The crystals grew to a maximum size of about 50×100 µm within 1 month after incubation at 20 °C.

X-ray data collection and structural determination

Preliminary X-ray diffraction was performed using an in-house X-ray source (MICROSTAR™, BRUKER, Thailand) at the MX end station at Synchrotron Light Research Institute (SLRI, Thailand), whereas high resolution data were collected at the beamline BL32XU of the SPring-8 synchrotron (Hyogo, Japan). Prior to data collection, a single crystal was briefly soaked in its reservoir solution containing 12 % sucrose as a cryoprotectant before being flash-frozen in the nitrogen stream at 100 K. A total of 560 images were collected with an oscillation angle of 0.2° at a crystal-to-detector distance of 150 mm. The exposure time per image was set to 1 s. The X-ray diffraction pattern images were processed and scaled using XDS program package [6]. The structure was solved with the molecular replacement program MOLREP in *CCP*4 program suite [7] using the crystal structure of human JNK1 [PDB:2XS0 [8]] as a search model. The electron density map was calculated and a model was built with the graphic software COOT [9]. Several rounds of refinement were performed using REFMAC5 [10]. Ligand and water molecules were added at nearly the final step of refinement and manually validated. The final model including water molecules was justified by monitoring the R_{factor} and R_{free}. The quality of protein structure was determined by MolProbity [11]. Statistics of data collection and refinement are summarized in Table 1. The residues falling outside the allowed region were modified by adjusting torsion angles, Phi (Φ), Psi (ψ). All images of the DJNK structure were prepared with PyMOL (Delano Scientific, Palo Alto, CA, http://www.pymol.org/).

Table 1 Data collection and refinement statistics of DJNK structure

Data collection	
Space group	$P2_12_12_1$
Unit-cell parameters (Å)	$a = 52.49$, $b = 55.33$, $c = 126.72$
High resolution limit (Å)	1.58
Total reflections	219226
Unique reflections	51133
Average mosaicity (°)	0.0
Completeness (%)	99.4 (97.6)[a]
Redundancy	4.3(4.2)[a]
[b]R_{merge} (%)	6.2 (79.4)[a]
$I/\sigma(I)$	11.8 (1.8)[a]
Refinement statistics	
Resolution range (Å)	28.28-1.79
Highest resolution shell (Å)	1.84-1.79
[c]R-factor (%)	18.9
[d]R_{free}-factor (%)	22.4
Averaged B-factor (Å²)	26.0
Number of non-hydrogen atoms	
Protein	2810
Water	143
ADPNP	1
Mg^{2+} ions	2
RMSD from ideal geometry	
Bond length (Å)	1.07
Bond angle (°)	1.06
[e]Ramachandran plot (%)	
Favored regions	97.0
Additional allowed regions	3.0
Outliers	0

[a]Number in parentheses refer to the outer resolution shell
[b]$R_{merge} = \Sigma_{hkl}\Sigma_i |I_{hkl,i} - \langle I_{hkl}\rangle| / \Sigma_{hkl}\Sigma_i I_{hkl,i}$, where $\langle I_{hkl}\rangle$ is the mean intensity of symmetry-equivalent reflection
[c]R-factor $= \Sigma |F_{obs}-F_{cal}|/\Sigma F_{obs}$, where F_{obs} and F_{cal} are observed and calculated structure factor amplitudes, respectively
[d]R_{free}-factor value was calculated as R-factor but using a subset (10 %) of reflections that were not used for refinement
[e]Ramachandran plot was calculated using MolProbity [11]

Accession number

Atomic coordinates and structure factor amplitudes of DJNK have been deposited into the Protein Data Bank (PDB) with the accession code 5AWM.

Results and discussion
Overall structure of DJNK

The crystal structure of *Drosophila* JNK (DJNK) has been solved to 1.79 Å resolution and consists of 354 amino acid residues (Gln7-Tyr362) with one chemical compound of AMP-PNP (ATP analog) bound at the active site cleft

(Fig. 1). The electron densities are not visible in four loop regions encompassing residues Ser32-Gln35, Thr181-Tyr183, Asn282-Asn283, Asp342-Glu343, likely due to structural disorder. The overall DJNK structure is composed of two distinct domains. The N-terminal domain (residues 7−109 and residues 338−362) consists of seven β strands (β1L0, β2L0, β1, β2, β3, β4, and β5) and two α helices (αC and αL16), while the C-terminal domain (residues 110−172 and residues 187−337) contains mostly α helices (αD, αE, αF, αG, αH, αI, αL12, α1L14, α2L14, α3L14 and αIL16,) with five short 3_{10}-helices and three β strands (βL5, β7, and β8) (Fig. 1). The two domains are connected by the so-called hinge regions (residues 107−111 and residues 330−348). A deep cleft at the domain interface contains the ATP-binding site (Fig. 1). DJNK also possesses extensions and insertions characteristic for MAP kinases including an N-terminal β-hairpin (β1L0 and β2L0), an extended loop between αG and α1L14, α1L14, α2L14, α3L14, and αL16.

Structural comparisons of DJNK with other known JNK structures was performed using the DALI server [12]. As expected, the overall structure of DJNK shows a

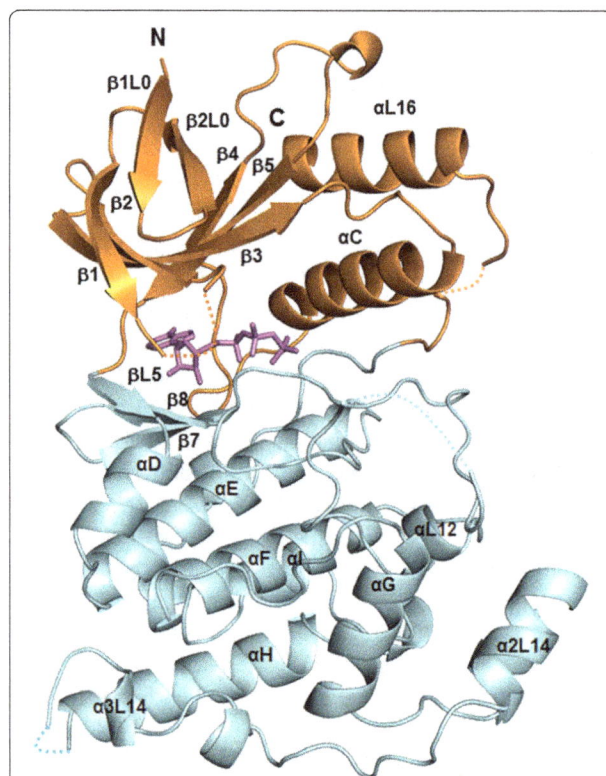

Fig. 1 Ribbon representation of *Drosophila* JNK complexed with AMP-PNP at a resolution of 1.79 Å. The DJNK structure is formed by two distinct domains: the N-terminal domain (orange) and the C-terminal domain (cyan). The bound AMP-PNP located in a deep cleft at the domain interface is shown in a magenta stick model. Secondary-structure elements are labeled as described in additional file 1. The disordered connecting loops are shown by dotted lines

similar protein kinase fold to that of mammalian JNK1 [PDB:1UKH [13]], mammalian JNK2 [PDB:3E7O [14]] and mammalian JNK3 [PDB:1JNK [15]] with amino acid sequence identities of 82, 77 and 79 %, respectively (see Additional file 1).

The highly-flexible activation loop

The activation loop of DJNK, also referred to as the phosphorylation lip, spans residues Leu166 to Ala191. This region contains the dual phosphorylation sites phosphorylated by specific upstream MAP kinase kinases. These residues, Thr181 and Tyr183, are found in the conserved Thr-X-Tyr motif. Despite the common regulatory mechanism of dual phosphorylation, the number of residues in the activation loops among protein kinase family members is variable [16]. No electron density, however, is visible for residues Thr181-Tyr183, including the two phosphorylation sites in the activation loop of DJNK, implying its high flexibility in the unphosphorylated form. The similar disordered phosphorylation sites are also found in the unphosphorylated forms of JNK2 and JNK3 [14, 15, 17]. On the other hand, one of the two protein chains of JNK2 reveals a well-ordered activation loop which is mainly stabilized by crystal packing interactions. Based on these observations, it could be assumed that both Thr181 and Tyr183 of DJNK are solvent-exposed and do not participate in crystal packing interaction similarly to the corresponding residues Thr221 and Tyr223 in the unphosphorylated JNK3 [15], thereby becoming readily accessible to the JNK upstream activating kinases. Although the activation loop of DJNK is assumed to adopt multiple conformations in solution, the restructuring of this loop to adopt an appropriate conformation upon phosphorylation to allow the enzyme to recognize the specific substrate may also occur. Hence, the inherent flexibility of DJNK activation loop may play a central role in the enzymatic regulation upon substrate binding.

The putative conformational changes of the catalytic site upon peptide binding

In the structure of DJNK complexed with AMP-PNP, the AMP-PNP is bound in a deep cleft between the N- and C-terminal domains. The glycine-rich sequence (Gly31-Ser–Gly-Ala-Gln-Gly-Ile-Val38) of DJNK, however, is not well defined, implicating that this loop is highly flexible and does not directly interact with AMP-PNP. The residues important for the ATP binding come from different parts of the DJNK structure. The orientation of AMP-PNP in DJNK is determined by the hydrogen bonds between the phosphate groups of AMP-PNP and the backbone amide of Met109 and backbone carbonyl groups of Glu107 and Ser153. The side chain of Lys53, a putative catalytic residue which is highly conserved within the MAP kinase family [15], is directly involved

in the formation of hydrogen bonds with the α- and β-phosphoryl groups of the nucleotide, thereby playing a key role in the proper positioning of ATP in MAP kinases. Glu107 and Asn112, found to be conserved among JNK isoforms, also make hydrogen bonds with the phosphate groups. Two fixed magnesium ions are observed in the DJNK. The side chain carbonyl groups of Asn154 and Glu71 appear to interact with the phosphate groups of AMP-PNP through the Mg^{2+} ions. An important role in metal chelation has been proposed for Asp184 in cAMP-dependent protein kinase (cAPK), corresponding to Asp167 of DJNK, which forms direct interactions with the metal ions [18]. The side chain of Asp167 of DJNK does not directly interact with the metal ion, however, it is connected to the metal ion through an interaction with the side chain of Asn154. A similar interaction is also observed between Asp207 and Asn194 of JNK3 (Fig. 2) [15].

To gain insights into the conformational changes upon peptide binding, the structure of JNK3 complexed with 11-mer JIP1 peptide (pepJIP1), a peptide of JNK scaffolding protein, was compared with the structure of peptide-free DJNK. In JNK3-pepJIP1 structure [PDB:4H39 [17]], part of the activation loop coils into a helix that partially occupies the ATP binding site. When the full structures of DJNK and JNK3-pepJIP1 complex were superimposed, the root-mean-square (r.m.s.) Cα distance was increased by ~1 Å compared with that from the superposition with peptide-free JNK3. Major structural changes upon peptide binding

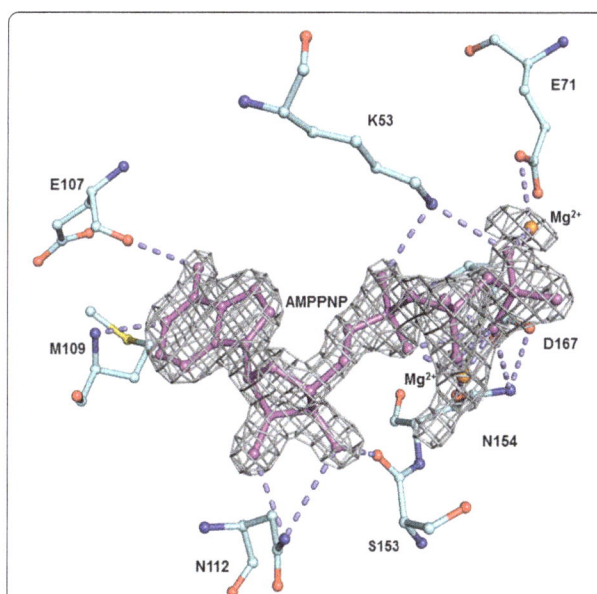

Fig. 2 AMP-PNP bound to *Drosophila* JNK. F_o-F_c omit electron density map (contoured at 4σ) is shown in grey for the bound DJNK-AMP-PNP (magenta stick model). Amino acid residues crucial for the ATP binding of DJNK are shown in cyan. Two Mg^{2+} ions are shown as orange balls. Hydrogen bonds are indicated as dashed lines

were observed in several regions especially in the N-terminal domain. In particular, the β1-β2 and glycine-rich loop of the JNK3-pepJIP1 was shifted by ~6 Å relative to the corresponding regions of the DJNK (Fig 3a). This substantial shift is required to allow space for the activation loop to dock into the ATP binding site of JNK3 (Fig. 3a). As a result of the shift in glycine-rich loop, the β1L0-β2L0 hairpin loop of JNK3-pepJIP1 is moved away by ~8 Å relative to the corresponding loop of DJNK to avoid the steric clashes.

The conserved aspartic acids in both HRDLKXXN and DFG motifs in the C-terminal domain, corresponding to Asp149 and Asp167 of DJNK, have been thought to be essential for protein kinase activity as previously described [19]. These two catalytic residues together with two other conserved key residues, Lys53 and Glu71 in the N-terminal domain of DJNK, are compared with the corresponding residues of JNK3-pepJIP1. Most of the catalytic residues of DJNK are in slightly different positions compared to their corresponding residues in JNK3-pepJIP1 (Fig. 3b). The disruption of hydrogen bond network formed between Asn194, Glu207 and Mg^{2+} ions, known as a requisite for catalysis, is apparent in JNK3-pepJIP1 structure (Fig. 3b). The peptide-induced

conformational changes also involve a shift in αC helix, consequently disrupting the hydrogen bonding interactions of residues in the αC helix with those in the activation loop (Fig. 3a). The shifted αC is also associated with a subtle move of the N-terminal MAPK insert αL16 (Fig. 3a).

Taken together, the conformational changes upon pepJIP1 binding potentially blocks the formation of active catalytic sites as a result of the failure to properly bind to ATP. This allosteric inhibition mechanism, nonetheless, is likely reversible once the activation loop is phophorylated, allowing the catalytically active conformation to form. Considering that the ATP binding sites and catalytically important residues are evolutionarily conserved among mammalian JNK isoforms and DJNK, the different JNKs thus likely undergo common structural transformations during peptide binding.

The putative peptide-binding and docking sites
In the MAPK signaling pathway, docking interactions are formed between the specific conserved regions on MAPKs, so called common docking groove or CD groove, and their interacting molecules. Specificity and enzymatic reactions are determined by the specific interactions of CD grooves

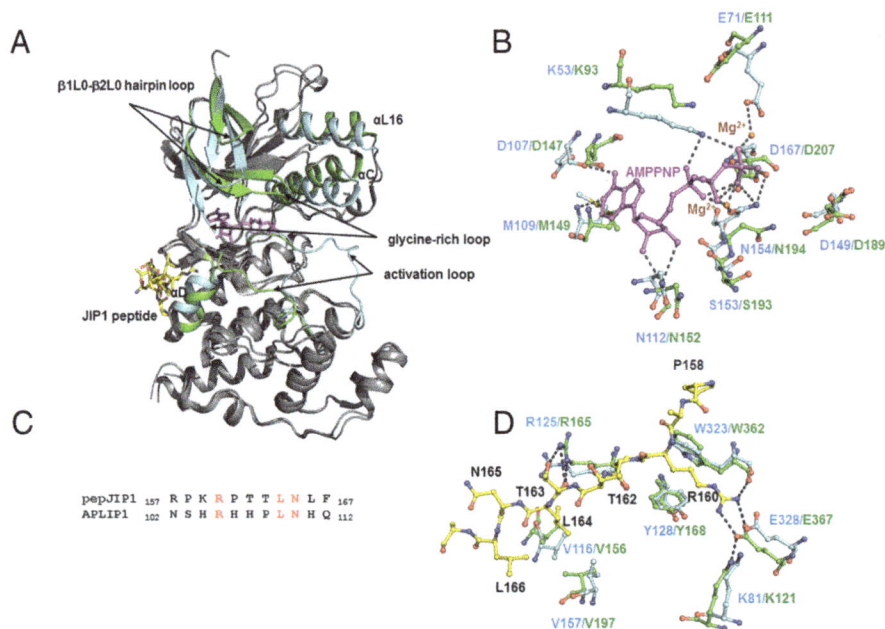

Fig. 3 The putative conformational changes of DJNK upon peptide binding. **a** Global superposition of the DJNK bound to AMP-PNP with the JNK3-pepJIP1 complex. The regions of major structural changes upon peptide binding including the glycine-rich loop, the activation loop, the N-terminal MAPK insert (β1L0-β2L0 hairpin), αL16, αC, and αD are highlighted in green for JNK3-pJIP1 and cyan for DJNK, respectively, whereas the homologous regions are in grey color. **b** Close-up view of conformational changes in the ATP-binding and catalytic sites. Amino acid residues crucial for the ATP binding and catalytic activity of DJNK and JNK3-pepJIP1 are shown in cyan and green, respectively. The bound DJNK-AMP-PNP is shown in a magenta stick model. Two Mg^{2+} ions are shown as orange balls. Hydrogen bonds are indicated as dashed lines. **c** Sequence alignment of the docking sites of the scaffolding proteins JIP1 and APLIP1. The conserved residues are highlighted in red. **d** Close-up view of superposition of DJNK and JNK3-pepJIP1 in the peptide-binding sites. Amino acid residues crucial for the scaffold protein binding of DJNK and JNK3-pepJIP1 are labeled as in panel **b**, whereas the residues of pepJIP1 are presented in yellow sticks and labeled in black

with specific docking sites or docking motifs commonly found in transcription factors, upstream activating kinases, phosphatases, scaffold proteins, and substrates [20]. The docking sites possess both basic and hydrophobic sequences in the arrangement of $(R/K)_{2-3}$-$(X)_{1-6}$-$\emptyset_A$-X-$\emptyset_B$ (where $\emptyset_A$ and $\emptyset_B$ are hydrophobic residues) [20]. Despite sharing common linear motifs, variations in sequences of CD grooves and docking sites appear to be an important determinant of pathway specificity for the different MAPKs.

JNK-interacting protein-1 (JIP1) was identified as a scaffold protein of the JNK module. The mammalian JIP homolog, namely APLIP1 (β-Amyloid Precursor-Like Interacting Protein 1), was identified in *Drosophila melanogaster* [21]. The similarities of APLIP1 to the mammalian homologs JIP1 and JIP2 include abundant expression in neural tissues, interactions with component(s) of the JNK signaling pathway and with the motor kinesin, and formation of homo-oligomers [21], reflecting their conserved functions and structures. Based on amino acid sequence alignment of mammalian JIP1 and *Drosophila* APLIP1 (Accession numbers Q9UQF2 and Q9W0K0, respectively), a putative docking motif could be mapped at positions 102–112 of APLIP1 (Fig. 3c, also see Additional file 2). Thus, although the structure of DJNK has been solved in the unbound state, it is still possible to understand some aspects of the docking specificity via both primary sequence and three-dimensional structure comparisons.

Superposition of DJNK and JNK3-pepJIP1 reveals that amino acid residues in their CD grooves, corresponding to those on the C-terminal domain surface covering αD, αE, and β7-β8 reverse turn, are conserved and homologous in positions (Fig. 3d). The side chain of Arg160 of pepJIP1, which is at equivalent position to Arg105 of APLIP1, rotates to form an extensive electrostatic network with Glu367 (in αIL16) and Lys121 (in loop following αC) of JNK3. These residues are well superimposed on their corresponding DJNK Glu328 and Lys81 residues, implying the presence of a conserved electrostatic network in DJNK (Fig. 3d). The pepJIP1 Arg160 also forms a hydrogen bond with the backbone carbonyl oxygen atom of JNK3 Trp362, conserved in position with Trp323 of DJNK. In fact, the critical importance of Arg160 in pepJIP1 was verified by ITC measurement as substitutions of this residue conferred significant lower affinity of JNK3-pepJIP1 binding [17]. Another conserved pepJIP1 Leu164 residue ($\emptyset_A$ in the $\emptyset_A$-X-$\emptyset_B$ motif), corresponding to Leu109 of APLIP1, makes van der Waal contacts with Val156 and Val197 of JNK3 which are also conserved in positions with Val116 and Val157 of DJNK, respectively. These two conserved basic and hydrophobic residues in the docking sites of JIP1 and APLIP1 thus serve as anchor points for interacting with the common surface features of JNKs. On the other hand, pepJIP1 Asn165, conserved with APLIP1 Asn110,

does not have any interaction with JNK3 but it forms a hydrogen bond with the backbone nitrogen atom of pepJIP1 Phe167, suggesting a role for this conserved residue in local folding of the peptide.

Despite the presence of consensus residues in the docking site, APLIP1 failed to interact with DJNK under experimental conditions *in vitro* [21]. This seems to be associated with the lower affinity of DJNK and APLIP1 binding, compared with the JNK3-pepJIP1 counterpart. A notable difference is that one of the key hydrophobic residues corresponding to $\emptyset_B$ in the $\emptyset_A$-X-$\emptyset_B$ motif is replaced by His111 in APLIP1, thus likely contributing to the weak hydrophobic interactions with the hydrophobic pocket of DJNK docking groove. In conjunction with the hydrophobic motif, the variable amino acids located between the conserved anchor points also could be important determinants of the ineffective docking interaction of DJNK and APLIP1. On the other hand, APLIP1 was found to bind to *Drosophila* Hep, a protein kinase upstream of DJNK, which is functionally similar to mammalian JIP1 [21]. Thus, APLIP1 shares most of the features in common with mammalian JIP1, except for the binding to JNK. Therefore, a structural basis of DJNK binding specificity with its binding partners remains to be elucidated via biochemical assays and co-crystallization with the peptide.

Conclusions

This is the first structural determination of *Drosophila* JNK, thereby serving as a representative of insect JNKs. The structure reveals common architectures with those of the mammalian JNK isoforms, allowing the identification of putative catalytic and substrate binding sites. Structural changes induced by peptide binding could be anticipated based on the comparison with the closely-related JNK3 structure in complex with pepJIP1. Although the evidence of specificity of DJNK with partner proteins is still limited, the structure and sequence alignments provide some clues for the docking interaction of DJNK with its putative scaffold protein.

Abbreviations
JNK: c-Jun N-terminal kinase; DJNK: *Drosophila* JNK.

Competing interests
The authors declare that they have no competing interests.

Author's contributions
SC collected the X-ray diffraction, determined the structure and carried out the coordinate deposition. JS carried out the protein expression, purification, crystallization, and structure determination. KS assisted with coordinate deposition. MS assisted with protein expression, purification and crystallization. AK assited with experimental design, data analysis and manuscript editing. PB conceived the study and developed the manuscript. All authors read and approved the final manuscript.

Acknowledgements

We are grateful to Yoshiaki Kawano and Yoshitaka Bessho for data collection at the BL32XU station of SPring-8 (Harima, Japan), and the Synchrotron Light Research Institute (Public Organization, Thailand) for providing the in-house X-ray machine (PX station). This work was supported by Mahidol University (to SC, AK and PB), the Naito Memorial Grant for Natural Science Research from the Naito Foundation, Japan (to SC) and the Thailand Research Fund (IRG5780009).

Author details

[1]Institute of Molecular Biosciences, Mahidol University, Salaya, Phuttamonthon, Nakhon Pathom 73170, Thailand. [2]Faculty of Allied Health Sciences, Burapha University, Mueang District, Saen Sook, Chonburi 20131, Thailand.

References

1. Johnson GL, Lapadat R. Mitogen-activated protein kinase pathways mediated by ERK, JNK, and p38 protein kinases. Science. 2002;298:1911–2.
2. Davis RJ. Signal transduction by the JNK group of MAP kinases. Cell. 2000;103:239–52.
3. Noselli S, Agnes F. Roles of the JNK signaling pathway in *Drosophila* morphogenesis. Curr Opin Genet Dev. 1999;9:466–72.
4. Sluss HK, Han Z, Barrett T, Goberdhan DC, Wilson C, Davis RJ, et al. A JNK signal transduction pathway that mediates morphogenesis and an immune response in *Drosophila*. Genes Dev. 1996;10:2745–58.
5. Udomsinprasert R, Bogoyevitch MA, Ketterman AJ. Reciprocal regulation of glutathione S-transferase spliceforms and the *Drosophila* c-Jun N-terminal kinase pathway components. Biochem J. 2004;383:483–90.
6. Kabsch W. Xds. Acta Crystallogr D Biol Crystallogr. 2010;66:125–32.
7. Vagin A, Teplyakov A. Molecular replacement with MOLREP. Acta Crystallogr D Biol Crystallogr. 2010;66:22–5.
8. Garai A, Zeke A, Gogl G, Toro I, Fordos F, Blankenburg H, et al. Specificity of linear motifs that bind to a common mitogen-activated protein kinase docking groove. Sci Signal. 2012;5:ra74.
9. Emsley P, Lohkamp B, Scott WG, Cowtan K. Features and development of Coot. Acta Crystallogr D Biol Crystallogr. 2010;66:486–501.
10. Skubak P, Murshudov GN, Pannu NS. Direct incorporation of experimental phase information in model refinement. Acta Crystallogr D Biol Crystallogr. 2004;60:2196–201.
11. Davis IW, Leaver-Fay A, Chen VB, Block JN, Kapral GJ, Wang X, et al. MolProbity: all-atom contacts and structure validation for proteins and nucleic acids. Nucleic Acids Res. 2007;35:W375–83.
12. Holm L, Rosenstrom P. Dali server: conservation mapping in 3D. Nucleic Acids Res. 2010;38:W545–9.
13. Heo YS, Kim SK, Seo CI, Kim YK, Sung BJ, Lee HS, et al. Structural basis for the selective inhibition of JNK1 by the scaffolding protein JIP1 and SP600125. EMBO J. 2004;23:2185–95.
14. Shaw D, Wang SM, Villasenor AG, Tsing S, Walter D, Browner MF, et al. The crystal structure of JNK2 reveals conformational flexibility in the MAP kinase insert and indicates its involvement in the regulation of catalytic activity. J Mol Biol. 2008;383:885–93.
15. Xie X, Gu Y, Fox T, Coll JT, Fleming MA, Markland W, et al. Crystal structure of JNK3: a kinase implicated in neuronal apoptosis. Structure. 1998;6:983–91.
16. Okano I, Hiraoka J, Otera H, Nunoue K, Ohashi K, Iwashita S, et al. Identification and characterization of a novel family of serine/threonine kinases containing two N-terminal LIM motifs. J Biol Chem. 1995;270:31321–30.
17. Laughlin JD, Nwachukwu JC, Figuera-Losada M, Cherry L, Nettles KW, LoGrasso PV. Structural mechanisms of allostery and autoinhibition in JNK family kinases. Structure. 2012;20:2174–84.
18. Knighton DR, Xuong NH, Taylor SS, Sowadski JM. Crystallization studies of cAMP-dependent protein kinase. Cocrystals of the catalytic subunit with a 20 amino acid residue peptide inhibitor and MgATP diffract to 3.0 Å resolution. J Mol Biol. 1991;220:217–20.
19. Gibbs CS, Knighton DR, Sowadski JM, Taylor SS, Zoller MJ. Systematic mutational analysis of cAMP-dependent protein kinase identifies unregulated catalytic subunits and defines regions important for the recognition of the regulatory subunit. J Biol Chem. 1992;267:4806–14.
20. Sharrocks AD, Yang SH, Galanis A. Docking domains and substrate-specificity determination for MAP kinases. Trends Biochem Sci. 2000;25:448–53.
21. Taru H, Iijima K, Hase M, Kirino Y, Yagi Y, Suzuki T. Interaction of Alzheimer's beta -amyloid precursor family proteins with scaffold proteins of the JNK signaling cascade. J Biol Chem. 2002;277:20070–8.

13

Designing and evaluating the MULTICOM protein local and global model quality prediction methods in the CASP10 experiment

Renzhi Cao[1], Zheng Wang[4] and Jianlin Cheng[1,2,3*]

Abstract

Background: Protein model quality assessment is an essential component of generating and using protein structural models. During the Tenth Critical Assessment of Techniques for Protein Structure Prediction (CASP10), we developed and tested four automated methods (MULTICOM-REFINE, MULTICOM-CLUSTER, MULTICOM-NOVEL, and MULTICOM-CONSTRUCT) that predicted both local and global quality of protein structural models.

Results: MULTICOM-REFINE was a clustering approach that used the average pairwise structural similarity between models to measure the global quality and the average Euclidean distance between a model and several top ranked models to measure the local quality. MULTICOM-CLUSTER and MULTICOM-NOVEL were two new support vector machine-based methods of predicting both the local and global quality of a single protein model. MULTICOM-CONSTRUCT was a new weighted pairwise model comparison (clustering) method that used the weighted average similarity between models in a pool to measure the global model quality. Our experiments showed that the pairwise model assessment methods worked better when a large portion of models in the pool were of good quality, whereas single-model quality assessment methods performed better on some hard targets when only a small portion of models in the pool were of reasonable quality.

Conclusions: Since digging out a few good models from a large pool of low-quality models is a major challenge in protein structure prediction, single model quality assessment methods appear to be poised to make important contributions to protein structure modeling. The other interesting finding was that single-model quality assessment scores could be used to weight the models by the consensus pairwise model comparison method to improve its accuracy.

Keywords: Protein model quality assessment, Protein model quality assurance program, Protein structure prediction, Support vector machine, Clustering

Background

Predicting protein tertiary structure from amino acid sequence is of great importance in bioinformatics and computational biology [1,2]. During the last few decades, a lot of protein tertiary structure prediction methods have been developed. One category of methods adopts a template-based approach [3-7], which uses experimentally determined structures as templates to build structural models for a target protein without known structure. Another category uses a template-free approach [8,9], which

tries to fold a protein from scratch without using known template structures. The two kinds of methods were often combined to handle a full spectrum of protein structure prediction problems ranging from relatively easy homology modeling to hard *de novo* prediction [10-13].

During protein structure prediction, one important task is to assess the quality of structural models produced by protein structure prediction methods. A model quality assessment (QA) method employed in a protein structure prediction pipeline is critical for ranking, refining, and selecting models [3]. A model quality assessment method can generally predict a global quality score measuring the overall quality of a protein structure model and a series of local quality scores measuring the local quality of each residue in the model. A global

* Correspondence: chengji@missouri.edu
[1]Computer Science Department, University of Missouri, Columbia, Missouri 65211, USA
[2]Informatics Institute, University of Missouri, Columbia, Missouri 65211, USA
Full list of author information is available at the end of the article

quality score can be a global distance test (GDT-TS) score [14-16] that is predicted to be the structural similarity between a model and the unknown native structure of a protein. A local quality score of a residue can be the Euclidean distance between the position of the residue in a model and that in the unknown native structure after they are superimposed.

In general, protein model quality assessment methods can be classified into two categories: multi-model methods [17-21] and single-model methods [13-17]. Multi-model methods largely use a consensus or clustering approach to compare one model with other models in a pool of input models to assess its quality. Generally, a model with a higher similarity with the rest of models in the pool receives a higher global quality score. The methods tend to work well when a large portion of models in the input pool are of good quality, which is often the case for easy to medium hard template-based modeling. Multi-model methods tend to work particularly well if a large portion of good models were independently generated by a number of independent, diverse protein structure prediction methods as seen in the CASP (the Critical Assessment of Techniques for Protein Structure Prediction) experiments, but they worked less well when being applied to the models generated by one single protein structure prediction method because they prefer the average model of the largest model cluster in the model pool. And multi-model methods tend to completely fail if a significant portion of low quality modes are similar to each other and thus dominate the pairwise model comparison as seen in some cases during the 10th CASP experiment (CASP10) held in 2012. Single-model methods strive to predict the quality of a single protein model without consulting any other models [22-26]. The performance of single-model methods is still lagging behind the multi-model methods in most cases when most models in the pool are of good quality [23,27]. However, because of their capability of assessing the quality of one individual model, they have potential to address one big challenge in protein structure modeling – selecting a model of good quality from a large pool consisting of mostly irrelevant models. Furthermore, as the performance of multi-model quality assessment methods start to converge, single-model methods appear to have a large room of improvement as demonstrated in the CASP10 experiment.

In order to critically evaluate the performance of multi-model and single-model protein model quality assessment methods, the CASP10 experiment was designed to assess them in two stages. On Stage 1, 20 models of each target spanning a wide range of quality were used to assess the sensitivity of quality assessment methods with respect to the size of input model pool and the quality of input models. On Stage 2, about top 150 models selected by a naïve consensus model quality

assessment method were used to benchmark model quality assessment methods' capability of distinguishing relatively small differences between more similar models. The new settings provided us a good opportunity to assess the strength and weakness of our multi-model and single-model protein model quality assessment methods in terms of accuracy, robustness, consistency and efficiency in order to identify the gaps for further improvement.

In addition to evaluating our four servers on the CASP10 benchmark, we compare our methods with three popular multi-model clustering-based methods (Davis-QAconsensus [28], Pcons [29], and ModFOLDclust2 [21]). Our clustering-based methods (MULTICOM-REFINE, MULTICOM-CONSTRUCT) performed comparably to the three external tools in most cases. Our single-model methods (MULTICOM-CLUSTER, MULTICOM-NOVEL) had a lower accuracy than the clustering-based methods, but performed considerably better than them on the models of hard template-free targets. Besides the reasonable performance and a comprehensive comparative study, our methods have some methodological innovations such as using single-model quality scores to weight models for clustering methods, repacking side chains before model evaluation, and improved machine learning methods for single-model quality assessment for template-free targets.

The rest of the paper is organized as follows. In the Results and discussions section, we analyze and discuss the performance of the methods on the CASP10 benchmark. In the Conclusion section, we summarize this work and conclude it with the directions of future work. In the Methods section, we introduce the methods in our protein model quality assessment servers tested in CASP10.

Results and discussions
Results of global quality predictions
We evaluated the global quality predictions using five measures (see the detailed descriptions of the evaluation methods in the Evaluation methods section). The results of the global quality evaluation on Stage 1 of CASP10 are shown in Table 1. The weighted pairwise model comparison method MULTICOM-CONSTRUCT performed best among all our four servers according to all the five measures, suggesting using single-model quality prediction scores as weights can improve the multi-model pairwise comparison based quality prediction methods such as MULTICOM-REFINE. The two multi-model global quality assessment methods had the better average performance than the two single-model global quality assessment methods (MULTICOM-NOVEL and MULTICOM-CLUSTER) on average on Stage 1, suggesting that the advantage of multi-model methods over single-model methods was not much affected by the relatively small size of input models (i.e. 20). Instead, the

Table 1 The average correlation (Ave. Corr.), overall correlation (Over. Corr.), average GDT-TS loss (Ave. Loss), average Spearman's correlation (Ave. Spearman), average Kendall tau correlation (Ave. Kendall) of MULTICOM servers, DAVIS-QAconsensus, Pcons, and ModFOLDclust2 on Stage 1 of CASP10

Servers	Ave. corr.	Over. corr.	Ave. loss	Ave. Spearman	Ave. Kendall
MULTICOM-REFINE	0.6494	0.8162	0.0615	0.5989	0.4908
MULTICOM-CLUSTER	0.5144	0.5946	0.0727	0.4364	0.3273
MULTICOM-NOVEL	0.5016	0.4848	0.0791	0.4483	0.3380
MULTICOM-CONSTRUCT	0.6838	0.8300	0.0613	0.6182	0.5043
DAVIS-QAconsensus	0.6403	0.7927	0.0537	0.5798	0.4745
Pcons	0.7501	0.7683	0.0327	0.6781	0.5457
ModFOLDclust2	0.6775	0.8301	0.0572	0.6206	0.5064

multi-model methods still work reasonably well on a small model pool that contains a significant portion of good quality models. It is worth noting that the average loss of the two single-model quality assessment methods (MULTICOM-CLUSTER and MULTICOM-NOVEL) is close to that of the two multi-model quality assessment methods (MULTICOM-REFINE and MULTICOM-CONSTRUCT) (i.e. +0.07 versus +0.06). We also compared our methods with three popular multi-model clustering-based methods (DAVIS-QAconsensus, Pcons, and ModFOLDclust2) on Stage 1. According to the evaluation, MULTICOM-CONSTRUCT performed slightly better than the naive consensus method DAVIS-QAconsensus and ModFOLDclust2, while Pcons performed best.

Table 2 shows the global quality evaluation results on Stage 2. Similarly as in Table 1, the weighted pairwise comparison multi-model method (MULTICOM-CONSTRUCT) performed better than the simple pairwise multi-model method (MULTICOM-REFINE) and both had better performance than the two single-model quality assessment methods (MULTICOM-CONSTRUCT and MULTICOM-NOVEL). That the two single-model quality prediction methods yielded the similar performance indicated that some difference in their input features (amino acid sequence versus sequence profile) did not significant affect their accuracy. In comparison with Stage 1, all our methods performed worse on

Stage 2 models. Since the models in Stage 2 are more similar to each other than in Stage 1 in most cases, the results may suggest that both multi-model and single-model quality assessment methods face difficulty in accurately distinguishing models of similar quality. On Stage 2 models, MULTICOM-CONSTRUCT delivered a performance similar with DAVIS-QAconsensus and Pcons, and had a higher average correlation than ModFOLDclust2.

We used the Wilcoxon signed ranked sum test to assess the significance of the difference in the performance of our four servers, DAVIS-QAconsensus, Pcons, and ModFOLDclust2. The p-values of the difference between these servers are reported in Table 3. On Stage 1 models, according to 0.01 significant threshold, the difference between clustering-based methods (MULTICOM-REFINE and MULTICOM-CONSTRUCT) and single-model methods (MULTICOM-CLUSTER and MULTICOM-NOVEL) is significant, but the difference between our methods in the same category is not significant. One Stage 2 models, the difference between all pairs of our servers except the two single-model methods is significant. Compared with the three external methods (DAVIS-QAconsensus, Pcons, and ModFOLDclust2), the difference between our multi-model method MULTICOM-REFINE and the three methods is not significant, while the difference between our single-model methods (MULTICOM-CLUSTER, MULTICOM-NOVEL) and the three methods is significant. The difference between

Table 2 The average correlation, overall correlation, average GDT-TS loss, average Spearman's correlation, average Kendall tau correlation of MULTICOM servers, DAVIS-QAconsensus, Pcons, and ModFOLDclust2 on Stage 2 of CASP10

Servers	Ave. corr.	Over. corr.	Ave. loss	Ave. Spearman	Ave. Kendall
MULTICOM-REFINE	0.4743	0.8252	0.0511	0.4763	0.3510
MULTICOM-CLUSTER	0.3354	0.6078	0.0675	0.3361	0.2343
MULTICOM-NOVEL	0.3350	0.5057	0.0654	0.3394	0.2358
MULTICOM-CONSTRUCT	0.4853	0.8272	0.0510	0.4824	0.3566
DAVIS-QAconsensus	0.5050	0.8383	0.0499	0.5031	0.3686
Pcons	0.4891	0.8194	0.0416	0.4843	0.3524
ModFOLDclust2	0.4489	0.8337	0.0470	0.4621	0.3393

Table 3 The p-value of pairwise Wilcoxon signed ranked sum test for the difference of correlation score between MULTICOM servers and three external methods (DAVIS-QAconsensus, Pcons, ModFOLDclust2) on Stage 1 and Stage 2 of CASP10

MULTICOM servers, DAVIS-QAconsensus, Pcons, and ModFOLDclust2 on Stage 1 or Stage 2	P-value
MULTICOM-REFINE and MULTICOM-CLUSTER on Stage 1	7.552e-05
MULTICOM-REFINE and MULTICOM-NOVEL on Stage 1	3.280e-05
MULTICOM-REFINE and MULTICOM-CONSTRUCT on Stage 1	0.031
MULTICOM-CLUSTER and MULTICOM-NOVEL on Stage 1	0.201
MULTICOM-CLUSTER and MULTICOM-CONSTRUCT on Stage 1	3.757e-06
MULTICOM-NOVEL and MULTICOM-CONSTRUCT on Stage 1	7.013e-07
MULTICOM-REFINE and Pcons on Stage 1	0.1723
MULTICOM-REFINE and ModFOLDclust2 on Stage 1	0.578
MULTICOM-REFINE and DAVIS-QAconsensus on Stage 1	0.6238
MULTICOM-CLUSTER and Pcons on Stage 1	2.872e-08
MULTICOM-CLUSTER and ModFOLDclust2 on Stage 1	5.517e-05
MULTICOM-CLUSTER and DAVIS-QAconsensus on Stage 1	0.002873
MULTICOM-NOVEL and Pcons on Stage 1	5.65e-09
MULTICOM-NOVEL and ModFOLDclust2 on Stage 1	2.116e-05
MULTICOM-NOVEL and DAVIS-QAconsensus on Stage 1	0.002066
MULTICOM-CONSTRUCT and Pcons on Stage 1	0.7492
MULTICOM-CONSTRUCT and ModFOLDclust2 on Stage 1	0.01223
MULTICOM-CONSTRUCT and DAVIS-QAconsensus on Stage 1	0.0002211
MULTICOM-REFINE and MULTICOM-CLUSTER on Stage 2	4.133e-05
MULTICOM-REFINE and MULTICOM-NOVEL on Stage 2	3.180e-05
MULTICOM-REFINE and MULTICOM-CONSTRUCT on Stage 2	2.439e-05
MULTICOM-CLUSTER and MULTICOM-NOVEL on Stage 2	0.658
MULTICOM-CLUSTER and MULTICOM-CONSTRUCT on Stage 2	7.75e-06
MULTICOM-NOVEL and MULTICOM-CONSTRUCT on Stage 2	5.276e-06
MULTICOM-REFINE and Pcons on Stage 2	0.2465
MULTICOM-REFINE and ModFOLDclust2 on Stage 2	0.08742
MULTICOM-REFINE and DAVIS-QAconsensus on Stage 2	0.4976
MULTICOM-CLUSTER and Pcons on Stage 2	1.114e-05
MULTICOM-CLUSTER and ModFOLDclust2 on Stage 2	0.001202
MULTICOM-CLUSTER and DAVIS-QAconsensus on Stage 2	7.495e-06
MULTICOM-NOVEL and Pcons on Stage 2	1.073e-05
MULTICOM-NOVEL and ModFOLDclust2 on Stage 2	0.001128
MULTICOM-NOVEL and DAVIS-QAconsensus on Stage 2	5.717e-06
MULTICOM-CONSTRUCT and Pcons on Stage 2	0.9807
MULTICOM-CONSTRUCT and ModFOLDclust2 on Stage 2	0.003362
MULTICOM-CONSTRUCT and DAVIS-QAconsensus on Stage 2	9.597e-05

MULTICOM-CONSTRUCT and Pcons is not significant, while the difference between MULTICOM-CONSTRUCT and the other two external methods (DAVIS-QAconsensus and ModFOLDclust2) is significant.

To elucidate the key factors that affect the accuracy of multi-model or single-model quality assessment methods, we plot the per-target correlation scores of each target on Stage 2 against the ratio of the average real quality of the largest model cluster in the pool and the average real quality of all the models in the pool in Figure 1. To get the largest model cluster for each target, we first calculate the GDT-TS score between each pair of models, and then use (1 – the GDT-TS score) as the distance measure to hierarchically cluster the models. Finally, we use a distance threshold to cut the hierarchical tree to get the largest cluster so that the total number of models in the largest cluster is about one third of the total number of models in the pool.

Figure 1 shows that the quality prediction accuracy (i.e. per-target correlation scores of each target) positively correlates with the average real quality of the largest model cluster divided by the average real quality of all models for two multi-model methods (MULTICOM-REFINE, MULTICOM-CONSTRUCT), whereas it has almost no correlation with single-model methods (MULTICOM-CLUSTER, MULTICOM-NOVEL). The results suggest that the performance of clustering-based multi-model methods depends on the relative real quality of the large cluster of models and that of single-model methods does not. This is not surprising because multi-model methods rely on pairwise model comparison, but single-model methods try to assess the quality from one model.

As CASP10 models were generated by many different predictors from around of the world, the side chains of these models may be packed by different modeling tools. The difference in side chain packing may result in difference in input features (e.g. secondary structures) that affect the quality prediction results of single-model methods even though they only try to predict the quality of backbone of a model. In order to remove the side-chain bias, we also tried to use the tool SCWRL [30] to rebuild the side chains of all models before applying a single-model quality prediction method - ModelEvaluator. Figure 2 compares the average correlation and loss of the predictions with or without side-chain repacking. Indeed, repacking side-chains before applying single-model quality assessment increased the average correlation and reduced the loss. We did a Wilcoxon signed ranked sum test on the correlations and losses of the predictions before and after repacking side-chains. The p-value for average correlation before and after repacking side-chains on Stage 1 is 0.18, and on Stage 2 is 0.02. The p-value for loss on Stage 1 is 0.42, and on Stage 2 is 0.38.

Since mining a few good models out of a large pool of low-quality models is one of the major challenges in protein structure prediction, we compare the performance of single-model methods and multi-model methods on the models of several hard CASP10 template-free

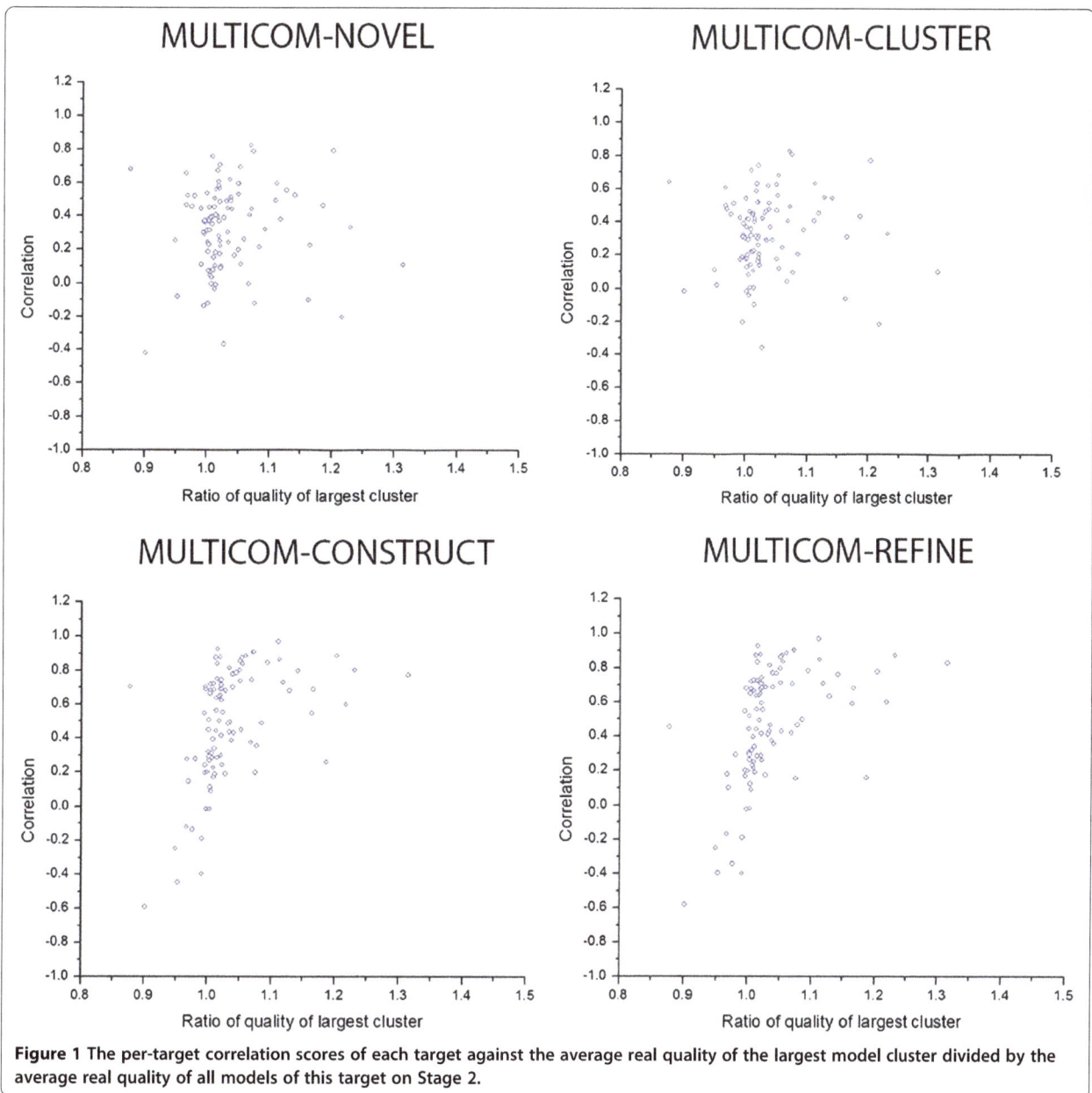

Figure 1 The per-target correlation scores of each target against the average real quality of the largest model cluster divided by the average real quality of all models of this target on Stage 2.

targets. Tables 4 and 5 report the evaluation results of our four servers, DAVIS-QAconsensus, Pcons, and ModFOLDclust2 on all standalone template-free modeling (FM) targets on Stages 1 and 2, i.e. the targets whose domains are all FM domains. The results show that the single-model methods (MULTICOM-CLUSTER and MULTICOM-NOVEL) clearly performed better than the multi-model methods (MULTICOM-REFINE and MULTICOM-CONSTRUCT) on both stages. They also performed better than the DAVIS-QAconsensus and ModFOLDclust2 on both stages, achieved the similar performance with Pcons on Stage 1, and the better performance than Pcons on Stage 2. For instance, the average Pearson's correlation score of MULTICOM-NOVEL on

Stage 1 is 0.539, which is much higher than 0.082 of MULTICOM-REFINE. The multi-model methods even get low negative correlation for some targets. For example, the Pearson's correlation score of MULTICOM-REFINE on target T0741 at Stage 1 is −0.615. We use the tool TreeView [31] to visualize the hierarchical clustering of the models of T0741 in Figure 3. The qualities of the models in the largest cluster are among the lowest, but they are similar to each other leading to high predicted quality scores when being assessed by multi-model methods. The example indicates that multi-model methods often completely fail (i.e. yielding negative correlation) when the models in the largest cluster are of worse quality, but similar to each other. Multi-model methods often perform

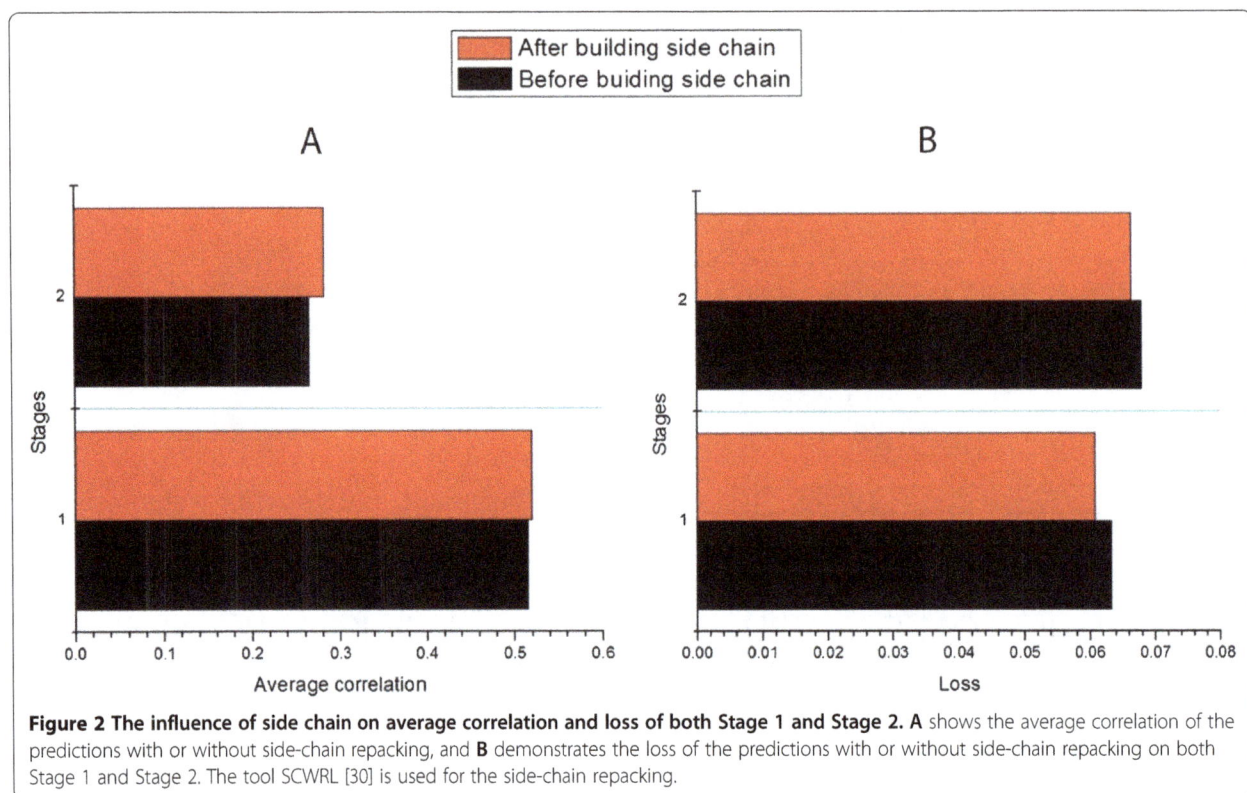

Figure 2 The influence of side chain on average correlation and loss of both Stage 1 and Stage 2. A shows the average correlation of the predictions with or without side-chain repacking, and **B** demonstrates the loss of the predictions with or without side-chain repacking on both Stage 1 and Stage 2. The tool SCWRL [30] is used for the side-chain repacking.

worse than single-model methods when all models in pool are of low quality and are different from each other. In this situation, the quality scores predicted by multi-model methods often do not correlate with the real quality scores, whereas those predicted by single-model methods still positively correlate with real quality scores to some degree. As an example, Figure 4 plots the real GDT-TS scores and predicted GDT-TS scores of a single-model predictor MULTICOM-NOVEL and a multi-model predictor MULTICOM-REFINE on the models of a hard target T0684 whose best model has quality score less than 0.2. It is worth noting that, since the quality of the models of the template-free modeling targets is rather low on average, the quality assessment on these models can be more arbitrary than on the template-based models of better quality.

Therefore, more cautions must be put into the interpretation of the evaluation results.

Based on the per-target correlation between predicted and observed model quality scores of the official model quality assessment results [28], the MULTICOM-CONSTRUCT was ranked 5th on Stage 2 models of CASP10 among all CASP10 model quality assessment methods. The performance of MULTICOM-CONSTRUCT was slightly better than the DAVIS-QAconsensus (the naïve consensus method that calculates the quality score of a model as the average structural similarity (GDT-TS score) between the model and other models in the pool) on Stage 2, which was ranked at 10th. The methods MULTICOM-REFINE, MULTICOM-NOVEL, and MULTICOM-CLUSTER were ranked at 11th, 28th,

Table 4 Pearson correlation of the FM (template-free modeling) targets on Stage 1 of CASP10

Targets	MULTICOM-NOVEL	MULTICOM-CLUSTER	MULTICOM-CONSTRUCT	MULTICOM-REFINE	DAVIS-QA consensus	Pcons	ModFOLDclust2
T0666	0.570	0.454	0.138	0.272	0.274	0.346	0.538
T0735	0.725	0.704	0.414	0.083	0.086	0.667	0.030
T0734	0.522	0.544	0.152	−0.099	−0.096	0.509	−0.014
T0737	0.878	0.878	0.221	0.118	0.124	0.565	0.421
T0740	0.558	0.512	0.710	0.732	0.726	0.684	0.770
T0741	−0.020	0.214	−0.659	−0.615	−0.611	0.475	−0.674
Average	0.539	0.551	0.163	0.082	0.084	0.541	0.179

Table 5 Pearson correlation of all FM (template-free modeling) targets on Stage 2 of CASP10

Targets	MULTICOM-NOVEL	MULTICOM-CLUSTER	MULTICOM-CONSTRUCT	MULTICOM-REFINE	DAVIS-QA consensus	Pcons	ModFOLDclust2
T0666	0.213	0.206	0.490	0.499	0.492	0.338	0.520
T0735	0.466	0.433	0.261	0.159	0.150	0.238	−0.070
T0734	0.459	0.44	−0.134	−0.342	−0.334	0.199	−0.363
T0737	0.787	0.806	0.200	0.155	0.147	0.583	0.525
T0740	0.490	0.451	0.487	0.412	0.411	0.434	0.478
T0741	−0.079	0.022	−0.444	−0.397	−0.397	0.125	−0.382
Average	0.389	0.393	0.143	0.081	0.078	0.320	0.118

and 29th, respectively. However, it was not surprising that the single-model methods such as MULTICOM-NOVEL and MULTICOM-CLUSTER were ranked lower than most clustering-based methods because the latter tended to work better on most CASP template-based targets with good-quality predicted models. But, among all single-model methods, MULTICOM-NOVEL and MULTICOM-CLUSTER were ranked at 3th and 4th.

Results of local quality

Table 6 shows the performance of local quality assessment of our four local quality assessment servers,

DAVIS-QAconsensus, Pcons, and ModFOLDclust2 on both Stage 1 and Stage 2. Among our four servers, the multi-model methods performed better than single-model methods on average for all the targets. We used the pairwise Wilcoxon signed ranked sum test to assess the significance of the difference between our four servers and the three external methods (Table 7). Generally speaking, the difference between multi-model local quality methods (MULTICOM-REFINE, DAVIS-QAconsensus, Pcons, and ModFOLDclust2) and single-model local quality methods (MULTICOM-NOVEL, MULTICOM-CLUSTER, MULTICOM-CONSTRUCT) on both stages

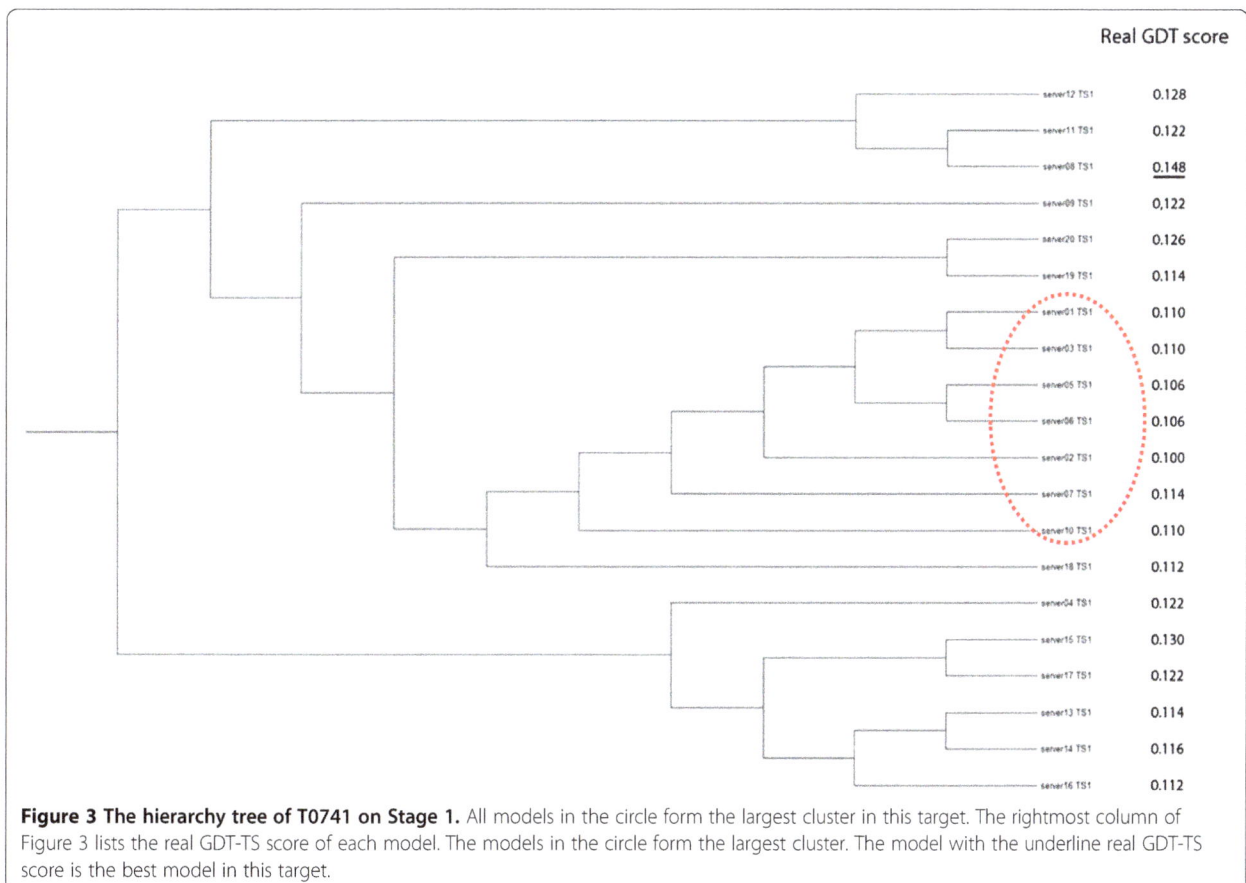

Figure 3 The hierarchy tree of T0741 on Stage 1. All models in the circle form the largest cluster in this target. The rightmost column of Figure 3 lists the real GDT-TS score of each model. The models in the circle form the largest cluster. The model with the underline real GDT-TS score is the best model in this target.

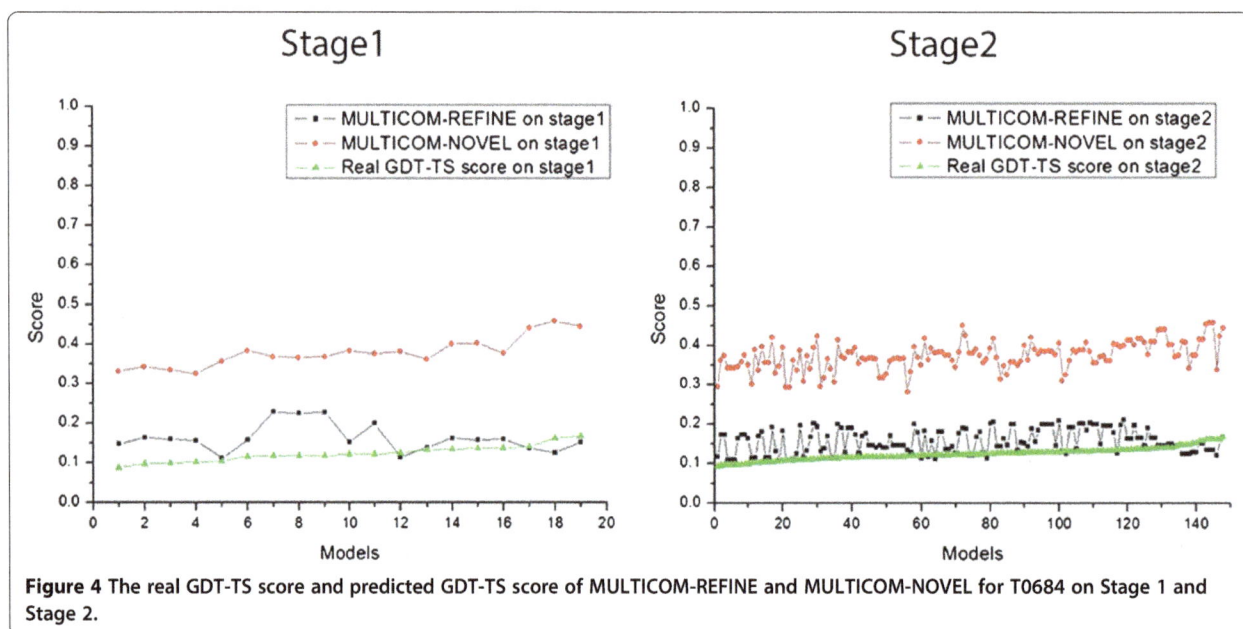

Figure 4 The real GDT-TS score and predicted GDT-TS score of MULTICOM-REFINE and MULTICOM-NOVEL for T0684 on Stage 1 and Stage 2.

is significant. The difference between MULTICOM-REFINE and Pcons is not significant on both stages according to a 0.01 threshold.

However, the single-model local quality prediction methods (MULTICOM-NOVEL, MULTICOM-CLUSTER, MULTICOM-CONSTRUCT) and the multi-model local quality prediction method (MULTICOM-REFINE) performed not very differently on FM targets as shown in Tables 8 and 9. This is not surprising because multi-model methods cannot select real good models as reference methods for evaluating the local quality of residues.

According to the CASP official evaluation [28], MULTICOM-REFINE performs best among all of our four servers for the local quality assessment on both Stage 1 and Stage 2 models of CASP10. Compared with DAVIS-QAconsensus, Pcons, and ModFOLDclust2, the multi-model local quality prediction method MULTICON-REFINE performed best on Stage 1, achieved the similar

performance with Pcons on Stage 2, but performed worse than DAVIS-QAconsensus and ModFOLDclust2 on Stage 2.

Conclusion

In this work, we rigorously benchmarked our multi-model and single-model quality assessment methods blindly tested in the Tenth Critical Assessment of Techniques for Protein Structure Prediction (CASP10). In general, the performance of our multi-model quality prediction methods (e.g., MULTICOM-REFINE) was comparable to the state-of-the-art multi-model quality assessment methods in the literature. The multi-model quality prediction methods performed better than the single-model quality prediction methods (e.g., MULTICOM-NOVEL, MULTICOM-CLUSTER), whereas the latter, despite in its early stage of development, tended to work better in assessing a small number of models of wide-range quality usually associated with a hard target. Our experiment demonstrated that the prediction accuracy of multi-model quality assessment methods is largely influenced by the proportion of good models in the pool or the average quality of the largest model cluster in the pool. The multi-model quality assessment methods performed better than single-model methods on easy modeling targets whose model pool contains a large portion of good models. However, they tend to fail on the models for hard targets when the majority of models are of low-quality and particularly when some low-quality models are similar to each other severely dominating the calculation of pairwise model similarity. The problem can be somewhat remedied by using single-model quality prediction scores as weights in calculating the average similarity scores

Table 6 Evaluation result of local quality score of four MULTICOM servers, DAVIS-QAconsensus, Pcons, and ModFOLDclust2 on Stage 1 and Stage 2 of CASP10

Servers	Ave. corr. on Stage 1	Ave. corr. on Stage 2
MULTICOM-REFINE	0.6102	0.6251
MULTICOM-CLUSTER	0.2604	0.2956
MULTICOM-NOVEL	0.2882	0.3289
MULTICOM-CONSTRUCT	0.2889	0.3095
DAVIS-QAconsensus	0.5841	0.6633
Pcons	0.5793	0.6226
ModFOLDclust2	0.5997	0.6526

Table 7 The P-value of pairwise Wilcoxon signed ranked sum tests for the difference of correlation scores for local model quality prediction methods (MULTICOM servers, DAVIS-QAconsensus, Pcons, and ModFOLDclust2)

MULTICOM servers, DAVIS-QAconsensus, Pcons, and ModFOLDclust2 and on Stage 1 or Stage 2	P-value
MULTICOM-REFINE and MULTICOM-CLUSTER on Stage 1	2.220e-16
MULTICOM-REFINE and MULTICOM-NOVEL on Stage 1	6.661e-16
MULTICOM-REFINE and MULTICOM-CONSTRUCT on Stage 1	6.661e-16
MULTICOM-CLUSTER and MULTICOM-NOVEL on Stage 1	0.0009948
MULTICOM-CLUSTER and MULTICOM-CONSTRUCT on Stage 1	0.0008437
MULTICOM-NOVEL and MULTICOM-CONSTRUCT on Stage 1	0.1781
MULTICOM-REFINE and Pcons on Stage 1	0.01575
MULTICOM-REFINE and ModFOLDclust2 on Stage 1	0.2678
MULTICOM-REFINE and DAVIS-QAconsensus on Stage 1	0.00699
MULTICOM-CLUSTER and Pcons on Stage 1	2.2e-16
MULTICOM-CLUSTER and ModFOLDclust2 on Stage 1	2.553e-16
MULTICOM-CLUSTER and DAVIS-QAconsensus on Stage 1	2.442e-15
MULTICOM-NOVEL and Pcons on Stage 1	2.2e-16
MULTICOM-NOVEL and ModFOLDclust2 on Stage 1	3.046e-16
MULTICOM-NOVEL and DAVIS-QAconsensus on Stage 1	4.885e-15
MULTICOM-CONSTRUCT and Pcons on Stage 1	2.2e-16
MULTICOM-CONSTRUCT and ModFOLDclust2 on Stage 1	3.137e-16
MULTICOM-CONSTRUCT and DAVIS-QAconsensus on Stage 1	4.78e-15
MULTICOM-REFINE and MULTICOM-CLUSTER on Stage 2	2.269e-16
MULTICOM-REFINE and MULTICOM-NOVEL on Stage 2	6.661e-16
MULTICOM-REFINE and MULTICOM-CONSTRUCT on Stage 2	3.137e-16
MULTICOM-CLUSTER and MULTICOM-NOVEL on Stage 2	0.00327
MULTICOM-CLUSTER and MULTICOM-CONSTRUCT on Stage 2	0.5493
MULTICOM-NOVEL and MULTICOM-CONSTRUCT on Stage 2	1.029e-14
MULTICOM-REFINE and Pcons on Stage 2	0.2498
MULTICOM-REFINE and ModFOLDclust2 on Stage 2	0.0005575
MULTICOM-REFINE and DAVIS-QAconsensus on Stage 2	2.443e-06
MULTICOM-CLUSTER and Pcons on Stage 2	2.220e-16
MULTICOM-CLUSTER and ModFOLDclust2 on Stage 2	2.220e-16
MULTICOM-CLUSTER and DAVIS-QAconsensus on Stage 2	2.220e-16
MULTICOM-NOVEL and Pcons on Stage 2	4.441e-16
MULTICOM-NOVEL and ModFOLDclust2 on Stage 2	2.220e-16
MULTICOM-NOVEL and DAVIS-QAconsensus on Stage 2	2.220e-16
MULTICOM-CONSTRUCT and Pcons on Stage 2	4.089e-16
MULTICOM-CONSTRUCT and ModFOLDclust2 on Stage 2	2.2e-16
MULTICOM-CONSTRUCT and DAVIS-QAconsensus on Stage 2	2.2e-16

coupling information, torsion angle information, and statistical contact potentials may be used to improve the discriminative power of single-model methods; on the other hand, new powerful machine learning and data mining methods such as deep learning, random forests and outlier detection methods may be developed to use existing quality features more effectively. Despite it may take years for single-model methods to mature, we believe that improved single-model quality prediction methods will play a more and more important role in protein structure prediction.

Methods
Protein model quality prediction methods
The methods used by the four automated protein model quality assessment servers are briefly described as follows.

MULTICOM-REFINE is a multi-model quality assessment method using a pairwise model comparison approach (APOLLO) [29] to generate global quality scores. The 19 top models based on the global quality scores and the top 1 model selected by SPICKER [32] formed a top model set for local quality prediction. After superimposing a model with each model in the top model set, it calculated the average absolute Euclidean distance between the position of each residue in the model and that of its counterpart in each model in the top model set. The average distance was used as the local quality of each residue.

MULTICOM-CLUSTER is a single-model, support vector machine (SVM)-based method initially implemented in [24]. The input features to the SVM include a window of amino acids encoded by a 20-digit vector of 0 and 1 centered on a target residue, the difference between secondary structure and solvent accessibility predicted by SCRATCH [33] from the protein sequence and that of a model parsed by DSSP [34], and predicted contact probabilities between the target residue and its spatially neighboring residues. The SVM was trained to predict the local quality score (i.e. the Euclidean distance between its position in the model and that in the native structure) of each residue. The predicted local quality scores of all the residues was converted into the global quality score of the model according to the formula [35] as follows:

$$Global\ quality\ score = \frac{1}{L}\sum_{i=1}^{t}\left(\frac{1}{1+\left(\frac{S_i}{T}\right)^2}\right)$$

In the formula, L is the total number of residues, S_i is the local quality score of residue i, and T is a distance threshold set to set to 5 Angstrom. Residues that did not have a predicted local quality score were skipped in averaging.

between models. However, to completely address the problem, more accurate single-model quality prediction methods that can assess the quality of a single model need to be developed. On one hand, more informative features such as sequence conservation information, evolutionary

Table 8 Local quality score of four MULTICOM servers, DAVIS-QAconsensus, Pcons, and ModFOLDclust2 for all FM (template-free modeling) targets on Stage 1 of CASP10

Targets	MULTICOM-NOVEL	MULTICOM-CLUSTER	MULTICOM-CONSTRUCT	MULTICOM-REFINE	DAVIS-QA consensus	Pcons	ModFOLDclust2
T0666	0.261	0.216	0.262	0.261	0.195	0.303	0.164
T0735	0.118	0.083	0.122	0.366	0.190	0.214	0.224
T0734	0.025	0.105	0.025	0.402	0.302	0.166	0.232
T0737	0.554	0.664	0.551	0	0.186	0.704	0.122
T0740	0.242	0.196	0.243	0.442	0.368	0.377	0.407
T0741	0.078	−0.035	0.084	0.227	0.108	−0.072	0.136
Average	0.213	0.205	0.215	0.283	0.225	0.282	0.214

MULTICOM-NOVEL is the same as MULTICOM-CLUSTER except that amino acid sequence features were replaced with the sequence profile features. The multiple sequence alignment of a target protein used to construct profiles was generated by PSI-BLAST [36].

MULTICOM-CONSTRUCT uses a new, weighted pairwise model evaluation approach to predict global quality. It uses ModelEvaluator [37] – an ab initio single-model global quality prediction method – to predict a score for each model, and uses TM-score [35] to get the GDT-TS score for each pair of models. The predicted global quality score of a model i is the weighted average GDT-TS score between the model and other models, calculated according to the formula: $S_i = \sum_{j=1}^{N}\left(X_{i,j} * \dfrac{W_J}{\sum_{j=1}^{N} W_j}\right)$. In this formula, S_i is the predicted global quality score for model i, N is the total number of models, $X_{i,j}$ is the GDT-TS score between model i and model j, W_j is the score for model j predicted by ModelEvaluator, which is used to weight the contribution of $X_{i,j}$ to S_i. In case that no score was predicted for a model by ModelEvaluator, the weight of the model is set to the average of all the scores predicted by ModelEvaluator. The local quality prediction of MULTICOM-CONSTRUCT is the same as MULTICOM-NOVEL

except that additional SOV (segment overlap measure of secondary structure) score features were used by the SVM to generate the local quality score.

Evaluation methods

CASP10 used two-stage experiments to benchmark for model quality assessment. Stage 1 had 20 models with different qualities for each target, and Stage 2 had 150 top models for each target selected from all the models by a naïve pairwise model quality assessment method. We downloaded the native structures of 98 CASP10 targets, their structural models, and the quality predictions of these models made by our four servers during the CASP10 experiment running from May to August, 2012 from the CASP website (http://predictioncenter.org/casp10/index.cgi).

We used TM-score [35] to calculate the real GDT-TS scores between the native structures and the predicted model as their real global quality scores. The predicted global quality scores of our four servers were used to compare with the real global quality scores. In order to calculate real local quality scores of residues in a model, we first used TM-score to superimpose the native structure and the model, and then calculate the Euclidean distance between each residue's coordinates in the superimposed native structure and the model as the real local

Table 9 Local quality score of four MULTICOM servers, DAVIS-QAconsensus, Pcons, and ModFOLDclust2 for all FM (template-free modeling) targets on Stage 2 of CASP10

Servers	MULTICOM-NOVEL	MULTICOM-CLUSTER	MULTICOM-CONSTRUCT	MULTICOM-REFINE	DAVIS-QA consensus	Pcons	ModFOLDclust2
T0666	0.244	0.226	0.227	0.310	0.322	0.282	0.337
T0735	0.125	0.122	0.127	0.288	0.290	0.150	0.351
T0734	0.129	0.151	0.122	0.172	0.330	0.255	0.305
T0737	0.426	0.578	0.419	0	0.202	0.583	0
T0740	0.268	0.197	0.257	0.270	0.422	0.377	0.425
T0741	0.105	−0.011	0.109	0.165	0.129	0.009	0.119
Average	0.216	0.211	0.210	0.200	0.283	0.276	0.256

quality score of the residue. The real local and global quality scores of a model were compared with that predicted by the model quality assessment methods to evaluate their prediction accuracy.

We evaluated the global quality of our predictions from five aspects: the average of per-target Pearson correlations, the overall Pearson's correlation, average GDT-TS loss, the average Spearman's correlation, and the average Kendall tau correlation. The average of per-target Pearson's correlations is calculated as the average of all 98 targets' Pearson correlations between predicted and real global quality scores of their models. The overall Pearson's correlation is the correlation between predicted and real global quality scores of all the models of all the targets pooled together. The average GDT-TS loss is the average difference between the GDT-TS scores of the real top 1 model and the predicted top 1 model of all targets, which measures how well a method ranks good models at the top. The Spearman's correlation is the Pearson's correlation of the ranked global quality scores. In order to calculate the Spearman's rank correlation, we first convert the global quality scores into the ranks. The identical values (rank ties or duplicate values) are assigned a rank equal to the average of their positions in the rank list. And then we calculate the Pearson's correlation between the predicted ranks and true ranks of the models. The Kendall tau correlation is the probability of concordance minus the probability of discordance. For two vectors x and y with global quality scores of n models of a target, the number of total possible model pairs for x or y is $N = \frac{n*(n-1)}{2}$. The number of concordance is the number of pairs and (X_j,Y_j) when $(x_i - x_j) * (y_i - y_i) > 0$, and the number of discordance is the number of pairs X_i,Y_i and (X_j,Y_j) when $(x_i - x_j) * (y_i - y_i) < 0$. The Kendall tau correlation is equal to the number of concordance minus the number of discordance divided by N. (http://en.wikipedia.org/wiki/Kendall_tau_rank_correlation_coefficient).

The accuracy of local quality predictions was calculated as the average of the Pearson's correlations between predicted local quality scores and real local quality scores of all the models of all the targets. For each model, we used TM-score to superimpose it with the native structure, and then calculated the Euclidean distance between Ca atom's coordinates of each residue in a superimposed model and the native structure as the real local quality score of each residue. The Pearson's correlation between the real quality scores and the predicted ones of all the residues in each model was calculated. The average of the Pearson's correlations of all the models for all 98 targets was used to evaluate the performance of the local quality prediction methods.

Competing interests
The authors declare that they have no competing interests.

Authors' contributions
JC conceived and designed the method and the system. RC, ZW implemented the method, built the system, carried out the CASP experiments. RC, ZW, JC evaluated and analyzed data. RC, JC wrote the manuscript. All the authors approved the manuscript.

Acknowledgements
The work was partially supported by an NIH grant (R01GM093123) to JC.

Author details
[1]Computer Science Department, University of Missouri, Columbia, Missouri 65211, USA. [2]Informatics Institute, University of Missouri, Columbia, Missouri 65211, USA. [3]Christopher S. Bond Life Science Center, University of Missouri, Columbia, Missouri 65211, USA. [4]School of Computing, University of Southern Mississippi, Hattiesburg, MS 39406-0001, USA.

References
1. Eisenhaber F, Persson B, Argos P: Protein structure prediction: recognition of primary, secondary, and tertiary structural features from amino acid sequence. Crit Rev Biochem Mol Biol 1995, 30(1):1–94.
2. Rost B: Protein structure prediction in 1D, 2D, and 3D. Encyclopaedia Comput Chem 1998, 3:2242–2255.
3. Cheng J: A multi-template combination algorithm for protein comparative modeling. BMC Struct Biol 2008, 8(1):18.
4. Zhang Y, Skolnick J: Automated structure prediction of weakly homologous proteins on a genomic scale. Proc Natl Acad Sci U S A 2004, 101(20):7594–7599.
5. Sali A, Blundell T: Comparative protein modelling by satisfaction of spatial restraints. Protein Structure Distance Anal 1994, 64:C86.
6. Peng J, Xu J: RaptorX: exploiting structure information for protein alignments by statistical inference. Proteins 2011, 79(S10):161–171.
7. Zhang J, Wang Q, Barz B, He Z, Kosztin I, Shang Y, Xu D: MUFOLD: a new solution for protein 3D structure prediction. Proteins 2010, 78(5):1137–1152.
8. Jones DT, McGuffin LJ: Assembling novel protein folds from super-secondary structural fragments. Proteins 2003, 53(S6):480–485.
9. Simons KT, Kooperberg C, Huang E, Baker D: Assembly of protein tertiary structures from fragments with similar local sequences using simulated annealing and Bayesian scoring functions. J Mol Biol 1997, 268(1):209–225.
10. Cheng J, Eickholt J, Wang Z, Deng X: Recursive protein modeling: a divide and conquer strategy for protein structure prediction and its case study in CASP9. J Bioinform Comput Biol 2012, 10(3). doi:10.1142/S0219720012420036.
11. Liu P, Zhu F, Rassokhin DN, Agrafiotis DK: A self-organizing algorithm for modeling protein loops. PLoS Comput Biol 2009, 5(8):e1000478.
12. Lee J, Lee D, Park H, Coutsias EA, Seok C: Protein loop modeling by using fragment assembly and analytical loop closure. Proteins 2010, 78(16):3428–3436.
13. Yang Y, Zhou Y: Ab initio folding of terminal segments with secondary structures reveals the fine difference between two closely related all-atom statistical energy functions. Protein Sci 2008, 17(7):1212–1219.
14. Zemla A, Venclovas Č, Moult J, Fidelis K: Processing and analysis of CASP3 protein structure predictions. Proteins 1999, 37(S3):22–29.
15. Zemla A, Venclovas Č, Moult J, Fidelis K: Processing and evaluation of predictions in CASP4. Proteins 2002, 45(S5):13–21.
16. Zemla A: LGA: a method for finding 3D similarities in protein structures. Nucleic Acids Res 2003, 31(13):3370–3374.
17. Wang Z, Eickholt J, Cheng J: APOLLO: a quality assessment service for single and multiple protein models. Bioinformatics 2011, 27(12):1715–1716.
18. McGuffin LJ: The ModFOLD server for the quality assessment of protein structural models. Bioinformatics 2008, 24(4):586–587.
19. McGuffin LJ: Prediction of global and local model quality in CASP8 using the ModFOLD server. Proteins 2009, 77(S9):185–190.
20. Wang Q, Vantasin K, Xu D, Shang Y: MUFOLD-WQA: a new selective consensus method for quality assessment in protein structure prediction. Proteins 2011, 79(Supplement S10):185–95. doi:10.1002/prot.23185.

21. McGuffin LJ, Roche DB: **Rapid model quality assessment for protein structure predictions using the comparison of multiple models without structural alignments.** *Bioinformatics* 2010, **26**(2):182–188.

22. Tress ML, Jones D, Valencia A: **Predicting reliable regions in protein alignments from sequence profiles.** *J Mol Biol* 2003, **330**(4):705.

23. Wallner B, Elofsson A: **Identification of correct regions in protein models using structural, alignment, and consensus information.** *Protein Sci* 2009, **15**(4):900–913.

24. Kalman M, Ben-Tal N: **Quality assessment of protein model-structures using evolutionary conservation.** *Bioinformatics* 2010, **26**(10):1299–1307.

25. Liithy R, Bowie JU, Eisenberg D: **Assessment of protein models with three-dimensional profiles.** *Nature* 1992, **356**:83–85.

26. Ray A, Lindahl E, Wallner B: **Improved model quality assessment using ProQ2.** *BMC Bioinformatics* 2012, **13**(1):224.

27. Kaján L, Rychlewski L: **Evaluation of 3D-Jury on CASP7 models.** *BMC Bioinformatics* 2007, **8**(1):304.

28. Kryshtafovych A, Barbato A, Fidelis K, Monastyrskyy B, Schwede T, Tramontano A: **Assessment of the assessment: evaluation of the model quality estimates in CASP10.** *Proteins* 2013, **82**(Suppl 2):112–26. doi:10.1002/prot.24347.

29. Larsson P, Skwark MJ, Wallner B, Elofsson A: **Assessment of global and local model quality in CASP8 using Pcons and ProQ.** *Proteins* 2009, **77**(S9):167–172.

30. Krivov GG, Shapovalov MV, Dunbrack RL Jr: **Improved prediction of protein side-chain conformations with SCWRL4.** *Proteins* 2009, **77**(4):778–795.

31. Page RD: **TreeView: an application to display phylogenetic trees on personal computer.** *Comp Appl Biol Sci* 1996, **12**:357–358.

32. Zhang Y, Skolnick J: **SPICKER: a clustering approach to identify near-native protein folds.** *J Comput Chem* 2004, **25**(6):865–871.

33. Cheng J, Randall A, Sweredoski M, Baldi P: **SCRATCH: a protein structure and structural feature prediction server.** *Nucleic Acids Res* 2005, **33**(suppl 2):W72–W76.

34. Kabsch W, Sander C: **Dictionary of protein secondary structure: pattern recognition of hydrogen-bonded and geometrical features.** *Biopolymers* 1983, **22**(12):2577–2637.

35. Zhang Y, Skolnick J: **Scoring function for automated assessment of protein structure template quality.** *Proteins* 2004, **57**(4):702–710.

36. Altschul SF, Madden TL, Schäffer AA, Zhang J, Zhang Z, Miller W, Lipman DJ: **Gapped BLAST and PSI-BLAST: a new generation of protein database search programs.** *Nucleic Acids Res* 1997, **25**(17):3389–3402.

37. Wang Z, Tegge AN, Cheng J: **Evaluating the absolute quality of a single protein model using structural features and support vector machines.** *Proteins* 2009, **75**(3):638–647.

14

Structure to function prediction of hypothetical protein KPN_00953 (Ycbk) from *Klebsiella pneumoniae* MGH 78578 highlights possible role in cell wall metabolism

Boon Aun Teh[1], Sy Bing Choi[2], Nasihah Musa[3], Few Ling Ling[4], See Too Wei Cun[4], Abu Bakar Salleh[3], Nazalan Najimudin[1], Habibah A Wahab[5*] and Yahaya M Normi[3*]

Abstract

Background: *Klebsiella pneumoniae* plays a major role in causing nosocomial infection in immunocompromised patients. Medical inflictions by the pathogen can range from respiratory and urinary tract infections, septicemia and primarily, pneumonia. As more *K. pneumoniae* strains are becoming highly resistant to various antibiotics, treatment of this bacterium has been rendered more difficult. This situation, as a consequence, poses a threat to public health. Hence, identification of possible novel drug targets against this opportunistic pathogen need to be undertaken. In the complete genome sequence of *K. pneumoniae* MGH 78578, approximately one-fourth of the genome encodes for hypothetical proteins (HPs). Due to their low homology and relatedness to other known proteins, HPs may serve as potential, new drug targets.

Results: Sequence analysis on the HPs of *K. pneumoniae* MGH 78578 revealed that a particular HP termed KPN_00953 (YcbK) contains a M15_3 peptidases superfamily conserved domain. Some members of this superfamily are metalloproteases which are involved in cell wall metabolism. BLASTP similarity search on KPN_00953 (YcbK) revealed that majority of the hits were hypothetical proteins although two of the hits suggested that it may be a lipoprotein or related to twin-arginine translocation (Tat) pathway important for transport of proteins to the cell membrane and periplasmic space. As lipoproteins and other components of the cell wall are important pathogenic factors, homology modeling of KPN_00953 was attempted to predict the structure and function of this protein. Three-dimensional model of the protein showed that its secondary structure topology and active site are similar with those found among metalloproteases where two His residues, namely His169 and His209 and an Asp residue, Asp176 in KPN_00953 were found to be Zn-chelating residues. Interestingly, induced expression of the cloned *KPN_00953* gene in lipoprotein-deficient *E. coli* JE5505 resulted in smoother cells with flattened edges. Some cells showed deposits of film-like material under scanning electron microscope.

(Continued on next page)

* Correspondence: habibahw@usm.my; normi_yahaya@upm.edu.my
[5]Malaysian Institute of Pharmaceuticals and Nutraceuticals, Ministry of Science, Technology and Innovation, Blok 5-A, Halaman Bukit Gambier, 11700 Pulau Pinang, Malaysia
[3]Enzyme and Microbial Technology Research Center (EMTECH), Faculty of Biotechnology and Biomolecular Sciences, Universiti Putra Malaysia, 43400 Serdang, Selangor, Malaysia
Full list of author information is available at the end of the article

(Continued from previous page)

Conclusions: We postulate that KPN_00953 is a Zn metalloprotease and may play a role in bacterial cell wall metabolism. Structural biology studies to understand its structure, function and mechanism of action pose the possibility of utilizing this protein as a new drug target against *K. pneumoniae* in the future.

Keywords: KPN_00953, Hypothetical protein, Homology modeling, Peptidase M15_3 superfamily, Cell wall metabolism

Background

Klebsiella pneumoniae is a Gram-negative, rod-shaped bacterium that is widely distributed in soil and water [1] as well as the intestine, urethra and respiratory tract of mankind and other animals [2]. This opportunistic pathogen has been regarded as one of the major causes of respiratory and urinary tract infections, septicemia and the third-most-common bacterial cause of hospital-acquired pneumonia in immunocompromised patients [3]. Studies in Taiwan showed that this pathogen has the capacity to cause pyogenic liver abscess in human [4,5]. Similar cases have been observed in other countries as well, indicating that such medical infliction is not only confined to Taiwan per se and may potentially emerge as a global problem [6]. To add to this problem, *K. pneumoniae* strains which produce extended-spectrum beta-lactamases and are highly resistant to a spectrum of antibiotics are emerging worldwide [6]. These strains, also known as *K. pneumoniae* carbapenemases (KPC)-encoding strains, are often associated with nearly complete antibiotic resistance whereby failure and mortality rates related to pneumonia caused by this pathogen can reach up to 50% even with antibiotic therapy [7]. This makes treatments for this bacterium more difficult and has certainly created obstacles, no less danger, to public health.

Many components of the bacteria have been identified as pathogenic factors such as its capsular polysaccharide [8], yersiniabactin [9,10] and enterobactin [11]. All these pathogenic factors are well characterized in which their mechanisms of action and effects are established. Little is known, however, on the roles of poorly characterized biomolecules such as hypothetical proteins (HPs) of this pathogen. As their sequences and structures remain largely non-similar with other known proteins, HPs are often regarded as proteins of unknown functions [12] or orphan proteins [13]. It is important to note that a substantial fraction (up to 30 – 40%) of any sequenced bacterial genomes consist of genes which encode HPs [13]. This is certainly the case even in model organisms such as *Escherichia coli*, *Bacillus subtilis* or *Saccharomyces cerevisiae* [14]. Efforts to gain basic understanding on the roles and possible functions of HPs are crucial to close the gap between the "knowns" and the "unknowns". This is especially important to fit and complete the genetic information puzzle of any living organisms, as well as to gain a 'complete' understanding of these organisms as biological systems as a whole [14].

As 25% of the complete genome sequence of *K. pneumoniae* MGH 78578 codes for HPs [15], it serves as a good mining pool for these proteins to be studied structurally and functionally. This effort is important particularly in substantiating the biological role and importance of HPs in the system of a pathogen. Improved understanding of these proteins may make them potential targets of antimicrobial drugs [14]. This present study highlights the *in silico* studies to characterize a HP, KPN_00953 (YcbK) from *K. pneumoniae* MGH 78578. The results revealed that KPN_00953 is a Zn-metalloprotease possibly related to the functions of the cell wall whereby its induced expression has interestingly changed the surface morphology of a lipoprotein-deficient *E. coli* JE5505 strain.

Results
Sequence analysis

The genome of *K. pneumoniae* MGH 78578 was obtained from NCBI website (Refseq: NC_009648) and thoroughly studied to identify the annotated proteins and HPs. A total of 1004 HPs were found in the genome of *K. pneumoniae* MGH 78578. Via pBLAST analysis [16] of the HPs against the non-redundant (NR) database, a particular hypothetical protein annotated as KPN_00953 (YcbK) gave more than 100 hits with values above the E-value threshold of 0.001. Majority of the top hits for this HP were also HPs and proteins with unknown functions (Table 1). Among these hits, a hypothetical lipoprotein from *Vibrio furnissii* NCTC 11218 and twin-arginine translocation pathway signal peptide showed high similarity to KPN_00953, up to 81% (data not shown) and 99%, respectively (Table 1). A search on the structures of these top hits in the Protein Data Bank (PDB) however, did not yield any result. In other words, no potential structural template among these top hits was found.

Conserved domains search on KPN_00953 using Uniprot [17] revealed that it contains a conserved domain found in the superfamily of M15_3 peptidases. Using this information, a similarity search was performed on KPN_00953 against all peptidases in the MEROPS Peptidase Database [18]. BLAST MEROPS results indicated that KPN_00953 shares similarity with many sub family M15A unassigned peptidases. This family

Table 1 Top 20 hits in BLAST search against NR database for KPN_00953

Accession number	Title	Organism	SI (%)	SS (%)	E-value
ZP_06550038.1 GI:290510668	Hypothetical protein HMPREF0485_02438	Klebsiella sp. 1_1_55	99	99	2e-113
YP_002918716.1 GI:238893982	Hypothetical protein KP1_1926	Klebsiella pneumoniae NTUH-K2044	99	100	4e-104
ZP_06014196.1 GI:262040974	Tat pathway signal sequence domain protein	Klebsiella pneumoniae subsp. rhinoscleromatis ATCC 13884	100	100	7e-103
YP_002239426.1 GI:206575875	Tat (twin-arginine translocation) pathway signal sequence domain/peptidase M15 family protein	Klebsiella pneumoniae 342	99	100	1e-102
ADO49107.1 GI:308749355	Protein of unknown function DUF882	Enterobacter cloacae SCF1	89	94	1e-100
NP_752993.1 GI:26246953	Hypothetical protein c1068	Escherichia coli CFT073	91	96	2e-95
ZP_05967118.1 GI:261339260	Hypothetical protein ENTCAN_05496	Enterobacter cancerogenus ATCC 35316	91	97	5e-95
YP_003613217.1 GI:296103071	Hypothetical protein ECL_02727	Enterobacter cloacae subsp. cloacae ATCC 13047	91	97	5e-95
CBK85464.1 GI:295096374	Uncharacterized protein conserved in bacteria	Enterobacter cloacae subsp. cloacae NCTC 9394	91	96	1e-94
YP_002382233.1 GI:218548442	Hypothetical protein EFER_1070	Escherichia fergusonii ATCC 35469	90	95	4e-94
NP_309036.1 GI:15830263	Hypothetical protein ECs1009	Escherichia coli O157:H7 str. Sakai	92	96	6e-94
ZP_03066641.1 GI:194434378	Tat (twin-arginine translocation) pathway signal sequence domain/peptidase M15 family protein	Shigella dysenteriae 1012	92	96	6e-94
YP_001438498.1 GI:156934582	Hypothetical protein ESA_02416	Cronobacter sakazakii ATCC BAA-894	87	95	8e-94
NP_459971.1 GI:16764356	Hypothetical protein Ent638_1445	Enterobacter sp. 638]	90	96	1e-93
ZP_07097188.1 GI:300816969	Putative outer membrane protein	Salmonella enterica subsp. enterica serovar Typhimurium str. LT2	91	96	5e-93
NP_455482.1 GI:16759865	Tat pathway signal sequence protein	Escherichia coli MS 107-1	91	96	6e-93
ZP_02346752.1 GI:167553002	Hypothetical protein STY0998	Salmonella enterica subsp. enterica serovar Typhi str. CT18	90	96	1e-92
YP_001588730.1 GI:161614765	Putative exported protein, Tat-dependent	Salmonella enterica subsp. enterica serovar Saintpaul str. SARA29	90	95	1e-92
ZP_04561370.1 GI:237730889	Hypothetical protein SPAB_02517	Salmonella enterica subsp. enterica serovar Paratyphi B str. SPB7	90	95	1e-92
ZP_04561370.1 GI:237730889	Conserved hypothetical protein	Citrobacter sp. 30_2	90	96	2e-92

of peptidase consists of metallopeptidases mostly specialized carboxypeptidases and dipeptidases such as Zn D-Alanyl-D-Alanine (D-Ala-D-Ala) carboxypeptidases. The biological functions of D-Ala-D-Ala carboxypeptidases are related to bacterial cell wall biosynthesis and metabolism [19-21].

To predict the possible function of KPN_00953, recently reported protein prediction methods comprising of FFPred [22], GOStruct [23], Argot2 [24], CombFunc [25] and PANNZER [26] were used. Out of this five prediction methods, GOStruct and PANNZER services were not available at the moment when the analyses were performed. Results obtained from FFPred indicated that KPN_00953 might be responsible for oxidation-reduction process with 0.952 probability. Since FFPred system is dedicated to assign gene ontology terms for eukaryotic protein sequences [22], the results obtained from the analysis of KPN_00953 using FFPred might not be accurate for a prokaryotic system. Analysis using CombFunc and Argot2 failed to predict any significant function for KPN_00953 although both highlighted that the protein consisted the Peptidase M_15 domain as well as a leucine rich domain. KEGG Orthology (KO) group analysis on KPN_00953 only annotated it as a hypothetical protein, unrelated to any KO group. Therefore, KPN_00953 could not be associated with any pathways based on the KO analysis.

A multiple sequence alignment (MSA) analysis of KPN_00953 with sequences from six other organisms containing similar domain from the BLAST result was performed using ClustalW [27]. 33 residues from KPN_00953 were identical with these proteins (indicated with *, Figure 1). There were also 34 residues (indicated as :) which are conserved suggesting that the same chemical properties are shared albeit differences in sequence identity. Two Histidine residues i.e. His169 and His209, and an Aspartate residue i.e. Asp176 which are postulated to be involved in Zn chelation in well-characterized D-Ala-D-Ala carboxypeptidases, are interestingly found to be conserved here.

Template selection
A search in the Protein Data Bank (PDB) on the potential structural template to be used to build the model of KPN_00953 was performed. No structure with high homology in PDB was detected for KPN_00953. However, since the MSA results confirmed the integrity of the putative Peptidase_M15_3 superfamily conserved domain in KPN_00953, this domain was used to search for such potential template in PDB instead. This resulted in the identification of one potential template, termed 1LBU with sequence identity of only 23%. 1LBU is a crystal structure of muramoyl-pentapeptide carboxypeptidase, a Zn^{2+} D-Ala-D-Ala carboxypeptidase from *Streptomyces albus* [20] which contain the particular Peptidase_M15_3

superfamily domain. 1LBU was also ranked top by Phyre2 search within CombFunc server as having the highest structural similarity KPN_0053. The results indicated that KPN_00953 might adopt similar fold and domain with 1LBU; namely the Hedgehog/D-Ala-D-Ala peptidase fold and Zn^{2+} D-Ala-D-Ala carboxypeptidase C-terminal catalytic domain. In fact, all the hits listed by Phyre2 contained the Hedgehog/D-Ala-D-Ala peptidase fold (Table 2). These results further stressed 1LBU as the best structural template for KPN_00953.

Phylogenetics analysis via SCOP search [28] was performed between KPN_00953 with other members of Peptidase_M15_3 superfamily to determine their degree of evolutionary relatedness. The analysis revealed that 1LBU was at a further clad from KPN_00953 (*Klebsiella sp*) in the cladrogram as compared to other organisms (Figure 2). Although the sequence identity of 1LBU compared with KPN_00953 is only 23%, it is evolutionary closer to *Klebsiella sp* based on its phylogenetic relationship. Moreover, the length of 1LBU is similar to KPN_00953 (Figure 1). Thus, 1LBU was selected as the template for homology modeling.

Homology modelling of KPN_00953 and model validation
Homology modeling of KPN_00953 using MODELLER 9v8 [29] with 1LBU as the template randomly generated 20 models. The best model (with the lowest DOPE score) was subsequently validated using PROCHECK [30]. The Ramachandran analysis revealed that 96.6% of the amino acid residues reside in the most favourable and additional allowed regions (Table 3). The built model was further verified using Verify3D [31] and ERRAT [32]. Verify3D indicated that the built protein model scored 79.91%, suggesting compatibility between the amino acid sequence and the environment of the amino acid side chains in the model. ERRAT analysis on the protein model gave forth score of 63.285, a relatively acceptable assessment value on the arrangement of atoms with respect to one another in the protein model. In addition to these analyses, the compactness of the built model was also validated using ProQ protein quality prediction tool [33]. The result showed LG score of 1.304 and MaxSub score of 0.130, indicating that the built model of KPN_00953 is within the range of an acceptable model. Calculations of the interaction energy and Z-score using ProSA-Web [34] energy plot for each residue of the model gave forth value of -3.5 kcalmol^{-1}. Based on these various structural evaluation results, the particular model can be accepted as a potential model for KPN_00953 (Figure 3).

Structural and motif analyses
Structural alignment of KPN_00953 with the template 1LBU and 2VO9 (crystal structure of the distantly related

```
KPN_00953     ----------------MPGNISVVASDSELILASRKAEKVYNDSLTVDLIIMDKFDANR  43
Enterobacter  MSYEWVKTGQFVCRACIYWGWATLTSAAMTVKVPFVRQKCINYDHLLVDLIIMDKFDANR  60
Escherichia   -------------------------------------------MIIMDKFDANR  11
Shigella      -------------------------------------------------MDKFDANR   8
Cronobacter   -----------------------MTPLRLAVKVPLVRQKCIVIVHLIVDLNIMDKIDAHR  37
Salmonella    -------------------------------------------------MDKFDANR   8
1LBU          -------------------------DGCYTWSGTLSEGSSGEAVRQLQIRVAGYPGTG  33
                                                                 :       .

KPN_00953     RRLLALGGAALGAAAILPAPAFATLS-TPRPRILTLNNLHTGESLRAEFFDGRGYIQDEL 102
Enterobacter  RKLLALGGVAFGAAAILPTPAFATLS-TPRPRILTLNNLHTGESIKAEFFDGRGYIQDEL 119
Escherichia   RKLLALGGVALG-AAILPTPAFATLS-TPRPRILTLNNLHTGESIKAEFFDGRGYIQEEL  69
Shigella      RKLLALGGVALG-AAILPTPAFATLS-TPRPRILTLNNLHTGESIKAEFFDGRGYIQEEL  66
Cronobacter   RKLLTIGGAALG-AAILPTPAFATLS-TPRPRILTLNNLHTGESIKAEFFDGRGYIQDEL  95
Salmonella    RKLLALGGVALG-AAILPAPAFATLS-TPRPRILTLNNLHTGESIKAEFFDGRAYIQDEL  66
1LBU          AQLAIDG--QFGPATKAAVQRFQSAYGLAADGIAGPATFNKIYQLQDDDCTPVNFTYAEL  91
               :*    *    :* *:   ..   * :      .     *    .::.  .:: :        :    **

KPN_00953     ARLNHFFRDYRANKIKSIDPNLFDHLYRLQGLLGTN--KPVQLISGYRSLDTNDELRARS 160
Enterobacter  AKLNHFFRDFRANKIKSIDPKLFDQLYRLQGLLGTN--KPVQLVSGYRSLDTNNELRERS 177
Escherichia   AKLNHFFRDYRANKIKSIDPGLFDQLYRLQGLLGTR--KPVQLISGYRSIDTNNELRARS 127
Shigella      AKLNHFFRDYRANKIKSIDPRLFDQLYRLQGLLGTR--KPVQLISGYRSIDTNNELRARS 124
Cronobacter   AKLNHFFRDYRANKVKAIDPRLFDQLFRLQGLLGTR--KPVQLISGYRSVDTNNELRSKS 153
Salmonella    AKLNHFFRDYRANKVRSIDPRLFDQLYRLQGLLGTR--KPVQLISGYRSLDTNNELRARS 124
1LBU          NRCNSDWSGGKVSAATARANALVT-MWKLQAMRHAMGDKPITVNGGFRSVTCNSNVGG-- 148
               : *   :  . :..    :       *.   :::**.:   :     **: :  .*:*:  *.::

KPN_00953     RGVAKHSYHTKGQAMDFHIEGISLSNIRKAALS---MRAGGVGYYPRSNFVHIDTGPVRH 217
Enterobacter  RGVAKHSYHTKGQAMDFHIEGISLSNVRKAALS---MRAGGVGYYPSSNFVHIDTGPTRH 234
Escherichia   RGVAKKSYHTKGQAMDFHIEGIALSNIRKAALS---MRAGGVGYYPRSNFVHIDTGPARH 184
Shigella      RGVAKKSYHTKGQAMDFHIEGIALSNIRKAALS---MRAGGVGYYPRSNFVHIDTGPARH 181
Cronobacter   RGVAKHSYHTKGQAMDFHIEGISLSNIRKAALS---LRAGGVGYYPSSNFVHIDTGPLRH 210
Salmonella    SGVAKKSYHTKGQAMDFHIEGVALSNIRKAALS---MRAGGVGYYPRSNFVHIDTGPARH 181
1LBU          ---ASNSRHMYGHAPDLGAGSQGFCALAQAARNHGFTEILGPGYPGHNDHTHVAGGDGRF 205
              *.:*  *    *:.:  *:       . .::  :  :**  .     *  * **   ..:..*:   *   *.
```

HXXXXXXD (H-x6-D motif)

```
KPN_00953     W-------  218
Enterobacter  W-------  235
Escherichia   W-------  185
Shigella      W-------  182
Cronobacter   W-------  211
Salmonella    W-------  182
1LBU          WSAPSCGI  213
              *
```

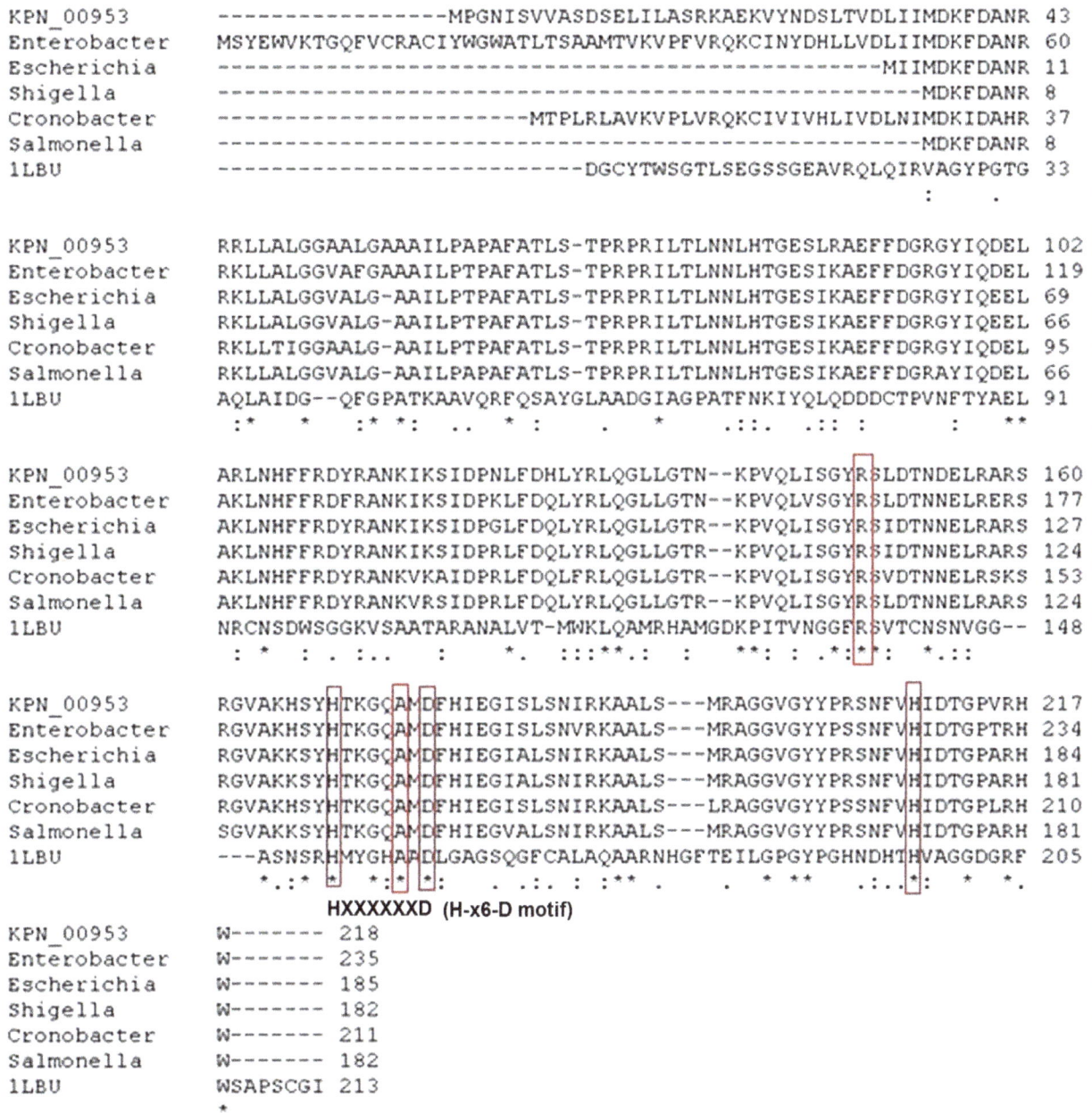

Figure 1 MSA of KPN_00953 with 6 other conserved hypothetical proteins. The two conserved His residues as well as Arg, Ala and Asp (highlighted in red) are believed to be responsible for Zn binding. The presence of the H-x6-D motif in the sequences is observed.

L-alanoyl-D-glutamate endopeptidase domain of *Listeria* bacteriophage endolysin Ply500) [35] showed the integrity of the conserved domain. Structural analysis showed that the secondary structure, in particular the four beta stranded region and one single helix region, are well aligned (Figure 4). This is a unique secondary structure topology shared among metalloproteases [36,37]. The average RMSD between KPN_00953 and these two other structures is 5.42 Å. Further analysis on the secondary structure elements of the built model with 1LBU and 2VO9 using STRIDE [38] lends further support that these

proteins share conserved secondary structure topologies (Figure 5).

Certain peptidases, particularly those of peptidoglycan hydrolases such as D-Ala-D-Ala metallopeptidases are believed to contain a Zn^{2+} ligand in most of the structures where the metal ion is coordinated by two histidines, an aspartate and a water molecule [36,37,39]. The presence of these active site residues is clearly observed in our built model, where His169, His209 and Asp176 are located exactly at Zn^{2+}-chelating positions (Figure 3). Interestingly, these residues are also found to be highly conserved in

Table 2 Results extracted from Phyre2 analysis from CombFunc server

Template	Confidence level (5)	Domain
1LBU	100.0	Hedgehog/D-Ala-D-Ala peptidase
2 V09	98.1	Hydrolase/D-Ala-D-Ala peptidase
4MUR	97.9	Hydrolase/D-Ala-D-Ala peptidase
4JID	97.2	Hydrolase/D-Ala-D-Ala peptidase
2R44	95.8	Hedgehog/D-Ala-D-Ala peptidase

certain hypothetical proteins from other organisms as well (Figure 1). It has been reported that other than these active site residues, there is another second conserved His residue two residues upstream of the His Zn^{2+} ligand [40]. This particular His residue, His166, was observed in the sequence of KPN_00953 where it is located two residues upstream of His169 (the Zn^{2+} ligand) (Figure 1).

The intactness of both the secondary structure topology and the three Zn^{2+} ligand-binding residues of the built model suggest that KPN_00953 may function as a cell wall (peptidoglycan)-hydrolyzing enzyme, similar to a few characterized Zn D-Ala-D-Ala metallopeptidases such as muramoyl-pentapeptide carboxypeptidase from *S. albus* [20] and VanX from *Enterococcus faecalis* [19]. Closer inspection on the sequence of KPN_00953 in comparison to the abovementioned proteins revealed that it does not contain the characteristic H-x-H motif which is predominantly found in nearly all D-Ala-D-Ala metallopeptidases, except VanX [40]. This motif is present in the sequence of 1LBU, the template used for the homology modeling of KPN_00953 (Figure 1). In the case of VanX, instead of the signature H-x-H motif it bears the E-x-x-H motif in its sequence [40]. This motif was absent also in the sequence of KPN_00953. However, KPN_00953 was found to possess the H-x (3–6)-D motif similar to MepA peptidase (Figure 1). Similar to the Zn D-Ala-D-Ala metallopeptidases stated above, MepA is a Zn-metalloprotein shown to be involved in cell wall related functions [41]. The only deviation to this similarity is the absence of the H-x-H motif in KPN_00953, which is reported to be present in MepA [40].

Amplification and cloning of KPN_00953

To characterize further the possible function of KPN_00953, its Open Reading Frame (ORF) was amplified from the genome of *K. pneumoniae* MGH 78578 using specifically designed primers. A specific amplicon of 657 bp was obtained (Figure 6(a)). Cloning of this amplicon into pGEM®-7zf (+) was subsequently achieved, as confirmed from blue-white screening (data not shown), colony PCR (Figure 6(b)) and sequencing (data not shown).

Altered cell surface morphology of *E. coli* JE5505 overexpressing KPN_00953

Since homology modeling results point to the possibility of KPN_00953 having cell wall related metabolic functions i.e. peptidoglycan degradation, the effect of overexpressing this HP on cell surface morphology was investigated. For this purpose, the cloned *KPN_00953* construct was introduced into the lipoprotein-deficient *E. coli* JE5505 strain [42] and subsequently overexpressed via IPTG induction. Cells which overexpressed KPN_00953 appeared to have different surface morphology than cells which do not expressed this protein. They appear to be slightly smoother with flattened edges (Figure 7(c)), and some of them seem to have deposits on their surfaces (Figure 7(b) and (c)). In contrast, the cells which contained only the expression vector (control) have more well-defined and rougher surface texture (Figure 7(a)). Such alterations and deposits observed on the surface of the cells may suggest possible cell wall degradation by KPN_00953.

Discussion

We have identified that HP KPN_00953 from *K. pneumoniae* MGH 78578 contains a well conserved domain belonging to the M15 superfamily of peptidases. Template identification based on this domain has led to the building of a 3D model of KPN_00953 via homology modeling using the crystal structure of muramoyl-pentapeptide carboxypeptidase (PDB id: 1LBU), a Zn D-Ala-D-Ala metallopeptidase from *S. albus* [20] as the template. The built model has been verified to be acceptable and topologically conserved with other available structures related to peptidases such as the L-alanoyl-D-glutamate endopeptidase domain of *Listeria* bacteriophage

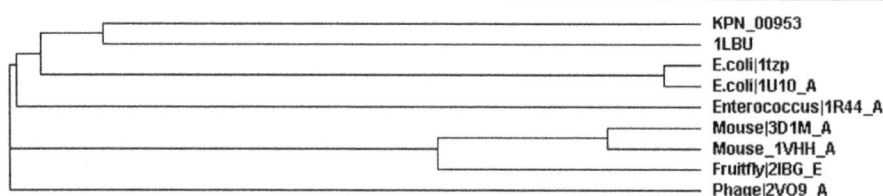

Figure 2 Phylogenetic analysis of KPN_00953 with other protein structures from 8 other organisms including 1LBU (selected template). The proteins were selected using SCOP hierarchy search. All the proteins contained the conserved Peptidase_M15_3 superfamily domain.

Table 3 Statistical result of Ramachandran plot analysis for the best model in homology modeling

Ramachandran plot (%)	Built KPN_00953 model (*K. pneumoniae*) (%)
Most favoured regions	84.2
Additional allowed regions	12.4
Generously allowed	2.6
Disallowed regions	1.1

endolysin Ply500 (PDB id: 2VO9) [35]. Two His residues, His169 and His209, as well as an Asp residue, Asp176 of this model are postulated to be involved in Zn chelation (Figure 3) and are interestingly found to be conserved in other well characterized Zn-metalloproteases [36,37,39,40]. It is important to note that several members of Zn-metalloproteases act as peptidoglycan hydrolases; in which they are involved in cell wall metabolism.

The cell wall of bacteria contains peptidoglycan which is important in preserving and maintaining the structural integrity of the cell by withstanding turgor. It is closely linked to several physiological processes such as cell growth and division. Inhibition of its biosynthesis via the action of antiobiotics for instance as well as its degradation by lysozyme will result in cell lysis [21]. Peptidoglycan in general is made of alternating units of N-acetyl-glucosamine and N-acetyl-muramic acid that are linked via 1,4-glycosidic bonds. The muramyl residues serve as platforms for the attachment of short polypeptides which contain both L- and D-amino acids and typically have two D-Ala residues at the C terminus. These peptide components on the muramyl residues can be crosslinked by transpeptidation, which subsequently will result in a loss of the terminal D-Ala and strengthening of the bacterial cell wall [40,43].

Members of Zn-metalloproteases which have been characterized to be involved in peptidoglycan (cell wall)

biosynthesis and metabolism include Zn D-Ala-D-Ala carboxypeptidases and dipeptidases [19]. Muramoyl-pentapeptide carboxypeptidase, which is used as a template in the homology modeling of KPN_00953, is a specific Zn D-Ala-D-Ala carboxypeptidase [20]. It was reported that this particular enzyme from *Streptomyces* hydrolyzes the C-terminal peptide bond of peptides of general structure R-D-Ala-D-Xaa. The lytic and extracellular characteristics of the enzyme brought about the suggestion that this enzyme is used by *Streptomyces* for fighting competitors in its ecological niche since it does not hydrolyze the *Streptomyces* peptidoglycan [20]. Another instance is the Zn-dependent D-Ala-D-Ala (amino) dipeptidase, VanX. This peptidase reduces the cellular pool of the D-Ala-D-Ala dipeptide so that only the D-Ala-D-lactate peptidoglycan chain precursors are produced and incorporated into the cell wall instead of the former. The modified peptidoglycan reportedly exhibited a 1,000-fold decrease in affinity for vancomycin. This feature is responsible in conferring antibiotic resistance to pathogenic bacteria such as vancomycin-resistant Enterococci (VRE) [19]. Both of these proteins were reported to be similar and structurally related [44,45] despite the differences in their sequences. The muramoyl-pentapeptide carboxypeptidase from *S. albus*, like nearly all Zn D-Ala-D-Ala metallopeptidases, contains the H-x-H motif. VanX, on the other hand, contains the E-x-x-H motif making it an exception among the D-Ala-D-Ala metallopeptidases [40].

In the case of KPN_00953, it does not possess both of these characteristic motifs. Instead, it contains the H-x (3–6)-D motif, similar to MepA peptidase [40,41]. However, it is important to stress that MepA contains as well the characteristic H-x-H motif [40,41] which KPN_00953 lacks. In terms of molecular function, MepA is reported to cleave D-alanyl-meso-2,6-diamino-pimelyl peptide bonds in *E. coli* peptidoglycan and is classified as a peptidase of unknown

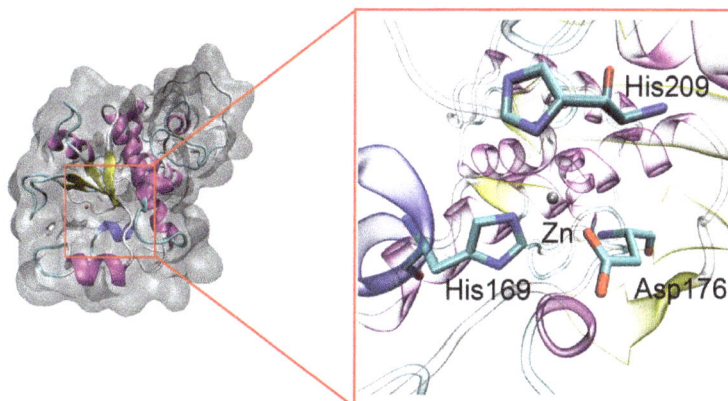

Figure 3 The best model for KPN_00953 built using Modeller. Conserved Zn-chelating residues such as His169, His209 and Asp176 are located within 4 Å from the Zn atom.

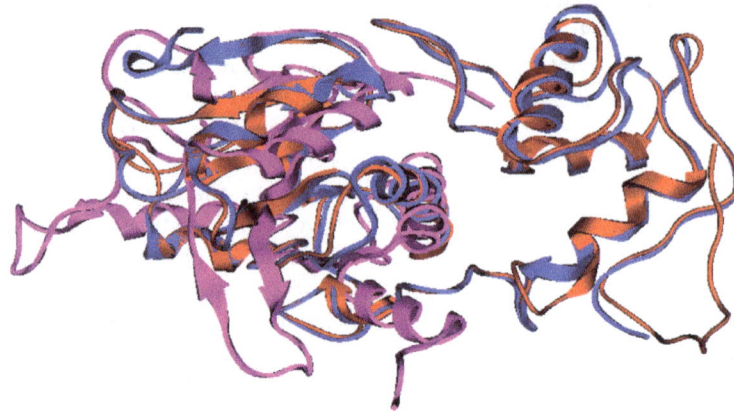

Figure 4 Structural alignment of KPN_00953, 1LBU and 2VO9. Four beta stranded regions are well aligned among KPN_00953 (red), 1LBU (blue) and 2VO9 (purple).

fold and catalytic class due to its low sequence similarity with other peptidases [41]. KPN_00953 resembles MepA in this respect in which KPN_00953 is shown to be related to the subfamily M15A of unassigned peptidases from BLAST MEROPS scan. In contrast to MepA however, KPN_00953 could be assigned to a subfamily or catalytic class of peptidases, namely the M15A subfamily of peptidase. As this subfamily of peptidases consists of a number of characterized members such as Zn D-Ala-D-Ala carboxypeptidases, functional inference of KPN_00953 could be made based on this information. Although KPN_00953 contains the H-x (3–6)-D motif of MepA, conserved domain analysis result seems to relate KPN_00953 more to Zn D-Ala-D-Ala metallopeptidase in general, excluding VanX. Hence, the presence of the H-x (3–6)-D motif of MepA in KPN_00953 may occur by chance.

It is important to note that whilst both domain analysis and homology modeling of KPN_00953 revealed the conservation of important domains, Zn-chelating residues and secondary structure topologies to D-Ala-D-Ala carboxypeptidase, sequence similarity search revealed that this particular HP is also related to the twin-arginine translocation (Tat) pathway signal sequence. Tat pathway is a protein transport system for the export of folded proteins [46]. Proteins which are targeted to the Tat pathway are exported to the cell envelope or to the

extracellular space by tripartite N-terminal signal peptides and Tat translocase, which are found in the cytoplasmic membrane [47]. However, it is important to note that KPN_00953 lacks the consensus twin-arginine motif, (S/T)-R-R-x-F-L-K, which is reported to be present in all types of bacterial signal peptides [48,49], despite the sequence similarity mentioned above. In addition to this, KPN_00953 does not contain signal peptide sequence as revealed by analyses using various signal peptide detection software such as SignalP 4.0 [50], Signal–3 L [51], iPSORT [52] and SOSUISignal [53]. Hence, this omits the possibility that KPN_00953 is functionally related to Tat pathway.

Further inference on the possible function of KPN_00953 was attempted by cloning and expressing its gene in *E. coli* JE5505 strain which is deficient in lipoprotein production. Induced production of KPN_00953 in this particular strain changed the morphology of the cells. They appeared to be smoother with less defined edges with some having deposits on their surfaces. Cells which did not produce the cloned KPN_00953 protein appeared to be rougher with defined edges (Figure 7). This observation highlights the possibility of KPN_00953 to be involved in the functions of the cell wall, similar to other characterized peptidoglycan-hydrolyzing Zn metallopeptidases.

Figure 5 Secondary structure comparison of KPN_00953, 1LBU and 2VO9.

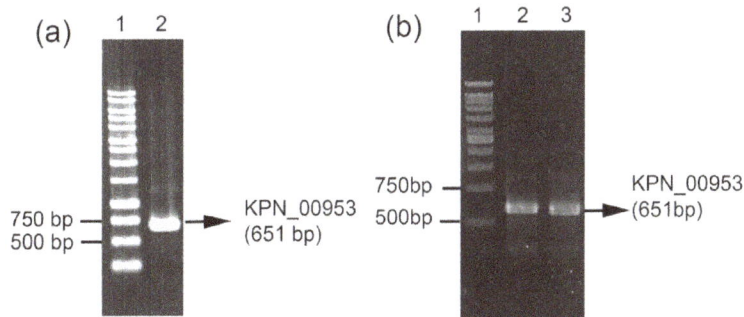

Figure 6 Amplification of KPN_00953. KPN_00953 amplicon (657 bp) amplified from **(a)** the genome of *K. pneumoniae* MGH 78578 (Lane 2) and **(b)** *E. coli* JM109 transformants via colony PCR (Lanes 2 and 3).

Conclusions

Based on the three-dimensional model, domain and residues conservation of KPN_00953 to D-Ala-D-Ala carboxypeptidase, we hypothesize that KPN_00953 adopts the functionality as a metallopeptidase with an important role in cell wall metabolism. This is further supported by the altered surface morphology of *E. coli* JE5505 cells overexpressing KPN_00953. The mechanism as to how KPN_00953 brings about these changes is worthy to be investigated in the near future. This can be achieved via gene-knockout and structural biology studies

to understand its structure, function and mechanism of action. Such efforts will undoubtedly address the possibility of utilizing this protein as a new drug target against *K. pneumoniae* in the future.

Methods

Bacterial strains and plasmids used

K. pneumoniae MGH 78578 was purchased from American Type Culture Collection (ATCC number: 700721). *Escherichia coli* JM109 [*end*A1, *rec*A1, *gyr*A96, *thi*, *hsd*R17 (r$_k$−, m$_k$+), *rel*A1, *sup*E44, Δ (lac-proAB)] was used for standard cloning purposes. For microbial plate assay of the expressed HP on cell surface morphology, a lipoprotein deletion mutant, *Escherichia coli* JE5505 [Δ (*gpt-proA*)62, *lacY1*, *tsx-29*, *glnV44*(AS), *galK2*(Oc), λ⁻, Δ*lpp-254*, *pps-6*, *hisG4*(Oc), *xylA5*, *mtl-1*, *argE3* (Oc), *thi-1*] [33] was used (purchased from *E. coli* Genetic Resource Center, Yale University). pGEM-7Zf (+) (Promega) was used as an expression vector to express the cloned *KPN_00953* gene.

Sequence analysis, model building and validation

The complete genome sequence of *K. pneumoniae* MGH 78578 was obtained from NCBI website [Refseq: NC_009648]. The sequences for HPs of the pathogen were selected and analyzed preliminarily using Uniprot [17]. KPN_00953 (YcbK) was selected based on the presence of the conserved M15 peptidase domain. KPN_00953 was subjected to a series of BLAST [16] search against non-redundant database (NR) and Protein Data Bank (PDB). FFPred [22], GOStruct [23], Argot2 [24], CombFunc [25] and PANNZER [26] were used to predict the possible function of KPN_00953. SignalP 4.0 [50], Signal–3 L [51], iPSORT [52] and SOSUISignal [53] were used to determine the possible presence of signal peptide in the sequence. Multiple sequence alignment (MSA) of KPN_00953 with six other proteins sequences from other organisms was later performed with ClustalW [27]. These proteins were selected from the list of potential hits in BLAST result

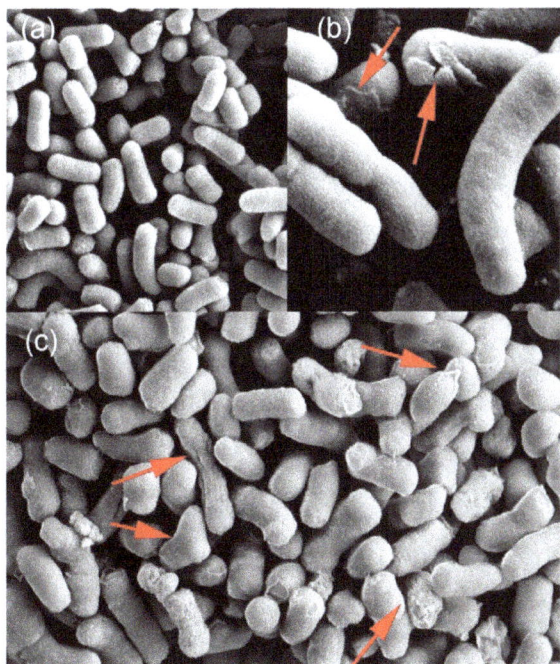

Figure 7 Cell surface morphology of *E. coli* JE5505 cells. Cell surface morphology of *E. coli* JE5505 cells **(a)** containing pGEM-7zf (+) plasmid (control) and those overexpressing KPN_00953 at **(b)** 200 nm and **(c)** 1 μm scale. Red arrows indicate material deposits on the surface of the cells.

which contained similar domain with KPN_00953. 1LBU [20] was selected as the template for homology modeling of KPN_00953 using MODELLER 9v8 [29]. Twenty models were generated randomly and model with the best Discrete Optimized Potential Energy (DOPE) score was selected and subsequently verified using PROCHECK [30], Verify3D [31], ERRAT [32], ProQ [33] and ProSA-Web [34] energy plot.

Genomic DNA extraction

Genomic DNA extraction of *K. pneumoniae* MGH 78578 was performed using Wizard® Genomic DNA Purification Kit (Promega). The integrity and quality of the genomic DNA extracted were analyzed via agarose gel electrophoresis.

Amplification of KPN_00953 Open Reading Frame (ORF)

KPN_00953 (*ycbK*) gene was amplified by Polymerase Chain Reaction (PCR) using: 5'-TTAAGCTTTTGCCGGG CAACATCTCG-3' (forward primer) and 5'-TCTTGGAT CCTTACCAGTGCCTTACGGG-3 (reverse primer). Under-lined sequences refer to incorporated *Hind*III and *Bam*HI restriction sites, respectively. The PCR reaction mixture (50 μl) contained Go Taq® Flexi Buffer (1X), dNTP mixture (0.2 mM), MgCl$_2$ (1.0 mM), forward and reverse primers (1.0 μM each), genomic DNA template (0.5 ng) and 0.25U Go Taq® polymerase (Promega). A cycle of 95°C denatur-ation for 5 minutes, followed by 30 cycles of denaturation at 95°C for 1 minute, annealing at 55°C for 1 minute and extension at 72°C for 1 minute, and lastly further extension at 72°C for 5 minutes were performed. The amplicon was subsequently analyzed and purified using the QIAquick PCR Purification Kit (QIAGEN).

Cloning of KPN_00953 amplicon into plasmid vector

The KPN_00953 amplicon and pGEM®-7zf (+) plasmid vector were subjected to *Bam*HI/*Hind*III double digestion at 37°C for 5 hours and subsequently analyzed by gel electrophoresis followed by purification. The products were later ligated using T4 DNA ligase (New England Biolabs) at 16°C, overnight. Ligated products were then transformed into *E. coli* JM109 via the TSS transformation method [54]. The presence of the recombinant plasmid constructs was verified via blue-white screening, colony PCR and sequencing. Upon positive verification of the de-sired construct, it was transformed into *E. coli* JE5505 (a lipoprotein-deficient strain).

Cell surface observation of E. Coli JE5505 overexpressing KPN_00953 hypothetical protein by scanning electron microscope (SEM)

E. coli JE5505 cells harboring the cloned *KPN_00953* gene were cultivated in 5 ml LB broth supplemented with 100 μg/ml ampicillin and 0.1 mM IPTG for 16 hours at 180 rpm and 37°C. As the pGEM®-7zf (+) plasmid con-tains a T7 promoter, IPTG was used to induce the expres-sion of *KPN_00953* in *E. coli* JE5505. Cells were harvested by centrifugation at 2000 rpm for 15 minutes at room temperature and were resuspended with 300 μl of McDowell-Trump fixative prepared in 0.1 M phosphate buffer (pH7.2) for at least 2 hours. The cells were centri-fuged at the same speed and duration before being resus-pended again in 500 μl 0.1 M phosphate buffer. This step was repeated once and followed by resuspension in 1% osmium tetroxide prepared in the phosphate buffer for 1 hour. The cells were centrifuged and finally resuspended in distilled water. These steps were repeated thrice to ensure the pellet was properly washed. Next, dehydration steps using increasing concentrations of ethanol were performed on the sample: 50% ethanol for 10 minutes, 75% ethanol for 10 minutes, 95% ethanol for 10 minutes and 100% ethanol for 10 minutes (twice). The final dehy-dration process was performed using hexamethyldisila-zane. The centrifugation speed and duration employed for each of the dehydration step were the same as described above. After hexamethyldisilazane was decanted, the cells in the tube were left in the desiccators to be air-dried at room temperature. They were then mounted onto an SEM specimen stub with a double-sided sticky tape and coated with gold to be viewed under SEM.

Competing interests
The authors declare that they have no competing interests.

Authors' contributions
TBA carried out the molecular cloning, expression and characterization studies. SBC carried the *in silico* studies related to sequence homology and evolutionary relatedness analyses as well as homology modeling. NM carried out the localization, signal peptide and sequence motifs analyses as well as sequencing analysis of the cloned plasmid. All three drafted the manuscript. FLL, STWC, NN, ABS, HAW and YMN revised and proofread the manuscript. YMN and HAW conceived the study and participated in its design and coordination. NN, ABS, FLL and STWC gave technical advice to the study. All authors read and approved the final manuscript.

Acknowledgements
We would like to thank the Electron Microscopy Unit of School of Biological Sciences, USM for the excellent technical support and service. This work was supported by USM and UPM RU Grant Schemes (No.:1001/PBIOLOGI/815014 and 05-01-11-1191RU, respectively). BA, Teh and Nasihah, M wish to thank the Ministry of Science, Technology and Innovation (MOSTI) for the National Science Fellowship and UPM Graduate Research Fellowship awarded.

Author details
[1]School of Biological Sciences, Universiti Sains Malaysia, 11800 USM Pulau Pinang, Malaysia. [2]School of Industrial Technology, Universiti Sains Malaysia, 11800 USM Pulau Pinang, Malaysia. [3]Enzyme and Microbial Technology Research Center (EMTECH), Faculty of Biotechnology and Biomolecular Sciences, Universiti Putra Malaysia, 43400 Serdang, Selangor, Malaysia. [4]School of Health Sciences, Health Campus, Universiti Sains Malaysia, 16150 Kubang Kerian, Kelantan, Malaysia. [5]Malaysian Institute of Pharmaceuticals and Nutraceuticals, Ministry of Science, Technology and Innovation, Blok 5-A, Halaman Bukit Gambier, 11700 Pulau Pinang, Malaysia.

References

1. Brisse S, Grimont F, Grimont PAD: The genus Klebsiella. *Prokaryotes* 2006, 6:159–196.

2. Podschun R, Ullmann U: Klebsiella spp. as nosocomial pathogens: epidemiology, taxonomy, typing methods, and pathogenicity factors. *Clin Microbiol Rev* 1998, 11(4):589–603.

3. Dhingra KR: A case of complicated urinary tract infection: klebsiella pneumoniae emphysematous cystitis presenting as abdominal pain in the emergency department. *West J Emerg Med* 2008, 9(3):171–173.

4. Fang CT, Chuang YP, Shun CT, Chang SC, Wang JT: A novel virulence gene in klebsiella pneumoniae strains causing primary liver abscess and septic metastatic complications. *J Exp Med* 2004, 199(5):697–705.

5. Fung CP, Chang FY, Lee SC, Hu BS, Kuo BI, Liu CY, Ho M, Siu LK: A global emerging disease of klebsiella pneumoniae liver abscess: is serotype K1 an important factor for complicated endophthalmitis? *Gut* 2002, 50(3):420–424.

6. Won SY, Munoz-Price LS, Lolans K, Hota B, Weinstein RA, Hayden MK: Emergence and rapid regional spread of klebsiella pneumoniae carbapenemase-producing enterobacteriaceae. *Clin Infect Dis* 2011, 53(6):532–540.

7. Bordow RA, Ries AL, Morris TA: *Manual of clinical problems in pulmonary medicine: with annotated key references, 6th edn.* USA: Lippincott Williams & Wilkins; 2005.

8. Lawlor MS, Hsu J, Rick PD, Miller VL: Identification of Klebsiella pneumoniae virulence determinants using an intranasal infection model. *Mol Microbiol* 2005, 58(4):1054–1073.

9. Lawlor MS, O'Connor C, Miller VL: Yersiniabactin is a virulence factor for Klebsiella pneumoniae during pulmonary infection. *Infect Immun* 2007, 75(3):1463–1472.

10. Mokracka J, Koczura R, Kaznowski A: Yersiniabactin and other siderophores produced by clinical isolates of Enterobacter spp. and Citrobacter spp. *FEMS Immunol Med Microbiol* 2004, 40(1):51–55.

11. Raymond KN, Dertz EA, Kim SS: Enterobactin: an archetype for microbial iron transport. *Proc Natl Acad Sci U S A* 2003, 100(7):3584–3588.

12. Galperin MY: Conserved 'hypothetical' proteins: new hints and new puzzles. *Comp Funct Genomics* 2001, 2(1):14–18.

13. Bork P: Powers and pitfalls in sequence analysis: the 70% hurdle. *Genome Res* 2000, 10(4):398–400.

14. Galperin MY, Koonin EV: Searching for drug targets in microbial genomes. *Curr Opin Biotechnol* 1999, 10(6):571–578.

15. Choi SB, Normi YM, Wahab HA: Revealing the functionality of hypothetical protein KPN00728 from Klebsiella pneumoniae MGH78578: molecular dynamics simulation approaches. *BMC Bioinforma* 2011, 12(13):S11.

16. Altschul SF, Madden TL, Schäffer AA, Zhang J, Zhang Z, Miller W, Lipman DJ: Gapped BLAST and PSI-BLAST: a new generation of protein database search programs. *Nucleic Acids Res* 1997, 25(17):3389–3402.

17. Magrane M: the UniProt consortium: UniProt Knowledgebase: a hub of integrated protein data. *Database* 2011, 2011:bar009.

18. Rawlings ND, Barrett AJ, Bateman A: MEROPS: the database of proteolytic enzymes, their substrates and inhibitors. *Nucleic Acids Res* 2012, 40(D1):D343–D350.

19. Lessard IA, Walsh CT: Mutational analysis of active-site residues of the enterococcal D-ala-D-Ala dipeptidase VanX and comparison with Escherichia coli D-ala-D-Ala ligase and D-ala-D-Ala carboxypeptidase VanY. *Chem Biol* 1999, 6(3):177–187.

20. Charlier P, Wery J, Dideberg O, Frere J: Streptomyces albus g d-ala-a-ala carboxypeptidase. *Handbook Of Metalloproteins* 2004, 3:164.

21. Vollmer W, Blanot D, De Pedro MA: Peptidoglycan structure and architecture. *FEMS Microbiol Rev* 2008, 32(2):149–167.

22. Thompson JD, Higgins DG, Gibson TJ: Clustal-W - Improving the Sensitivity of Progressive Multiple Sequence Alignment through Sequence Weighting, Position-Specific Gap Penalties and Weight Matrix Choice. *Nucl Acids Res* 1994, 22(22):4673–4680.

23. Minneci F, Piovesan D, Cozzetto D, Jones DT: FFPred 2.0: improved homology-independent prediction of gene ontology terms for eukaryotic protein sequences. *PLoS One* 2013, 8:e63754.

24. Sokolov A, Funk C, Graim K, Verspoor K, Ben-Hur A: Combining heterogeneous data sources for accurate functional annotation of proteins. *BMC Bioinforma* 2013, 14(3):S10.

25. Falda M, Toppo S, Pescarolo A, Lavezzo E, Di Camillo B, Facchinetti A, Cilia E, Velasco R, Fontana P: Argot2: a large scale function prediction tool relying on semantic similarity of weighted Gene Ontology terms. *BMC Bioinforma* 2012, 13(4):S14.

26. Wass MN, Barton G, Sternberg MJE: CombFunc: predicting protein function using heterogeneous data sources. *Nucleic Acids Res* 2012, 40:W466–W470.

27. Radivojac P, *et al*: A large-scale evaluation of computational protein function prediction. *Nat Methods* 2013, 10:221–227.

28. Murzin AG, Brenner SE, Hubbard T, Chothia C: SCOP: A structural classification of proteins database for the investigation of sequences and structures. *J Mol Biol* 1995, 247(4):536–540.

29. Sali A, Blundell TL: Comparative protein modelling by satisfaction of spatial restraints. *J Mol Biol* 1993, 234(3):779–815.

30. Laskowski RA, Macarthur MW, Moss DS, Thornton JM: Procheck - a program to theck the stereochemical quality of protein structures. *J Appl Crystallogr* 1993, 26:283–291.

31. Eisenberg D, Lüthy R, Bowie JU: VERIFY3D: assessment of protein models with three-dimensional profiles. *Meth Enzymol* 1997, 277:396–404.

32. Colovos C, Yeates TO: Verification of protein structures: patterns of nonbonded atomic interactions. *Protein Sci* 1993, 2(9):1511–1519.

33. Wallner B, Elofsson A: Can correct protein models be identified? *Protein Sci* 2003, 12(5):1073–1086.

34. Wiederstein M, Sippl MJ: ProSA-web: interactive web service for the recognition of errors in three-dimensional structures of proteins. *Nucl Acids Res* 2007, 35:W407–W410.

35. Korndörfer IP, Kanitz A, Danzer J, Zimmer M, Loessner MJ, Skerra A: Structural analysis of the L-alanoyl-D-glutamate endopeptidase domain of Listeria bacteriophage endolysin Ply500 reveals a new member of the LAS peptidase family. *Acta Cryst* 2008, D64:644–650.

36. Hooper NM: Families of zinc metalloproteases. *FEBS Lett* 1994, 354(1):1–6.

37. Makarova KS, Grishin NV: The Zn-peptidase superfamily: functional convergence after evolutionary divergence. *J Mol Biol* 1999, 292(1):11–17.

38. Heinig M, Frishman D: STRIDE: a web server for secondary structure assignment from known atomic coordinates of proteins. *Nucl Acids Res* 2004, 32:W500–W502.

39. Zhang J, Yang W, Piquemal J-P, Ren P: Modeling Structural Coordination and Ligand Binding in Zinc Proteins with a Polarizable Potential. *J Chem Theory Comput* 2012, 8(4):1314–1324.

40. Bochtler M, Odintsov SG, Marcyjaniak M, Sabala I: Similar active sites in lysostaphins and D-Ala-D-Ala metallopeptidases. *Protein Sci* 2004, 13(4):854–861.

41. Marcyjaniak M, Odintsov SG, Sabala I, Bochtler M: Peptidoglycan amidase MepA is a LAS metallopeptidase. *J Biol Chem* 2004, 279(42):43982–43989.

42. Hirota Y, Suzuki H, Nishimura Y, Yasuda S: On the process of cellular division in Escherichia coli: a mutant of E. coli lacking a murein-lipoprotein. *Proc Natl Acad Sci U S A* 1977, 74(4):1417–1420.

43. Schleifer KH, Kandler O: Peptidoglycan types of bacterial cell walls and their taxonomic implications. *Bacteriol Rev* 1972, 36(4):407–477.

44. McCafferty DG, Lessard IA, Walsh CT: Mutational analysis of potential zinc-binding residues in the active site of the enterococcal D-Ala-D-Ala dipeptidase VanX. *Biochemistry* 1997, 36(34):10498–10505.

45. Bussiere DE, Pratt SD, Katz L, Severin JM, Holzman T, Park CH: The structure of VanX reveals a novel amino-dipeptidase involved in mediating transposon-based vancomycin resistance. *Mol Cell* 1998, 2(1):75–84.

46. Widdick DA, Dilks K, Chandra G, Bottrill A, Naldrett M, Pohlschroder M, Palmer T: The twin-arginine translocation pathway is a major route of protein export in Streptomyces coelicolor. *Proc Natl Acad Sci U S A* 2006, 103(47):17927–17932.

47. Sargent F, Berks BC, Palmer T: Pathfinders and trailblazers: a prokaryotic targeting system for transport of folded proteins. *FEMS Microbiol Lett* 2006, 254(2):198–207.

48. Berks BC: A common export pathway for proteins binding complex redox cofactors? *Mol Microbiol* 1996, 22(3):393–404.

49. Berks BC, Sargent F, Palmer T: The Tat protein export pathway. *Mol Microbiol* 2000, 35(2):260–274.

50. Petersen TN, Brunak S, von Heijne G, Nielsen H: SignalP 4.0: discriminating signal peptides from transmembrane regions. *Nat Methods* 2011, 8(10):785–786.

51. Shen HB, Chou KC: Signal-3 L: A 3-layer approach for predicting signal peptides. *Biochem Bioph Res Co* 2007, 363(2):297–303.

52. Bannai H, Tamada Y, Maruyama O, Nakai K, Miyano S: **Extensive feature detection of N-terminal protein sorting signals.** *Bioinformatics* 2002, **18**(2):298–305.
53. Gomi MSM, Mitaku S: **High performance system for signal peptide prediction: SOSUIsignal.** *Chem-Bio Informatics Journal* 2004, **4**(4):142–147.
54. Chung CT, Niemela SL, Miller RH: **One-step preparation of competent Escherichia coli: transformation and storage of bacterial cells in the same solution.** *Proc Natl Acad Sci U S A* 1989, **86**(7):2172–2175.

15

Buried chloride stereochemistry in the Protein Data Bank

Oliviero Carugo[1,2]

Abstract

Background: Despite the chloride anion is involved in fundamental biological processes, its interactions with proteins are little known. In particular, we lack a systematic survey of its coordination spheres.

Results: The analysis of a non-redundant set (pairwise sequence identity < 30%) of 1739 high resolution (<2 Å) crystal structures that contain at least one chloride anion shows that the first coordination spheres of the chlorides are essentially constituted by hydrogen bond donors. Amongst the side-chains positively charged, arginine interacts with chlorides much more frequently than lysine. Although the most common coordination number is 4, the coordination stereochemistry is closer to the expected geometry when the coordination number is 5, suggesting that this is the coordination number towards which the chlorides tend when they interact with proteins.

Conclusions: The results of these analyses are useful in interpreting, describing, and validating new protein crystal structures that contain chloride anions.

Background

It is sufficient to open any biochemistry or bioinorganic chemistry book to verify that, although chloride is essential for any form of life, its biological chemistry receives less attention than other small ions like for example sodium(I), calcium(II) or magnesium(II).

Together with sodium(I) and potassium(I) ions, chloride is responsible for the osmotic and charge stability of cells [1]. Chloride is essential to maintain cellular and whole body pH, which is mainly buffered by the CO_2/HCO_3^- equilibrium, since the exchange of bicarbonate across the plasma membrane is coupled with chloride exchange [2]. A well-known channel with specificity for anions, in particular chloride and bicarbonate, is the cystic fibrosis transmembrane conductance regulator [3].

Chloride is also present in some proteins. For example in many amylases, a chloride anion coordinated by arginine and lysine side-chains is bound close the active site, where it may assist the reaction [4]. In photosystem II two chloride anions have been identified close to the Mn_4CaO_6 cluster and may contribute to it stability/reactivity [5,6]. The two chloride binding sites are at the start of hydrogen bond networks that connect the cluster with the bulk solution. These networks may function as proton exit channels or water inlet channels [5,6].

In this manuscript, the chloride anions observed in the protein crystal structures deposited in the Protein Data Bank [7,8] are analyzed systematically, with the aim of finding their preferred coordination numbers, the amino acids/atoms that tend to interact with them, and their coordination stereochemistry. The results of these analyses should prove useful as benchmarks against which it is possible to critically validate new experimental results.

Results and discussion

Solvent exposure of the chlorides

The solvent accessibility of a chloride anion is defined here as the fraction of surface that is accessible to a solvent probe (radius = 1.4 Å) and that is not already covered by other atoms, including water molecules that were experimentally positioned. This is an uncommon definition of solvent accessibility. However, it is adopted here to identify the chloride anions with a well defined first coordination sphere.

On average, the chloride anions have 29 Å² exposed to the solvent, which means about 22% of their surface (the surface of a sphere of radius equal to 1.81 + 1.40 Å is 129 Å²). Several chloride anions are well packed (Figure 1).

Correspondence: oliviero.carugo@univie.ac.at
[1]Department of Structural and Computational Biology, Max F. Perutz Laboratories, Vienna University, Vienna, Austria
[2]Department of Chemistry, University of Pavia, Pavia, Italy

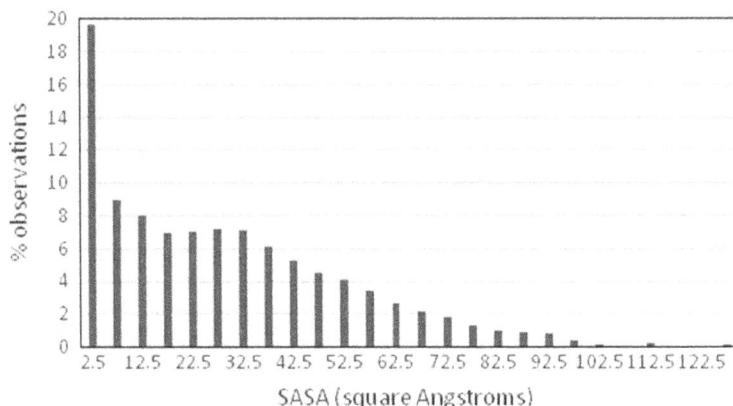

Figure 1 Distribution of the SASA values of the chloride anions observed in a non-redundant set of chloride-containing protein crystal structures.

Nearly 20% of them have a residual SASA lower than 5 Å². The attention was focused on them. The other chloride anions, located at the protein surface and considerably exposed to the solvent were disregarded since their first coordination sphere is incomplete.

Quite frequently two or more very similar chloride anions are present in an individual PDB file, for example, when one of them interacts with a polypeptide chain and another interacts with another polypeptide chain of identical sequence. These chloride anions are in the same asymmetric unit and are therefore not equivalent from a symmetry perspective. However, they are chemically analogous and therefore only one of them must be considered to avoid data redundancy. For this reason, only one chloride anion (the first in the list) was analyzed in each PDB file, though this implies that some chloride anions are arbitrarily disregarded.

Coordination numbers

The distance between a chloride anion and the atoms that are in its first coordination sphere is within 3.4 Å. Chloride coordination by urea derivatives, molecules that mimic polypeptides, has been studied in detail and it has been observed that the distances between the chloride anions and the urea nitrogen atoms range from 3.26 to 3.35 Å [9]. In chloride supramolecular salts of diammonium-bis-pyridinium cations, the distances between the chloride anions and the cationic nitrogen atoms range from 3.03 to 3.04 Å [10].

In the data set of the chloride-containing PDB files that are examined here, it is possible to examine these distances from a statistical perspective. Figure 2 shows the distribution of the distances between the chloride anion and the atoms (any type of atom) around it. Obviously, there are nearly no atoms within 2.5 Å, and the

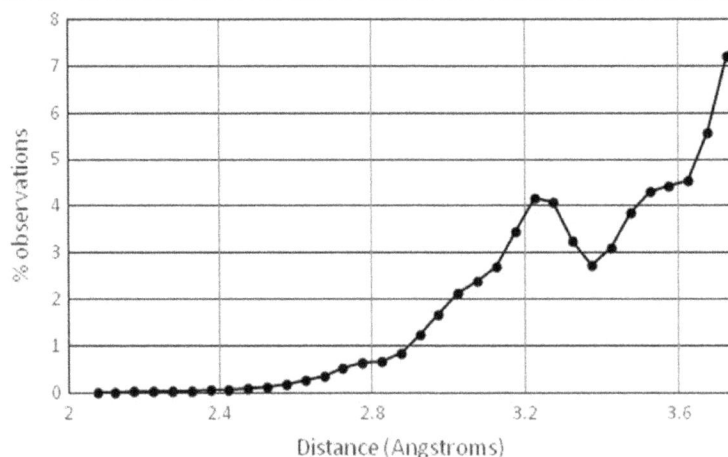

Figure 2 Distribution of the distances between the buried chloride anion and the atoms around it observed in a non-redundant set of chloride-containing protein crystal structures.

number of atoms increases at larger distances. However, this increase is discontinuous. There is a maximum around 3.2 Å, followed by a minimum around 3.4 Å.

This type of trend indicates a considerable clustering tendency of the atoms around the chloride anions, according to the Lacey-Cole statistics, which is routinely used (i) to verify if the objects are randomly distributed or have a natural tendency to cluster into well separate groups and (ii) to estimate the natural separation between adjacent groups [11]. In other words, the results shown in Figure 2 suggest that (i) the atoms are not randomly distributed around chloride anions (at any distance from the anion) and that (ii) the atoms that surround a chloride anion tend to be within 3.4 Å from the chloride anion.

The atoms are not randomly distributed but tend to be within a sphere of radius 3.4 Å, centred on the chloride anion.

On the basis of these observations, an atom was considered to be within the first coordination sphere of a chloride anion if it falls in a sphere of radius equal to 3.4 Å. Figure 3 shows the distribution of the coordination numbers. The most common coordination numbers are 3, 4, and 5. Only about 10% of the chlorides have a coordination number lower than 3 or higher than 5.

Only one case of coordination number equal to 0 was observed (Figure 4): it is the atom Cl 601 B, located in between the chains B and D of a tetrameric glycoside hydrolase from *Parabacteroides distasonis ATCC 8503* (PDB file 1fj6), which is naked, in the sense that its closest neighbour is the main chain nitrogen atom of Lys 50 B, at 3.48 Å [12].

The atomic displacement parameters (ADPs), normalized to zero mean and unit variance (as it is usually done when it is necessary to compare B-factors of different protein crystal structures [13,14]), of the chloride anions

and their donor atoms are independent of the coordination number and do not present any particular trend. The average ADP of the chloride anions is slightly larger than zero (0.10), while the average ADP of the ligands is slightly smaller than zero for protein donor atoms (−0.40) and for donor atoms of small molecules (−0.16); it is on the contrary slightly larger than zero for water molecules coordinated to the chlorides (0.10).

Residues in the first coordination sphere

Table 1 summarizes the frequencies of each type of residue (including water) in the first coordination sphere of the chloride anions.

Water is very frequent, between one fourth and one third, indicating that it is rare that chlorine interacts only with atoms of protein.

Arginine is also very frequent, in the range 11-17%, considerably more than expected – only about 6% of the amino acids are arginines in proteins [15]. Also histidine is more frequent than expected: 3-9% of the residues in the chloride first coordination sphere are histidines, while only about 2-3% of the residues are histidines in proteins [15]. Curiously, lysine is less frequent, in the range 3-6%, roughly as it is expected (about 6% of the residues are lysines in proteins [15]).

Polar residues (Asn, Gln, Ser, and Thr) are also quite frequent, in the range 17-26%. Glycine is also quite frequent, in the range 3-6%, as expected [15].

Despite the electrostatic repulsion between their anionic side-chains and the chloride anion, aspartate and glutamate are not totally absent (range 6-9%). Usually they contact the chloride anion with their main-chain nitrogen atom and this interaction is reinforced if the chloride anion interacts also with cations, which can be coordinated by the carboxylate moiety of Glu and Asp. The chloride coordination via the main-chain amido

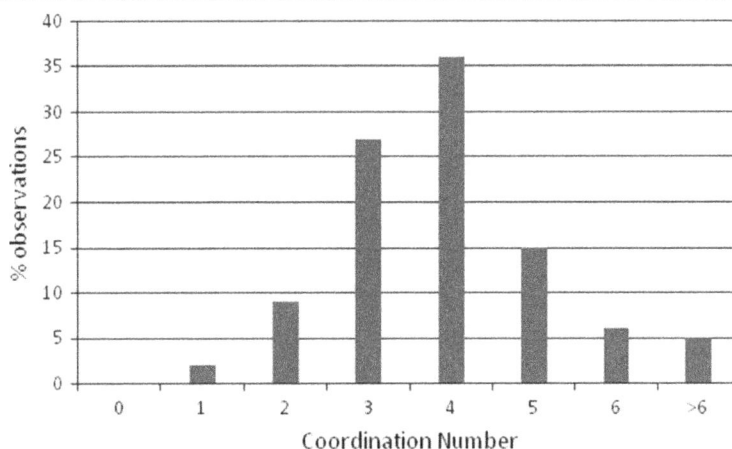

Figure 3 Distribution of the coordination numbers of the buried chloride anions observed in a non-redundant set of chloride-containing protein crystal structures.

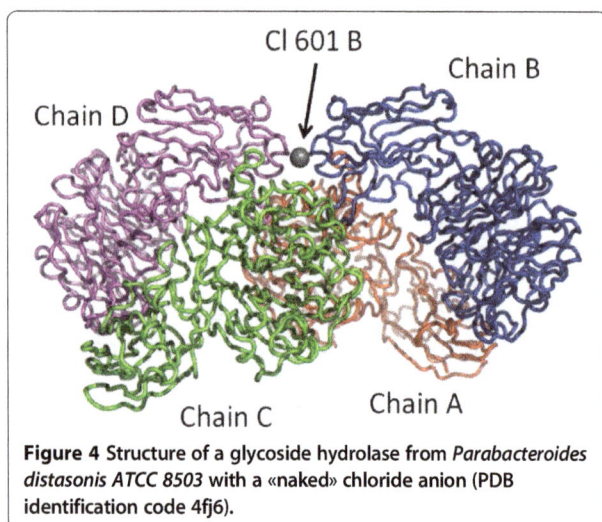

Figure 4 Structure of a glycoside hydrolase from *Parabacteroides distasonis ATCC 8503* with a «naked» chloride anion (PDB identification code 4fj6).

Table 2 Percentage with which the various types of chemical elements are found in the first coordination sphere of chloride anions, for various coordination numbers (CN)

Element	CN = 2	CN = 3	CN = 4	CN = 5	CN = 6
N	61.1	56.0	46.6	38.9	39.1
O	34.4	40.4	48.3	47.3	37.9
C	4.4	3.3	4.7	12.7	20.1
Others	0.0	0.2	0.4	1.1	2.9

At lower coordination numbers, nitrogen atoms are considerable more frequent than oxygen atoms, while the two types of elements have comparable frequencies at higher coordination numbers. Other types of chemical elements, typically carbon, are present in considerable amount only when the coordination number is high. For example, only 4% of the atoms in the first coordination sphere are carbons when the coordination number is 2, while up to 20% of the atoms in the first coordination sphere are carbons when the coordination number is 6.

In general, the presence of carbon atoms in the first coordination sphere of a chloride anion may seem surprising, since most of the carbon atoms that can be present in protein crystals are rather apolar and therefore little prone to interact with a charged ion. However, the fact that carbon atoms are rather abundant only at high coordination numbers suggests that in many cases their presence is purely accidental. They are in general covalently bound to another atom that can be attracted electrostatically by the chloride anion, for example, a carboxylic carbon atom of the side-chain of an asparagine, which is bound to an amide NH2 moiety, which in turn may form a hydrogen bond with the chloride anion.

A deeper analysis of the composition of the first coordination sphere of the chloride anions was conducted only on the protein atoms, excluding cofactors/prosthetic-groups, since only for the protein atoms it is easy to classify automatically the type of electronic structure and covalent context of each non-hydrogen atom.

Table 3 shows a brief summary of the results. As expected, most of the atoms of the chloride first coordination sphere are potential hydrogen bond donors: the main-chain nitrogen atom, other N-H groups of the side chains of some residues (glutamine, asparagine, histidine, and tryptophan), ammonium and guanidinium groups of lysine and arginine, and O-H groups of the side-chains of some amino acids (serine, threonine, and tyrosine).

Unexpected atoms are also observed in the chloride first coordination spheres (carbon atoms and even negatively charged oxygen atoms of the side chains of aspartate and glutamate). However, their presence is particularly evident only at high coordination numbers and it is reasonable to suppose that their presence in the chloride first coordination sphere is simply accidental.

group explains also the presence of apolar residues in the chloride first coordination sphere (range 11-24%).

Atoms in the first coordination sphere

The large majority of the atoms in the chloride first coordination sphere are oxygen and nitrogen atoms (Table 2).

Table 1 Percentage with which the various types of amino acids are found in the first coordination sphere of chloride anions, for various coordination numbers (CN)

Residue	CN = 2	CN = 3	CN = 4	CN = 5	CN = 6
Ala	3.5	1.5	2.7	2.7	4.2
Arg	17.4	11.9	13.2	10.9	12.5
Asn	2.3	6.5	6.6	5.6	6.2
Asp	4.7	3.4	1.8	5.0	0.0
Cys	0.0	1.0	0.3	1.5	0.0
Gln	4.7	4.8	2.7	3.0	2.8
Glu	4.7	2.7	3.0	2.4	6.2
Gly	4.7	2.7	4.8	6.2	6.2
His	9.3	5.1	3.1	5.0	4.2
Ile	0.0	1.7	1.8	1.5	1.4
Leu	1.2	2.7	1.9	2.4	4.9
Lys	3.5	5.8	3.1	3.0	2.8
Met	0.0	0.5	0.6	0.3	2.1
Phe	0.0	1.7	0.9	0.9	2.8
Pro	0.0	0.2	0.3	1.5	0.7
Ser	9.3	8.2	7.2	9.8	3.5
Thr	4.7	6.3	5.7	3.3	4.9
Trp	2.3	1.2	1.3	0.9	2.8
Tyr	1.2	4.4	3.3	1.5	1.4
Val	3.5	3.6	1.5	1.5	3.5
HOH	23.0	24.1	34.2	31.1	26.9

Table 3 Percentage with which the various types of protein atoms are found in the first coordination sphere of chloride anions, for various coordination numbers (CN)

Atom	CN = 2	CN = 3	CN = 4	CN = 5	CN = 6
Main-chain N-H	47.0	39.0	35.2	33.6	27.6
Other N-H	13.6	16.7	14.6	8.6	12.5
N charged	19.7	18.9	22.3	15.5	17.2
Main-chain O	1.5	3.5	3.6	8.6	9.5
O-H	10.6	15.0	15.5	12.1	5.8
C = O	3.0	2.6	0.9	2.6	3.8
Charged O	0.0	0.0	0.7	1.3	1.9
Main-chain C	0.0	0.3	0.7	2.6	1.0
CA	0.0	0.0	1.8	4.7	3.8
CB	0.0	1.3	1.1	4.3	6.7
Other C	4.5	2.4	3.5	5.5	9.6
S	0.0	0.3	0.2	0.0	1.0

Analysis of the coordination polyhedra

The interactions between the chloride anion and the atoms that surround it are expected to be essentially electrostatic and thus non-directional. Consequently, it is possible to predict the most stable conformation for each coordination number. When two, identical atoms are in contact with the chloride anion, they should occupy two positions diametrically opposite on a sphere centred on the chloride anion. This results in a linear stereochemistry with an angle X-Cl-X equal to 180 degrees.

Analogously, if the coordination number is three, the stereochemistry is expected to be trigonal planar with the chloride lying at the centre of an equilateral triangle, at the vertices of which there are three, identical atoms. There are then three, identical angles X-Cl-X equal to 120 degrees.

If the coordination number is equal to four, two, nearly degenerate stereochemistries are possible. In the tetrahedral stereochemistry, the chloride anion is at the centre of a tetrahedron, the vertices of which are occupied by four, identical atoms; there are then six X-Cl-X angles, all equal to 109.5 degrees. In the square planar stereochemistry, four, identical atoms are at the vertices of a square, at the centre of which lies the chloride anion; in this case there are then two X-Cl-X angles of 180 degrees and four X-Cl-X angles of 90 degrees.

Also for the coordination number five, two stereochemistries have comparable energies. In the trigonal bipyramidal stereochemistry, two trigonal pyramids share a triangular face and the five vertices are occupied by five, identical atoms, while the chloride anion lies in the centre of the polyhedron. There are then ten X-Cl-X angles, one of which is equal to 180 degrees, three of which are equal to 120 degrees, and six of which are equal to 90 degrees. In the alternative, square pyramidal

stereochemistry, the chloride anion lies at the centre of the square base of the pyramid and five, identical atoms occupy the vertices of the polyhedron. Two of the ten X-Cl-X angles are then equal to 180 degrees while all the other eight are equal to 90 degrees.

If the coordination number is equal to six, the most stable stereochemistry is octahedral, with six, identical atoms at the vertices of the octahedron and the chloride anion at the centre of the octahedron. Three of the 15 X-Cl-X angles are then equal to 180 degrees, while all the other 12 are equal to 90 degrees. Another possible stereochemistry is the trigonal prismatic (or anti-prismatic), where the chloride anion is at the centre of a trigonal prism, the vertices of which are occupied by six, identical atoms. In this case, the amplitudes of the X-Cl-X angles have not fixed values; the only constraint is that six of them must be equal to each other and that the other nine must be equal to each other, with the two amplitudes being independent of each other.

Obviously, deviations from these ideal stereochemistries are possible. They might be due to the fact the atoms that surround the chloride anion are not identical and can have different dimensions. It is also necessary to consider that the interactions between the chloride anion with some of its neighbours can be different from the interactions with other neighbours. In addition, one cannot ignore the sterical role of the chemical groups that may influence the atoms of the first coordination sphere even if they do not interact directly with the chloride anion.

The degree of the distortion from the ideal geometries was measured by means of the distortions of the angles centred on the chloride anion. For example, for a chloride anion surrounded by two atoms X and Y, with the angle X-Cl-Y being equal to A, the quantity $D = |A-180|$ was computed and the stereochemistry was considered to be linear if $D < 20$ degrees. For a chloride anion surrounded by the three atoms X, Y, and Z, the amplitudes of the three angles X-Cl-Y (indicated by A1), X-Cl-Z (A2), and Y-Cl-Z (A3) was computed and the stereochemistry was considered to be trigonal planar if the quantity $D = (|A1-120| + |A2-120| + |A3-120|)/3$ was less than 20 degrees.

An analogous quantity D was computed for larger coordination numbers, by considering that different angle permutations are possible and must then be inspected. For example, in a square planar stereochemistry, two angles centred on the chloride anion must be equal to 180 degrees and four angles must be equal to 90 degrees. There are numerous ways to write this list of these six angle amplitudes: one might be 180, 180, 90, 90, 90, 90; a second might be 180, 90, 180, 90, 90, 90; a third might be 180, 90, 90, 180, 90, 90; and so on. In total, there are 15 permutations, and all of them must be inspected to measure the minimal distortion from the ideal, square planar stereochemistry of a real molecular moiety constituted by a chloride anion surround by four atoms. Analogously, there are 45 permutations of the angles centred on the chloride anion in a square planar geometry, 840 in a trigonal bipyramidal geometry, 455 in a octahedral geometry, and 5005 in a trigonal prismatic geometry.

The analysis of the deviations of the real stereochemistry around the chloride anions from the ideal stereochemistries is rather surprising (Table 4). The cases with coordination numbers two are severely distorted, on average, from the ideal stereochemistry: the average distortion is larger than 70 degrees. This clearly suggests that the second coordination sphere play a relevant role and, in other words, that the coordination number two is often just a misinterpretation of the structural data: other atoms should have been included into the first coordination sphere, despite they are too distant from the chloride anion or despite their positions were not determined with sufficient accuracy.

The distortions from the ideal geometry are less pronounced for coordination numbers higher than two. They are still relatively large for coordination number three (27 degrees), four (26–27 degrees) and six (31 degrees) and they are smaller for coordination number five (18–25 degrees). Notably, nearly all (91%) the chlorides surrounded by five atoms can be described, within 20 degrees, as regular trigonal bipyramids or square pyramids. A so high percentage is rather unexpected, since the actual coordination sphere of a chloride anion inaccessible to the solvent is expected to be influenced severely by the overall molecular packing requirements. In other words, the type and the shape of the molecules and of the molecular fragments that can be present in a protein crystal are expected to restrain the natural tendency to occupy optimally the space, because of their covalent constraints. It is then possible to hypothesize that five is the more appropriate coordination number for a chloride anion in a protein crystal structure. In other words, a chloride anion seems to have a natural tendency to surround itself with five atoms, when it co-crystallizes with proteins and other molecules that might be present in the crystallization medium.

Packing bridges involving chlorides

About 10–20% of the chlorides examined in the present paper are bound to residues of different polypeptide chains, behaving as packing bridges [16]. This percentage varies with the coordination number: it is 9% for coordination number 2, 21% for 3, 19% for 4, 12% for 5, and 7% for 6. About one-half of these packing bridges connect polypeptide chains in the same asymmetric unit and the other half connect chains that are related by a symmetry operation in the crystallized material.

This shows that the chloride anions share a feature with other ions and small molecules: they have a marked tendency to connect adjacent molecules in the solid state, stabilizing nascent crystals. Given that several charged and polar residues form the surface of globular proteins, it is not surprising that chloride anions can approach the protein surface in solution and then connect a protein molecule with an adjacent protein in the nucleation phase of the crystallization.

Conclusions

A non-redundant set (pairwise sequence identity < 30%) of 1739 high resolution (<2 Å) crystal structures that contain chloride anions was analyzed. The first coordination sphere of the chlorides is essentially constituted by hydrogen bond donors: main-chain N-H groups and side-chains N-H and O-H groups. Amongst the side-chains positively charged, arginine interacts with chlorides much more frequently than lysine. The most common coordination number is 4, though the coordination stereochemistry is closer to the expected geometry when the coordination number is 5, suggesting that this is the coordination number towards which the chlorides are predisposed when they interact with proteins.

Table 4 Distortions from the ideal geometries of the chloride first coordination spheres

Coordination number	Stereochemistry	Average distortions with standard error in parentheses (degrees)	% of chlorides that assume the ideal stereochemistry (within 20 degrees)
2	Linear	75.0 (5.4)	0
3	Trigonal	27.5 (1.1)	29
4	Tetrahedral	25.8 (0.7)	25
	Square planar	27.5 (0.6)	10
5	Trigonal bipyramidal	18.4 (0.6)	70
	Square pyramidal	24.7 (0.7)	21
6	Octahedral	34.8 (0.9)	0
	Trigonal prismatic	30.6 (1.1)	7

These results should prove useful to examine new protein crystal structures that contain chloride anions, especially to validate, on the basis of the chemical and geometrical features, the chloride binding sites.

Methods

All the files of the Protein Data Bank that contain at least one chloride ion were identified with help of the server PLI (http://bioinformatics.istge.it/pli/; [17]) and only the crystal structures refined at a resolution better than 2.0 Å were retained. Sequence redundancy was reduced to 30% of pairwise sequence identity with the service provided by the PDB advanced search service and 1739 PDB files were retained (see Additional file 1).

It is important to observe that this dataset probably underestimate the number of chlorides anions in protein crystal structures. It was in fact observed that chlorides and other ions, for example phosphate and metal cations, tend to be overlooked in macromolecular crystallography and misinterpreted as water molecules, unless soft x-rays diffraction data are used [18]. However, routinely, these data are not collected yet. However, if it is true that there are some false negatives (chlorides misinterpreted as water molecules) it is also true that at high resolution there should be few false positives (electron density peaks that are interpreted as chlorides despite they are not chlorides). Moreover, the data refined with soft X-rays are still not sufficiently numerous to allow a statistical survey of the chloride coordination chemistry in biological macromolecules.

The solvent accessible surface areas were computed with Naccess (http://www.bioinf.manchester.ac.uk/naccess/), with a probe radius of 1.4 Å, in the modality that includes the hetero-atoms in the computations. The Shannon's ionic radius of the Chloride (1.81 Å) was used [19]. The solvent accessibility of a chloride anion is the fraction of surface that is accessible to a solvent probe and that is not already covered by other atoms, including water molecules that were experimentally positioned.

Packing contacts and packing bridges were identified as described previously [16].

Competing interests
The author declares that he has no competing interests.

Author's contributions
OC designed and executed the project and wrote the manuscript.

Acknowledgements
I would like to acknowledge Kristina Djinovic-Carugo for helpful discussions.

References
1. da Silva JJR F, Williams RJP: *The biological chemistry of the elements.* Oxford: Oxford University Press; 2001.
2. Bonar PT, Casey JR: **Plasma membrane Cl⁻/HCO₃ exchangers: structure, mechanism and physiology.** *Channels (Austin)* 2008, **2**:337–345.
3. Cant N, Pollock N, Ford RC: **CFTR structure and cystic fibrosis.** *Int J Biochem Cell Biol* 2014, **52**:15–25.
4. Buisson G, Duee E, Haser R, Payan F: **Three dimensional structure of porcine pancreatic cl-amylase at 2.9 A resolution.** *EMBO J* 1987, **6**:3909–3916.
5. Bricker TM, Roose JL, Fagerlund RD, Frankel LK, Eaton-Rye JJ: **The extrinsic proteins of Photosystem II.** *Biochim Biophys Acta* 1817, **2012**:121–142.
6. Umena Y, Kawakami K, Shen J-R, Kamiya N: **Crystal structure of oxygen-evolving Photosystem II at a resolution of 1.9 Å.** *Nature* 2011, **473**:55–60.
7. Bernstein FC, Koetzle TF, Williams GJ, Meyer EF Jr, Brice MD, Rodgers JR, Kennard O, Shimanouchi T, Tasumi M: **The Protein Data Bank: a computer-based archival file for macromolecular structures.** *J Mol Biol* 1977, **112**(3):535–542.
8. Berman HM, Westbrook J, Feng Z, Gilliland G, Bhat TN, Weissig H, Shindyalov IN, Bourne PE: **The Protein Data Bank.** *Nucleic Acids Res* 2000, **28**(1):235–242.
9. Wu B, Jia C, Wang X, Li S, Huang X, Yang X-J: **Chloride coordination by oligoureas: From mononuclear crescents to dinuclear foldamers.** *Org Lett* 2012, **14**:684–687.
10. Keegan J, Kruger PE, Nieuwenhuyzen M, O'Brien J, Martin N: **Anion directed assembly of a dinuclear double helicate.** *Chem Cummun* 2001, 2192–2193.
11. Carugo O: **Clusteriung crietria and algorithms.** *Methods Mol Biol* 2010, **609**:175–196.
12. Genomics JCS: *Author of the PDB file*; 2012.
13. Carugo O, Argos P: **Reliability of atomic displacement parameters in protein crystal structures.** *Acta Crystallogr Sect D: Biol Crystallogr* 1999, **55**(Pt 2):473–478.
14. Carugo O, Argos P: **Accessibility to internal cavities and ligand binding sites monitored by protein crystallographic thermal factors.** *Proteins* 1998, **31**:201–213.
15. Carugo O: **Amino acid composition and protein dimension.** *Protein Sci* 2008, **17**:2187–2191.
16. Caugo O, Djiovic-Carugo K: **Packing bridges in protein crystal structures.** *J Appl Cryst* 2014, **47**:458–461.
17. Gallina AM, Bisignano P, Bergamino M, Bordo D: **PLI: a web-based tool for the comparison of protein-ligand interactions observed on PDB structures.** *Bioinformatics* 2013, **29**:395–397.
18. Mueller-Dieckmann C, Panjikar S, Schmidt A, Mueller S, Kuper J, Geerlof A, Wilmanns M, Singh RK, Tucker PA, Weiss MS: **On the routine use of soft X-rays in macromolecular crystallography. Part IV. Efficient determination of anomalous substructures in biomacromolecules using longer X-ray wavelengths.** *Acta Crystallogr* 2007, **D63**:366–380.
19. Shannon R: **Revised effective ionic radii and systematic studies of interatomic distances in halides and chalcogenide.** *Acta Crystallogr* 1976, **32**:751–767.

16

Molecular analysis of hyperthermophilic endoglucanase Cel12B from *Thermotoga maritima* and the properties of its functional residues

Hao Shi[1,2,3], Yu Zhang[1,2], Liangliang Wang[1,2], Xun Li[1,2], Wenqian Li[1,2,3], Fei Wang[1,2*] and Xiangqian Li[3*]

Abstract

Background: Although many hyperthermophilic endoglucanases have been reported from archaea and bacteria, a complete survey and classification of all sequences in these species from disparate evolutionary groups, and the relationship between their molecular structures and functions are lacking. The completion of several high-quality gene or genome sequencing projects provided us with the unique opportunity to make a complete assessment and thorough comparative analysis of the hyperthermophilic endoglucanases encoded in archaea and bacteria.

Results: Structure alignment of the 19 hyperthermophilic endoglucanases from archaea and bacteria which grow above 80°C revealed that Gly30, Pro63, Pro83, Trp115, Glu131, Met133, Trp135, Trp175, Gly227 and Glu229 are conserved amino acid residues. In addition, the average percentage composition of residues cysteine and histidine of 19 endoglucanases is only 0.28 and 0.74 while it is high in thermophilic or mesophilic one. It can be inferred from the nodes that there is a close relationship among the 19 protein from hyperthermophilic bacteria and archaea based on phylogenetic analysis. Among these conserved amino acid residues, as far as Cel12B concerned, two Glu residues might be the catalytic nucleophile and proton donor, Gly30, Pro63, Pro83 and Gly227 residues might be necessary to the thermostability of protein, and Trp115, Met133, Trp135, Trp175 residues is related to the binding of substrate. Site-directed mutagenesis results reveal that Pro63 and Pro83 contribute to the thermostability of Cel12B and Met133 is confirmed to have role in enhancing the binding of substrate.

Conclusions: The conserved acids have been shown great importance to maintain the structure, thermostability, as well as the similarity of the enzymatic properties of those proteins. We have made clear the function of these conserved amino acid residues in Cel12B protein, which is helpful in analyzing other undetailed molecular structure and transforming them with site directed mutagenesis, as well as providing the theoretical basis for degrading cellulose from woody and herbaceous plants.

Keywords: Cellulose, Conserved amino acid residues, Endoglucanase, Phylogenetic analysis, Thermostability

Background

Cellulose is the most abundant organic compound and renewable carbon resource on earth [1]. Biodegradation of cellulose, an abundant plant polysaccharide, is a complex process that requires the coordinate action of three enzymes, among which endoglucanases (EC 3.2.1.4), are able to break the internal bonds of cellulose, and disrupt its crystalline structure, exposing the individual cellulose polysaccharide chains, playing in most important role [2-4]. The degradation is mainly carried out by bacteria, fungi, and protozoa, commensals in the guts of herbivorous animals, as well as the termite *Reticulitermes speratus* [5], from which, there are variety of endoglucanases. The complex chemical nature and heterogeneity of cellulose account for the multiplicity of endoglucanases produced by microorganisms. The activity of different endoglucanases with subtle differences in substrate specificity and mode of action contributes to improvement of the degradation of plant cellulose in natural habitats. There are fourteen families of glycoside hydrolases (GHF) that are used

* Correspondence: lixq2002@126.com; hgwf@njfu.edu.cn
[1]College of Chemical Engineering, Nanjing Forestry University, Nanjing 210037, China
[3]Department of Life Science and Chemistry, Huaiyin Institute of Technology, Huaian 223003, China
Full list of author information is available at the end of the article

for cellulose hydrolysis [6]. More and more extremophiles have been studied in recent years, especially the hyperthermophilic enzymes. Based on amino acid sequence homologies and hydrophobic cluster analysis, hyperthermophilic endoglucanases obtained from extremophiles, which are widely distributed in terrestrial and marine hydrothermal areas, as well as in deep subsurface oil reservoirs, have been classified into GHF12 [7-14]. As described above, there are hyperthermophilic endoglucanases from archaea, most of which were chosen for sequencing on the basis of their physiology [15]. In addition, many hyperthermophilic endoglucanases gene which have been cloned were found in some heat-tolerant bacteria [16]. Those hyperthermophilic endoglucanases have a common feature that the amino acid sequences are mostly relatively short (less than 400 amino acid residues).

Although many hyperthermophilic endoglucanases of GHF12 amino acids have been reported from archaea and bacteria, a complete survey and classification of all sequences in these species from disparate evolutionary groups, and the relationship between their molecular structures and functions are lacking. The completion of several high-quality gene or genome sequencing projects provided us with the unique opportunity to make an unprecedented assessment and thorough comparative analysis of the hyperthermophilic endoglucanases encoded in archaea and bacteria. The analysis of the full set of hyperthermophilic endoglucanases genes in genomes from diverse species allows a definitive classification of hyperthermophilic endoglucanases and an assessment of their origins, evolutionary relations, patterns of differentiation, and proliferation in the various phylogenetic groups. We are interested in finding answers to the following questions: 1) What are the evolutionary relations among these hyperthermophilic endoglucanases?; 2) What is the common feature between these conserved amino acid residues and 3D topological structure?; 3) What the mechanism of the heat tolerance among these hyperthermophilic endoglucanases?

The broad analysis in this study provided a comprehensive classification scheme and proposed a molecular structure applicable to all hyperthermophilic endoglucanases. A clear picture of the patterns of endoglucanases classes in different species groups was provided. We identified and classified in this study a higher number of hyperthermophilic endoglucanase amino acids from the GHF12 than previously reported, allowing us to identify their relationships based on the phylogenetic clustering. We found that, similar to archaea, amino acids from hyperthermophilic bacteria are also quite different from the other sequences in GHF12. We characterized several conserved amino acid sites from these endoglucanases and predicted their functionality based on the amino acids similarity among the proteins available in databases. The

resulting rich data set of hyperthermophilic endoglucanases from GHF12, comprising 19 sequences, is available downloaded from NCBI (Table 1).

Results
Protein sequences characteristics
GenBank has grown fast in recent years and offer us with much better taxonomic sampling for such BLAST-based analysis [17]. We performed similar BLAST-based analysis for the 19 thermophilic endoglucanase protein sequences (which included the *T. maritima* endoglucanase sequences), using the nonredundant (*nr*) database as a reference and recording highest ranking matches. We also searched endoglucanase sequences in several plants, bacteria, fungi and algae sequences including the sequences of the *R. speratus*, using the protein BLAST search engine with a variety of endoglucanase amino acid sequences as queries for most of the thermophilic endoglucanase, else using endoglucanase as a keyword for searching other amino acid sequences of endoglucanase (Table 1). In most cases, whenever significant similarity to an endoglucanase sequence was identified, the amino acid sequence was excised and homology based protein predictions were performed using the most similar query as a guide. All of these 40 protein sequences range from 252 to 438 amino acid residues in length. Of these sequences, those from archaea and bacteria showed similar lengths, especially for those 19 thermophilic endoglucanase protein sequences where the average percentage composition of the residues cysteine and histidine is only 0.28 and 0.74, which are less frequent in thermophilic proteins according to the statistics of amino acid composition based on MEGA 5 (Table 2).

Phylogenetic analysis
Phylogenetic analysis based on the Maximum-parsimony (MP) and Neighbour-joining (NJ) procedure implemented in PAUP 4.0 [18] and other approaches (see Materials and Methods), indicated that all endoglucanase proteins can be reliably grouped into 3 distinct classes except for the outgroup *R. speratus*, which belongs to the insect family (Figure 1). Furthermore, from the multiple sequence alignments, the hyperthermophilic endoglucanase proteins belong to the class I, and others belong to class II and III. No obvious differentiations are implied in these 19 protein sequences. It was not surprising that there was a close relationship among 19 protein sequences from bacteria and archaea supported with good bootstrap values based on Maximum-likelihood (ML) tree by using MEGA 5 (Figure 2). It was inferred that the endoglucanases of *Dictyoglomus turgidum*, *Thermotoga naphthophila* and *Thermotoga maritima* which are currently studied in our research group are closely related compared to the others, although the identity of the amino acid sequences

Table 1 The phylogenetic distribution of endoglucanases from glycoside hydrolase family 12

	Organism	Length	*GenBank number
Euryarchaeota	*Acidilobus saccharovorans*	396	ADL19785
	Ignisphaera aggregans	360	ADM27702
	Metallosphaera cuprina	326	AEB95090
	Pyrococcus furiosus	319	AAD54602
	Sulfolobus acidocaldarius	311	AAY81158
	Sulfolobus islandicus	332	ADX81754
	Sulfolobus islandicus	334	ACP37717
	Sulfolobus islandicus	334	ACR41545
	Sulfolobus islandicus	334	ADX84872
	Sulfolobus solfataricus	334	AAK42142
	Thermococcus sp.	319	EEB73588
	Thermoproteus tenax	263	CCC81038
	Thermoproteus uzoniensis	252	AEA12777
	Vulcanisaeta distributa	330	ADN509821
Bacteria	*Acidobacterium sp.*	439	ZP_07030982
	Bacillus licheniformis	261	AAP44491
	Dictyoglomus turgidum	288	YP_002352530
	Paenibacillus mucilaginosus	266	AEI43442
	Spirochaeta thermophila	438	ADN02999
	Spirochaeta thermophila	433	AEJ62362
	Teredinibacter turnerae	278	ACR14297
	Thermobispora bispora	393	ADG87082
	Thermotoga naphthophila	274	YP_003346783
	Thermotoga maritima	275	Z69341
	Lysobacter enzymogenes	383	ABI54135
	Bacillus megaterium	345	ADE69644
	Streptococcus dysgalactiae	366	BAH80742
	Streptococcus dysgalactiae	366	YP_002995956
	Bacillus thuringiensis	349	ZP_04083086
Fungi	*Stachybotrys echinata*	237	AF435067
	Aspergillus fumigatus	378	EDP50688
	Aspergillus fumigatus	378	XP_751495
	Neosartorya fischeri	381	XP_001266710
	Aspergillus niger	396	XP_001400178
	Penicillium marneffei	379	XP_002147625
	Talaromyces stipitatus	503	XP_002481822
	Ajellomyces dermatitidis	357	XP_002621187
Planta	*Arabidopsis thaliana*	484	BAB11001
	Thalassiosira pseudonana	499	XP_002287341
Insect	*Reticulitermes speratus*	448	AB019095

*All the sequences are downloaded from GenBank (http://www.ncbi.nlm.nih. gov/protein/).

were shown less than 30% (Figure 1, Figure 2). Therefore, it was postulated that they may have a common origination based on protein evolution. Class II comprises of other 12 proteins from plant, fungi and bacteria, and class III comprises of 8 proteins from bacteria.

Analysis of conserved and catalytic amino acid residues

For the further analysis of the relationship among 19 hyperthermophilic endoglucanases from bacteria and archaea, those 19 amino acid sequences were aligned again with Clustal X2 (Figure 3). We found that the conserved amino acids of hyperthermophilic endoglucanase in Cel12B (for instance) include Gly30, Pro63, Pro83, Trp115, Glu131, Met133, Trp135, Trp175, Gly227 and Glu229 which are highlighted in red (Figure 3), which is very different from the previously reported data [19,20]. Among these conserved amino acids, two glutamic acid residues might be the catalytic nucleophile and proton donor like lysozyme with acid base catalysis [21], other eight conserved amino acids might be necessary to the thermostability of protein and binding of the substrate.

Hyperthermophilic protein homology modeling

All the hyperthermophilic protein sequences were rendered using SWISS-MODEL database for protein modeling, but only one good model, Cel12B protein model from *T. maritima*, can be used to describe conserved amino acids in which sites of secondary structure and enzymatic center of protein. As described with Cel12B protein model, Glu131, Glu229, Trp115, Trp135, Trp175 and Met133 residues, comprised the active center of the protein (Figure 4a). Cel12B protein is primarily composed of β-sheet (Figure 4a,b,c,d). Trp115, Glu131, Met133, Trp135 and Gly227 residues are in the β-sheet; Pro63 and Trp175 residues are in the turn; and Gly30, Pro83 and Glu229 residues are in the random coil (Figure 4b,d).

Analysis of site-directed mutagenesis

Base on the homology modeling, the functional amino acid residues Glu64, Pro63, Pro83 and Met133 of Cel12B were selected to be mutated. The results showed that the P63K, P83K, M133W, E64H, E64T and E64I mutant enzymes dramaticlly inhibited the enzyme activity of Cel12B toward CMC-Na, while E64S mutant protein apparently increased the enzyme activity (Table 3).

Discussion

Endoglucanases isolated from hyperthermophilic organisms are more active and stable at higher temperatures than their counterparts from mesophiles. In addition, they may be more appropriate for degradation of the cellulose. Since the enzyme activity of those hyperthermophilic endoglucanases is not high for degradation, the

Table 2 The frequencies of nineteen endoglucanases amino acids

	Ala	Cys	Asp	Glu	Phe	Gly	His	Ile	Lys	Leu	Met	Asn	Pro	Gln	Arg	Ser	Thr	Val	Trp	Tyr	Total
ADX81754	4.74	0.00	2.55	3.65	5.84	6.20	0.36	6.93	2.19	6.93	2.92	9.49	7.30	3.28	1.82	7.66	10.58	6.93	4.74	5.84	274.00
ACP37717	4.74	0.00	2.55	3.65	5.84	6.20	0.73	6.93	2.19	6.93	2.92	9.49	7.30	2.92	1.82	7.66	10.58	6.93	4.74	5.84	274.00
ADX84872	5.08	0.00	2.97	3.81	6.36	7.20	0.42	7.20	2.54	6.36	3.39	9.75	6.78	3.39	2.12	5.08	9.32	6.78	5.51	5.93	236.00
ACR41545	5.08	0.00	2.97	3.81	6.36	7.20	0.85	7.20	2.97	6.36	3.39	9.75	6.78	2.97	2.12	5.08	8.90	6.78	5.51	5.93	236.00
AAK42142	5.51	0.00	2.97	3.39	6.36	7.20	0.42	7.20	2.97	5.51	3.39	10.17	6.78	3.39	2.12	5.08	8.90	7.20	5.51	5.93	236.00
ADM27702	6.52	0.72	6.16	3.26	3.62	8.70	0.72	9.42	2.90	5.43	1.45	6.16	7.25	3.62	4.35	5.80	4.71	8.70	3.62	6.88	276.00
ADN02999	5.15	0.00	8.46	5.51	5.51	7.72	0.74	4.78	0.74	6.62	1.47	5.51	5.88	4.78	5.51	6.62	8.82	7.72	4.41	4.04	272.00
AEJ62362	5.15	0.00	8.09	6.25	5.51	7.72	0.74	4.04	0.74	6.62	1.10	5.51	5.88	4.41	5.51	6.62	8.82	8.82	4.41	4.04	272.00
AF181032	5.54	0.00	4.43	7.01	3.32	7.01	1.11	9.59	4.43	7.75	0.74	7.38	7.01	1.85	2.21	5.17	9.59	6.27	4.06	5.54	271.00
EEB73588	6.42	0.00	6.04	8.68	5.28	8.68	1.89	3.40	2.64	7.55	4.15	6.04	6.42	1.13	4.15	4.91	5.66	9.43	3.77	3.77	265.00
YP 003346783	3.97	0.40	6.35	7.94	7.14	6.75	1.19	4.37	6.75	5.56	1.98	5.95	4.76	1.98	1.59	4.76	7.94	10.32	4.76	5.56	252.00
Z69341	4.37	0.40	6.35	7.94	7.14	6.75	1.19	4.37	6.75	5.16	2.38	5.95	4.76	1.98	1.59	4.76	7.54	10.32	4.76	5.56	252.00
YP 002352530	5.43	0.00	4.26	8.53	4.26	5.04	1.16	9.69	9.30	5.81	1.94	7.75	4.65	1.55	2.33	5.43	5.04	6.59	4.26	6.98	258.00
AEA12777	10.71	0.40	4.37	6.35	4.76	7.54	0.00	5.16	3.57	5.95	3.97	3.17	7.14	2.78	3.57	8.33	4.76	6.75	4.37	6.35	252.00
AAY81158	2.90	0.36	5.07	2.54	5.80	7.97	0.72	8.33	2.90	7.97	2.90	9.78	3.99	3.26	1.45	7.61	7.97	8.70	2.17	7.61	276.00
AEB95090	2.89	0.00	4.33	4.33	5.42	7.94	0.36	6.14	3.25	8.66	4.33	7.22	5.42	2.89	2.17	10.11	5.78	7.58	2.53	8.66	277.00
ADN509821	3.90	1.42	2.48	3.90	3.19	7.09	0.71	9.22	3.19	9.22	3.19	11.35	6.74	1.42	1.77	7.80	4.61	6.03	4.61	8.16	282.00
ADL19785	3.96	0.00	3.24	3.60	3.24	12.59	0.36	6.12	0.72	11.51	4.32	7.19	5.76	2.16	2.88	7.91	5.76	8.63	3.96	6.12	278.00
CCC81038	9.13	1.66	4.98	4.56	3.32	9.13	0.41	2.49	1.24	9.96	1.24	3.32	7.47	2.49	6.22	7.88	4.15	9.96	3.32	7.05	241.00
Avg.	5.28	0.28	4.68	5.18	5.14	7.63	0.74	6.49	3.23	7.19	2.69	7.43	6.20	2.75	2.91	6.59	7.33	7.91	4.24	6.10	262.11

hyperthermophilic modification by using genetic engineering is essential. Few structures on databases have been reported so far for transforming those enzymes. In this paper, nineteen sequences of hyperthermophilic endoglucanases were aligned and used for phylogenetic tree construction and molecular modeling to illustrate the relationship between structure and themostability.

The features of the nature environment of ancestral organism can be inferred by reconstructing phylogenetic tree using amino acid sequences of these organisms [22]. From the alignment of the amino acids sequences, the hyperthermophilic proteins from bacteria and archaea are clustered together based on the phylogenetic tree (Figure 1). Archaea, known to be an ancient organisms on earth, grow in strictly anaerobic environment (terrestrial solfataric springs, hydrothermal areas, and deep subsurface oil reservoirs) at high temperature (generally above 80°C), and hyperthermophilic bacteria also live in the same conditions [13,23]. Therefore, it is inferred that endoglucanases from hyperthermophilic microorganisms from GHF12 could share the similar enzymatic properties and catalytic mechanism.

The stability of thermophilic proteins depend on several amino acid residues and structural factors [24]. Specific amino acid composition plays a critical role in the thermostability of hyperthermophilic endoglucanase, with the fewest cysteine and histidine residues that are thermal stability among the whole protein sequences by using statistical comparison of the amino acid composition [25,26], Consistent with this feature, the average content of cysteine and histidine in our reserach is only 0.24 and 0.72 respectively (Table 2).

Ten conserved amino acids were found by the alignment of nineteen hyperthermophilic protein sequences (Figure 3), that we hypothesize may play a significant role in proton donation, substrate binding as well as the high thermostability. Among these nineteen amino acid sequences, only thethree-dimensional structure of endoglucanase from *T. maritima* could be obtained (Figure 4), since there is no suitable template for other proteins homologous modeling. Thus, the relationship between the ten amino acid residues of these endoglucanases and their molecular structures will be illustrated in Cel12B protein from *T. maritima*. The substitution of non-Gly residue with Gly residue can be used as one of the general strategies to enhance the protein stability [27,28]. In our study, residues Gly30 and Gly227 located in random coil and β-sheet, respectively, might contribute to the thermostability of the protein (Figure 4b,d).

It is believed that loop and turn are the weak connections among the protein secondary structure elements, but recently it was demonstrated that they played a key role in thermostability of protein, especially for the proteins that proline is located in loop or turn region [29].

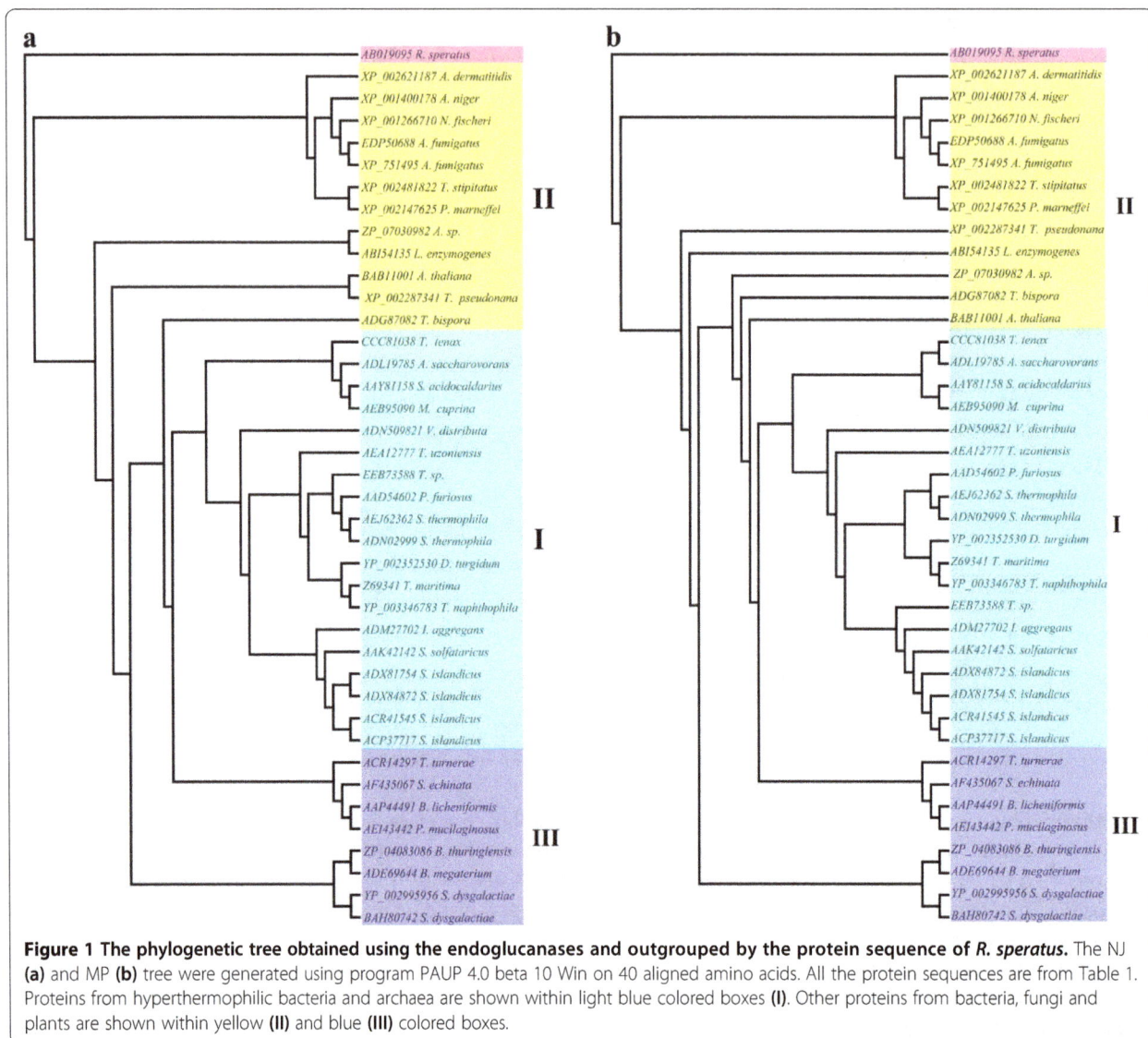

Figure 1 The phylogenetic tree obtained using the endoglucanases and outgrouped by the protein sequence of *R. speratus*. The NJ (a) and MP (b) tree were generated using program PAUP 4.0 beta 10 Win on 40 aligned amino acids. All the protein sequences are from Table 1. Proteins from hyperthermophilic bacteria and archaea are shown within light blue colored boxes (**I**). Other proteins from bacteria, fungi and plants are shown within yellow (**II**) and blue (**III**) colored boxes.

Proline in the polypeptide chain possesses less conformational freedom than other amino acids, as the pyrrolidine ring of proline imposes rigid constrains on the N-C rotation and restricts the available conformational space of the preceding residue. Therefore it can bend the polypeptide chain on itself so as to prepare the backbone much more easily to form the hydrogen bonds with the polar side chains of other turns; meanwhile, the hydrophobic part of proline can interact with the adjacent hydrophobic cavity [30,31]. Compared to mesophilic proteins, thermophilic proteins contain more proline residues especially occurring at the turn, with higher frequency, as well as the shorter loop region of the glucosidase. As the consequence of the flexibility reduction of the polypeptide chain, the protein thermostability can be increased by introducing prolines at specific sites based on the facts that illustrated above [29,31,32]. Hence, residues Pro63 and Pro83, located in the turn and random

coil respectively (Figure 4c,d), could provide closer packing of each region, as assumed for thermostability of protein. And then, it was finally confirmed by experimental results. Compared to other amino acids, lysine has longer side-chain groups and more vibrational degree of freedom, and it is more sensitive to the temperature. When the proline is substituted with lysine, the vibration of side-chain groups rises up at high temperature, and then the thermostability of the Cel12B decrease dramatically. Therefore, it is confirmed that residues Pro63 and Pro83 play an important role in stabilizing the Cel12B.

The crystal structure and protein molecular simulation supported that two glutamic acid residues are the catalytic nucleophile and proton donor that have been reported in many enzymes, lysozyme, xylanase as well as endoglucanase [33]. So, Glu131 (in β-sheet) and Glu229 (in random coil) residues are the proton donor and

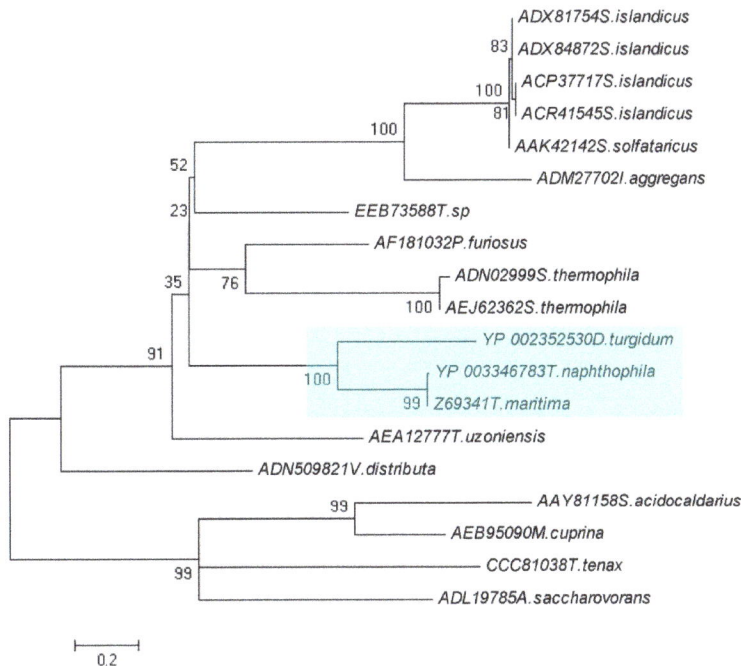

Figure 2 The ML tree obtained using the 19 endoglucanases amino acids using program MEGA 5. Numbers on nodes correspond to percentage bootstrap values for 1000 replicates.

catalytic nucleophile repectively (Figure 4b,d). Although the chemical nature of the tryptophan residue in the catalytic center does not significantly affect the conformational properties of lysozyme, it exhibited a pronounced effect on the binding of substrate and the enhancement of the total enzyme activity [34]. It was reported that structural changes at the active site (W95L) of alcohol dehydrogenase from *Sulfolobus solfataricus* are consistent with the reduced activity on substrates and decreased coenzyme binding [35]. Therefore, we propose that three tryptophan residues (Trp115, 135 and 175, Figure 4b,c) of Cel12B protein may be essential in mediating the total cooperativity of the response of the enzyme to substrate. Met133, located in the middle of Trp135 and Glu131 in β-sheet (Figure 4b), is predicted to be related to the binding of substrate and also finally confirmed by experimental results. When it is replaced by tryptophan residue, the enzyme activity is significantly decreased. With the homology modeling result (data not shown), it is inferred that Glu64 is probably another functional acid amino located near the catalytic center. It is supposed that residue Glu64 might contribute to stabilizing the intermediate product. Maintaining the intermediate product may be caused by the interaction of side-chain group of Glu64. Polar amino acids, histidine and threonine are able to stabilize the intermediate product to some extent. However, their side-chain groups are relatively large, and possess larger

steric hindrance, thus lead to decrease of the enzyme activity. Compared to glutamic acid, histidine and threonine, serine has smaller side-chain group and steric hindrance, so it can easily form hydrogen bond with product and stabilize it, and then increase the enzyme activity.

Conclusions

Nineteen hyperthermophilic homologous protein sequences from GHF12 were aligned and used for constructing phylogenetic tree. It was inferred from the nodes that there is a close relationship among these nineteen homologous endoglucanases from hyperthermophilic bacteria and archaea. We have made clear the function of these conserved amino acids in Cel12B protein, which is helpful in analyzing other molecular structure and transforming them with site directed mutagenesis.

Methods
Extraction of sequences from databases

Thorough BLASTP searches for several divergent endoglucanases of plants, animals, bacteria, fungi, alga and archaea were performed to retrieve endoglucanases genes through NCBI, PDB (http://www.rcsb.org/pdb/home/home.do), UniProt (http://www.uniprot.org/) database server. Hyperthermophilic endoglucanase amino acid sequence was used (GenBank No: Z6934) [16] as a BLAST query for seeking hyperthermophilic endoglucanases from bacteria and

```
ADN02999      EPNLWNVVGG-SGSVTMTFDDADGFDLDVQIDLSNIQQEDPSGWVHAYPEIWYGIK-IWNTVGPAQDG--PVPLPRRLSEL----
AEJ62362      EPNLWNVVGG-SGSVTMTFDDAEGFDLDVQIDLSNIQQEDPSGWVHAYPEIWYGIK-IWNTVGPAQDG--PVPLPRKLSEL----
AF181032      EINLWNILNA-TGFAEMTYNLTSGVLHYVQQ-LDNIVLRDRSNWVHGYPEIFYGNK-PWN-ANYATDG--PIPLPSKVSNL----
EEB73588      EIDPWNVKRA-SGFQRMTYDPESGRVEFVSN-LSDEVLVDPESWVHGYPEVYFGTK-PWN-GNSAPGF--GVELPVKVSEM----
YP_003346783  ELNFWNVKSY-EGETWLKFDGEK--VEFYAD-LYNIVLQNPDSWVHGYPEIYYGYK-PWAGHNSGVE-----FLPVKVKDL----
Z69341        ELNFWNVKSY-EGETWLKFDGEK--VEFYAD-LYNIVLQNPDSWVHGYPEIYYGYK-PWAGHNSGVE-----FLPVKVKDL----
YP_002352530  ELNFWNIAKY-EGNTWMAFYKEEDAVEYYAD-IKNIILKDKNSWVHGYPEVYYGYK-PWSAHGNSIEK---LVLPRKVLEF----
ADX81754      EVNMWNAKTW-NGNYTMVFNPLTRTLSVSFN----LTQVNPLQWTNGYPEIYVGRK-PWDTSYAG------NIFPMRIGNM----
ACP37717      EVNMWNAKTW-NGNYTMVFNPLTRTLSVSFN----LTQVNPLQWTNGYPEIYVGRK-PWDTSYAG------NIFPMRIGNM----
ADX84872      EVNMWNAKTW-NGNYTMVFNPLTRTLSVSFN----LTQVNPLQWTNGYPEIYVGRK-PWDTSYAG------NIFPMRIGNM----
ACR41545      EVNMWNAKTW-NGNYTMVFNPLTRTLSVSFN----LTQVNPLQWTNGYPEIYVGRK-PWDTSYAG------NIFPMRIGNM----
AAK42142      EVNMWNAKTW-NGNYTMVFNPLTRTLSVSFN----LTQVNPLQWTNGYPEIYVGRK-PWDTSYAG------NIFPMRIGNM----
ADM27702      SLNLYGVNSA-LGYQRMYIYIHNLTIKIVSD----LHSIQPVQWVNGYPEIYVGRK-PWDTRYIDGYG---VAFPINVDNP----
AEA12777      QINMWNIKSA-SGTAAMRYCDGVFYYEQALK---DIAEANPDAWVAGYPEIWLGYK-PWAGAASPNS-----PFPIKISDAES--
AEB95090      SPFLWNVKEG-EGQVTMNFS-NYLKVVINMS---KVNKITPSIPVDGYPGLMYGRE-MWFPPVAQTET-YRLNLPEIVNDL----
AAY81158      SPFLWNLKTA-LGYTNLTYRGNTLVVNVNFT---NFEKINSNLQVDGYPGVMYGQE-DWFPFAGRTLMPSCFVLPVKVISL----
ADN509821     YINMWNLANFSKGYAKMTYNPNKGVLCFYAL-LSNAVLKSPESQVWGYPEIFLVGKSPWFKSPVN--E--IINLPQRINDLLSSY
CCC81038      APYPWNAKEW-TGVVQVAYDGREVTASVNMS---YTAKFNPYAPVLGYPSVRYGCDPLFYYCSGRAQP---LELPEPASKAGG--
ADL19785      SPMLWNLVGG-TGNVTMVLSGGRLYVYINVT---GVSRAIKFTPVVGFPDIMYGWY-SWGPFYTRTSSYGFLSLPMPASEV----
                        30                              63               83

ADN02999      NDFYTTVDFSIQRLDPQLPFNFPFETWLTR---DTSRGRDVRSDEVEIMIWFNYYGLQGAGSQVD------TLTVPIEVNGQMRD
AEJ62362      NDFYTTVDFSIQRLDPELPFNFPFETWLTR---DTSRGRDVRSDEVEIMVWFNYYGLQGAGSQVD------TLTVPIEVNGQTRD
AF181032      TDFYLTISYKLEPKN-GLPINFAIESWLTR---EAWRTTGINSDEQEVMIWIYYDGLQPAGSKVK------EIVVPIIVNGTPVN
EEB73588      RHFLVHVEYSINLTD-PIPFNLAMETWLTK---EKNRSTGVFPGEAEIMVWLYYSNLTPAGEKIG------EAHVPLLVNGSLVN
YP_003346783  PDFYVTLDYSIWYEN-NLPINLAMETWITR---SPDQTSVSSGDAEIMFYNNVLMPGGQKVD------EFTTTVEINGVKQE
Z69341        PDFYVTLDYSIWYEN-NLPINLAMETWITR---SPDQTSVSSGDAEIMFYNNVLMPGGQKVD------EFTTTVEINGVKQE
YP_002352530  PDVLFNLKYNIWYER-NLSINFAMETWITK---EPYQKTVTAGDIEMMVWLYANRLSPAGRKVA------EVKIPIILNGNQKD
ADX81754      TPFMVSFYINLTKLDPSINFDIASDAWIVRPQIAFSPGTAPGNGDIEIMVWLFSQNLQPAGQQVG------ELVIPIYINHTLVN
ACP37717      TPFMVSFYINLTKLDPSINFDIASDAWIVRPQIAFSPGTAPGNGDIEIMVWLFSQNLHPAGQQVG------ELVIPIYINHTLVN
ADX84872      TPFMVSFYINLTKLDPSINFDIASDAWIVRPQIAFSPGTAPGNGDIEIMVWLFSQNLQPAGQQVG------ELVIPIYINHTLVN
ACR41545      TPFMVSFYINLTKLDPSINFDIASDAWIVRPQIAFSPGTAPGNGDIEIMVWLFSQNLHPAGQQVG------ELVIPIYINHTLVN
AAK42142      TPFMVSFYINLTKLDPSINFDIASDAWIVRPQIAFSPGTAPGNGDIEIMVWLFSQNLQPAGQQVG------EVVIPIYINHTLVN
ADM27702      RQFVVSFYVCIEDLDPTMNFNIAADAWIVRESVARAPGTPPGKGDIEIMVWLFSQNLGPAGDRVG------EEIIPIVINGTRID
AEA12777      SNFTISVDYSVEVPDPTLPLDFAFDLWVTR----STGERSVGQGEQEIMIWLYYQQLMPAGEKVG------EVRIPLVVNGSPAE
AEB95090      PSFYSILNYSIFVNE-GTVDDFSYDIWLSQ---NPNITYLKYGDFEVMIWLYWHENFSSDKYMIYT---GEMTIPVEVNGTFEP
AAY81158      PNFNSTLSYKINDNR-GIIDDFSYDIWLTQ---NPNTTYIQFPDVEIMIWLYHNETLS--DYFVKA---GVMSVNIMVNGTVIQ
ADN509821     PNLGIYVNYTLIHSP-STPMDWAYDIWFLR----DPNVTGVGPGDAEMMIWLYYSGYNTQWAYTGI------NVNIPIYVNGTLIN
CCC81038      ---VLVLSVSPGD-CNVTDFSYDIWFNR------GGFRLGAGDLELMLWLYYNVDPSASLPAPYWRYLGERRLRIAVDGAWQT
ADL19785      PDLWSVVNYSLSLSR-GAYNDFSYDIWLVR----SPGVTSLGPSDVELMLWMFANQSLAGLPYWVTWR---PITMPTLINGDIEN
                        115                          131 133 135

ADN02999      MTFEVWRSDAV--GNGGWEYFAFRPT--------TPISEGTVRFNWAPFIQKARSLSNRA---------DWENLYFTSVELGTEFG
AEJ62362      MTFEVWRSDAV--GNGGWEYFAFRPT--------TPVSEGTVRFNWAPFIQRARSLSNRA---------DWENLYFTSVELGTEFG
AF181032      ATFEVWK------ANIGWEYVAFRIK--------TPIKEGTVTIPYGAFISVAANISSLP---------NYTELYLEDVEIGTEFG
EEB73588      ATFEVWLDGN---MGDGWQYMAFMIA--------EPMREADVTLDPTLFVTAAENFS-RV---------DLKNLYLQDWEMGTEFG
YP_003346783  TKWDVYFAP------WGWDYLAFRLT--------TPMKEGKVKINVKDFVQKAAEVVKKHSTRI----DNFEELYFCVWEIGTEFG
Z69341        TKWDVYFAP------WGWDYLAFRLT--------TPMKEGKVKINVKDFVQKAAEVVKKHSTRI----DNFEELYFCVWEIGTEFG
YP_002352530  IVWEVYFSP------GSWDYIAYKSK--------ENIIQGEVKIPIKDFIKHLRTVIANNSSRIT---AEKYDQMYVTVWEIGTEFG
ADX81754      ATFQVWEMKSV--PWGGWEYIAFRPDG-------WKVTNGYVAYEPNLFIK-ALNNFTSY---------NITNYYLTDWEFGTEWG
ACP37717      ATFQVWEMKSV--PWGGWEYIAFRPDG-------WKVTNGYVAYEPNLFIK-ALNNFTSY---------NITNYYLTDWEFGTEWG
ADX84872      ATFQVWEMKSV--PWGGWEYIAFRPDG-------WKVTNGYVAYEPNLFIK-ALNNFTSY---------NITNYYLTDWEFGTEWG
ACR41545      ATFQVWEMKSV--PWGGWEYIAFRPDG-------WKVTNGYVAYEPNLFIK-ALNNFTSY---------NITNYYLTDWEFGTEWG
AAK42142      ATFQVWKMKNV--PWGGWEYIAFRPDG-------WKVTNGYVAYEPNLFIK-ALNNFASY---------NITNYYLTDWEFGTEWG
ADM27702      AKWDVYLQRSV--PWGGWDYIAFAPSG-------WSVRCGSVAYDPTLFIQ-AAKKYVS----------MSGYYLLNWEIGTEWG
AEA12777      AVFYVYRKEG-----MPWEYIAFALS--------KPMRSGSVYFRLADFIRAAAAYTALP---------NYSDMWLNDVELGSEFG
AEB95090      MNFSVYVLPRT-GSADGWTGVYLLAP--------RNLQG-SVGVPIAYVLNNMSPYLSKVKINIYN----TSKYYLDAIQVGMEFN
AAY81158      DDFTVYILPHT-GSSNGWIGVYYISQ--------LELSAGNITVPMSTLIKDSFNYIRGVFPDLQ-----TSAYYLNAIQVGMEFN
ADN509821     ETFMVLINCN---HGAGWTYIAFVPVN-------GGYRNGNIGVWLAPFLNYMVTLLPQKCPSIWKSPDNVSNLWLMDIELGSEFD
CCC81038      AEASAYVHLS----QGSWSVLVFVLR--------RPVRSGTVAVDLGQLVRAAQETLNGTP-------IDLERLNLTSIDVGMEFD
ADL19785      VTYQVFILPRN-GGPSGWMLIILIPEVNTSGGQYHGLLKGEYGVNLGELMNETLNIIGEFNGTR-----WEQGLYLSVIGLGAEVD
                        175                              227 229
```

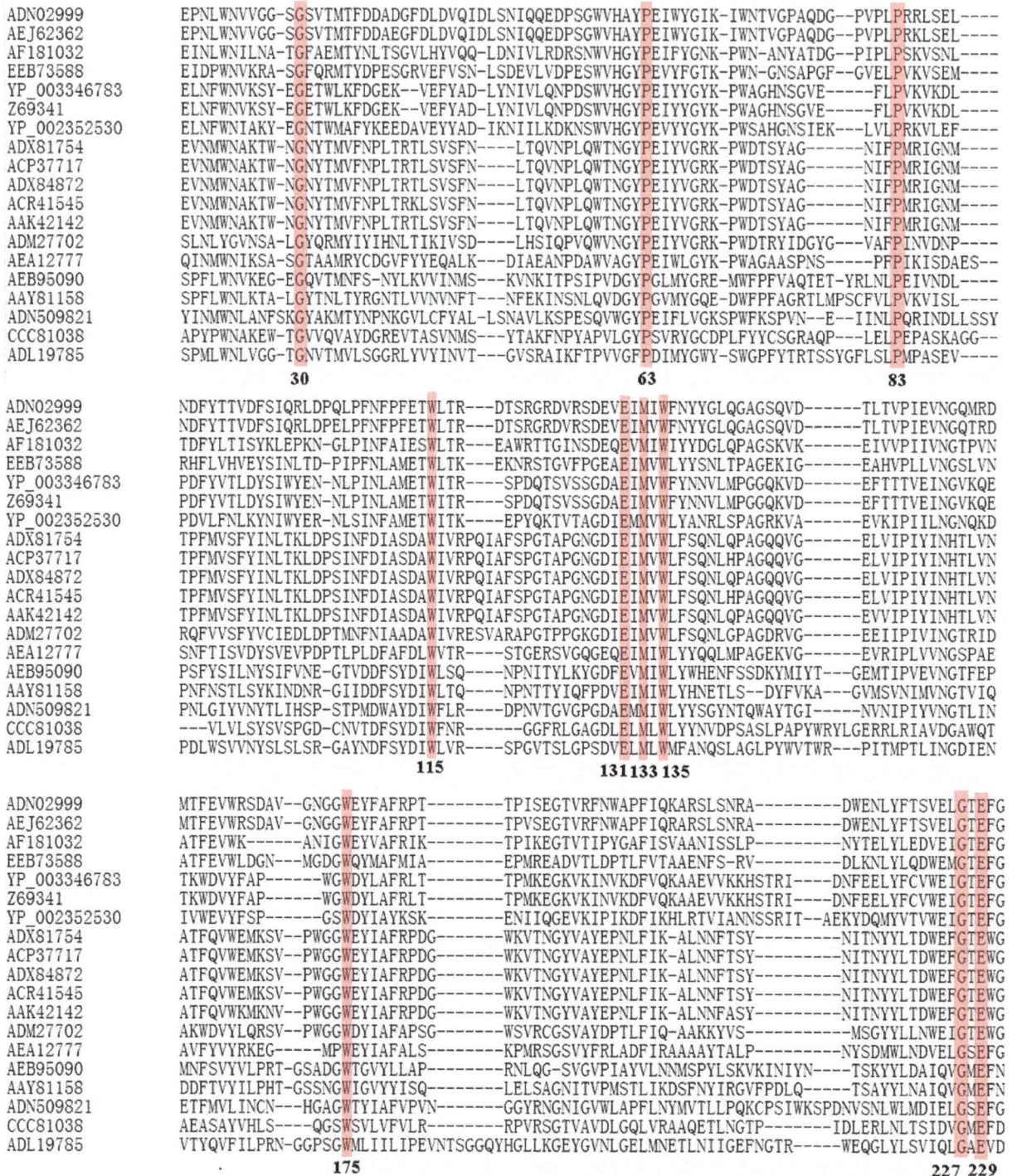

Figure 3 Alignment of 19 endoglucanases amino acids sequences using CLUSTAL X2.0. The highly conserved amino acids are colored in red.

archaea. New rounds of BLASTP searches for the nr protein and GenBank databases at NCBI restricted to plant or other organisms were carried out using representative endoglucanase from different classes of plants, bacteria, fungi and alga as queries.

Multiple sequence alignment and phylogenetic analysis

One of the most widely used bioinformatics analysis is multiple sequences alignment, and it needs several widely used software packages to analysis. In this study, the multiple sequence alignment tool Clustal X2 was

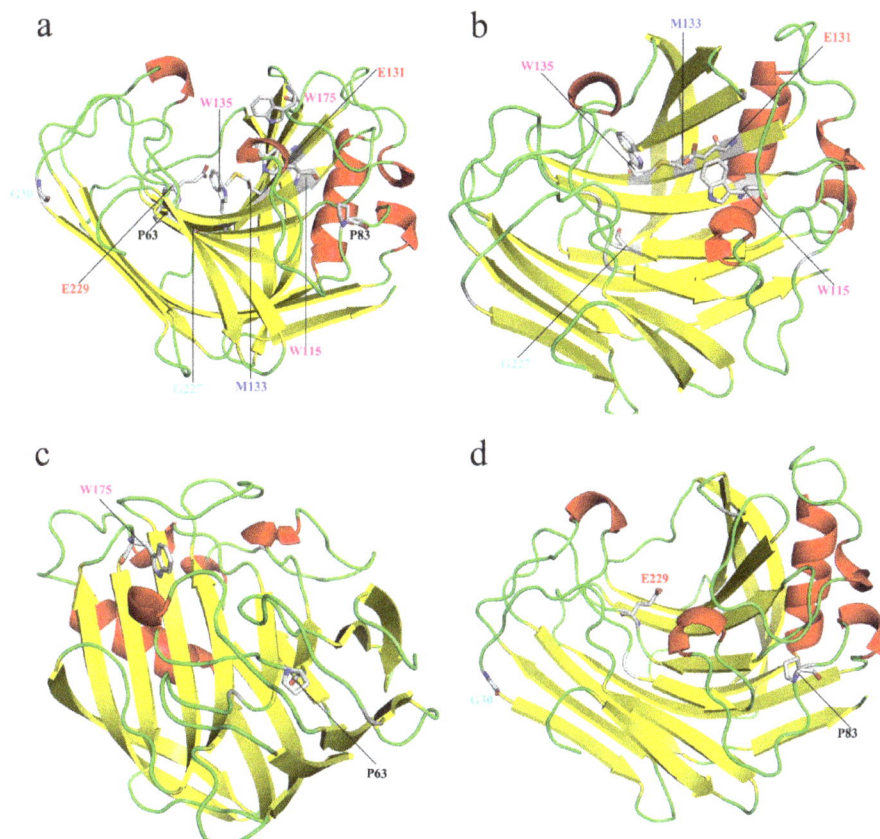

Figure 4 Structure modeling of the protein Cel12B. Different segments of the protein secondary structure are colored accordingly. The catalytic amino acids (Glu131 and Glu229) locating in the center of the structure were labeled in red **(a, b, d)**. The amino acids Trp115, Trp135 and Trp175 were labeled in magenta **(a, b, c)**, Met133 was labeled in blue **(a, b)**, where these four amino acids show a great importance in the substrate binding. The amino acids Pro63 and Pro83 were labeled in black **(a, c, d)**, Gly30 and Gly227 were labeled in cyan **(a, b, d)**, where these four amino acids are well related to the thermostability of the enzyme.

used for sequence alignment [36]. Sequences were further edited using the MEGA 5 when necessary and aligned manually [37]. In the phylogenetic analysis, sequences were trimmed so that only the relevant conserved domains were remained in the alignment. Phylogenetic relationships were inferred using the NJ and MP methods as

implemented in PAUP 4.0 [18] while the Maximum-Likelihood method as implemented in MEGA 5 [37]. The NJ, MP and ML trees, displayed using TREEVIEW 1.6.6 (http://taxonomy.zoology.gla.ac.uk/rod/treeview.html), were evaluated with 1000 bootstrap replicates.

Secondary structure prediction

For homology modeling, the crystal structure of the thermophilic endoglucanase (PDB ID: 3AAM) obtained

Table 3 Effect of site-directed mutagenesis on enzyme activity

Strain	Optimum temp (°C)	Specific activity (U mg^{-1})	Relative activity (%)
Control	90	105 ± 3.4	100 ± 3.2
E64T	85	53 ± 1.3	50 ± 1.2
E64H	85	25 ± 1.0	24 ± 1.0
E64L	ND	0	0
E64S	90	133 ± 2.5	127 ± 2.4
P63K	ND	0	0
P83K	ND	0	0
M133W	ND	0	0

ND: not determined. Values shown were the mean of triplicate experiments.

Table 4 Nucleotide sequences of used primers

Primers	Nucleotide sequence
Forward 1	5′-AGTAGAT**NNN**TGGATATCCATGCACCCAGC -3′
Reverse 1	5′-ACGGTTACAAGCCCTGGGCG -3′
Forward 2	5′- CATGGATAT**AAG**GAGATCTACTACGGTTACAAG -3′
Reverse 2	5′-CACCCAGCTGTCTGGATTCTGAAG -3′
Forward 3	5′-GAATTTCTT**AAG**CTGAAGGTGAAAGATCTTCC -3′
Reverse 3	5′-AACACCGCTGTTGTGCCCCG -3′
Forward 4	5′-CGGAGATC**TGG**GTTTGGTTCTACAACAACGTTC-3′
Reverse 4	5′-CGTCACCCGAAGAAACAGAGGTC -3′

from Protein Data Bank (PDB) was used as a template. The aligned sequences were submitted to SWISS-MODEL (http://www.expasy.org/swissmod/) to obtain the 3D structure of the endoglucanases [38-40]. The model was viewed using Swiss-PDB Viewer [41], and the quality of the model was evaluated by the local model quality estimation on SWISS-MODEL. The 3D structure of the protein was further modified by PyMOL (version 1.4.1, http://www.pymol.org/).

Test of functional residues

Site-directed mutagenesis was used to analyze the related functional amino acid residues using reverse PCR. Restriction enzymes, DNA polymerase, *Dpn*I, T4 polynucleotide kinase and T4 ligase were purchased from Takara (Dalian, China) and used according to the manufacturer's instructions. The sequence of *cel12B* gene (GenBank Protein No. Z69341) based on the *T. maritima* genomic DNA was amplified using primers 5′-GGAATTCCATATGAGGTGGG-CAGTTCTTCTGA-3′, and 5′-CCGCTCGAGTTATTACT CGAGTTTTACACCTTCGACAGAGAAGTC-3′ (primers with the added compatible restriction sites of *Nde*I and *Xho*I, respectively). PCR was performed as follows: 94°C, 5 min; 30 cycles of 94°C for 30 s, 55°C for 30 s and 72°C for 50 s; and 72°C, 10 min. The recombinant vector was constructed as follows: the amplified PCR products were purified, digested with *Nde*I and *Xho*I, and then ligated into pET-20b vector at the corresponding sites. Reverse PCR amplifications were conducted by high-fidelity *Pyrobest* DNA polymerase using recombinant pET-20b-*cel12B* as templates, and primers were shown in Table 4. The templates were cleaned away from the products using *Dpn*I. Then, the resulting products were purified with BIOMIGA PCR Purification Kit (Shanghai, China), followed by phosphorylation using T4 polynucleotide kinase and finally ligated with T4 ligase. DNA sequencing was performed with ABI 3730 (Applied Biosystems, USA).

E. coli BL21 (DE3) cells harboring recombinants were grown at 37°C and 200 rpm in 200 mL of Luria-Bertani (LB) with appropriate antibiotic selection. When the OD_{600} reached 0.6-0.8, the expression of mutated enzymes were induced by the addition of 0.5 mM isopropyl β-D-1-thiogalactopyranoside (IPTG) and the culture was incubated at 37°C and 200 rpm for 5 h. Cells were harvested by centrifugation at 4°C (10000 rpm, 5 min), washed twice with 20 mM Tris-HCl buffer (pH 8.0), and re-suspended in 5 mL of 5 mM imidazole, 0.5 M NaCl, and 20 mM Tris-HCl buffer (pH 7.9). All subsequent steps were carried out at 4°C. The cell extracts after sonication were heat treated at 50°C for 30 min, cooled in an ice bath, and then centrifuged (15000 g, 4°C, 20 min). The resulting supernatants were loaded onto a 1 ml Ni^{2+} affinity column (Novagen, USA) and the bounded proteins were eluted by discontinuous imidazole gradient.

Enzyme activity was determined using 5-dinitrosalicylic acid (DNS) method [42]. The reaction mixture, containing 50 mM imidazole-potassium buffer (pH 6.0), 0.5% sodium carboxymethyl cellulose (CMC-Na), and a certain amount of endoglucanase (0.1 µg) in 0.2 mL, was incubated for 10 min at 85°C. The reaction was stopped by the addition of 0.3 mL DNS. The absorbance of the mixture was measured at 520 nm. One unit of enzyme activity was defined as the amount of enzyme necessary to liberate 1 µmol of reducing sugars per min under the assay conditions. All the values of enzymatic activities shown in figures were averaged from three replicates.

Competing interests
The authors declare that they have no competing interests.

Authors' contributions
HS conceived the project, carried out phylogenetic phylogenetic analysis, LW, XL, YZ and WL carried out database searches and protein modeling, and FW and XL supervised the work. HS, XL, and FW wrote the manuscript. All authors read and approved the final manuscript.

Acknowledgements
This work was financially supported by the National Natural Science Foundation of China (No. 31170537), Jiangsu Provincial Government (CXZZ11_0526), Doctorate Fellowship Foundation of Nanjing Forestry University, as well as A Project Funded by the Priority Academic Program Development of Jiangsu Higher Education Institutions (PAPD).

Author details
[1]College of Chemical Engineering, Nanjing Forestry University, Nanjing 210037, China. [2]Jiangsu Key Lab of Biomass-Based Green Fuels and Chemicals, Nanjing 210037, China. [3]Department of Life Science and Chemistry, Huaiyin Institute of Technology, Huaian 223003, China.

References
1. Wang T, Liu X, Yu Q, Zhang X, Qu Y, Gao P: **Directed evolution for engineering pH profile of endoglucanase III from *Trichoderma reesei*.** *Biomol Eng* 2005, **22**(1–3):89–94.
2. Liang C, Fioroni M, Rodriguez-Ropero F, Xue Y, Schwaneberg U, Ma Y: **Directed evolution of a thermophilic endoglucanase (Cel5A) into highly active Cel5A variants with an expanded temperature profile.** *J Biotechnol* 2011, **154**(1):46–53.
3. Anbar M, Lamed R, Bayer EA: **Thermostability enhancement of *Clostridium thermocellum* cellulosomal endoglucanase Cel8A by a single glycine substitution.** *Chemcatchem* 2010, **2**(8):997–1003.
4. Nakazawa H, Okada K, Onodera T, Ogasawara W, Okada H, Morikawa Y: **Directed evolution of endoglucanase III (Cel12A) from trichoderma reesei.** *Appl Microbiol Biotechnol* 2009, **83**(4):649–657.
5. Watanabe H, Noda H, Tokuda G, Lo N: **A cellulase gene of termite origin.** *Nature* 1998, **394**(6691):330–331.
6. Davison A: **Ancient origin of glycosyl hydrolase family 9 cellulase genes.** *Mol Biol Evol* 2005, **22**(5):1273–1284.
7. Mardanov AV, Svetlitchnyi VA, Beletsky AV, Prokofeva MI, Bonch-Osmolovskaya EA, Ravin NV, Skryabin KG: **The genome sequence of the crenarchaeon *Acidilobus saccharovorans* supports a new order, *Acidilobales*, and suggests an important ecological role in terrestrial acidic hot Springs.** *Appl Environ Microbiol* 2010, **76**(16):5652–5657.
8. Reno ML, Held NL, Fields CJ, Burke PV, Whitaker RJ: **Biogeography of the *Sulfolobus islandicus* pan-genome.** *Proc Natl Acad Sci* 2009, **106**(21):8605–8610.
9. Guo L, Brugger K, Liu C, Shah SA, Zheng H, Zhu Y, Wang S, Lillestol RK, Chen L, Frank J, *et al*: **Genome analyses of Icelandic strains of *Sulfolobus islandicus*, model organisms for genetic and virus-host interaction studies.** *J Bacteriol* 2011, **193**(7):1672–1680.

10. Göker M, Held B, Lapidus A, Nolan M, Spring S, Yasawong M, Lucas S, Glavina Del Rio T, Tice H, Cheng J-F, *et al*: Complete genome sequence of *Ignisphaera aggregans* type strain (AQ1.S1T). *Stand Genomic Sci* 2010, **3**(1):66–75.

11. Angelov A, Liebl S, Ballschmiter M, Boemeke M, Lehmann R, Liesegang H, Daniel R, Liebl W: Genome sequence of the polysaccharide-degrading, thermophilic anaerobe *Spirochaeta thermophila* DSM 6192. *J Bacteriol* 2010, **192**(24):6492–6493.

12. Mardanov AV, Gumerov VM, Beletsky AV, Prokofeva MI, Bonch-Osmolovskaya EA, Ravin NV, Skryabin KG: Complete genome sequence of the thermoacidophilic crenarchaeon *Thermoproteus uzoniensis* 768–20. *J Bacteriol* 2011, **193**(12):3156–3157.

13. Chen LM, Brugger K, Skovgaard M, Redder P, She QX, Torarinsson E, Greve B, Awayez M, Zibat A, Klenk HP, *et al*: The genome of *Sulfolobus acidocaldarius*, a model organism of the *Crenarchaeota*. *J Bacteriol* 2005, **187**(14):4992–4999.

14. Liu L-J, You X-Y, Zheng H, Wang S, Jiang C-Y, Liu S-J: Complete genome sequence of *Metallosphaera cuprina*, a metal sulfide-oxidizing archaeon from a hot spring. *J Bacteriol* 2011, **193**(13):3387–3388.

15. Wu D, Hugenholtz P, Mavromatis K, Pukall R, Dalin E, Ivanova NN, Kunin V, Goodwin L, Wu M, Tindall BJ, *et al*: A phylogeny-driven genomic encyclopaedia of bacteria and archaea. *Nature* 2009, **462**(7276):1056–1060.

16. Liebl W, Ruile P, Bronnenmeier K, Riedel K, Lottspeich F, Greif I: Analysis of a *Thermotoga maritima* DNA fragment encoding two similar thermostable cellulases, CelA and CelB, and characterization of the recombinant enzymes. *Microbiol (Reading, England)* 1996, **142**(Pt 9):2533–2542.

17. Zhaxybayeva O, Swithers KS, Lapierre P, Fournier GP, Bickhart DM, DeBoy RT, Nelson KE, Nesbo CL, Doolittle WF, Gogarten JP, *et al*: On the chimeric nature, thermophilic origin, and phylogenetic placement of the thermotogales. *Proc Natl Acad Sci U S A* 2009, **106**(14):5865–5870.

18. Wilgenbusch JC, Swofford D: Inferring evolutionary trees with PAUP. *Curr Protoc Bioinformatics* 2003. Chaper 6, unit 6.4. http://www.currentprotocols.com/protocol/bi0604.

19. Chhabra SR, Shockley KR, Ward DE, Kelly RM: Regulation of endo-acting glycosyl hydrolases in the hyperthermophilic bacterium *Thermotoga maritima* grown on glucan- and mannan-based polysaccharides. *Appl Environ Microbiol* 2002, **68**(2):545–554.

20. Wang Y, Wang X, Tang R, Yu S, Zheng B, Feng Y: A novel thermostable cellulase from *Fervidobacterium nodosum*. *J Mol Catal B Enzym* 2010, **66**(3–4):294–301.

21. Sinnott ML: Catalytic mechanisms of enzymatic glycosyl transfer. *Chem Rev* 1990, **90**(7):1171–1202.

22. Gaucher EA, Thomson JM, Burgan MF, Benner SA: Inferring the palaeoenvironment of ancient bacteria on the basis of resurrected proteins. *Nature* 2003, **425**(6955):285–288.

23. Mardanov AV, Ravin NV, Svetlitchnyi VA, Beletsky AV, Miroshnichenko ML, Bonch-Osmolovskaya EA, Skryabin KG: Metabolic versatility and Indigenous origin of the archaeon *Thermococcus sibiricus*, isolated from a siberian oil reservoir, as revealed by genome analysis. *Appl Environ Microbiol* 2009, **75**(13):4580–4588.

24. Kumar S, Tsai CJ, Nussinov R: Factors enhancing protein thermostability. *Protein Eng* 2000, **13**(3):179–191.

25. Warren GL, Petsko GA: Composition analysis of alpha-helices in thermophilic organisms. *Protein Eng* 1995, **8**(9):905–913.

26. Kumar S, Bansal M: Dissecting alpha-helices: position-specific analysis of alpha-helices in globular proteins. *Proteins* 1998, **31**(4):460–476.

27. Kimura S, Kanaya S, Nakamura H: Thermostabilization of *Escherichia coli* ribonuclease HI by replacing left-handed helical Lys95 with Gly or Asn. *J Biol Chem* 1992, **267**(31):22014–22017.

28. Kawamura S, Kakuta Y, Tanaka I, Hikichi K, Kuhara S, Yamasaki N, Kimura M: Glycine-15 in the bend between two alpha-helices can explain the thermostability of DNA binding protein HU from *Bacillus stearothermophilus*. *Biochemistry* 1996, **35**(4):1195–1200.

29. Watanabe K, Kitamura K, Suzuki Y: Analysis of the critical sites for protein thermostabilization by proline substitution in oligo-1,6-glucosidase from *Bacillus coagulans* ATCC 7050 and the evolutionary consideration of proline residues. *Appl Environ Microbiol* 1996, **62**(6):2066–2073.

30. Suzuki Y, Oishi K, Nakano H, Nagayama T: A strong correlation between the increase in mumber of proline resdues and the rise in thermostability of 5 *Bacillus* oligo-1,6-glucsidases. *Appl Microbiol Biotechnol* 1987, **26**(6):546–551.

31. Zhu GP, Xu C, Teng MK, Tao LM, Zhu XY, Wu CJ, Hang J, Niu LW, Wang YZ: Increasing the thermostability of D-xylose isomerase by introduction of a proline into the turn of a random coil. *Protein Eng* 1999, **12**(8):635–638.

32. Suzuki Y: A general principle of increasing protein thermostability. *Proc Japan Acad Series B-Physl and Bio Sci* 1989, **65**(6):146–148.

33. Derewenda U, Swenson L, Green R, Wei Y, Morosoli R, Shareck F, Kluepfel D, Derewenda ZS: Crystal structure, at 2.6-A resolution, of the *streptomyces lividans* xylanase a, a member of the F family of beta-1,4-D-glycanases. *J bio chem* 1994, **269**(33):20811–20814.

34. Churakova NI, Cherkasov IA, Kravchenko NA: The role of the tryptophan-62 residue in the structure and function of lysozyme. *Biokhimiĩa (Moscow, Russia)* 1977, **42**(2):274–276.

35. Pennacchio A, Esposito L, Zagari A, Rossi M, Raia CA: Role of Tryptophan 95 in substrate specificity and structural stability of *Sulfolobus solfataricus* alcohol dehydrogenase. *Extremophiles* 2009, **13**(5):751–761.

36. Larkin MA, Blackshields G, Brown NP, Chenna R, McGettigan PA, McWilliam H, Valentin F, Wallace IM, Wilm A, Lopez R, *et al*: Clustal W and clustal X version 2.0. *Bioinformatics* 2007, **23**(21):2947–2948.

37. Tamura K, Peterson D, Peterson N, Stecher G, Nei M, Kumar S: MEGA5: Molecular evolutionary genetics analysis using maximum likelihood, evolutionary distance, and maximum parsimony methods. *Mol Biol Evol* 2011, **28**(10):2731–2739.

38. Schwede T, Kopp J, Guex N, Peitsch MC: SWISS-MODEL: an automated protein homology-modeling server. *Nucleic Acids Res* 2003, **31**(13):3381–3385.

39. Guex N, Peitsch MC: SWISS-MODEL and the Swiss-PdbViewer: an environment for comparative protein modeling. *Electrophoresis* 1997, **18**(15):2714–2723.

40. Arnold K, Bordoli L, Kopp J, Schwede T: The SWISS-MODEL workspace: a web-based environment for protein structure homology modelling. *Bioinformatics* 2006, **22**(2):195–201.

41. Kaplan W, Littlejohn TG: Swiss-PDB viewer (deep view). *Brief Bioinform* 2001, **2**(2):195–197.

42. Miller GL: Use of dinitrosalicylic acid reagent for determination of ruducing sugar. *Anal Chem* 1959, **31**(3):426–428.

17

Crystal structure and functional implications of the tandem-type universal stress protein UspE from *Escherichia coli*

Yongbin Xu[1,2*], Jianyun Guo[1], Xiaoling Jin[1], Jin-Sik Kim[3], Ying Ji[1], Shengdi Fan[1], Nam-Chul Ha[3] and Chun-Shan Quan[1]

Abstract

Background: The universal stress proteins (USP) family member UspE is a tandem-type USP that consists of two Usp domains. The UspE expression levels of the *Escherichia coli* (*E. coli*) become elevated in response to oxidative stress and DNA damaging agents, including exposure to mitomycin C, cadmium, and hydrogen peroxide. It has been shown that UspA family members are survival factors during cellular growth arrest. The structures and functions of the UspA family members control the growth of *E. coli* in animal hosts. While several UspA family members have known structures, the structure of *E. coli* UspE remains to be elucidated.

Results: To understand the biochemical function of UspE, we have determined the crystal structure of *E. coli* UspE at 3.2 Å resolution. The asymmetric unit contains two protomers related by a non-crystallographic symmetry, and each protomer contains two tandem Usp domains. The crystal structure shows that UspE is folded into a fan-shaped structure similar to that of the tandem-type Usp protein PMI1202 from *Proteus mirabilis*, and it has a hydrophobic cavity that binds its ligand. Structural analysis revealed that *E. coli* UspE has two metal ion binding sites, and isothermal titration calorimetry suggested the presence of two Cd^{2+} binding sites with a K_d value of 38.3–242.7 µM. Structural analysis suggested that *E. coli* UspE has two Cd^{2+} binding sites (Site I: His117, His 119; Site II: His193, His244).

Conclusion: The results show that the UspE structure has a hydrophobic pocket. This pocket is strongly bound to an unidentified ligand. Combined with a previous study, the ligand is probably related to an intermediate in lipid A biosynthesis. Subsequently, sequence analysis found that UspE has an ATP binding motif (Gly^{269}- X_2-Gly^{272}-X_9-Gly^{282}-Asn) in its C-terminal domain, which was confirmed by *in vitro* ATPase activity monitored using Kinase-Glo® Luminescent Kinase Assay. However, the residues constituting this motif were disordered in the crystal structure, reflecting their intrinsic flexibility. ITC experiments revealed that the UspE probably has two Cd^{2+} binding sites. The His117, His 119, His193, and His244 residues within the β-barrel domain are necessary for Cd^{2+} binding to UspE protein. As mentioned above, USPs are associated with several functions, such as cadmium binding, ATPase function, and involvement in lipid A biosynthesis by some unknown way.

Keywords: UspE, UspA superfamily, Tandem-type USP

* Correspondence: yongbinxu@dlnu.edu.cn
[1]Department of Bioengineering, College of Life Science, Dalian Nationalities University, Dalian 116600Liaoning, China
[2]Laboratory of Biomedical Material Engineering, Dalian Institute of Chemical Physics, Chinese Academy of Sciences, Dalian 116023Liaoning, China
Full list of author information is available at the end of the article

Background

The universal stress proteins (USP) superfamily is a group of conserved proteins that play an important role in *E. coli*. USPs' expression levels become elevated in response to a bewildering variety of stress conditions, such as heat shock, nutrient starvation, the presence of oxidants, DNA-damaging agents (including exposure to mitomycin C, cadmium, and hydrogen peroxide), as well as others, that may arrest cell growth. Proteins in the UspA family constitute a natural biological defense mechanism [1, 2]. Despite considerable research on the behavior of UspA family members, the biological and biochemical roles of these proteins remain largely uncharacterized. Very few details were available to help decipher their roles in the aforementioned cellular processes [1]. A better understanding of the molecular mechanisms of *E. coli's* UspA proteins is important for establishing effective therapeutic strategies. In particular, establishing the three-dimensional structural model of the UspE protein can provide hints to explore the function(s) of the UspA family.

E. coli has six small UspA superfamily genes: *uspA*, *-C* *-D*, *-E*, *-F*, and *-G*. To date, these proteins have been extensively investigated. Previous studies have shown that UspA family members show immaculate similarity. They encode either a small USP protein (approximately 14 to 15 kDa) that consists of two USP domains in tandem or a larger version (approximately 30 kDa) that consists of two peptides attached as a single functional protein [3, 4]. UspA, UspC, and UspD belong to class I; UspF and UspG belong to class II; two Usp domains of UspE belong to class II and IV based on the sequence and structural analysis [3, 5, 6]. Previous works have found that while Usp family members have partially overlapping functions, the functions of class I, II, and IV Usps are distinct [7]. UspA proteins differ in their responses to protect cells from oxidative stress and DNA damage agents; UspA, UspC, UspD, and UspE are induced by exposure to mitomycin C, cadmium, and hydrogen peroxide. However, class II proteins, UspG and UspF, were associated with iron scavenging in the cell [4]. As mentioned before, UspE is a tandem-type USP. When UspE proteins are split apart and treated separately, the UspE2 domain is more closely related to UspF and UspG. This is clearly visible in both the clustering analysis and the reconstructed cladogram. In contrast, UspE1 groups are more closely related to class I UspA proteins (UspACD) [1].

This paper includes structural and functional studies on UspE from *E. coli*. Specifically, it presents the three-dimensional X-ray crystal structure of the recombinantly produced UspE from *E. coli* at 3.2 Å resolution. Additionally, through the use of structural biochemical analyses, the UspE mechanisms were determined. In terms of its overall structure, UspE was found to be similar to the tandem-type Usp protein PMl1202 from *Proteus*

mirabilis, which has a hydrophobic cavity that binds an unidentified ligand. It was also observed that UspE has an ATP binding motif $(Gly^{269}\text{-}Thr\text{-}Val\text{-}Gly^{272}\text{-}X_9\text{-}Gly^{282}\text{-}Asn)$ in its C-terminal domain. The ATPase activity was then measured to determine if UspE had ATPase activity and to characterize UspE activity. Because previous research found that UspE is critical for Cd^{2+} defense, we characterized the role of UspE as part of the Cd^{2+} binding process by ITC and structural analysis and found that UspE has two Cd^{2+} binding sites in its tandem USP domain. These observations suggest that UspE performs several distinct functions, such as ATP hydrolysis and cadmium defense. Although the molecular function of this protein remains unknown, our three-dimensional structures of UspE offer valuable clues to understand its potential biochemical mechanisms.

Methods

Structure determination, refinement and protein data bank accession number of UspE

We have previously reported the crystallization and preliminary X-ray analysis of *E. coli* UspE [8]. Data collection and refinement statistics are summarized in Table 1. The structure of the *E. coli* UspE protein was determined by molecular replacement (MR) using the program CCP4 package [9]. We used the coordinates of the structures of *P. mirabilis* PMl1202 (78.5 % sequence identity; PDB code: 3OLQ) as a reference. To build our protein model, we first removed model bias by rounds of simulated annealing performed with the program PHENIX [10], followed by calculating the differences using Fourier maps. Then, the UspE model was rebuilt in the graphic program COOT [11]. The model was finally refined using the same programs by iterative rounds of energy minimization, B-factor, and anisotropic refinements. Then, the composite omit and differences were calculated by Fourier maps. UspE coordinates and structure have been deposited in the Protein Data Bank [12] under accession code: 5CB0 (www.rcsb.org/pdb).

Isothermal titration calorimetry

Isothermal titration calorimetry (ITC) measurements were performed on a Microcal iTC 200 (GE Healthcare) VP-ITC microcalorimeter at 298 K. The protein was dialyzed against 20 mM Hepes (pH 7.0) and 150 mM NaCl. The titration $CdCl_2$ solution was prepared with 20 mM Hepes (pH 7.0) and 150 mM NaCl by adding 2 mM $CdCl_2$. Both the protein and the titrant $CdCl_2$ solutions were thoroughly degassed in a ThermoVac apparatus (Microcal). The titration reaction was performed by sequential injections of 40 μl $CdCl_2$ solution into the sample cell. The duration of the injection was 120 s. The

Table 1 Diffraction statistics

X-ray source	Beamline 5C, Pohang Accelerator Laboratory
PDB code	5CB0
Wavelength (Å)	1.000
Space group	$I4_122$
Resolution (Å)	19.9-3.2
Parameters (Å)	$a = b = 121.1$ Å, $c = 241.7$ Å, $\alpha = \beta = \gamma = 90°$
Rsym (%)	15.3 % (33.4 %)
Completeness (%)	92.7 (90.2)
Redundancy	5.3 (3.6)
Average I/σ (I)	9.2 (3.3)
R-factor (%)	24.31
Rfree (%)	30.06
Rmsd for bonds (Å)	0.007
Rmsd for angles ( )	1.393
Ramachandran plot (%)	
Favored regions	93.64 %
Allowed regions	5.12 %
Disallowed regions	1.24 %
Number of atoms	
Protein	4630
Ligand	34
Average B factor (Å^2)	70.54

†$R_{merge} = \Sigma_{hkl}\Sigma_i \mid I_i(hkl)-<I(hkl)> \mid \Sigma_{hkl}\Sigma_i \mid I_i(hkl)$, where I(hkl), where I(hkl) is the intensity of reflection hkl, Σ_{hkl} is the sum over all reflections and Σ_i is the sum over i measurements of reflection hkl. ‡$R_{work} = \Sigma_{hkl} \mid F_o\text{-}F_c \mid /\Sigma_{hkl} \mid F_o \mid$ for all data with $F_o > 2\sigma(F_o)$, excluding data used to calculate R_{free}. §$R_{free} = \Sigma_{hkl} \mid F_o\text{-}F_c \mid /\Sigma_{hkl} \mid F_o \mid$ for all data with $F_o > 2\sigma(F_o)$ that were excluded from refinement

syringe was rotated at 600 rev min^{-1}. Triplet measurements were collected in each case.

Kinase-Glo® luminescent kinase assay

The *in vitro* ATPase activity of UspE was measured by quantifying the amount of ATP remaining in the solution following a kinase reaction using a Kinase-Glo® Luminescent Kinase Assay Kit (Promega, Fitchburg, WI, USA). The assay was performed in a 96-well plate in a kinase reaction volume of 50 μl containing 10 mM $MgCl_2$,5 μM ATP,10 mM HEPES (pH 8.0) and 150 mM NaCl. The reaction was initiated by adding the protein to a final concentration of 0.4 mg/ml-3.2 mg/ml. The reaction mixture was kept at 310 K for 20 min in a water bath. Reaction mixtures containing no UspE were used as negative controls. The kinase reaction mixture was incubated with 50 μl of ATP detection reagent. The plates were then incubated for another 10 min at 310 K. The Synergy2 Multi-Mode Microplate Reader (BioTek, Winooski, VT, USA) was used to collect the relative light unit (RLU) signal. The luminescent signal was positively

correlated with the amount of remaining ATP and inversely correlated with the amount of kinase activity.

Results and discussion
Overall structure
To examine the biochemical mechanisms responsible for UspE function, we determined the crystal structure of the UspE by the molecular replacement method with the synchrotron data set at a resolution of 3.2 Å. The final model refined to a R-factor of 0.24 (R_{free} = 0.30). The initial solution suggested the presence of two monomers per asymmetric unit, which is consistent with the Matthews' coefficient of 3.1 Å^3 Da^{-1} (60.37 % solvent). The tertiary structure of the UspE is very similar to that of the previously described *P. mirabilis* Usp protein PMI1202 (PDB code: 3OLQ), which was used as a search model in molecular replacement [13, 14]. UspE exists as a monomer, and the structure reveals a compact and 2-fold symmetric dimer in the crystal. Each monomer consists of two USP domains, and the final model contains two homologous subunits related by a non-crystallographic symmetry (Fig. 1a). The *E. coli* UspE has a high structural similarity compared with the *P. mirabilis* USP (PDB code: 4WY2) (Fig. 1b). The crystal structure shows that UspE is folded into a fan-shaped structure similar to that of the tandem-type Usp protein USP from *P. mirabilis* (Fig. 1c). In a Ramachandran plot, 93.82 % of the model residues were found in favored regions, 5.72 % in allowed regions, and 1.06 % in the disallowed regions. Their structures are virtually identical with a root-mean-square deviation (RMSD) value of 0.40 Å for a 244 Cα atom. UspE is composed of ten β-stranded mixed β-sheet and nine α-helices. In the core structure, ten β-strands form a central parallel β-sheet (Fig. 1d). Significant density was observed for all residues in the final electron density map except 163–170, 202–215, and 270–282. Table 1 provides refinement statistics and structure solution for all structures.

E. coli UspE accommodates an unidentified ligand
More importantly, we found an unambiguous stick-like electron density in the hydrophobic pocket of *E. coli* UspE throughout the refinement process. It looks like that the UspE carries the unidentified ligand. The crystal structure of *P. mirabilis* USP suggested that Uridine-5′-diphosphate-3-O-(R-3-hydroxymyristoyl)-N-acetyl-D-glucosamine was tightly bound to *P. mirabilis* USP. We found that the Uridine-5′-diphosphate-3-O-(R-3-hydroxymyristoyl)-N-acetyl-D-glucosamine binding pocket of *P. mirabilis* USP was very similar to the hydrophobic pocket of *E. coli* UspE (Fig. 2a) and that the hydrophobic pocket of *E. coli* UspE was appropriate for binding Uridine-5′-diphosphate-3-O-(R-3-hydroxymyristoyl)-N-acetyl-D-glucosamine. In addition, the stick-like electron densities were very similar to a 3-

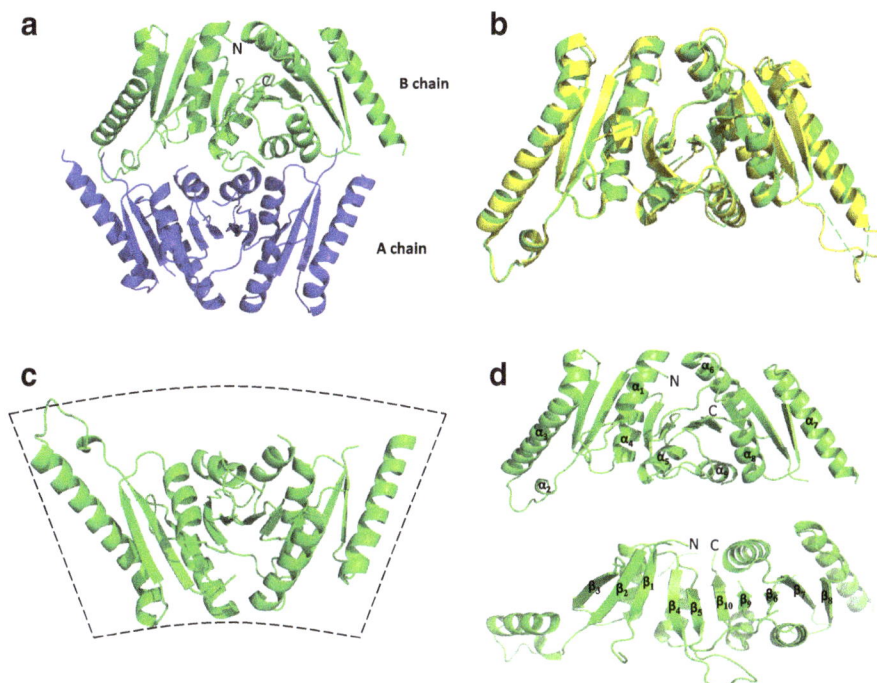

Fig. 1 Overall structure of the UspE. **a** Structure of *Ec*UspE in the tetragonal crystal form, displayed as ribbons. The asymmetric unit contains two protomers colored blue and green. **b** Structural comparison of the cartoon traces of *Ec*UspE and *P. mirabilis* USP (PDB code: 4WY2). *Ec*UspE is colored green, and the *P. mirabilis* USP is colored yellow. The disordered regions are shown with dashed lines. **c** The monomer structure of *Ec*UspE. **d** Secondary structural elements of UspE are numbered

hydroxymyristoyl group of Uridine-5′-diphosphate-3-O-(R-3-hydroxymyristoyl)-N-acetyl-D-glucosamine. Thus, this group was placed at density result. Despite the fact that *Ec*UspE crystals were obtained in the absence of 3-hydroxymyristoyl group molecules in both media and buffers, these positions could be successfully refined with no significant residual difference density and with associated B-factors comparable with those of the surrounding

atoms. The characteristic hydrophobic environment in the pocket indicates that UspE can bind unidentified ligand with a 3-hydroxymyristoyl group in the cavity. This pocket was surrounded by hydrophobic residues in β6, β9, β10, and α6 (Fig. 2b and c). This finding indicates that hydrophobic interactions are involved in the binding of this ligand. A previous study using mass spectrometric and surface analyses showed that the UspE homologue protein

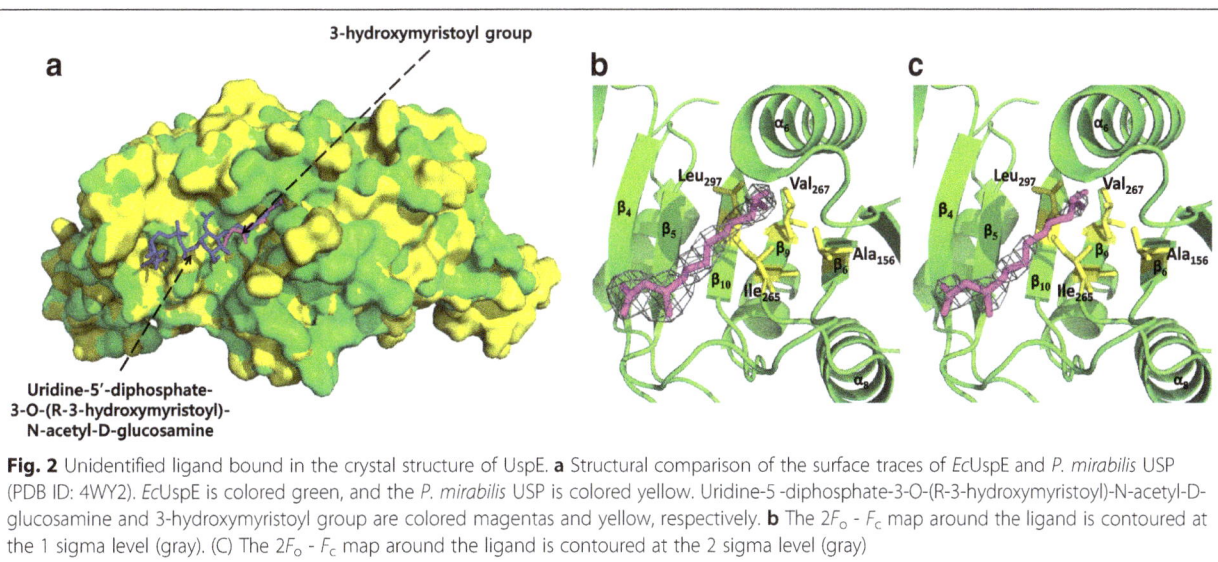

Fig. 2 Unidentified ligand bound in the crystal structure of UspE. **a** Structural comparison of the surface traces of *Ec*UspE and *P. mirabilis* USP (PDB ID: 4WY2). *Ec*UspE is colored green, and the *P. mirabilis* USP is colored yellow. Uridine-5′-diphosphate-3-O-(R-3-hydroxymyristoyl)-N-acetyl-D-glucosamine and 3-hydroxymyristoyl group are colored magentas and yellow, respectively. **b** The $2F_o - F_c$ map around the ligand is contoured at the 1 sigma level (gray). (C) The $2F_o - F_c$ map around the ligand is contoured at the 2 sigma level (gray)

YdaA from *Salmonella enterica* serovar Typhimurium might bind a large, non-polar ligand in its N-terminal domain; however, YdaA was not bound to any ligand in crystal structure [15]. The ligand bound to *P. mirabilis* USP was tentatively identified as UDP-(3-O-(R-3-hydroxymyristoyl))-N-acetylglucosamine. The published crystal structures of the LpxA and LpxC from *E. coli* contain UDP-(3-O-(R-3-hydroxymyristoyl))-N-acetylglucosamine and its deacetylated product, respectively [16, 17]. These proteins catalyze the first committed step of lipid A biosynthesis [18]. Combined with these data, the unidentified ligand bound to *E. coli* UspE is probably related to an intermediate in lipid A biosynthesis, similar to UDP-(3-O-(R-3-hydroxymyristoyl))-N-acetylglucosamine deacetylases.

E. coli UspE has an ATP binding motif and ATPase activity

Previous studies show that USPs can be divided into groups: those that bind ATP (UspFG-type), those that do not bind ATP (UspAs and UspA-like group), and those that hydrolyze adenine nucleotide substrates. USPs that bind ATP may function as an ATP-dependent signaling intermediate in a pathway that promotes persistent infection. Furthermore, it was suggested that USPs contain a conserved-sequence Gly-X_2-Gly-X_9-Gly (Ser/Thr/Asn) motif that is needed for binding ATP [3, 5, 6, 15, 19, 20]. However, the ATP binding motif of UspE was disordered in the crystal, due to the intrinsic flexibility of these regions (Fig. 3a). Therefore, it is likely that the bound ATP is disordered or the protein devoid of bound ATP was preferentially crystallized. In this structure, we found the presence of this motif (Gly^{269}- X_2-Gly^{272}-X_9-Gly^{282}-Asn) in the C-terminal domain of UspE, which is similar to other USPs that bind ATP (Fig. 3b). The *in vitro* ATPase activity of UspE was determined by measuring the amount of ATP left in solution following a kinase reaction using the Kinase-Glo® Luminescent Kinase Assay. As expected, the decline in luminescent signal depended on the increasing concentration of UspE (Fig. 3c). These results indicate that ATPase activity from UspE decreases the remaining ATP levels.

UspE has putative Cd^{2+} binding sites

Cadmium, in a variety of chemical forms, is toxic for the proper growth of microbial cells. Previous studies showed that cadmium (273 μM) can cause complete but transient inhibition of growth accompanied by the synthesis of cadmium-induced proteins (CDPs) [21]. The *E. coli* increase synthesis of CDPs (e.g., H-NS, UspA, UspC, UspD, UspE), which together make up the cadmium stress stimulon [7]. The UspE can sequester Cd from the cytosol to protect themselves. To analyze the relationship between UspE and cadmium, we investigated the cadmium binding ability of UspE by an ITC experiment.

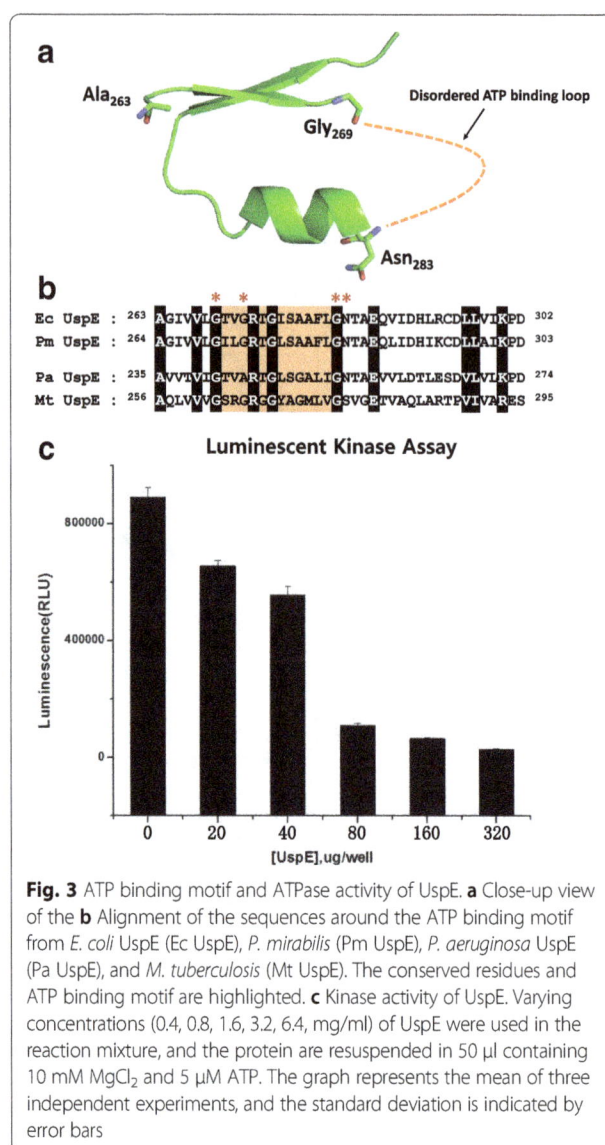

Fig. 3 ATP binding motif and ATPase activity of UspE. **a** Close-up view of the **b** Alignment of the sequences around the ATP binding motif from *E. coli* UspE (Ec UspE), *P. mirabilis* (Pm UspE), *P. aeruginosa* UspE (Pa UspE), and *M. tuberculosis* (Mt UspE). The conserved residues and ATP binding motif are highlighted. **c** Kinase activity of UspE. Varying concentrations (0.4, 0.8, 1.6, 3.2, 6.4, mg/ml) of UspE were used in the reaction mixture, and the protein are resuspended in 50 μl containing 10 mM $MgCl_2$ and 5 μM ATP. The graph represents the mean of three independent experiments, and the standard deviation is indicated by error bars

The ITC experiment was carried out using cadmium as titrant at pH 7.0. Initial attempts to fit the data to a double site-binding model were not successful. The isotherm was best fit when a sequential binding model with three binding sites was applied. Through analysis of these ITC data, we observed tight Cd^{2+} binding to UspE with $n = 3$. This suggests a binding stoichiometry of 3 moles of Cd^{2+} to 1 mole of UspE. The two have moderate binding affinities (K_d of 33.7 and 94.3 μM, respectively), whereas the other one has low affinity (K_d of 242.7 μM). The K_d value of site 1 and site 2 are high compared with site 3. At site 1 and site 2, the binding of Cd^{2+} with UspE is favorable, with an exothermic enthalpy (ΔH of -18.2 and -9.384 kcal mol^{-1}, respectively) and negative entropy (ΔS = -12.67 and -3.25 kcal mol^{-1}, respectively). At site 3, the binding of Cd^{2+} with UspE is unfavorable, with an exothermic enthalpy (ΔH = -8.131 kcal mol^{-1}) and negative entropy

Table 2 ITC experiment Cd^{2+} binding to UspE

	Kd	H	T S	G
	µM	kcal/mol	kcal/mol	kcal/mol
Site 1	94.3	-18.2	-12.67	-5.5
Site 2	33.7	-9.383	-3.25	-6.1
Site 3	242.7	-8.131	-3.19	-5.0

(ΔS = -3.19 kcal mol^{-1}). The detailed thermodynamic parameters are listed in Table 2 and Fig. 4a. Evidence has been reported that to obtain lethal effects in an exponentially growing culture, 600 µM CdCl$_2$ is required. In the lag phase before growth commenced, 3 µM CdCl$_2$ inhibited cell proliferation and 10 µM was lethal [13, 14]. To regulate the concentration of Cd^{2+}, E. coli might induce CDPs such as UspE. These results suggest that apoUspE and Cd^{2+} have a direct relationship. To understand the relationship between apoUspE and Cd^{2+}, we observed the ability of cadmium to interact with UspE through the ITC experiment. Our structural analysis indicates that E. coli has two putative binding sites (Site I: His117, His 119; Site II: His193, His244) (Fig. 4b). Additionally, sequence alignment showed that residues His117, His119, His193, and His244 within the β-barrel domain are highly conserved among the UspE proteins (Fig. 4c). Recently, a study has been performed to observe site I,

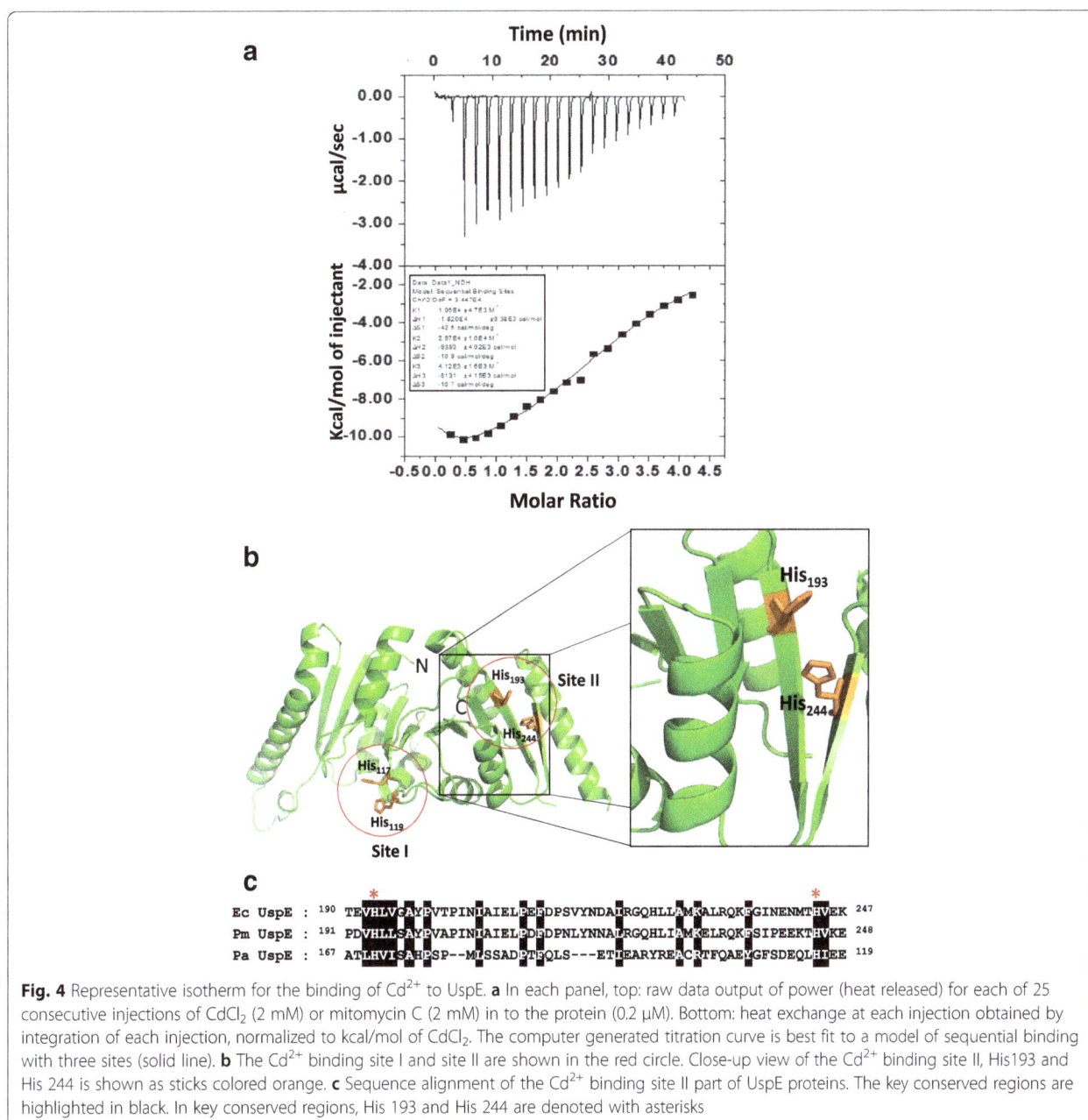

Fig. 4 Representative isotherm for the binding of Cd^{2+} to UspE. **a** In each panel, top: raw data output of power (heat released) for each of 25 consecutive injections of CdCl$_2$ (2 mM) or mitomycin C (2 mM) in to the protein (0.2 µM). Bottom: heat exchange at each injection obtained by integration of each injection, normalized to kcal/mol of CdCl$_2$. The computer generated titration curve is best fit to a model of sequential binding with three sites (solid line). **b** The Cd^{2+} binding site I and site II are shown in the red circle. Close-up view of the Cd^{2+} binding site II, His193 and His 244 is shown as sticks colored orange. **c** Sequence alignment of the Cd^{2+} binding site II part of UspE proteins. The key conserved regions are highlighted in black. In key conserved regions, His 193 and His 244 are denoted with asterisks

which is known to be crucial for zinc binding in the crystal structure of YdaA from *S. enterica* serovar Typhimurim (PDB ID: 4R2J) [15]. This provides further support for the original conclusion that cadmium binds at two locations. Our results clearly demonstrate that *E. coli* UspE has two different sets of binding sites and the protein may provide additional confirmation for the cadmium binding to these two sites.

In summary, the crystal structure of UspE from *E. coli* is representative of a tandem-type USP. The crystal structure of *E. coli* UspE reveals a hydrophobic pocket that moderately binds an unidentified ligand. Combined with previous studies, we can conclude that UspE is probably related to an intermediate in lipid A biosynthesis. We subsequently found through the sequence analysis that UspE has an ATP binding motif (Gly^{269}-X_2-Gly^{272}-X_9-Gly^{282}-Asn) in the C-terminal domain and has ATPase activity, though this did not appear in the crystal structure. We were also able to perform an ITC experiment which revealed that UspE probably has two Cd^{2+} binding sites and that the His117, His119, His193, and His244 residues within the β-barrel domain are critical for binding Cd^{2+}. We believe that this information is a significant contribution to understanding the molecular mechanisms of *E. coli* UspE.

Conclusions

In this study, we have determined the crystal structure of UspE of *E. coli* as a representative of a tandem-type USP. The UspE consists of two tandem USP domains that are highly conserved in this protein family. We found a hydrophobic pocket in the UspE structure, which was strongly bound to unidentified ligand. Combined with a previous study, evidence suggests that the UspE is related to an intermediate in lipid A biosynthesis. We subsequently found that sequence analysis suggests that UspE has an ATP binding motif (Gly^{269}-X_2-Gly^{272}-X_9-Gly^{282}-Asn) in the C-terminal domain of UspE and has ATPase activity, but this was not confirmed by the crystal structure. We were also able to perform the ITC experiment, which revealed that the UspE probably has two Cd^{2+} binding sites, comprised of the His117, His 119, His193, and His244 residues within the β-barrel domain. Both of them are essential for Cd^{2+} binding to UspE protein. As discussed before, USPs might be associated with several functions, such as cadmium binding, ATPase activity, and an intermediate in lipid A biosynthesis.

Abbreviations

E. coli: *Escherichia coli*; ITC: isothermal titration calorimetry; MR: molecular replacement; *P. mirabilis*: *Proteus mirabilis*; PDB: protein data bank; RLU: relative light unit; RMSD: root-mean-square deviation; *S. enterica* serovar Typhimurim: *Salmonella enterica* serovar Typhimurium; USP: universal stress protein.

Competing interests
The authors declare that they have no competing interests.

Authors' contributions
JG, XJ, and YJ carried out the protein expression, purification, crystallization, and biochemical experiments. YX and JK performed structural modeling and analyzed data. YX, SF, NH, and CQ helped write the manuscript. All authors read and approved the final manuscript.

Acknowledgements
We gratefully acknowledge the access to the beamline 5C at Pohang Light Source (PLS) (Pohang, South Korea). This study was supported by the National Natural Science Foundation of China (Grant No. 31200556 to Y. Xu, Grant No.31301447 to Y. Ji, and Grant No. 21172028 to Shengdi Fan), the China Postdoctoral Science Foundation (Grant No. 2013 M540229 to Y. Xu), and the Fundamental Research Funds for the Central Universities (Grant No. DC201502020203 to Y. Xu).

Author details
[1]Department of Bioengineering, College of Life Science, Dalian Nationalities University, Dalian 116600Liaoning, China. [2]Laboratory of Biomedical Material Engineering, Dalian Institute of Chemical Physics, Chinese Academy of Sciences, Dalian 116023Liaoning, China. [3]Department of Agricultural Biotechnology, College of Agriculture and Life Sciences, Seoul National University, Gwanak-gu, Seoul 151-742, Republic of Korea.

References
1. Tkaczuk KL, Shumilin IA, Chruszcz M, Evdokimova E, Savchenko A, Minor W. Structural and functional insight into the universal stress protein family. Evol Appl. 2013;6(3):434–49.
2. Gustavsson N, Diez A, Nystrom T. The universal stress protein paralogues of Escherichia coli are co-ordinately regulated and co-operate in the defence against DNA damage. Mol Microbiol. 2002;43(1):107–17.
3. Kvint K, Nachin L, Diez A, Nystrom T. The bacterial universal stress protein: function and regulation. Curr Opin Microbiol. 2003;6(2):140–5.
4. Nachin L, Nannmark U, Nystrom T. Differential roles of the universal stress proteins of Escherichia coli in oxidative stress resistance, adhesion, and motility. J Bacteriol. 2005;187(18):6265–72.
5. Sousa MC, McKay DB. Structure of the universal stress protein of Haemophilus influenzae. Structure. 2001;9(12):1135–41.
6. Zarembinski TI, Hung LW, Mueller-Dieckmann HJ, Kim KK, Yokota H, Kim R, et al. Structure-based assignment of the biochemical function of a hypothetical protein: a test case of structural genomics. Proc Natl Acad Sci U S A. 1998;95(26):15189–93.
7. Siegele DA. Universal stress proteins in Escherichia coli. J Bacteriol. 2005;187(18):6253–4.
8. Xu Y, Quan CS, Jin X, Jin X, Zhao J, Li X, et al. Crystallization and preliminary X-ray diffraction analysis of UspE from Escherichia coli. Acta Crystallogr F Struct Biol Commun. 2014;70(Pt 12):1640–2.
9. McCoy AJ, Grosse-Kunstleve RW, Adams PD, Winn MD, Storoni LC, Read RJ. Phaser crystallographic software. J Appl Crystallogr. 2007 Aug 1; 40(Pt 4): 658–74.
10. Adams PD, Afonine PV, Bunkoczi G, Chen VB, Davis IW, Echols N, et al. PHENIX: a comprehensive Python-based system for macromolecular structure solution. Acta Crystallogr D Biol Crystallogr. 2010;66(Pt 2):213–21.
11. Emsley P, Cowtan K. Coot: model-building tools for molecular graphics. Acta Crystallogr D Biol Crystallogr. 2004;60(Pt 12 Pt 1):2126–32.
12. Berman HM, Westbrook J, Feng Z, Gilliland G, Bhat TN, Weissig H, et al. The protein data bank. Nucleic Acids Res. 2000;28(1):235–42.
13. Lo YC, Lin SC, Shaw JF, Liaw YC. Crystal structure of Escherichia coli thioesterase I/protease I/lysophospholipase L1: consensus sequence blocks constitute the catalytic center of SGNH-hydrolases through a conserved hydrogen bond network. J Mol Biol. 2003;330(3):539–51.
14. Lo YC, Lin SC, Shaw JF, Liaw YC. Substrate specificities of Escherichia coli thioesterase I/protease I/lysophospholipase L1 are governed by its switch loop movement. Biochemistry. 2005;44(6):1971–9.
15. Bangera M, Panigrahi R, Sagurthi SR, Savithri HS, Murthy MR. Structural and functional analysis of two universal stress proteins YdaA and YnaF from

Salmonella typhimurium: possible roles in microbial stress tolerance. J Struct Biol. 2015;189(3):238–50.

16. Williams AH, Raetz CR. Structural basis for the acyl chain selectivity and mechanism of UDP-N-acetylglucosamine acyltransferase. Proc Natl Acad Sci U S A. 2007;104(34):13543–50.

17. Clayton GM, Klein DJ, Rickert KW, Patel SB, Kornienko M, Zugay-Murphy J, et al. Structure of the bacterial deacetylase LpxC bound to the nucleotide reaction product reveals mechanisms of oxyanion stabilization and proton transfer. J Biol Chem. 2013;288(47):34073–80.

18. Gao N, McLeod SM, Hajec L, Olivier NB, Lahiri SD, Bryan Prince D, et al. Overexpression of Pseudomonas aeruginosa LpxC with its inhibitors in an acrB-deficient Escherichia coli strain. Protein Expr Purif. 2014;104C:57–64.

19. Weber A, Jung K. Biochemical properties of UspG, a universal stress protein of Escherichia coli. Biochemistry. 2006;45(6):1620–8.

20. Saveanu C, Miron S, Borza T, Craescu CT, Labesse G, Gagyi C, et al. Structural and nucleotide-binding properties of YajQ and YnaF, two Escherichia coli proteins of unknown function. Protein Sci. 2002;11(11):2551–60.

21. Ferianc P, Farewell A, Nystrom T. The cadmium-stress stimulon of Escherichia coli K-12. Microbiology. 1998;144(Pt 4):1045–50.

18

REFOLDdb: a new and sustainable gateway to experimental protocols for protein refolding

Hisashi Mizutani[1], Hideaki Sugawara[1*] (ORCID), Ashley M. Buckle[2], Takeshi Sangawa[3], Ken-ichi Miyazono[4], Jun Ohtsuka[4], Koji Nagata[4], Tomoki Shojima[4], Shohei Nosaki[4], Yuqun Xu[4], Delong Wang[4], Xiao Hu[4], Masaru Tanokura[4] and Kei Yura[1,5,6,7]

Abstract

Background: More than 7000 papers related to "protein refolding" have been published to date, with approximately 300 reports each year during the last decade. Whilst some of these papers provide experimental protocols for protein refolding, a survey in the structural life science communities showed a necessity for a comprehensive database for refolding techniques. We therefore have developed a new resource – "REFOLDdb" that collects refolding techniques into a single, searchable repository to help researchers develop refolding protocols for proteins of interest.

Results: We based our resource on the existing REFOLD database, which has not been updated since 2009. We redesigned the data format to be more concise, allowing consistent representations among data entries compared with the original REFOLD database. The remodeled data architecture enhances the search efficiency and improves the sustainability of the database. After an exhaustive literature search we added experimental refolding protocols from reports published 2009 to early 2017. In addition to this new data, we fully converted and integrated existing REFOLD data into our new resource. REFOLDdb contains 1877 entries as of March 17th, 2017, and is freely available at http://p4d-info.nig.ac.jp/refolddb/.

Conclusion: REFOLDdb is a unique database for the life sciences research community, providing annotated information for designing new refolding protocols and customizing existing methodologies. We envisage that this resource will find wide utility across broad disciplines that rely on the production of pure, active, recombinant proteins. Furthermore, the database also provides a useful overview of the recent trends and statistics in refolding technology development.

Keywords: Solubilization, Inclusion body, Refolding, Renaturation, Crystallization

Background

Establishment of heterologous expression technology of recombinant proteins has revolutionized protein purification such that it is performed with cloned, recombinant proteins expressed in a suitable host. The predominant host is *Escherichia coli*. However, many overexpressed proteins in *E. coli* are found in an insoluble form called inclusion bodies (IBs). Since the target protein is often highly pure in washed IBs, the challenge is not so much to purify the target, but rather to solubilize IBs and refold the protein into its native, biologically active state [1, 2]. While many of the operations to prepare IBs are quite general—expression, cell disruption, IB isolation and washing, the precise conditions that are required to achieve efficient refolding vary for each protein.

The refolding experiments consist of two steps: (1) the solubilization of IBs by adding a denaturant and (2) the renaturation of the denatured protein by lowering the denaturant concentration. The solubilization step is relatively easily, performed by adding a denaturant, typically urea or guanidinium chloride at a final concentration of 6–8 M or 6 M, respectively. The renaturation step is often difficult. In order to maximize the refolding yield,

* Correspondence: hsugawar@nig.ac.jp
[1]Center for Information Biology, National Institute of Genetics, 1111 Yata Mishima, Shizuoka 411-8540, Japan
Full list of author information is available at the end of the article

the optimization of the following experimental methods/conditions of this process is required:

(a) refolding method: dilution and dialysis [3], gel filtration column chromatography, column adsorption and desorption [4], and high pressure [5] represent the most common methods. These methods are used to lower the denaturant concentration and allow protein refolding in an aqueous buffer.

(b) pH: In general, pI should not be used for refolding experiment to avoid isoelectric point precipitation [6].

(c) temperature: Temperature has an effect on the stability and mobility of the refolding intermediates [7].

(d) protein concentration: Protein concentration determines the degree of "crowding" and thus the frequency of molecular collisions between the unfolded molecules as well as folding intermediates, which can promote aggregation [8].

(e) additive(s): Some compounds may stabilize the refolding intermediates and avoid aggregation [9].

Because the suitable refolding methods/conditions differ from protein to protein, a knowledge database of optimized refolding methods/conditions for each protein is an important resource for many biochemists and molecular biologists. Thus, the REFOLD database established and published by Monash University in 2006 played an important role as the sole information source for refolding experiments [10–13]. This database, however, suspended its updates in 2009. We carried out a preliminary study in 2013 for the development of a sustainable database on protein refolding technologies. We decided that a new database, REFOLDdb, was required as a gateway to experimental methodologies that describe experimental refolding in detail. We therefore designed a simple data format and consistent data representation among entries so that users are able to easily interrogate the database and painlessly retrieve and understand search results. The design also allows straightforward maintenance, allowing the database to be sustainable over a long period.

The sustainability of biological databases is a serious issue [14, 15] and database developers have to analyze cost-effectiveness in advance. In the case of databases relating to technologies (Tech_db), the data volume will not expand as rapidly as in the case of molecular databases, for example the International Sequence Database [16], the Worldwide Protein Data Bank [17], UniProt [18], SUPERFAMILY database [19]. Nevertheless, developer of Tech_db must be sensitive to the direct and indirect cost of data extraction from the primary articles, curation and updating. REFOLDdb is designed to balance both cost and usefulness.

We have captured the refolding data from up-to-date literature as well as retrospectively from articles published since 2009. We also updated, converted and integrated the data stored in the REFOLD database into REFOLDdb. As of March 17th, 2017, REFOLDdb provides users with data on 1877 experimental methods for refolding 1628 proteins. Most of these data were extracted from 1232 publications.

Construction and content

We searched the NCBI PubMed database by a keyword search of "(refolding[All Fields] OR renaturation[All Fields]) AND ("proteins"[MeSH Terms] OR "proteins"[All Fields] OR "protein"[All Fields])" to find 2606 research reports published between 2009-early 2017 that might be relevant to REFOLDdb. Manual inspection of the results identified 420 reports that contained experimental protocols for the refolding of 650 proteins. These data were then integrated in REFOLDdb along with the data stored in the REFOLD database. REFOLDdb refers to 1232 publications in total (Full list available via a menu "List of publications referred by REFOLDdb" in "About" page at http://p4d-info.nig.ac.jp/refolddb/about.cgi?lang=EN).

Due to the standardization and other extension of the data format, the database now contains the following functionality: (1) it is searchable by sequence similarity; (2) it is equipped with statistics that enables the discovery of trends in refolding techniques; and (3) it is easy to upload/submit new data to the database manager. Specifically, the database has the following three sections: Article [title/abstract/PubMed ID/Author/Journal/Date], Protein [Protein name/Amino acid sequence/Comment/UniProt ID/Function/Domain] and Experiment [Refolding methods/pH/Temperature/Validation]. We did not itemize "protein concentration" and "additive(s)", because "protein concentration" is often missing in articles and the description of "additive(s)" is quite heterogeneous. REFOLDdb is composed of 12 tables in a relational database system.

REFOLDdb was created using open-source PostgreSQL relational database server software version 9.2.14 (https://www.postgresql.org/), running under CentOS 7 Server (version 7.2-1511) on a virtual machine based on VMware ESXi (http://www.vmware-e.com/products/esxi-and-esx.html). The system complies with the security policy of the National Institute of Genetics, Japan. A web-based query interface to the database was developed using the Perl programming language and PDO database abstraction classes (http://jp2.php.net/manual/en/book.pdo.php), and is hosted on the same virtual machine running the Apache 2.4.6 web server.

Fig. 1 The horizontal menu bar of REFOLDdb

Utility and discussion

The top page of REFOLDdb is composed of (a) a horizontal bar menu and (b) a large main search window.

(a) The horizontal bar menu

The menu includes 7 icons, "REFOLDdb", "About", "Statistics", "Blast Search", "Help", "Download" and "REFOLD", corresponding to options of database operations and 2 icons for language selection as shown in Fig. 1.

- "REFOLDdb" at the left end of the menu is a back button for the REFOLDdb user to return to the top page after several operations. "About" refers to a brief introduction of REFOLDdb and the REFOLD database. "Statistics" introduces the anatomy of REFOLDdb by graphically displaying the numbers of experiments by journals, refolding methods developed/used, pH, temperature, protein size, methods for validation, and also the number of refolding experiments by year. In addition, a "Statistics" page provides a search function, which is explained below.

- "Blast Search" page accepts a an amino acid sequence (AAseq) in FASTA format to search for "similar" proteins that were successfully refolded. A sub-menu "set sample" placed just above the blast search window toggles short, medium and long AAseqs for a quick trial (Fig. 2a). Figure 2b introduces the blast search result in a table format by choosing the medium length AAseq in the

Fig. 2 Results of "Blast Search" by one of the three sample amino acids sequences (the middle length). **a**) The blast search window with a sample AAseq. **b**) The result in a table format and the alignment of AAseqs of the query and the top hit in the table. **c**) Details of the top hit

sample set. By clicking "Blast Results" in the table, the alignment between the query sequence and AAseqs in the database is displayed as shown by the box that overlaps the table. "Detail" button navigates the user to the full record as shown in Fig. 2c. The full record includes a link to the corresponding record
in the REFOLD database, if available.

- "HELP", "DOWNLOAD", and "REFOLD" allow: browsing a compact manual for the utilization of REFOLDdb featuring screen captures, downloading the data contents of REFOLDdb in a tab-separated values (TSV) file, and accessing the previous REFOLD database respectively.

A "Statistics" page provides the user with an overview on REFOLDdb records and also a search interface. The pie and bar charts are clickable to retrieve the relevant data entries from REFOLDdb. The histogram of refolding methods developed/used is exemplified in Fig. 3a. It is obvious in the bar chart that "dilution" is the most popular methods and "high_pressure" is rarely used. It is straightforward to become familiar with

"high_pressure" method by clicking the bar in Fig. 3a. The data entries that contribute to the bar are displayed in a sortable table. The top 5 ~ 65 records in the table are shown in Fig. 3b. The user is able to directly reach the full description of the method in the database using the "Detail" button in the table and then the original articles, e.g. "A class-A GPCR solubilized under high hydrostatic pressure retains its ligand binding ability" [20].

"Statistics" on experimental conditions such as pH (Fig. 4) and temperature (Fig. 5) might be useful for protein crystallographers: the histogram in Fig. 4 suggests that protein refolding experiments are most successful in a pH range of 7 to 10 regardless of other factors; the histogram in Fig. 5 shows that protein refolding experiments have been mainly performed at two temperature ranges of 0.0–4.9° Celsius (~55%) and 25.0–29.9° Celsius (~23%). We envisage that the database may allow the identification of certain refolding conditions, such as low or high temperature, which may aid downstream crystallization attempts.

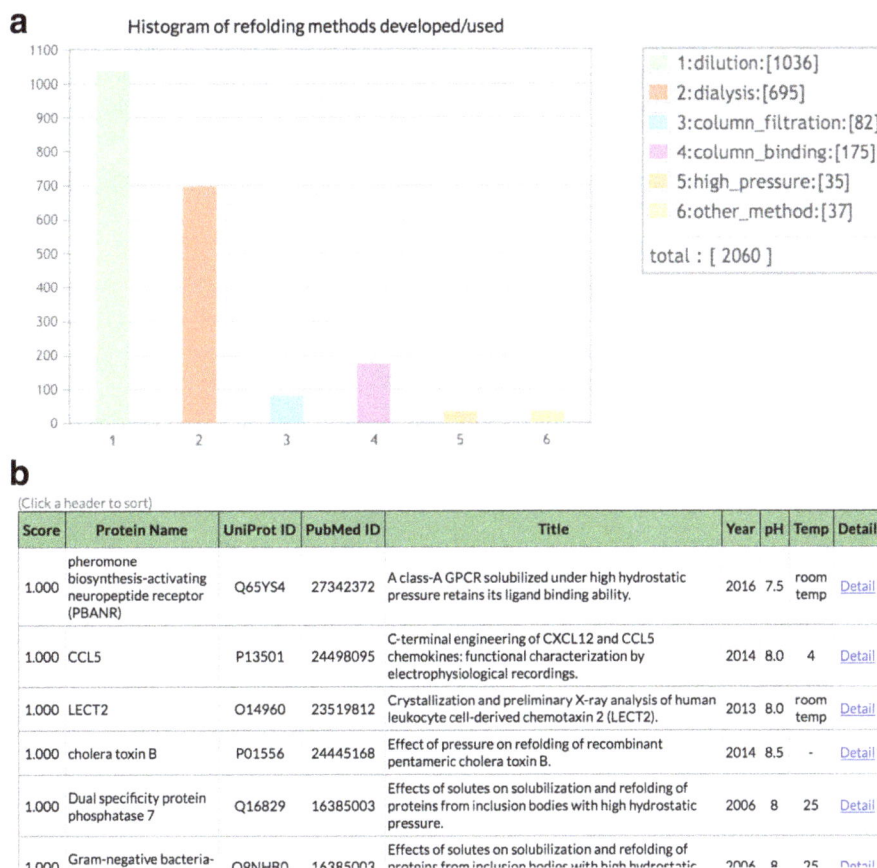

Fig. 3 Statistics for data retrieval. **a**) Histogram of refolding methods developed/used. **b**) A part of 35 REFOLDdb entries that compose the "high-pressure" bar in the Fig. 3a

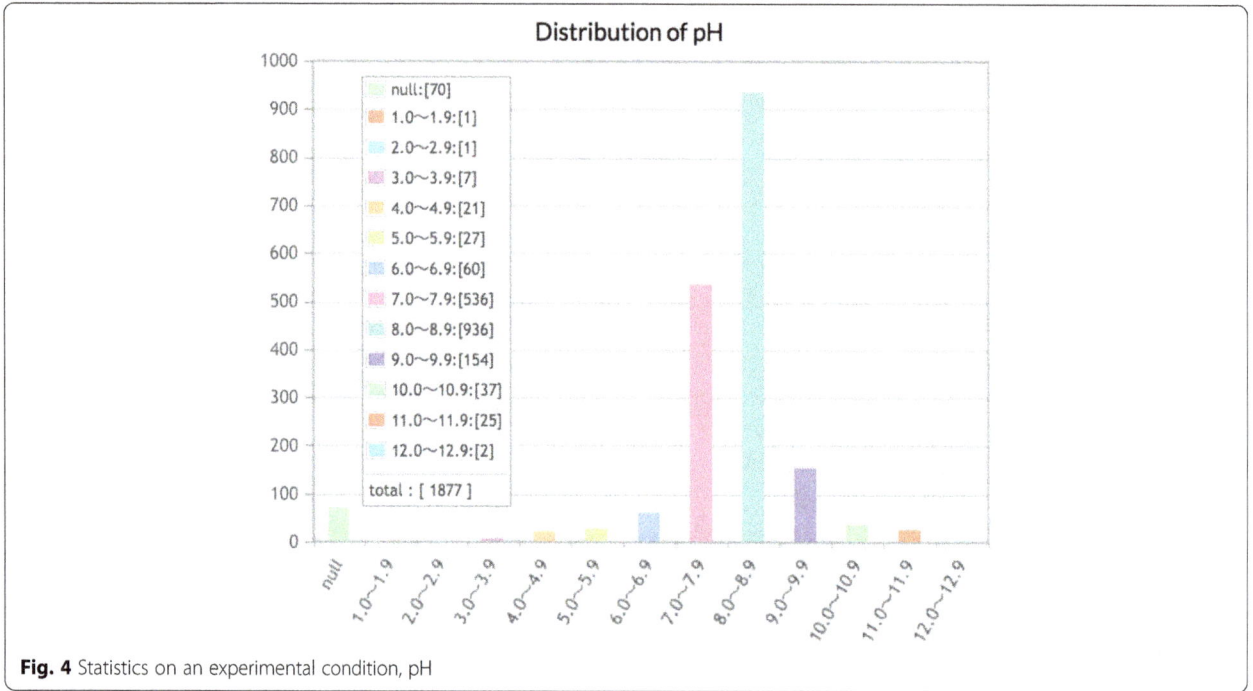

Fig. 4 Statistics on an experimental condition, pH

"Statistics" on properties of proteins could also be informative for optimal experimental design, e.g. disulfide bonds, domains, isoelectric point, metal ions, size and species. Performing this analysis shows that refolding techniques are available for a protein size range comparable to that in the PDB (Fig. 6) [21]. Both graphs imply that proteins of 100–400

amino acids are frequently analyzed. It is to be noted that REFOLDdb contains a diverse cross-section of protein architectures, including extracellular domains, subunits, whole proteins and multiple proteins. It is possible in theory to transform statistics to an inference engine based on AAseqs. However, it requires stringent cleansing of AAseqs in research articles and databases

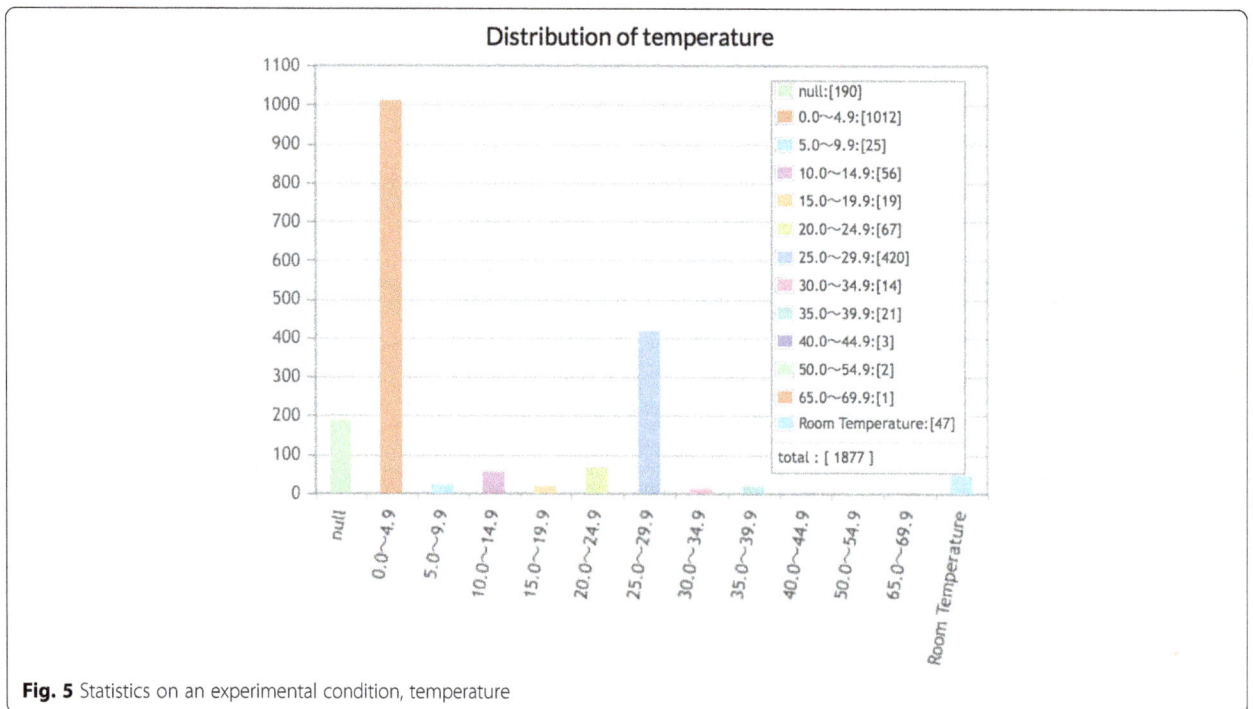

Fig. 5 Statistics on an experimental condition, temperature

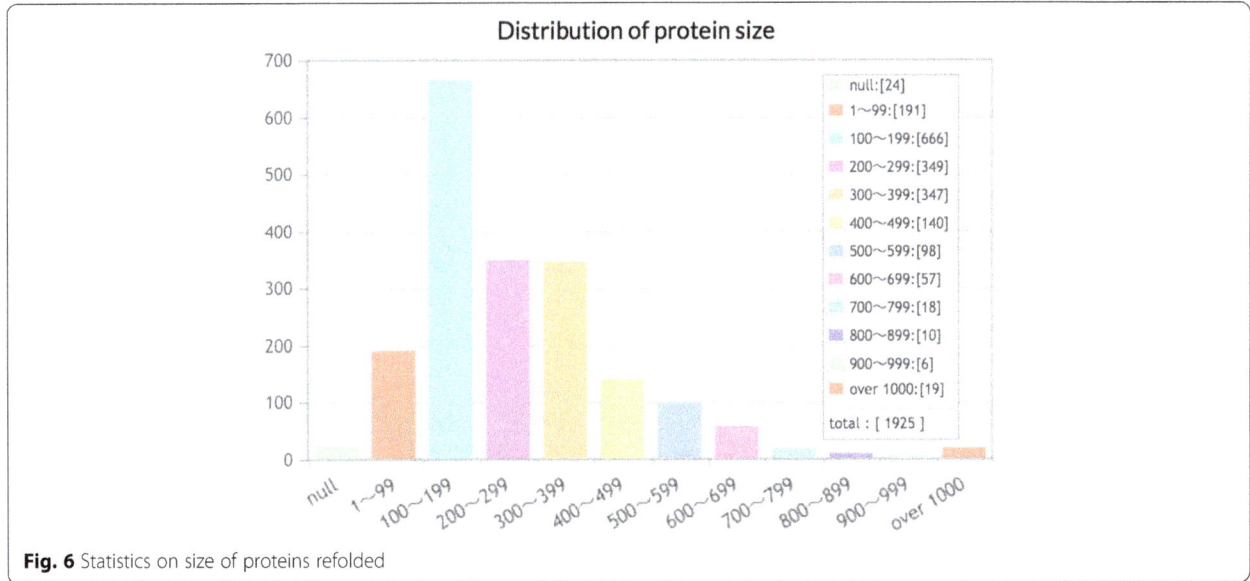

Fig. 6 Statistics on size of proteins refolded

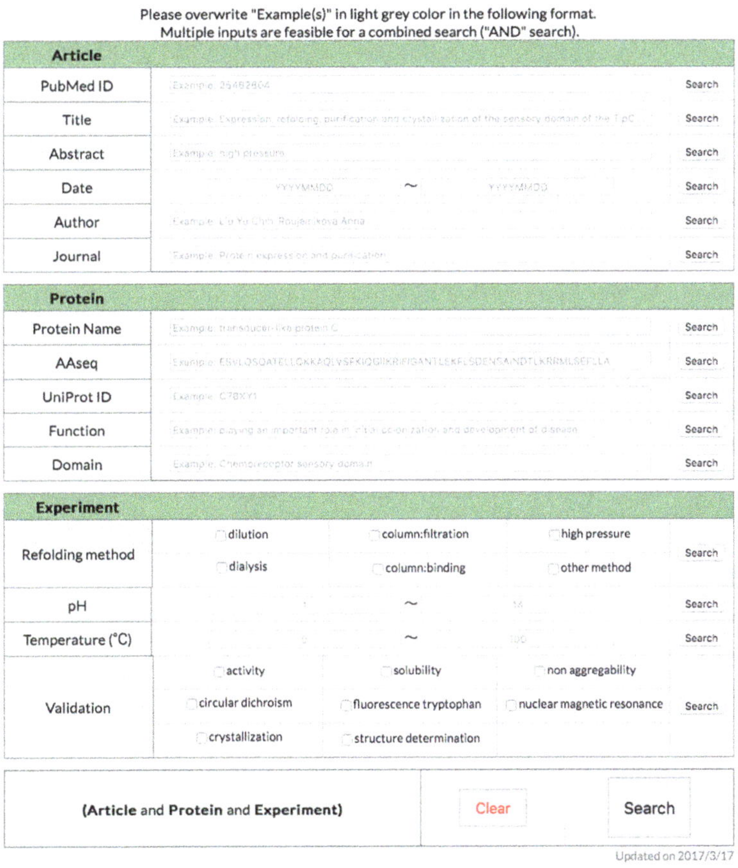

Fig. 7 The main window for searching REFOLDdb

that often implicitly include tags and linkers. It is also a difficult task to collect negative data that are prerequisite for the development of a reliable inference engine. Nevertheless, REFOLDdb is a good starting point for data mining in order to customize experimental conditions for a given protein in the future.

(b) The main search window (Fig. 7).

The window is located just under the horizontal bar menu. A combined search of REFOLDdb can be carried out in the following two steps:

1) Overwrite "Example(s)" in light grey color and/or check boxes as many as needed. In the case of the data items of "pH" and "Temperature", lower limit and/or upper limit can be specified.

2) Click a "Search" button in the line of a data item to go through the specified data item, or click the "Search" button at the bottom-right corner of the search window to perform a combined search, namely, "AND" search of multiple data items.

Multiple hits to a query will be displayed in a table format that is composed of sortable columns of "Protein Name", "UniProt ID", "PubMed ID"/"Title"/ "Year" of the publication, "pH", "Temp(erature)". "Detail" buttons in the table navigates the user to detailed information on proteins and experimental conditions.

Conclusions

The resources, including human resources, required for running and updating REFOLDdb is kept to a minimum. A team of one annotator who is knowledgeable about structural biology and a part time system engineer will be able to keep REFOLDdb up-to-date as far as collecting data from research papers on a monthly basis. The database system based on the virtual machine is almost autonomous and also flexible enough to allow future expansion.

In the future, we will evaluate new data sources other than research articles, such as patents, that might make the database more comprehensive. In addition, we will investigate the implementation of data-mining functionality to allow the prediction of suitable refolding methods based on chemical, physical and/or genetic features of proteins that have been successfully refolded.

Abbreviations
REFOLD database: The database on refolding technologies developed by Monash University; REFOLDdb: The database on refolding technologies developed by the authors; Tech_db: Databases on technologies

Acknowledgements
The authors are grateful to Professor Junichi Takagi (Institute for Protein Research, Osaka University) for his suggestions on the needs of the research communities to databases in structural life sciences.

Funding
This work was supported by the 'Platform for Drug Discovery, Informatics, and Structural Life Science' grant from the Ministry of Education, Culture, Sports, Science and Technology of Japan (MEXT) and the Japan Agency for Medical Research and Development (AMED). The funding body did not play any role in the design or conclusion of the study. Funding for open access charge: Waseda University, Japan.

Authors' contributions
HM, HS, TS, JO, KN and MT designed the REFOLD db. HM, KM, JO, TS, SN, YX, DW and HU contributed to the production of the data set based on research papers on refolding technologies. AB contributed to the conversion of the REFOLD database to REFOLD db. HS, KN, AB and KY wrote the manuscript. All authors read and approved the final manuscript.

Authors' information
Not applicable.

Competing interests
Not applicable.

Author details
[1]Center for Information Biology, National Institute of Genetics, 1111 Yata Mishima, Shizuoka 411-8540, Japan. [2]The Department of Biochemistry and Molecular Biology, Biomedicine Discovery Institute, Monash University, Clayton, VIC 3800, Australia. [3]Laboratory of Protein Synthesis and Expression, Institute for Protein Research, Osaka University, Suita, Osaka 565-0871, Japan. [4]Laboratory of Structural Biology and Food Biotechnology, Department of Applied Biological Chemistry, Graduate School of Agricultural and Life Sciences, The University of Tokyo, 1-1-1 Yayoi, Bunkyo-ku, Tokyo 113-8657, Japan. [5]Graduate School of Humanities and Sciences, Ochanomizu University, 2-1-1 Otsuka, Bunkyo, Tokyo 112-8610, Japan. [6]Center for Simulation Science and Informational Biology, Ochanomizu University, 2-1-1 Otsuka, Bunkyo, Tokyo 112-8610, Japan. [7]School of Advanced Science and Engineering, Waseda University, 3-4-1 Okubo, Shinjyuku, Tokyo 169-8555, Japan.

References
1.   Singh A, Upadhyay V, Upadhyay AK, Singh SM, Panda AK. Protein recovery from inclusion bodies of Escherichia coli using mild solubilization process. Microb Cell Fact. 2015;14:41.
2.   Eiberle MK, Jungbauer A. Technical refolding of proteins: Do we have freedom to operate? Biotechnol J. 2010;5(6):547–59.
3.   Yamaguchi H, Miyazaki M. Refolding techniques for recovering biologically active recombinant proteins from inclusion bodies. Biomolecules. 2014;4(1):235–51.
4.   Li M, Su ZG, Janson JC. In vitro protein refolding by chromatographic procedures. Protein Expr Purif. 2004;33(1):1–10.
5.   Okai M, Ohtsuka J, Asano A, Guo L, Miyakawa T, Miyazono K, Nakamura A, Okada A, Zheng H, Kimura K, Nagata K, Tanokura M. High pressure refolding, purification, and crystallization of flavin reductase from Sulfolobus tokodaii strain 7. Protein Expr Purif. 2012;84(2):214–8.
6.   Coutard B, Danchin EG, Oubelaid R, Canard B, Bignon C. Single pH buffer refolding screen for protein from inclusion bodies. Protein Expr Purif. 2012;82(2):352–9.
7.   Xie Y, Wetlaufer DB. Control of aggregation in protein refolding: the temperature-leap tactic. Protein Sci. 1996;5(3):517–23.
8.   Gupta P, Hall CK, Voegler AC. Effect of denaturant and protein concentrations upon protein refolding and aggregation: a simple lattice model. Protein Sci. 1998;7(12):2642–52.
9.   Yamaguchi S, Yamamoto E, Mannen T, Nagamune T, Nagamune T. Protein refolding using chemical refolding additives. Biotechnol J. 2013;8(1):17–31.
10. Phan J, Yamout N, Schmidberger J, Bottomley SP, Buckle AM. Refolding your protein with a little help from REFOLD. Methods Mol Biol. 2011;752:45–57.
11. Buckle AM, Devlin GL, Jodun RA, Fulton KF, Faux N, Whisstock JC, Bottomley SP. The matrix refolded. Nat Methods. 2005;2(1):3.

12. Chow MK, Amin AA, Fulton KF, Fernando T, Kamau L, Batty C, Louca L, Ho S, Whisstock JC, Bottomley SP, Buckle AM. The REFOLD database: a tool for the optimization of protein expression and refolding. Nucleic Acids Res. 2006;34(Database issue):D207–12.

13. Chow MK, Amin AA, Fulton KF, Whisstock JC, Buckle AM, Bottomley SP. REFOLD: an analytical database of protein refolding methods. Protein Expr Purif. 2006;46(1):166–71. Epub 2005 Aug 15.

14. Editorial. Database under maintenance. Nature Methods. 2016;13(9):699

15. Oliver SG, Lock A, Harris MA, Nurse P, Wood V. Model organism databases: essential resources that need the support of both funders and users. BMC Biology. 2016;14:49.

16. International Nucleotide Sequence Database. http://www.insdc.org/ Accessed 2 Dec 2016.

17. Worldwide Protein Data Bank (wwPDB). http://www.wwpdb.org/ Accessed 2 Dec 2016.

18. UniProt. http://www.uniprot.org/ Accessed 2 Dec 2016.

19. Oates ME, Stahlhacke J, Vavoulis DV, Smithers B, Rackham OJ, Sardar AJ, Zaucha J, Thurlby N, Fang H, Gough J. The SUPERFAMILY 1.75 database in 2014: a doubling of data. Nucleic Acids Res. 2015;43(Database issue):D227–33.

20. Katayama Y, Suzuki T, Ebisawa T, Ohtsuka J, Wang S, Natsume R, Lo YH, Senda T, Nagamine T, Hull JJ, Matsumoto S, Nagasawa H, Nagata K, Tanokura T. A class-A GPCR solubilized under high hydrostatic pressure retains its ligand binding ability. Biochim Biophys Acta. 2016;1858(9):2145–51.

21. RCSB Protein Data Bank. Residue Count Histogram http://www.rcsb.org/pdb/static.do?p=general_information/pdb_statistics/index.html Accessed 28 Feb 2017.

19

Structuprint: a scalable and extensible tool for two-dimensional representation of protein surfaces

Dimitrios Georgios Kontopoulos[1]* ⓘ, Dimitrios Vlachakis[2]*, Georgia Tsiliki[3] and Sofia Kossida[4]

Abstract

Background: The term 'molecular cartography' encompasses a family of computational methods for two-dimensional transformation of protein structures and analysis of their physicochemical properties. The underlying algorithms comprise multiple manual steps, whereas the few existing implementations typically restrict the user to a very limited set of molecular descriptors.

Results: We present Structuprint, a free standalone software that fully automates the rendering of protein surface maps, given - at the very least - a directory with a PDB file and an amino acid property. The tool comes with a default database of 328 descriptors, which can be extended or substituted by user-provided ones. The core algorithm comprises the generation of a mould of the protein surface, which is subsequently converted to a sphere and mapped to two dimensions, using the Miller cylindrical projection. Structuprint is partly optimized for multicore computers, making the rendering of animations of entire molecular dynamics simulations feasible.

Conclusions: Structuprint is an efficient application, implementing a molecular cartography algorithm for protein surfaces. According to the results of a benchmark, its memory requirements and execution time are reasonable, allowing it to run even on low-end personal computers. We believe that it will be of use - primarily but not exclusively - to structural biologists and computational biochemists.

Keywords: Molecular cartography, Protein surfaces, Visualization, Surface comparison, Structural biology

Background

Over the last two decades, the growth rate of the Protein Data Bank has been exponential. As structural data for biomolecules are increasingly made available, the study of homologous proteins can be performed not only at the level of sequence, but also at the level of three-dimensional structure. This has led to the development of numerous sophisticated methods, concerning, among others, the analysis of structural evolution [1] and the structure-based design of new drugs [2].

For the comparison of protein surfaces in particular, a family of methods is based on the reduction of the dimensionality of the system. The concept of projecting a three-dimensional protein structure to two dimensions was first introduced by Fanning et al. under the term 'molecular cartography' [3]. They presented this notion as a novel method for studying the entire surface of a protein, emphasizing on the topography of antigenic sites. It involved conversion of the protein structure into a triaxial ellipsoid, followed by its transformation into a graticule (a latitude/longitude grid). Pawłowski and Godzik later expanded on this approach by annotating protein surface maps according to the physicochemical properties of the exposed residues (e.g., charge or hydrophobicity), as a means to compare evolutionarily related proteins [4].

Even though a number of modifications to the aforementioned methodologies for two-dimensional protein representation have been proposed [5–7], molecular cartography has not found much use in the literature. This may be partly due to the significant amount of effort that is required to manually convert the atomic coordinates of a PDB file first into a spherical structure and then into a

* Correspondence: d.kontopoulos13@imperial.ac.uk; dvlachakis@bioacademy.gr
[1]Department of Life Sciences, Imperial College London, Silwood Park Campus, Ascot, UK
[2]Bioinformatics & Medical Informatics Team, Biomedical Research Foundation, Academy of Athens, Athens, Greece
Full list of author information is available at the end of the article

map. Visualizing the distribution of a particular physico-chemical property on the surface further increases the complexity and the overall approach becomes increasingly tedious. A few applications that implement molecular cartography algorithms are available (SURF'S UP! [8], PST [9], Udock [10]), but the range of supported physicochemical descriptors for visualization is typically limited to charge and hydrophobicity. Integrating other predictors is either unfeasible or not straightforward for the end user, creating an obstacle for specialized analyses. Moreover, an application that harnesses the power of multiprocessor systems to simultaneously render multiple protein surface maps is not to this day available. This would be very useful, for example, when visualizing entire molecular dynamics simulations or comparing the members of a large protein family.

To fill these gaps, we introduce Structuprint, a new tool for visualization of protein surfaces in two dimensions. Its name is a combination of the terms 'structure' and 'fingerprint', alluding to the fingerprint-like figures that it generates (see Fig. 1 for an example). Structuprint can produce single 2D maps starting from a PDB file, or GIF animations from multiple files. It is designed with a focus on scalability and extensibility. The tool can utilize

multiple CPU cores on GNU/Linux and OS X machines and can easily incorporate any physicochemical predictors provided by the user, other than those in its own default set. The following sections describe the design choices behind its algorithm, present the results from a benchmark and show three characteristic examples of use.

Implementation
Amino acid properties database
Values for 328 properties/descriptors were calculated for the 20 common amino acids with MOE 2010.10 [11] and were stored within an SQLite database. In particular, the database contains 11 categories of descriptors: i) 33 adjacency and distance matrix descriptors [12–16] (e.g., Balaban's connectivity topological index [14]); ii) 41 atom/bond count descriptors [17, 18] (e.g., the number of double bonds); iii) 18 conformation dependent charge descriptors [19] (e.g., the water accessible surface area of polar atoms); iv) the 16 Kier and Hall connectivity and kappa shape indices [20, 21] (e.g., the Zagreb index); v) 21 MOPAC descriptors [22] (e.g., the ionization potential); vi) 48 partial charge descriptors (e.g., the total positive partial charge); vii) 12 pharmacophore feature descriptors (e.g., the number of hydrophobic atoms);

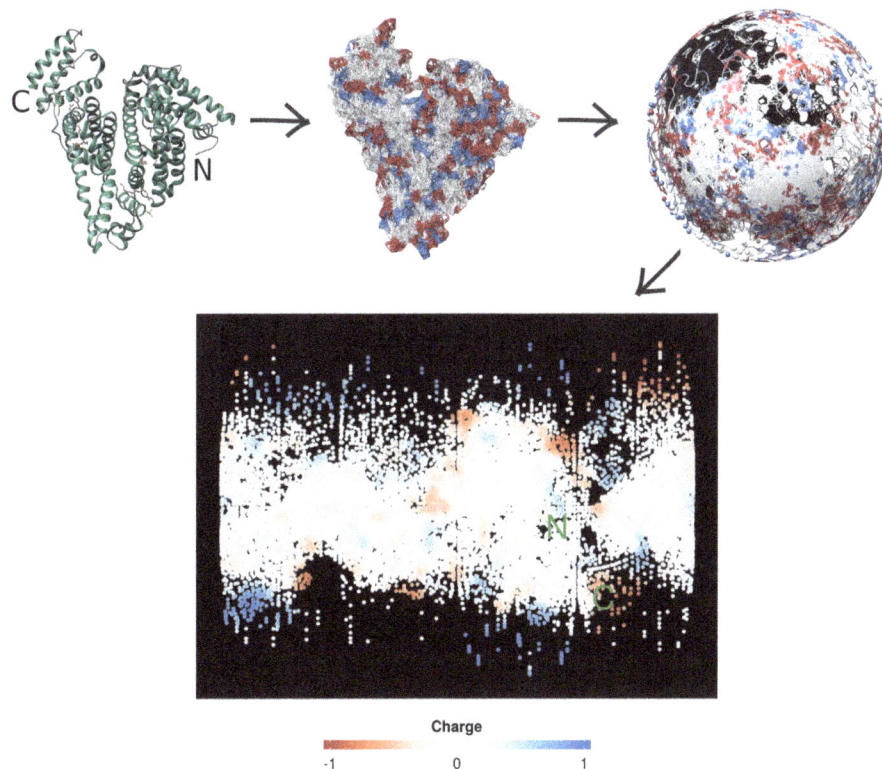

Fig. 1 The main steps of the algorithm executed by Structuprint. Here, a mould of the surface of the 3D structure of the leporine serum albumin ([PDB: 4F5V]) is first generated. The property values (e.g., charge) of the amino acids below the mould are retained. Then, the dummy atoms consisting the mould are mapped onto a sphere. Finally, the sphere is projected onto a map using the Miller cylindrical transformation and a smoothing of the property values is performed. The elements of the upper half of the figure were rendered with UCSF Chimera

viii) 11 potential energy descriptors (e.g., the solvation energy); ix) 16 physical properties [18, 23–27] (e.g., the molecular weight); x) 18 subdivided surface areas; xi) 94 surface area, volume, and shape descriptors (e.g., globularity). A detailed explanation of each descriptor is provided in the properties codebook which accompanies the tool. By drawing values from this database, Structuprint can visualize the distribution of a property across protein surfaces. Users can extend it by adding measurements for more chemical components or provide their own custom SQLite database in order to incorporate novel descriptors.

Algorithm

Generation of a mould of the surface of a protein

The main steps of the algorithm implemented by Structuprint are shown in Fig. 1. The tool first produces a mould of the protein structure's surface in two steps. The structure is initially placed within a 3D grid with cell dimensions of $1 \times 1 \times 1$ Å. Then, one dummy atom is inserted in each empty grid cell that neighbours a single protein atom. This process was previously described by Vlachakis et al. [28] and is extended here, with dummy atoms being assigned the identity of the amino acid to which their neighbouring protein atom belongs. This results to a quite accurate approximation of the underlying protein surface at the level of residue atoms.

Transformation of the mould into a sphere

The next step involves the conversion of the dummy atoms mould to a sphere. To this end, the algorithm calculates the coordinates of the centre of mass of the mould c - i.e., the average position of all atoms -, and the maximum distance of any atom v_i from the centre of mass ($radius$):

$$\mathbf{c} = (x_c, \ y_c, \ z_c)$$
$$= \left(\frac{\sum_{i=1}^{n} x_i}{n}, \ \frac{\sum_{i=1}^{n} y_i}{n}, \ \frac{\sum_{i=1}^{n} z_i}{n} \right) \quad (1)$$

$$radius = \max_{1 \leq i \leq n} \sqrt{(x_i - x_c)^2 + (y_i - y_c)^2 + (z_i - z_c)^2} \quad (2)$$

The coordinates of each atom are normalized with respect to the centre of mass:

$$\mathbf{v_i'} = \left(x_i', \ y_i', \ z_i' \right) = (x_i - x_c, \ y_i - y_c, \ z_i - z_c) \quad (3)$$

Then, to transfer the dummy atoms onto the surface of a sphere, each vector $\mathbf{v_i}$ is scaled to a length equal to the $radius$:

$$\mathbf{w_i} = \left(x_i'', \ y_i'', \ z_i'' \right) = \frac{radius}{\sqrt{x_i'^2 + y_i'^2 + z_i'^2 \cdot \mathbf{v_i'}}} \quad (4)$$

Projection of the sphere onto a map

The Cartesian coordinates of each $\mathbf{w_i}$ are converted to latitude/longitude values (in units of radians) using the following set of equations:

$$latitude_i = \tan^{-1} \frac{z_i''}{\sqrt{x_i''^2 + y_i''^2}} 2longitude_i = \tan^{-1} \frac{y_i''}{x_i''} \quad (5)$$

For the two-dimensional projection, several techniques were initially tested (e.g., the sinusoidal projection [29] and the Hammer projection [29, 30]), before deciding on the Miller cylindrical projection [29, 31]:

$$\mathbf{m_i} = \left(longitude_i, \ \frac{5}{4} \cdot \ln \left[\tan \left(\frac{\pi}{4} + \frac{2}{5} \cdot latitude_i \right) \right] \right) \quad (6)$$

This projection was selected on the basis of its simplicity and ease of understanding. It is one of the most popular projections in cartography, as it can depict the entirety of the sphere, including the poles. Latitude and longitude lines are parallel and straight. Projection-induced distortion is zero at the equator, increases gradually towards higher latitudes, and becomes maximal at the poles. This leads to significant overestimation of the distance among atoms at the upper and lower parts of the figure (Fig. 1), similarly to the areal exaggeration of Greenland and Antarctica. Nevertheless, the Miller cylindrical projection introduces less polar distortion than the Mercator projection, on which it is based.

Map smoothing

The previous step resulted in a map of the protein surface with data points coloured by a property of choice. However, this 'primary' map is not suitable for detecting areas with an overall concentration of atoms with high or low property values, which is one of the main benefits of this cartographic approach. For instance, a small area with both negatively and positively charged residue atoms would not appear as almost neutrally charged, but as a tiny dipole. To prevent the appearance of small 'hot spots' and redistribute the property values among neighbouring data points, the algorithm includes a smoothing step. The map is iteratively divided in grid squares of varying dimensions, from $0.001° \times 0.001°$ to $0.5° \times 0.5°$, with a step increase of $0.001°$. In each iteration of this process, grid cells are assigned the average value of all data points within them. Finally, the value of every data point is defined as the average value of its corresponding grid cell across all iterations. This smoothing method ensures that areas with pronounced accumulation of high or low values are easily discernible from those with a mixed population.

User interfaces

The default interface of Structuprint is a cross-platform, command-line interface (CLI). It consists of two executables: structuprint_frame and structuprint. The structuprint_frame executable produces a TIFF figure from a single input PDB file, using the R package ggplot2 [32] for plotting. The structuprint executable is responsible for processing multiple superimposed PDB files - either serially or in a parallel manner -, generating a TIFF figure per input file and a final GIF animation, rendered with the Imager Perl module [33]. Most parameters of the underlying algorithms can be modified by the user, such as the delay between animation frames, the background colour, and the appearance of ID numbers on final figures. A full descriptive list of the available parameters for both executables can be found in Structuprint's manual, distributed along with the application and also available from its website.

Other than the CLI, Structuprint also comes with a Graphical User Interface (GUI), available by default only on GNU/Linux systems. The GUI is built with the Gtk2 toolkit and offers a user-friendly interface to all the command line arguments and options. As an example of its capabilities, in Fig. 2 Structuprint's GUI is producing an animation on a multiprocessor machine using 30 cores.

Parallelism

On Unix-like systems (e.g., GNU/Linux, OS X), Structuprint supports task parallelism when generating animations. Using the Parallel::ForkManager Perl module [34], Structuprint can take advantage of multiple CPU cores by assigning each input PDB file to a different processor. The simultaneous rendering of multiple individual frames considerably reduces the total execution time, allowing for visualization of entire molecular dynamics simulations within a reasonable time frame.

Results and discussion
Benchmark

To understand how execution time and memory consumption scale with the number of atoms in an input PDB file, we ran Structuprint against 700 randomly selected structures from the Protein Data Bank (Additional file 1). For simplification purposes, multi-model PDB entries were excluded, as a large proportion of the atoms would overlap in 3D space, being essentially indistinguishable. The benchmark was performed on a GNU/Linux system with an Intel Xeon E5-1650 v2 CPU at 3.50 GHz and 31.4 GB of memory. Structuprint was launched 10 times per PDB file and the execution time was measured as the median time for completion. Memory usage was measured similarly. We then performed linear regressions using execution time and memory consumption as dependent variables and number of atoms as the independent variable. In both regressions, we applied a Box-Cox transformation [35] to the dependent variable to ensure that the residuals were normally distributed. The final fitted models are shown in Fig. 3. Execution time increases linearly with the number of atoms, whereas

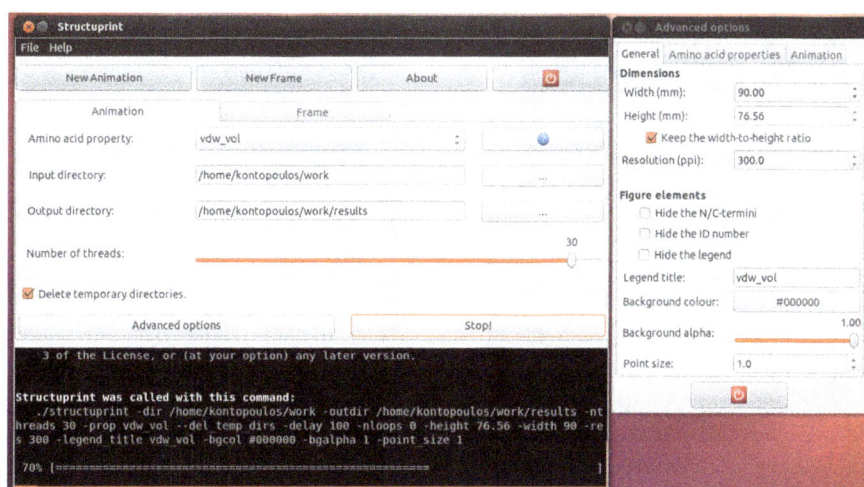

Fig. 2 Structuprint's Graphical User Interface. The main window is split between two tabs for preparation of 1) animations and 2) single static maps. The default parameters of the algorithm can be modified using the 'Advanced options' popup window. When Structuprint is rendering a figure, its progress is shown in a temporary terminal

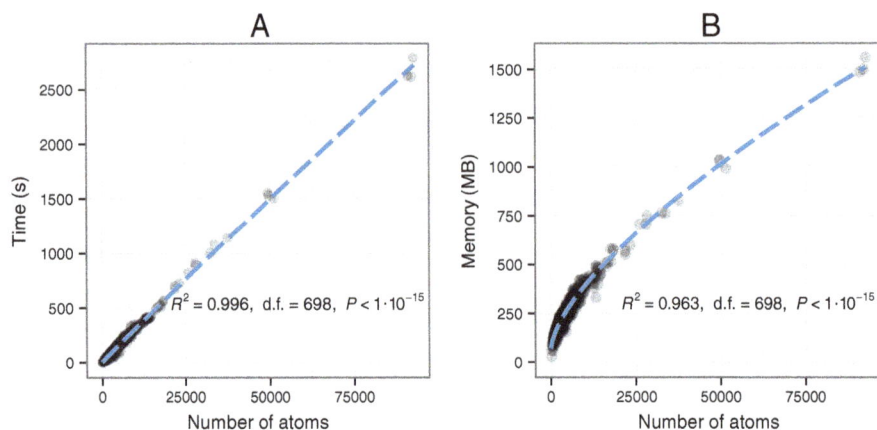

Fig. 3 Execution time (**a**) and memory consumption (**b**) of Structuprint, as a function of the atom count (n). The runtime complexity is O(n), whereas the memory complexity is O($\sqrt{n}$). The uneven distribution of atom counts reflects the composition of the Protein Data Bank. As of March 2015, ~99 % of entries in the PDB had an atom count of 61,000 or less, with the overall mean being 9,006 atoms

memory consumption only increases with the square root of the atom count. For example, on the aforementioned system it took 88 seconds and 211 MB of RAM to generate a Structuprint figure for a relatively small protein with 2,461 atoms ([PDB:1YLP]).

Examples of usage

To illustrate the utility of this tool, we present three different examples of usage in this section. Two-dimensional visualization with Structuprint enhances the representation of protein surfaces and facilitates the interpretation of the results in all three cases.

Visualization of molecular dynamics simulations

A seldom explored application of molecular cartography involves the generation of 2D animations from a series of PDB files. Here, we visualized a portion of a folding

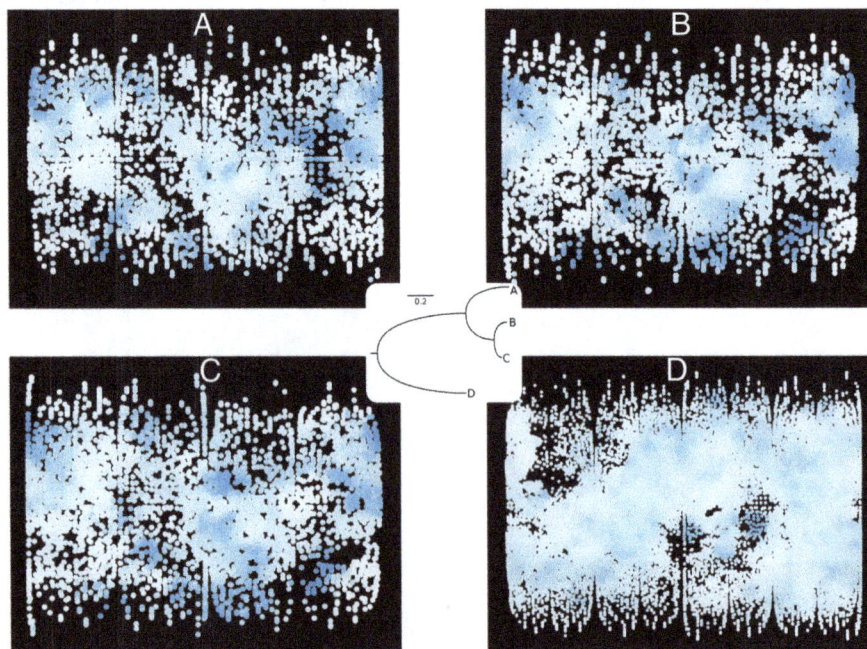

Fig. 4 Evolution of protein surfaces, as represented via Structuprint figures. **a–c**: Plastocyanin orthologs from *Spinacia oleracea*, *Ulva pertusa*, and *Ulva prolifera*, respectively. **d** Chloroplastic fructose 1,6-biphosphatase from *Spinacia oleracea*. The colour depth denotes the FASA_H value across each map, with darker areas having higher values of the descriptor. Despite the obvious conservation of surface shape and hydrophobicity, 2D maps can distinguish even slight differences among evolutionarily related proteins. Inset: The maximum likelihood phylogenetic tree of the proteins in panels **a–d**

simulation of a variant of the chicken villin headpiece subdomain (HP-35 NleNle) from the Folding@Home project [36]. The part of the input simulation was 50 ps long, with one frame being extracted every 0.25 ps. Each frame was structurally superimposed to the previous one with UCSF Chimera's MatchMaker tool [37]. Then, two separate animations were produced: one of the simulation frames in ribbon representation and one of the corresponding 2D maps, with the topological polar surface area - a measure of polarity - as the property of choice. For comparison purposes, these two animations are jointly shown in Additional file 2. This approach simplifies the detection of conformational changes during the course of the simulation, along with fluctuations in the distribution of physicochemical variables.

Depiction of surface conservation

The evolution of protein surfaces and the conservation - or lack - thereof is another domain in which Structuprint can be applied. As an example, we performed a brief phylogenetic analysis of three orthologs of plastocyanin - a protein involved in electron transfer in photosynthesis [38] - for which crystallographic structures were available. The amino acid sequences of spinach plastocyanin (*Spinacia oleracea* [Swiss-Prot:P00289]) and those of two green algal species (*Ulva pertusa* [Swiss-Prot:P56274], *Ulva prolifera* [Swiss-Prot:P07465]) were retrieved from the UniProt database, along with the sequence of the spinach chloroplastic fructose 1,6-biphosphatase ([Swiss-Prot:P22418]) that would be later used as an outgroup. The sequences were aligned with ProbCons 1.12 [39] and the best model of amino acid substitution was determined with RAxML 8.1.16 [40]. Ten maximum likelihood trees were then inferred with RAxML using the biphosphatase as the outgroup sequence, and the best scoring tree was selected. Next, 2D protein surface maps of the corresponding 3D structures ([PDB:1AG6, 1IUZ, 7PCY, 1SPI]) were produced with Structuprint, after performing a structural superposition. For this example we used a more complex descriptor, FASA_H:

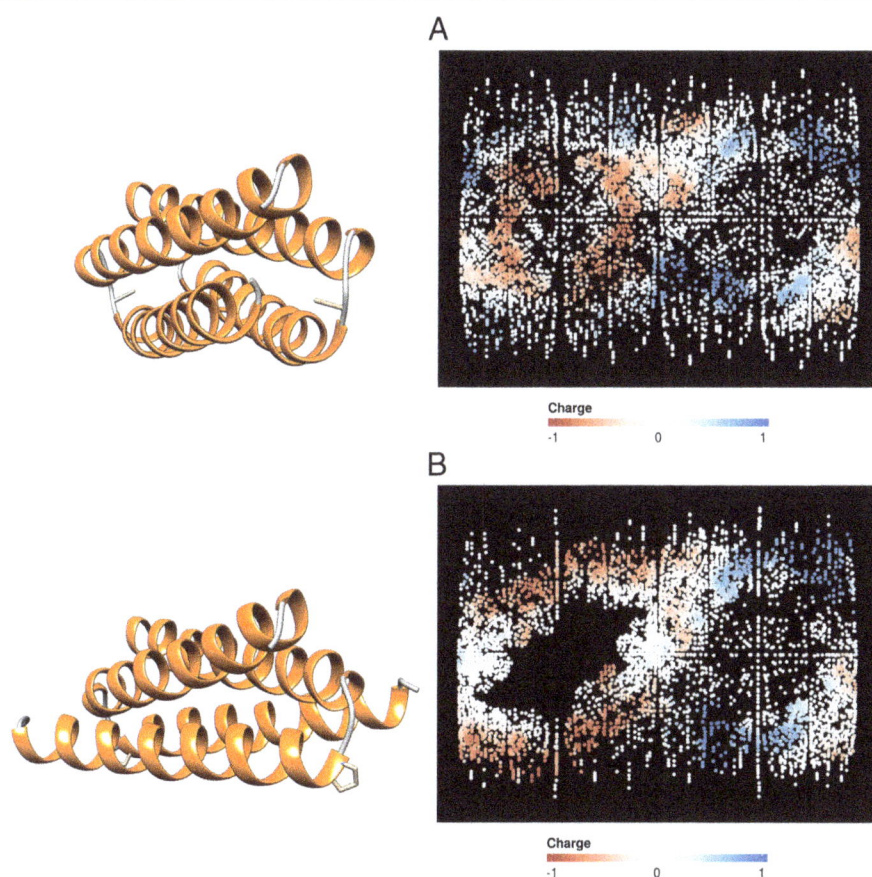

Fig. 5 Three- and two-dimensional depiction of the native Rop structure (**a**) and the A31P mutant (**b**). In the 3D representation, the amino acid side chain of the 31st residue - in the turn region - is shown in stick style. Positively charged residues are shown with blue colour, negatively charged ones with red, and non-charged residues with white. With the 2D representation generated by Structuprint, large differences can be observed not only in the shape of the surface, but also in the location of exposed negatively charged residues

$$FASA\_H = \frac{water\ accessible\ surface\ area\ of\ hydrophobic\ atoms}{water\ accessible\ surface\ area\ of\ all\ atoms}$$

$$(7)$$

The results are shown in Fig. 4. There is significant conservation of both surface structure and hydrophobicity patterns among all three species, with the algal orthologs (Fig. 4b, c) exhibiting greater similarity, as expected. Finally, the representation of the chloroplastic fructose 1,6-biphosphatase (Fig. 4d) is vastly different from the others, highlighting the long sequence distance among these proteins.

Comparison of conformational changes, e.g., due to mutations

A third proposed application of Structuprint involves visually contrasting protein surfaces before and after events such as mutations, ligand binding, pH or temperature alterations. We exemplify this case using a mutant of Rop, a small regulatory protein from *Escherichia coli* with a native tertiary structure of a homodimeric four-helix bundle. The native structure has been shown to be disrupted by a single amino acid substitution (Ala31 → Pro) in the turn region [41]. To show the consequences of this mutation, we generated Structuprint maps of the wild type protein ([PDB:1ROP]) and the A31P mutant ([PDB:1B6Q]) after superposition. Figure 5 illustrates the mutation-induced conformation change, comprising different surface shape and grouping of negatively charged residues.

Conclusions

We have developed a user-friendly application for two-dimensional visualization of protein surfaces, optionally supporting multicore processing and user-provided physicochemical descriptors. Structuprint provides an alternative view of molecular surfaces, which - as shown in the previous section - could be of great use to a variety of researchers, including biochemists, structural biologists, and biophysicists.

Availability and requirements
Project name: Structuprint

Project home page: http://dgkontopoulos.github.io/Structuprint/

Operating systems: Prebuilt packages and installers are available for GNU/Linux distributions (Ubuntu 14.04, Debian 8, Fedora 22, CentOS 7, openSUSE 13.2), Windows, and OS X. For all other operating systems, installation from the source code is required. The GUI is available by default only for GNU/Linux systems.

Programming languages: Perl 5, R

License: GNU GPLv3+

Any restrictions to use by non-academics: None

Additional files

> **Additional file 1:** Table of PDB entries used in the benchmark. Accession codes and atom counts of 700 random, non-multimodel PDB entries that were included in the benchmark.
>
> **Additional file 2:** Conventional and molecular cartographic visualizations of a molecular dynamics simulation of the chicken villin headpiece subdomain (HP-35 NleNle). Comparison between animations produced with conventional rendering methods (UCSF Chimera), and with 2D maps generated by Structuprint. The right half shows the movement of exposed amino acids with high topological polar surface area values (blue) during the course of the simulation.

Abbreviations
2D: two-dimensional; 3D: three-dimensional; A31P: ala31 → pro mutant; CLI: command-line interface; CPU: central processing unit; FASA_H: fractional water accessible surface area of hydrophobic atoms over all atoms; GB: gigabyte; GUI: graphical user interface; HP-35 NleNle: villin headpiece subdomain double norleucine mutant (Lys24Nle/Lys29Nle); MB: megabyte; MOE: molecular operating environment; MOPAC: molecular orbital package; PDB: protein data bank; RAM: random-access memory; RAxML: randomized axelerated maximum likelihood.

Competing interests
The authors declare that they have no competing interests.

Authors' contributions
DGK contributed to the conception of the software, designed, developed and packaged the software, ran benchmarks, prepared the figures, and drafted the manuscript. DV, GT, and SK conceived of and designed the software, reviewed and revised the draft. All authors read and approved the final manuscript.

Acknowledgements
The authors express their gratitude to two anonymous reviewers for helpful comments, and to all researchers who made their data publicly available on the Protein Data Bank, the UniProt database, or on Simtk.org. No funding was received for this project.

Author details
[1]Department of Life Sciences, Imperial College London, Silwood Park Campus, Ascot, UK. [2]Bioinformatics & Medical Informatics Team, Biomedical Research Foundation, Academy of Athens, Athens, Greece. [3]School of Chemical Engineering, National Technical University of Athens, Athens, Greece. [4]IMGT®, The International ImMunoGeneTics Information System®, Université de Montpellier, Laboratoire d'ImmunoGénétique Moléculaire LIGM, UPR CNRS 1142, Institut de Génétique Humaine, Montpellier, France.

References
1. Orengo CA, Thornton JM. Protein families and their evolution - a structural perspective. Annu Rev Biochem. 2005;74:867–900. doi:10.1146/annurev.biochem.74.082803.133029.
2. Cheng T, Li Q, Zhou Z, Wang Y, Bryant SH. Structure-based virtual screening for drug discovery: a problem-centric review. AAPS J. 2012;14(1):133–41. doi:10.1208/s12248-012-9322-0.
3. Fanning DW, Smith JA, Rose GD. Molecular cartography of globular proteins with application to antigenic sites. Biopolymers. 1986;25(5):863–83. doi:10.1002/bip.360250509.
4. Pawłowski K, Godzik A. Surface map comparison: studying function diversity of homologous proteins. J Mol Biol. 2001;309(3):793–806. doi:10.1006/jmbi.2001.4630.

5. Chirgadze Y, Kurochkina N, Nikonov S. Molecular cartography of proteins: surface relief analysis of the calf eye lens protein gamma-crystallin. Protein Eng. 1989;3(2):105–10. doi:10.1093/protein/3.2.105.

6. Badel-Chagnon A, Nessi J, Buffat L, Hazout S. "Iso-depth contour map" of a molecular surface. J Mol Graph. 1994;12(3):162–8. doi:10.1016/0263-7855(94)80082-0.

7. Yang H, Qureshi R, Sacan A. Protein surface representation and analysis by dimension reduction. Proteome Sci. 2012;10(Suppl 1):S1. doi:10.1186/1477-5956-10-S1-S1.

8. Sasin JM, Godzik A, Bujnicki JM. SURF'S UP! - protein classification by surface comparisons. J Biosci. 2007;32(1):97–100. doi:10.1007/s12038-007-0009-0.

9. Koromyslova AD, Chugunov AO, Efremov RG. Deciphering fine molecular details of proteins' structure and function with a Protein Surface Topography (PST) method. J Chem Inf Model. 2014;54(4):1189–99. doi:10.1021/ci500158y.

10. Levieux G, Montes M. Towards real-time interactive visualization modes of molecular surfaces: examples with Udock. IEEE VR 2015 Workshop on Virtual and Augmented Reality dedicated to Molecular Science (VARMS). 2015.

11. Molecular Operating Environment (MOE). 2010.10. 1010 Sherbooke St. West, Suite #910, Montreal, QC, Canada, H3A 2R7: Chemical Computing Group Inc; 2010. https://www.chemcomp.com/MOE-Molecular_Operating_Environment.htm. Accessed 19 Feb 2016.

12. Wiener H. Structural determination of paraffin boiling points. J Am Chem Soc. 1947;69(1):17–20.

13. Balaban AT. Five new topological indices for the branching of tree-like graphs. Theor Chim Acta. 1979;53:355–75.

14. Balaban AT. Highly discriminating distance-based topological index. Chem Phys Lett. 1982;89(5):399–404. doi:10.1016/0009-2614(82)80009-2.

15. Petitjean M. Applications of the radius-diameter diagram to the classification of topological and geometrical shapes of chemical compounds. J Chem Inf Comput Sci. 1992;32(4):331–7. doi:10.1021/ci00008a012.

16. Pearlman RS, Smith KM. Novel software tools for chemical diversity. In: Kubinyi H, Folkers G, Martin YC, editors. 3D QSAR in drug design: three-dimensional quantitative structure activity relationships. Volume 2. Netherlands: Springer; 1998. p. 339–53. doi:10.1007/0-306-46857-3_18.

17. Lipinski CA, Lombardo F, Dominy BW, Feeney PJ. Experimental and computational approaches to estimate solubility and permeability in drug discovery and development settings. Adv Drug Deliv Rev. 1997;23(1–3):3–25. doi:10.1016/S0169-409X(96)00423-1.

18. Oprea TI. Property distribution of drug-related chemical databases. J Comput Aided Mol Des. 2000;14(3):251–64. doi:10.1023/A:1008130001697.

19. Stanton DT, Jurs PC. Development and use of charged partial surface area structural descriptors in computer-assisted quantitative structure-property relationship studies. Anal Chem. 1990;62(21):2323–9. doi:10.1021/ac00220a013.

20. Kier LB, Hall LH. The nature of structure-activity relationships and their relation to molecular connectivity. Eur J Med Chem. 1977;12:307–12.

21. Hall LH, Kier LB: The molecular connectivity chi indexes and kappa shape indexes in structure-property modeling. In: Lipkowitz KB, Boyd DB, editors. Reviews in Computational Chemistry. Volume 2. Hoboken, New Jersey: John Wiley & Sons, Inc.; 1991. p. 367–422. doi:10.1002/9780470125793.ch

22. Stewart JJP. MOPAC manual. 7th ed. 1993.

23. Lide DR, editor. CRC handbook of chemistry and physics. Boca Raton: CRC Press; 1994.

24. Wildman SA, Crippen GM. Prediction of physicochemical parameters by atomic contributions. J Chem Inf Comput Sci. 1999;39(5):868–73. doi:10.1021/ci990307l.

25. Ertl P, Rohde B, Selzer P. Fast calculation of molecular polar surface area as a sum of fragment-based contributions and its application to the prediction of drug transport properties. J Med Chem. 2000;43(20):3714–7. doi:10.1021/jm000942e.

26. Hou TJ, Xia K, Zhang W, Xu XJ. ADME evaluation in drug discovery. 4. Prediction of aqueous solubility based on atom contribution approach. J Chem Inf Comput Sci. 2004;44(1):266–75. doi:10.1021/ci034184n.

27. Kazius J, McGuire R, Bursi R. Derivation and validation of toxicophores for mutagenicity prediction. J Med Chem. 2005;48(1):312–20. doi:10.1021/jm040835a.

28. Vlachakis D, Kontopoulos DG, Kossida S. Space constrained homology modelling: the paradigm of the RNA-dependent RNA polymerase of dengue (type II) virus. Comput Math Methods Med. 2013;2013:108910. doi:10.1155/2013/108910.

29. Snyder JP. Map projections - a working manual, U.S. Geological survey professional paper 1395. Washington, DC: United States Government Printing Office; 1987.

30. Hammer E. Über die Planisphäre von Aitow und verwandte Entwürfe, insbesondere neue flächentreue iihnlicher Art. Petermanns Geogr Mitt. 1892;38(4):85–7.

31. Miller OM. Notes on cylindrical world map projections. Geogr Rev. 1942;32(3):424–30.

32. Wickham H. ggplot2: elegant graphics for data analysis. New York: Springer; 2009.

33. Cook T. Imager - Perl extension for generating 24 bit images. https://metacpan.org/pod/Imager. Accessed 27 Sep. 2015.

34. Champoux Y. Parallel::ForkManager - A simple parallel processing fork manager. https://metacpan.org/pod/Parallel::ForkManager. Accessed 27 Sep. 2015.

35. Box GEP, Cox DR. An analysis of transformations. J R Stat Soc Series B Stat Methodol. 1964;26(2):211–52.

36. Ensign DL, Kasson PM, Pande VS. Heterogeneity even at the speed limit of folding: large-scale molecular dynamics study of a fast-folding variant of the villin headpiece. J Mol Biol. 2007;374(3):806–16. doi:10.1016/j.jmb.2007.09.069.

37. Pettersen EF, Goddard TD, Huang CC, Couch GS, Greenblatt DM, Meng EC, Ferrin TE. UCSF Chimera - a visualization system for exploratory research and analysis. J Comput Chem. 2004;25(13):1605–12. doi:10.1002/jcc.20084.

38. Shibata N, Inoue T, Nagano C, Nishio N, Kohzuma T, Onodera K, Yoshizaki F, Sugimura Y, Kai Y. Novel insight into the copper-ligand geometry in the crystal structure of Ulva pertusa plastocyanin at 1.6-Å resolution: structural basis for regulation of the copper site by residue 88. J Biol Chem. 1999;274(7):4225–30. doi:10.1074/jbc.274.7.4225.

39. Do CB, Mahabhashyam MSP, Brudno M, Batzoglou S. ProbCons: Probabilistic consistency-based multiple sequence alignment. Genome Res. 2005;15(2):330–40. doi:10.1101/gr.2821705.

40. Stamatakis A. RAxML version 8: a tool for phylogenetic analysis and post-analysis of large phylogenies. Bioinformatics. 2014;30(9):1312–3. doi:10.1093/bioinformatics/btu033.

41. Glykos NM, Cesareni G, Kokkinidis M. Protein plasticity to the extreme: changing the topology of a 4-α-helical bundle with a single amino acid substitution. Structure. 1999;7(6):597–603. doi:10.1016/S0969-2126(99)80081-1.

20

A lack of peptide binding and decreased thermostability suggests that the CASKIN2 scaffolding protein SH3 domain may be vestigial

Jamie J. Kwan[1,2] and Logan W. Donaldson[1*]

Abstract

Background: CASKIN2 is a neuronal signaling scaffolding protein comprised of multiple ankyrin repeats, two SAM domains, and one SH3 domain. The CASKIN2 SH3 domain for an NMR structural determination because its peptide-binding cleft appeared to deviate from the repertoire of aromatic enriched amino acids that typically bind polyproline-rich sequences.

Results: The structure demonstrated that two non-canonical basic amino acids (K290/R319) in the binding cleft were accommodated well in the SH3 fold. An K290Y/R319W double mutant restoring the typical aromatic amino acids found in the binding cleft resulted in a 20 °C relative increase in the thermal stability. Considering the reduced stability, we speculated that the CASKIN2 SH3 could be a nonfunctional remnant in this scaffolding protein.

Conclusions: While the NMR structure demonstrates that the CASKIN2 SH3 domain is folded, its cleft has suffered two substitutions that prevent it from binding typical polyproline ligands. This observation led us to additionally survey and describe other SH3 domains in the Protein Data Bank that may have similarly lost their ability to promote protein-protein interactions.

Keywords: SH3 domain, NMR spectroscopy, Molecular modeling

Background

A pair of neuronal scaffolding proteins, represented in humans by CASKIN1 and CASKIN2, participate in signaling pathways involved in axon guidance and the creation of neuromuscular junctions. While mammals possess two CASKIN homologs, *Drosophila* only possesses one form (*Ckn*) that differs in domain composition and *C. elegans* has no homolog at all leading to the idea that the general role of CASKINs may be to provide more complex signaling outcomes in higher organisms by recruiting additional sets of protein partners [1]. Typical for many scaffolding proteins [2, 3], CASKIN1 and CASKIN2 are composed entirely of protein-protein interaction domains including six ankyrin repeats, an SH3 domain and tandem SAM domain (Fig. 1a). The C-terminal half of the protein, spanning over 600 residues and only partially conserved, is characterized by a proline-rich segment of unknown function [4].

CASKIN1, and CASKIN2 by homology, were named by their ability to serve as ligands of the MAGUK protein, CASK, a prominent signaling protein linked to calcium mediated signaling events, actin microfilament assembly, and synaptic communication through the neurexin cell surface receptor [4, 5]. Later structural studies of CASKIN1 revealed that the CASK interaction domain (CID), shared with X11/Mint and located in a region between the SH3 and SAM domains, was not present in CASKIN2. Thus, the namesake of CASKIN2 is misleading because it is unable to bind the calmodulin kinase domain of CASK [6]. This distinction lead us to speculate that despite their organizational similarity,

* Correspondence: logand@yorku.ca
[1]Department of Biology, York University, 4700 Keele Street, Toronto, ON M3J 1P3, Canada
Full list of author information is available at the end of the article

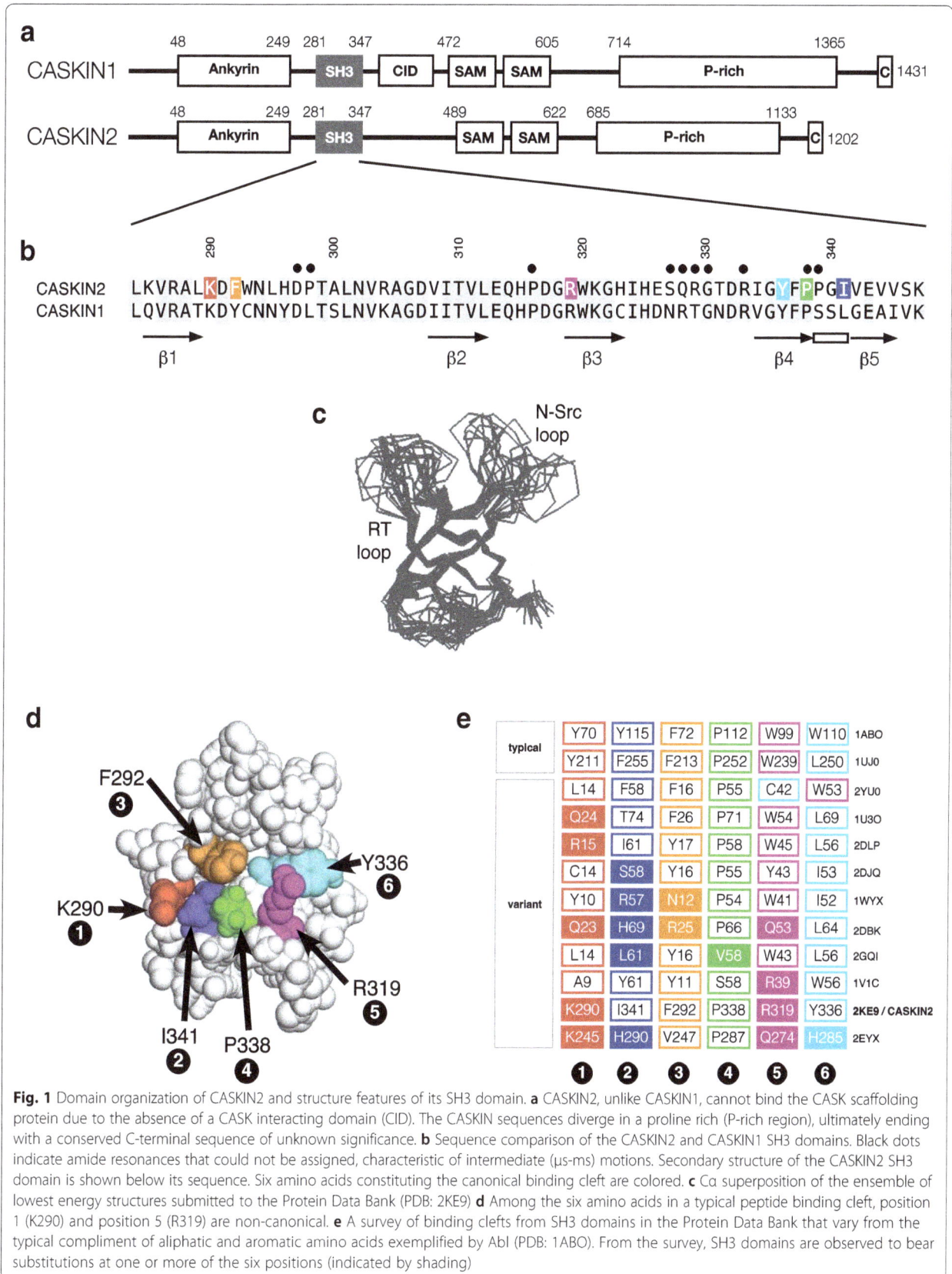

Fig. 1 Domain organization of CASKIN2 and structure features of its SH3 domain. **a** CASKIN2, unlike CASKIN1, cannot bind the CASK scaffolding protein due to the absence of a CASK interacting domain (CID). The CASKIN sequences diverge in a proline rich (P-rich region), ultimately ending with a conserved C-terminal sequence of unknown significance. **b** Sequence comparison of the CASKIN2 and CASKIN1 SH3 domains. Black dots indicate amide resonances that could not be assigned, characteristic of intermediate (μs-ms) motions. Secondary structure of the CASKIN2 SH3 domain is shown below its sequence. Six amino acids constituting the canonical binding cleft are colored. **c** Cα superposition of the ensemble of lowest energy structures submitted to the Protein Data Bank (PDB: 2KE9) **d** Among the six amino acids in a typical peptide binding cleft, position 1 (K290) and position 5 (R319) are non-canonical. **e** A survey of binding clefts from SH3 domains in the Protein Data Bank that vary from the typical compliment of aliphatic and aromatic amino acids exemplified by Abl (PDB: 1ABO). From the survey, SH3 domains are observed to bear substitutions at one or more of the six positions (indicated by shading)

CASKIN1 and CASKIN2 may have diverged with respect to their scaffolding functions in neurons [7].

Owing to the prominence of CASK in the development of neurological disease [8, 9], research to date has concentrated on CASKIN1 [10–12]. In this report, we have concentrated on the CASKIN2 SH3 domain to explore the question of whether functional differences between CASKIN1 and CASKIN2 extend to the structural level. Building upon the NMR solution structure of the CASKIN2 SH3 domain, we demonstrate that the typical binding cleft has two hydrophobic amino acid substitutions that do not affect the overall fold, but do affect thermostability and the ability to interact with peptides. These observations led us to consider that the CASKIN2 SH3 domain was non-functional. When hydrophobic amino acids were reintroduced, thermostability increased and high affinity binding was restored towards a peptide ligand of one of CASKIN2's most homologous family members.

Results and discussion
Domain organization and sequence similarity
The CASKIN1 and CASKIN2 SH3 domains are located at the same position within each protein and are approximately 60 % similar (Fig. 1b). Following the SH3 domain, CASKIN1 and CASKIN2 diverge in sequence, the notable differences being the lack of a CASK interaction domain (CID) in CASKIN2, as well as the length and composition of a carboxy terminal proline-rich region. Thus, depending on the perspective, CASKIN1 and CASKIN2 are simultaneously similar and different, and therefore warrant some caution when making functional assumptions about one protein in the absence of knowledge from the other.

Structure of the CASKIN2 SH3 domain
We began this study with a NMR solution structure determination of the CASKIN2 SH3 domain to provide a high-resolution framework for exploring how the SH3 domain may interact with ligands and react to post-translational modifications. Backbone atom precision for the ensemble of structures was 0.60 ± 0.14 Å, consistent with the number of experimental observations used in the structure calculation. Complete statistics pertaining to the experimental observations and structural quality are presented in Table 1. From the ensemble of the lowest energy structures, the RT and N-Src loops presented a high degree of disorder (Fig. 1c), presumably due to the absence of a ligand. Experimentally, the decrease in backbone atom precision is manifested by missing backbone assignments (D297, S327-D330, and R333) and short-range NOE observations (depicted as dots above the sequence in Fig. 1). A survey of the PDB for similar structures revealed the SH3 domains from human

Table 1 Restraints and statistics for the ensemble of 15 structures

NOE restraints	
Total	413
Intraresidue ($|i - j| = 0$)	204
Sequential ($|i - j| = 1$)	102
Medium range ($1 < |i - j| < 5$)	8
Long range ($|i - j| \geq 5$)	99
Additional restraints	
Hydrogen bond distance restraints	32
Backbone angle torsion angle restraints	74
RMS deviations[a]	
Bonds	0.0123 ± 0.0003
Angles	1.2827 ± 0.0363
Improper angles	1.6905 ± 0.1112
Dihedral angles	0.2015 ± 0.1411
RMS violations	
NOE restraints > 0.5 Å	0.0 ± 0.0
NOE restraints > 0.3 Å	2.3 ± 1.2
NOE restraints > 0.1 Å	25.5 ± 4.1
Dihedral angles > 5°	6.2 ± 0.6
Ramachandran analysis for ordered residues[b]	
Most favored regions	95.2 %
Additional allowed regions	4.8 %
Generously allowed regions	0.0 %
Disallowed regions	0.0 %

[a] As reported by XPLOR-NIH 2.30 using the standard protein force field
[b] As reported by PROCHECK for residues 284–292, 298–312, 320–324, 334–343

KIAA17833 and yeast NBP2 (both from structural proteomics studies) along with the well characterized SH3 domain from STAM2 [13] (Table 2).

The CASKIN2 SH3 domain peptide binding cleft
A typical SH3 domain presents a binding cleft comprised of six, nearly linearly arranged, hydrophobic amino acids. As shown in Fig. 1d, the six amino acids in CASKIN2 are K290, I341, F292, P338, R319, and Y336. Notably, there is strong deviation at the first and fifth positions, denoted by K290 and R319. Since the side chains of the SH3 domain binding cleft are surface exposed, K290 and R319 do not affect the protein fold, and presumably, this is also the case for CASKIN1.

Table 2 Structurally similar proteins to the CASKIN2 SH3 domain

PDB	Protein	Source	RMSD	Aligned	Identity
2DLP	KIAA17833	NMR	1.1 Å	49 aa	29 %
1YN8	NBP2	X-ray	1.2 Å	49 aa	21 %
1UJ0	STAM2 + UBPY peptide	X-ray	1.6 Å	57 aa	33 %

Given that the binding cleft deviates at two positions from the typical set of amino acids observed in PxxP type SH3 domains like Abl kinase [14] and RxxK type SH3 domains like STAM2 [13], we hypothesized that the CASKIN2 SH3 domain might bind an unconventional peptide ligand. To answer this question, we performed four successive rounds of panning a 6xHis _-tagged CASKIN2 SH3 domain against a commercial bacteriophage 12-mer peptide display library (PHD-12; Novagen). A comparison of eight peptides from the last round of panning revealed a Pxx[L/M/W] motif in many of the candidates [presented in Additional file 1]. However, upon testing this motif with two chemically synthesized peptides and one reversed peptide as a control, we did not observe any binding in NMR based titrations suggesting that the library may have identified a very weak interaction. Since the phage display method failed to identify a ligand for the CASKIN2 SH3 domain, it led us to consider that the CASKIN2 SH3 domain might be a non-functional remnant of the protein.

Figure 1e presents a comparison of high resolution SH3 domain structures from the Protein Data Bank with an emphasis on representatives having peptide binding clefts that deviate from the usual complement of aliphatic and aromatic amino acids, and consequently, have no ligands reported for them. The Abl SH3 domain leads a set of sequences in Fig. 1e as a point of reference and prototype for the typical array of aromatic and aliphatic amino acids that populate the six positions of the binding cleft. Next to Abl in the presentation is the RUN/TBC1 domain containing protein-3 SH3 domain (PDB: 2YUO) that swaps an aromatic/aliphatic pair at positions 5 and 6 of the cleft with a compensatory aliphatic/aromatic pair. A survey of the remaining representatives in the figure demonstrate that it is possible for each of the six binding cleft positions to tolerate a hydrophilic or charged amino acid substitution. The Crk-II SH3(2) domain represents the most extreme case with substitutions at positions 1, 2, 5 and 6. Loss of ligand binding in the Crk-II SH3(2) domain in the context of the full adaptor protein is consistent with a stabilizing function rather than a regulatory function in this domain [15].

Substitution mutagenesis of the CASKIN2 peptide binding cleft

A CASKIN2 SH3 domain double mutant (K290Y, R319W) that we termed SH3-2x, was expressed to reintroduce a full complement of aromatic amino acids in the peptide binding cleft. Using circular dichroism (CD) spectroscopy, the SH3-2x protein demonstrated a 20 °C increase in its thermal denaturation midpoint (Table 3). This observation suggests that aromatics in these two positions contribute to the stability of the SH3 fold. Given the high degree of

Table 3 Thermal denaturation midpoints of CASKIN2 SH3 domain mutants

SH3	Mutation(s)	T_m (°C)
WT	none	50
2x	K290Y, R319W	70
6x	K290Y, H296E, A300E, R319W, Y336L, I341F	70

structural similarity to the STAM2 SH3 domain in complex with a UBPY peptide [16], we introduced four more substitutions into the SH3-2x framework to produce a full mimic of the STAM2 interaction in the CASKIN SH3 domain. Overall, this new mutant termed SH3-6x, contains two substitutions to reintroduce aromatics (K290Y, R319W), two minor substitutions that maintain hydrophobicity (Y336L, I341F) and two substitutions in the RT-loop (H296E, A300E) that provide important complementary ionic contacts to the UBPY peptide ligand in STAM2. Since the SH3-2x mutant was a subset of the SH3-6x mutant, the expressed protein had a similar thermal denaturation midpoint of 70 °C (Table 3). When tested against a Class-III UBPY peptide containing an RxxKP motif, high affinity binding (0.46 ± 0.11 μM) was observed (Fig. 2) in the same order of magnitude as another Class-III SH3 domain GADS with an SLP-76 peptide ligand [17]. As expected, the wild type CASKIN2 SH3 domain had an over 100 fold less affinity for the UBPY peptide. While ultralow affinity SH3 domains with biological relevance have been reported, there is no evidence to suggest if CASKIN2 is among this group [18, 19].

Conclusions

The NMR structure of the CASKIN2 SH3 domain demonstrates a variant ligand binding cleft populated with charged amino acids. Based on this unique feature, we hypothesized that the CASKIN2 SH3 domain may bind a ligand that does not belong to any of the canonical, mainly proline-rich sequences that SH3 domains typically bind. Since a phage display failed to identify a ligand for the CASKIN2 SH3 domain, we concluded that it may indeed be non-functional. Increased thermal stability and ligand binding could be engineered back into the CASKIN2 SH3 domain sequence with relatively few amino acid substitutions. This results presented in the report emphasize the considerable versatility that SH3 domains, and signaling adaptor proteins, in general, have at their disposal.

Methods
Cloning, expression and protein purification
The human CASKIN2 SH3 domain (aa. 284–348; Uniprot Q8WXE0) was amplified by PCR from a human cDNA and inserted into the *BamHI* and *XhoI* restriction

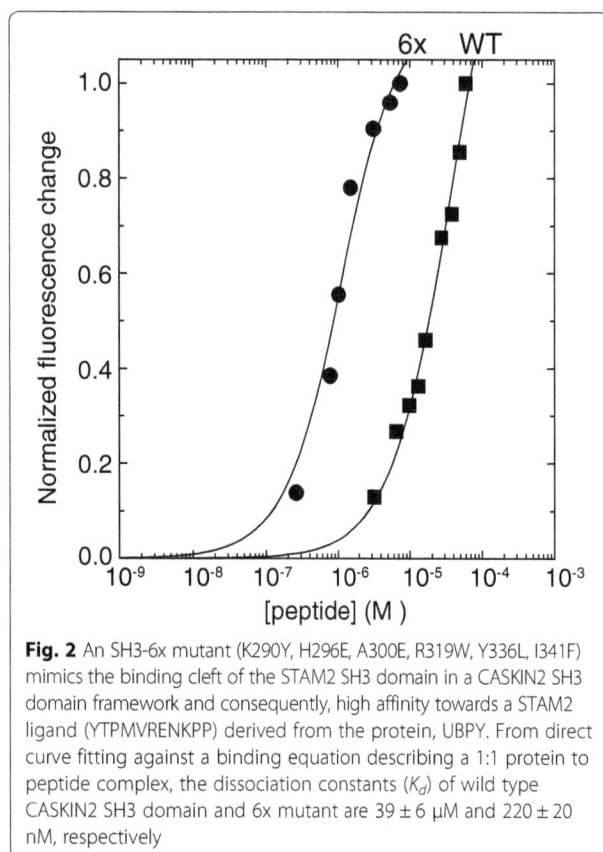

Fig. 2 An SH3-6x mutant (K290Y, H296E, A300E, R319W, Y336L, I341F) mimics the binding cleft of the STAM2 SH3 domain in a CASKIN2 SH3 domain framework and consequently, high affinity towards a STAM2 ligand (YTPMVRENKPP) derived from the protein, UBPY. From direct curve fitting against a binding equation describing a 1:1 protein to peptide complex, the dissociation constants (K_d) of wild type CASKIN2 SH3 domain and 6x mutant are 39 ± 6 μM and 220 ± 20 nM, respectively

sites of pGEX4T2 (GE Life Sciences) followed by transformation into *E. coli* BL21:DE3 to produce a glutathione S-transferase (GST) tagged fusion protein with intervening thrombin protease cleavage site. Two CASKIN2 SH3 domain mutants, termed 2x and 6x, contained two aromatic substitutions to reconstitute a typical hydrophobic binding cleft and six substitutions mimicking the canonical binding cleft of the STAM2 SH3 domain. These mutants were made using Quikchange method (Agilent). All constructs were verified by sequencing at the York University Core Facility. Isotopic labeling of wild type GST-CASKIN2 SH3 domain and mutants for NMR spectroscopy was achieved by performing a 1.5 L fermentation in a minimal medium containing 1 g of $^{15}NH_4Cl$ as the sole nitrogen source and/or 3 g of ^{13}C-glucose as the sole carbon source. The cell pellet was suspended in T300 buffer (20 mM Tris-HCl, 300 mM NaCl, 0.05 % NaN_3) and lysed by French press. Purification of the GST-CASKIN2 SH3 protein was achieved by glutathione affinity chromatography (GE Biosciences). After an 8-h digestion with activated human thrombin (Sigma-Aldrich), the SH3 domain was resolved from the GST carrier protein by gel filtration chromatography (Sephacryl-100, HiLoad 16/60; GE Life Sciences). The final buffer for all analyses was phosphate buffered saline (PBS; 20 mM sodium phosphate, pH 7.8, 0.15 M NaCl, 0.05 % (w/v) sodium azide).

Circular dichroism

Far UV spectra (190–260 nm) of CASKIN SH3 proteins at 20 μM in PBS were obtained on a Jasco J-810 spectropolarimeter and processed with Spectra Analysis 1.54.04 software. A midpoint thermal denaturation curve (T_m) was generated by monitoring the ellipticity signal at 220 nm as the temperature was ramped from 20 to 90 °C at a rate of 1 °C/min.

Peptide binding

A solution of peptide YTPMVRENKPP at > 90 % purity (Canpeptide; Montreal, QC) corresponding to the UBPY ligand in the STAM2 SH3 domain structure (PDB: 1UJ0) was titrated into 1 μM wild type CASKIN2 SH3 domain, a 6x CASKIN-SH3 domain mutant, or a reference cell containing buffer only (10 mM Tris-HCl, pH 7.5, 150 mM NaCl, 0.05 % sodium azide). Upon excitation at 280 nm, intrinsic fluorescence emission was measured at 340 nm corresponding to one partially buried tryptophan in the CASKIN2 SH3 domain. Each measurement was made three times and averaged.

NMR spectroscopy and structure determination

All experiments were performed on a uniformly $^{13}C,^{15}N$ labeled sample of the CASKIN2 SH3 domain sample at 0.6 mM plus 10 % D_2O. A conventional heteronuclear, triple-resonance strategy was employed with all experiments being acquired on an Agilent 600 MHz spectrometer equipped with a 5 mm cryoprobe. Backbone directed experiments : HNCACB and CBCAcoNH ($^{15}N_{sw} = 1450$ Hz, $^{15}N_{pts} = 28$; $^{13}C_{sw}$, $= 11309$ Hz, $^{13}C_{pts} = 40$), HNCO and HNcaCO ($^{15}N_{sw} = 1450$ Hz, $^{15}N_{pts} = 28$; $^{13}C_{sw}$, $= 2262$ Hz, $^{13}C_{pts} = 28$). Side chain directed experiments: HccoNH ($^{15}N_{sw} = 1450$ Hz, $^{15}N_{pts} = 28$; $^{1}H_{sw}$, $= 6596$ Hz, $^{1}H_{pts} = 80$), hCcoNH ($^{15}N_{sw} = 1450$ Hz, $^{15}N_{pts} = 28$; $^{13}C_{sw}$, $= 11309$ Hz, $^{13}C_{pts} = 28$), HCCH-TOCSY ($^{13}C_{sw}$, $= 11309$ Hz, $^{13}C_{pts} = 28$), aromatic HBcbcgCD and HBcbcgcdCE ($^{13}C_{sw}$, $= 4524$ Hz, $^{13}C_{pts} = 32$). Distance restrains were measures from peak volumes from a ^{15}N-edited NOESY ($^{15}N_{sw} = 1450$ Hz, $^{15}N_{pts} = 28$; $^{1}H_{sw}$, $= 6596$ Hz, $^{1}H_{pts} = 80$, 100 ms mixing time) and ^{13}C-edited aliphatic ($^{1}H_{sw}$, $= 6596$ Hz, $^{1}H_{pts} = 80$, $^{13}C_{sw}$, $= 2800$ Hz, $^{13}C_{pts} = 36$, 100 ms mixing time) and aromatic 3D-NOESYs aliphatic ($^{1}H_{sw}$, $= 6596$ Hz, $^{1}H_{pts} = 80$, $^{13}C_{sw}$, $= 2800$ Hz, $^{13}C_{pts} = 24$, 100 ms mixing time). Datasets were processed with NMRpipe [20] and interpreted with NMRView [21]. Distance restraints were obtained using CYANA. Backbone torsion angles were predicted from backbone chemical shift data with TALOS+ [22]. From an initial set of 500 structures calculated with CYANA 2.0 [23], the top 25 structures were selected with no NOE violations > 0.3 Å and no torsion angle violations < 5°. This ensemble was then subjected to additional refinement in explicit solvent [24] with XPLOR-NIH 2.30

[25] and 15 final lowest energy structures were selected for deposition to the Protein Data Bank.

Abbreviations

CASK: Calcium/calmodulin-dependent serine protein kinase; CASKIN2: CASK interacting protein-2; CD: Circular dichroism; CID: CASK interaction domain; GST: glutathione S-transferase; NMR: Nuclear magnetic resonance; PDB: Protein data bank

Acknowledgements
Not applicable.

Funding
This research was funded by an operating grant (MOP-81250) from the Canadian Institutes for Health Research to LWD.

Authors' contributions

JJK and LWD performed experiments and analyzed the data. LWD wrote the manuscript. JJK and LWD approved the manuscript.

Competing interests

The authors declare they have no competing interests.

Author details

[1]Department of Biology, York University, 4700 Keele Street, Toronto, ON M3J 1P3, Canada. [2]Present Address: McEwen Centre for Regenerative Medicine, Ontario Cancer Institute, 101 College Street, Toronto, ON M5G 1L7, Canada.

References

1. Weng Y-L, Liu N, DiAntonio A, Broihier HT. The cytoplasmic adaptor protein Caskin mediates Lar signal transduction during Drosophila motor axon guidance. J Neurosci. 2011;31:4421–33.
2. Good MC, Zalatan JG, Lim WA. Scaffold proteins: hubs for controlling the flow of cellular information. Science. 2011;332:680–6.
3. Bornberg-Bauer E. Signals: tinkering with domains. Sci Signal. 2010;3:pe31–1.
4. Tabuchi K, Biederer T, Butz S, Sudhof TC. CASK participates in alternative tripartite complexes in which Mint 1 competes for binding with caskin 1, a novel CASK-binding protein. J Neurosci. 2002;22:4264–73.
5. Butz S, Okamoto M, Sudhof TC. A tripartite protein complex with the potential to couple synaptic vesicle exocytosis to cell adhesion in brain. Cell. 1998;94:773–82.
6. Stafford RL, Ear J, Knight MJ, Bowie JU. The molecular basis of the Caskin1 and Mint1 interaction with CASK. J Mol Biol. 2011;412:3–13.
7. Ernst A, Sazinsky SL, Hui S, Currell B, Dharsee M, Seshagiri S, et al. Rapid evolution of functional complexity in a domain family. Sci Signal. 2009;2:ra50.
8. Hsueh Y-P. The role of the MAGUK protein CASK in neural development and synaptic function. CMC. 2006;13:1915–27.
9. Hsueh Y-P. A versatile player. J Mol Biol. 2011;412:1–2.
10. Stafford RL, Hinde E, Knight MJ, Pennella MA, Ear J, Digman MA, et al. Tandem SAM domain structure of human Caskin1: a presynaptic, self-assembling scaffold for CASK. Structure. 2011;19:1826–36.
11. Pesti S, Balázs A, Udupa R, Szabó B, Fekete A, Bőgel G, et al. Complex formation of EphB1/Nck/Caskin1 leads to tyrosine phosphorylation and structural changes of the Caskin1 SH3 domain. Cell Commun Signal. 2012;10:36.
12. Balázs A, Csizmok V, Buday L, Rakács M, Kiss R, Bokor M, et al. High levels of structural disorder in scaffold proteins as exemplified by a novel neuronal protein, CASK-interactive protein1. FEBS J. 2009;276:3744–56.
13. Kaneko T, Kumasaka T, Ganbe T, Sato T, Miyazawa K, Kitamura N, et al. Structural insight into modest binding of a non-PXXP ligand to the signal transducing adaptor molecule-2 Src homology 3 domain. J Biol Chem. 2003;278:48162–8.
14. Musacchio A, Saraste M, Wilmanns M. High-resolution crystal structures of tyrosine kinase SH3 domains complexed with proline-rich peptides. Nat Struct Biol. 1994;1:546–51.
15. Kobashigawa Y, Sakai M, Naito M, Yokochi M, Kumeta H, Makino Y, et al. Structural basis for the transforming activity of human cancer-related signaling adaptor protein CRK. Nat Struct Mol Biol. 2007;14:503–10.
16. Kaneko T, Huang H, Zhao B, Li L, Liu H, Voss CK, et al. Loops govern SH2 domain specificity by controlling access to binding pockets. Sci Signal. 2010;3:ra34–4.
17. Liu Q, Berry D, Nash P, Pawson T, McGlade CJ, Li SS-C. Structural basis for specific binding of the Gads SH3 domain to an RxxK motif-containing SLP-76 peptide: a novel mode of peptide recognition. Mol Cell. 2003;11:471–81.
18. Vaynberg J, Qin J. Weak protein-protein interactions as probed by NMR spectroscopy. Trends Biotech. 2006;24:22–7.
19. Vaynberg J, Fukuda T, Chen K, Vinogradova O, Velyvis A, Tu Y, et al. Structure of an ultraweak protein-protein complex and its crucial role in regulation of cell morphology and motility. Mol Cell. 2005;17:513–23.
20. Delaglio F, Grzesiek S, Vuister GW, Zhu G, Pfeifer J, Bax A. NMRPipe: a multidimensional spectral processing system based on UNIX pipes. J Biomol NMR. 1995;6:277–93.
21. Johnson BA. Using NMRView to visualize and analyze the NMR spectra of macromolecules. Methods Mol Biol. 2004;278:313–52.
22. Shen Y, Delaglio F, Cornilescu G, Bax A. TALOS+: a hybrid method for predicting protein backbone torsion angles from NMR chemical shifts. J Biomol NMR. 2009;44:213–23.
23. Güntert P, Buchner L. Combined automated NOE assignment and structure calculation with CYANA. J Biomol NMR. 2015;62:453-471.
24. Linge JP, Williams MA, Spronk CAEM, Bonvin AMJJ, Nilges M. Refinement of protein structures in explicit solvent. Proteins. 2003;50:496–506.
25. Schwieters CD, Kuszewski JJ, Tjandra N, Clore GM. The Xplor-NIH NMR molecular structure determination package. J Magn Reson. 2003;160:65–73.

21

Controlled dehydration improves the diffraction quality of two RNA crystals

HaJeung Park[1]* (ID), Tuan Tran[2], Jun Hyuck Lee[3,4], Hyun Park[3,4] and Matthew D. Disney[2]*

Abstract

Background: Post-crystallization dehydration methods, applying either vapor diffusion or humidity control devices, have been widely used to improve the diffraction quality of protein crystals. Despite the fact that RNA crystals tend to diffract poorly, there is a dearth of reports on the application of dehydration methods to improve the diffraction quality of RNA crystals.

Results: We use dehydration techniques with a Free Mounting System (FMS, a humidity control device) to recover the poor diffraction quality of RNA crystals. These approaches were applied to RNA constructs that model various RNA-mediated repeat expansion disorders.

Conclusion: The method we describe herein could serve as a general tool to improve diffraction quality of RNA crystals to facilitate structure determinations.

Keywords: Post-crystallization modification, RNA, Crystal dehydration, Free mounting system, Repeat expansion

Background

Even with sufficient size and volume, RNA molecules often yield poorly diffracting crystals for X-ray diffraction analysis. The nature of RNA, such as its repetitive negatively charged phosphate backbone, scarce functional groups, and high structural flexibility, results in poor packing of RNA molecules in crystals, which consequently leads to poor diffraction. For the same reasons, various crystal optimization efforts to improve diffraction quality often result in only a modest response. A common method to overcome this problem is to re-design the construct and repeat the crystallization screening until a perfect construct is obtained for diffraction experiments [1, 2].

Dehydration methods were utilized for many protein crystals, where the diffraction quality improved dramatically both in the resolution limit and mosaicity [3–5]. Crystal dehydration can be achieved by either vapor diffusion or humidity control instrumentations. In vapor diffusion, the main precipitants, used at a slightly higher concentration than used in the crystallization condition, are applied as the dehydration agents. The subject crystals are incubated for hours to days with a gradual increase of the dehydration agents [6]. A more rapid way of dehydration can be performed with humidity controlled devices such as Free Mounting System (FMS) and H1Cb, and an in-house made device [4, 5, 7]. The basic concept of the humidity controlled device is to encapsulate a bare crystal within a humidity- and temperature- controlled air stream and then gradually change the humidity to the desired level while observing the effect of dehydration by measuring the diffractions of the subject crystal. The advantage of the device over the vapor diffusion method is that the relative humidity (Rh) of a crystal can be fine-tuned over a desired time frame and the outcome of the dehydration is observed in real time. However, mounting a crystal onto a dehydration device requires some practice and only one crystal can be tested at a time.

Repeat expansion diseases are human genetic disorders affecting the nervous and muscular systems and are caused by the expansion of repeated microsatellite sequences in the coding or noncoding regions of the gene [8]. The repeat modules are generally three to six nucleotides in length [8, 9]. Crystal structures of these pathogenic repeat sequences could give insight into disease

---

* Correspondence: hajpark@scripps.edu; disney@scripps.edu
[1]X-ray Core Facility, The Scripps Research Institute, Scripps Florida, 130 Scripps Way, Jupiter, FL 33458, USA
[2]Department of Chemistry, The Scripps Research Institute, Scripps Florida, 130 Scripps Way, Jupiter, FL 33458, USA
Full list of author information is available at the end of the article

mechanisms and also give insights into the development of therapeutics [10]. Therefore, we have crystallized and determined structures of RNAs containing CCUG and AUUCU repeat sequences. The RNAs containing these repeats yielded a number of hits rather quickly in screening. However, none of the crystals diffracted beyond a ~15 Å resolution. We subjected these crystals to dehydration before designing and testing new constructs. Dehydration dramatically improved the diffraction limit of the crystals, and crystal structures were determined successfully and reported elsewhere [11, 12].

Although dehydration methods have been practiced to rescue many protein crystals of poor diffraction quality, there is only one report that describes this technique in detail for nucleic acid crystals in addition to an anecdotal account of *glmS* ribozyme crystals [13–15]. The dehydration approach used by Zhang and Ferre-D'Amare was to soak the poorly diffracting RNA crystals in a higher concentration of precipitants while exchanging cations to induce better contact among the RNAs in the crystal lattice. In *glmS* ribozyme crystals, the dehydration was unintentionally introduced by a stabilization solution for cryocooling. Our approach of crystal dehydration, however, was completed using FMS exclusively. The technique is also appropriate for the quick evaluation of RNA crystals with poor diffraction, even after they have grown to a sufficient size. Herein, we describe the dehydration method we have used to improve the diffraction quality of RNA crystals and our thoughts about the technique in general. This approach could have broad utility for structural studies of RNA crystals.

Methods
Crystallization screening
RNA samples containing three repeats of CCUG and two repeats of AUUCU were screened against the Nucleix Suite (Qiagen, Valencia, CA, USA) using a Gryphon crystallization robot (Art Robinsons, Sunnyvale, CA, USA) at room temperature. Crystals from hit conditions were tested for diffraction using the in-house X-ray diffraction system equipped with a Mar345dtb (Rayonix, Evanstone, IL, USA) and a Micromax 007 HFM (Rigaku Americas, The Woodlands, TX, USA). None of the tested crystals showed a promising diffraction pattern. Crystals obtained from precipitants containing 100 mM ammonium acetate, 5 mM $MgSO_4$, 50 mM 2-(N-morpholino)ethanesulfonic acid, pH 6.0, and 600 mM NaCl were used to test our dehydration method.

Dehydration protocol and diffraction imaging
Dehydration of the crystals was completed on the FMS (Rigaku Americas). The relative humidity (Rh) of the precipitant was determined to be ~96 %. To test the response of the crystals to dehydration, single crystals

were mounted on a Litholoop (Molecular Dimensions, Altamonte Springs, FL, USA) and placed in a goniometer head. Diffraction images of the crystals were collected every 5 min, while the Rh of the crystals was reduced to 70 % at a gradient of 0.25 % Rh change per min. The best diffracting crystals with a Bragg spacing of 3.0 Å and 3.3 Å from the CCUG and AUUCU crystals, respectively, were harvested by following the established dehydration protocol. The crystals were coated with perfluoropolyether cryo oil (Hampton Research, Aliso Viejo, CA, USA) to prevent any change in humidity and were then immediately cryocooled by submersion in liquid nitrogen.

Data collection and structure refinement
The diffraction dataset of a CCUG crystal in the space group of $P4_12_12$ was obtained on the PILATUS detector at beam line 11–1 of Stanford Synchrotron Radiation Lightsource, SLAC. The dataset of an AUUCU crystal in the space group of $P4_1$ was obtained on the MAR-300 detector at beam line ID-G of LS-CAT, Advanced Photon Source, Argonne National Laboratory. Datasets were processed with iMOSFLM [16]. Structures were determined by molecular replacement using Phaser [17] with the tetraloop-tetraloop receptor of PDB ID 4FNJ [18] as the search model. The refined final structure was deposited under the PDB IDs 4 K27 and 5BTM for CCUG and AUUCU, respectively, and the research papers featuring the structures were published separately elsewhere [11, 12].

Results and discussion
Construct design and crystallization
To overcome the inherent limitation of intermolecular crystal contact in RNAs, the GAAA tetraloop and the tetraloop receptor have been utilized as a general module to promote RNA crystallization [18, 19]. We applied this strategy and designed RNA constructs containing the target repeat sequences (Fig. 1a). Constructs were screened against the Nucleix suite (Qiagen), yielding a number of crystal hits with a tetragonal bipyramidal shape (Fig. 1b). However, none of the crystals tested showed a diffraction pattern. Before proceeding to redesign RNA sequences for new crystallization trials, we tested whether the non-diffracting crystals could be rescued by the dehydration technique.

Overall structure and crystal packing
The structure of the CCUG crystal showed that all bases were well ordered, with an overall B value of 35.1 $Å^2$, and the electron density map correlated with the model. Detailed structural analysis can be found elsewhere [11]. On the contrary, approximately 20 % of the bases in the AUUCU crystal were not modeled owing to poor

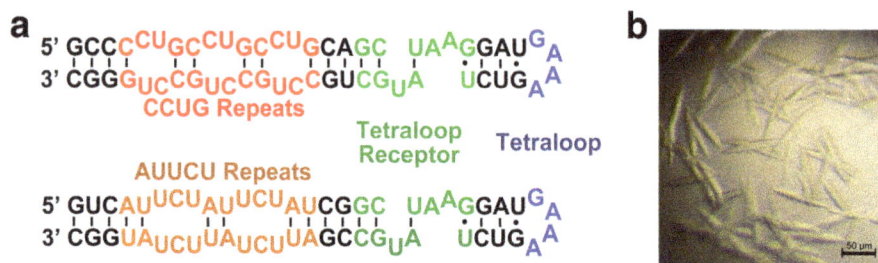

Fig. 1 Design and crystallization of RNAs with repeat expansions. **a** Representations of the secondary structure of the RNAs used for crystallization and structure determination. **b** Picture showing the typical morphology of the crystals used for dehydration experiments

electron density. The affected residues were located at the stem end [12]. The most prominent crystal packing interactions of both crystal forms were mediated through tetraloop and tetraloop receptor interactions. Coaxial stacking and phosphate backbone interactions between neighboring RNA molecules were also observed in both crystals. Although the symmetry-related RNA molecules aligned coaxially in AUUCU crystals, base stacking between them could not be observed owing to disorder in the stem ends (Fig. 2).

Comparison of the molecular packing of the two crystal forms revealed that the overall molecular arrangements are very similar between the two crystals (Fig. 2). Distinct differences appeared to be caused by a slight rearrangement of two molecules in parallel; that is, the two asymmetric molecules in the AUUCU crystal and their equivalent molecules in the CCUG crystal. The asymmetric molecules slid along the c-axis compared with the equivalent molecules in the CCUG crystal, resulting in tighter interaction between the two asymmetric unit molecules, shorter unit cell values in the a- and b-axes, and the lower solvent content of the crystal. These observations also suggest the possibility that the space groups had diverged

during the crystal transformation and the content of the RNA sequences could have dictated each space group formation. Space group transition during dehydration has been reported in monoclinic lysozyme crystals [20].

Crystal dehydration and diffraction analyses

Initial tests showed that the crystals responded to an Rh change through diffraction patterns in the 10–15 Å Bragg spacing range (Fig. 3a). Generally, these changes were too insignificant to affect the diffraction limits. Occasionally, however, crystals underwent a dramatic improvement in the diffraction quality (Fig. 3b). In such cases, the improvements in diffraction quality were noticeable at approximately 85 % Rh and improved continuously until 75 % Rh (Fig. 3 and Additional file 1: Movie). Lowering the Rh further reduced the resolution limit; thus, the crystals were prepared at 75 % Rh for synchrotron data collection.

Diffraction analyses after the dehydration experiments revealed that the CCUG and AUUCU crystals were in different space groups in a primitive tetragonal lattice, even though the two crystals were morphologically identical and had emerged from the same precipitant. The

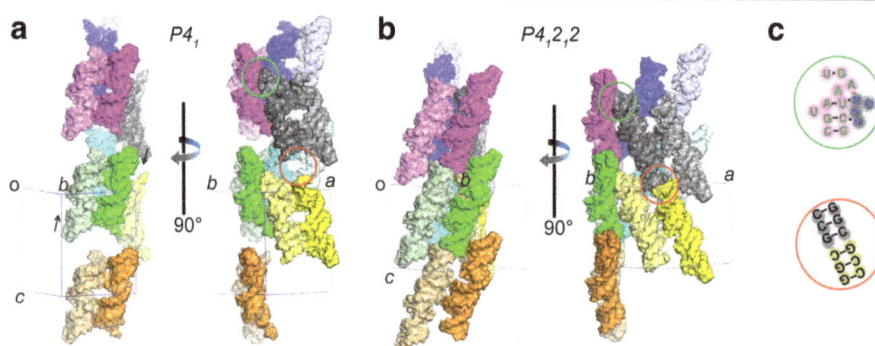

Fig. 2 Comparison of the crystal packing between the AUUCU (**a**) and CCUG (**b**) crystals. The two asymmetric unit molecules of the AUUCU crystal are shown as pure and tinted colors in panel (**a**), and the equivalently arranged molecules in the CCUG crystal are shown with the same colors in panel (**b**). Tetraloop-tetraloop receptor interactions (e.g., green circled area of magenta and grey molecules in panels (**a**) and (**b**)) and the coaxial arrangement (e.g., red circled area of grey and yellow molecules in panels (**a**) and (**b**)) that appear invariant between the two crystals are shown as secondary structural drawings in panel (**c**). The two asymmetric molecules in the AUUCU crystal slid along the c-axis (*arrows*), making the interaction of the two molecules tighter than the equivalent interface in the CCUG crystal (e.g., *green* and *light green*)

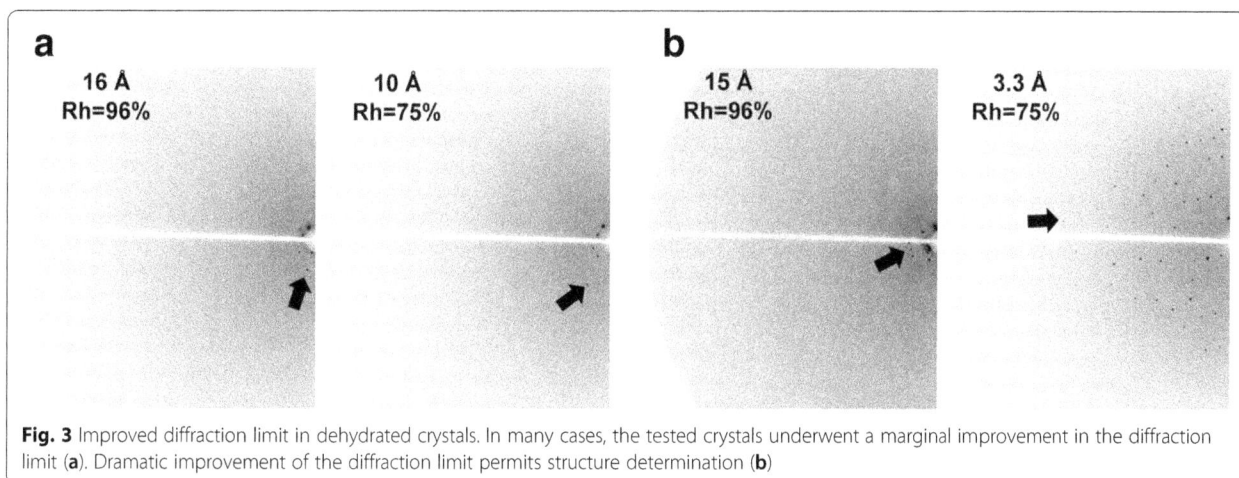

Fig. 3 Improved diffraction limit in dehydrated crystals. In many cases, the tested crystals underwent a marginal improvement in the diffraction limit (a). Dramatic improvement of the diffraction limit permits structure determination (b)

CCUG crystal was in the $P4_12_12$ space group, with one molecule in the asymmetric unit. The unit cell values of the crystal were as follows: a = b = 75.08 Å, c = 59.90 Å, $\alpha = \beta = \gamma = 90°$. The best crystal diffracted to 2.35 Å in a synchrotron radiation source. There was one molecule in the asymmetric unit with a Matthew's coefficient value of 2.53 (51.37 % solvent content). The apparent space group of the AUUCU crystal was also point group 422 ($P422$). However, detailed analysis of the diffraction revealed that the crystal was merohedrally twined (twin fraction = 0.43). Therefore, the space group was lowered to point group 4. The crystal was also severely anisotropic. Molecular replacement followed by structural refinements confirmed the space group to be $P4_1$. The final unit cell values of the crystal were a = b = 63.12 Å, c = 72.95 Å, and $\alpha = \beta = \gamma = 90°$. The best crystal diffracted to 2.78 Å in a synchrotron radiation source. There were two molecules in the asymmetric unit with a Matthew's coefficient value of 2.19 (43.75 % solvent content).

Analysis of the individual diffraction images during dehydration of the AUUCU crystal ($P4_1$) using HKL2000 [21] revealed that the dehydration process had introduced lattice shrinkage. The unit cell values calculated are shown in Fig. 4 and Table 1, along with the approximate humidity. Lattice shrinkage is a common observation in the dehydration process and has been reported by others [4, 22–24]. For example, dehydration of bovine mitochondrial F_1-ATPase with orthorhombic crystals using FMS resulted in a more dramatic shrinkage of 12 and 6 % in the a- and c-axes, respectively, during the Rh change of 96 to 90 %. The lattice change in our crystal between Rh of 80 and 75 % was a 2.3 % reduction of the a- and b-axes and a 1.9 % reduction of the c-axis. The contractions are relatively minor compared to F_1-ATPase even though Rh value decrease was about 15 %. Large contraction in protein crystals are related to domain motions and crystal contact improvement. Considering no diffraction in Rh range of 96 ~ 80 %, it is

safe to assume that intermolecular contacts of RNAs in our crystals at this state are poor. The crystals are held together through tetraloop-tetraloop receptor interactions and possibly through unorganized phosphate backbone interactions of neighboring helices with coarse co-axial and parallel packing. The small changes in contractions beyond the Rh 80 % indicate the overall arrangements of the RNA molecules in the untreated crystals may not be much different than that in the dehydrated crystals. Also, the small changes of contractions after the dehydration unlikely influenced the RNA structures. Helical parameters of the RNAs from the two dehydrated crystals are comparable to those of RNAs without dehydration [11, 12]. The variation in the degree of shrinkage can be explained by variables such as crystal packing interactions, the flexibility of subject molecules (and subdomain movements), the size of the unit cell, and the solvent content.

An earlier crystal dehydration study by Dobrianov et al., using tetragonal lysozyme, reported lattice shrinkage

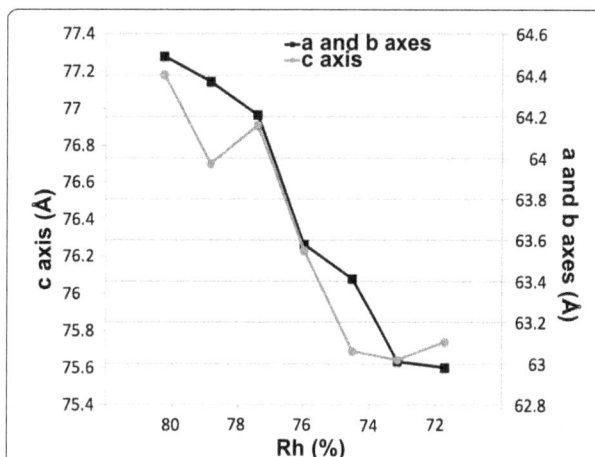

Fig. 4 Relative humidity versus unit cell values during the dehydration experiment of an AUUCU crystal

Table 1 Effect of dehydration on unit cell values and diffraction limit of the AUUCU crystal ($P4_1$)

Image No.	Rh (%)	a or b axis (Å)	c axis (Å)	Denzo distortion index (%)	Diffraction limit (Å)	Solvent content (%)
1 ~ 11	96 - 82	N/A	N/A	N/A	16 – 5.5	N/A
12	80	64.49	77.18	0.28	4	49.07
13	79	64.37	76.7	0.17	3.5	48.56
14	77	64.21	76.91	0.18	3.3	48.44
15	76	63.58	76.23	0.15	3	46.94
16	75	63.41	75.69	0.12	3	46.28
17	73	63.01	75.64	0.06	3.3	45.56
18	72	62.98	75.74	0.08	3.3	45.58
19	73	63.42	75.98	0.17	3	46.50
20	75	63.58	76.28	0.13	3	46.98

featuring nonlinear contraction of the crystal lattice and reproducible lattice transition points, beyond which irreversibility occurred (Rh value of 88 % for tetragonal lysozyme crystals) [23]. Such nonlinear contraction was also observed during the dehydration experiments of the bovine mitochondrial F_1-ATPase [24]. Although inconclusive as a result of coarse data points, our RNA crystal also showed a similar trend, where the lattice shrinkage rate decreased towards the end of the dehydration experiment (Fig. 4 and Table 1).

Points to consider for dehydration experiments

Although the response to dehydration was evident, it did not always guarantee high-quality diffraction at the end of the dehydration experiment. Within the same batch of crystals, there were crystal-to-crystal variations, where certain crystal diffracted well and others did not (Fig. 3). Visual inspection of the crystal could not identify or predict whether the crystal would yield quality diffraction after dehydration. Therefore, the identification of well-diffracting crystals was solely dependent on screening through dehydration experiments. Only about 13 %, or 2 out of 15 for CCUG crystals and 1 out of 8 AUUCU crystals, were of good diffraction quality. Bowler et al. also observed variability between crystals and their reaction to dehydration [24]. Therefore, if the diffraction quality of a tested crystal did not improve past the transition point (~80 % in our case), then the testing was terminated, and we moved on to a new crystal to save screening time. It needs to be further investigated whether the low reproducibility could be improved by modifying Rh gradient rate or other dehydration schemes.

Reported Rh value of saturated NaCl at 20 °C is 75 % [25]. Therefore, we also tested vapor diffusion dehydration following the methods published by Heras and Martin [6], where the concentration of reservoir NaCl, the main precipitant, was increase gradually from 600 mM to saturation point over a 5 day period. However, treatment had no effect on improving poor diffraction quality. The negative result could be because of an insufficient sample number (10 crystals were tested for diffraction), as we did not pursue the method extensively. Furthermore, the process of crystal dehydration using a higher concentration of precipitant takes longer to complete, as the sealed crystallization chamber needs to be equilibrated over hours to days, whereas dehydration by the FMS takes <2 h to complete. On the other hand, vapor diffusion dehydration can test multiple crystals with multiple variables simultaneously; controlled dehydration using the FMS must be completed serially.

Conclusion

RNA molecules with multiple motifs are in general less likely to produce well-ordered crystals owing to the inherent flexibility of their structures and therefore require significant effort in the design and screening of constructs. The dehydration method presented here can be used as a routine technique to test RNA crystals with poor diffraction quality as an alternative to new construct screenings, provided that suitable instrumentation is available.

Abbreviation

FMS: Free mounting system

Acknowledgements

We thank Dr. Jessica Childs-Disney for discussions and critical review of the manuscript and the staff at the LS-CAT, APS, and beam line 11–1, SSRL, SLAC for synchrotron supports.

Funding

This work was funded by the National Institutes of Health (R01-GM097455 to MDD), The Scripps Research Institute, and the Antarctic Organisms: Cold-Adaptation Mechanisms and Its Application grant funded by the Korea Polar Research Institute (PE15070 and PE16070 to HJP and HP).

Authors' contributions

HJP designed and conducted experiments. TT prepared RNA samples. HJP, JHL, HP and MDD interpreted the data, wrote and revised the manuscript. All authors have read and approved this manuscript.

Competing interests
The authors declare that they have no competing interests.

Author details
[1]X-ray Core Facility, The Scripps Research Institute, Scripps Florida, 130 Scripps Way, Jupiter, FL 33458, USA. [2]Department of Chemistry, The Scripps Research Institute, Scripps Florida, 130 Scripps Way, Jupiter, FL 33458, USA. [3]Unit of Polar Genomics, Korea Polar Research Institute, Incheon 21990, Republic of Korea. [4]Department of Polar Sciences, University of Science and Technology, Incheon 21990, Republic of Korea.

References
1. Reyes FE, Garst AD, Batey RT. Strategies in RNA crystallography. Methods Enzymol. 2009;469:119–39.
2. Oubridge C, Ito N, Teo CH, Fearnley I, Nagai K. Crystallisation of RNA-protein complexes. II. The application of protein engineering for crystallisation of the U1A protein-RNA complex. J Mol Biol. 1995;249(2):409–23.
3. Heras B, Edeling MA, Byriel KA, Jones A, Raina S, Martin JL. Dehydration converts DsbG crystal diffraction from low to high resolution. Structure. 2003;11(2):139–45.
4. Sanchez-Weatherby J, Bowler MW, Huet J, Gobbo A, Felisaz F, Lavault B, Moya R, Kadlec J, Ravelli RB, Cipriani F. Improving diffraction by humidity control: a novel device compatible with X-ray beamlines. Acta Crystallogr D Biol Crystallogr. 2009;65(Pt 12):1237–46.
5. Kiefersauer R, Than ME, Dobbek H, Gremer L, Melero M, Strobl S, Dias JM, Soulimane T, Huber R. A novel free-mounting system for protein crystals: transformation and improvement of diffraction power by accurately controlled humidity changes. J Appl Cryst. 2000;33(5):1223–30.
6. Heras B, Martin JL. Post-crystallization treatments for improving diffraction quality of protein crystals. Acta Crystallogr D Biol Crystallogr. 2005;61(Pt 9): 1173–80.
7. Baba S, Hoshino T, Ito L, Kumasaka T. Humidity control and hydrophilic glue coating applied to mounted protein crystals improves X-ray diffraction experiments. Acta Crystallogr D Biol Crystallogr. 2013;69(Pt 9):1839–49.
8. La Spada AR, Taylor JP. Repeat expansion disease: progress and puzzles in disease pathogenesis. Nat Rev Genet. 2010;11(4):247–58.
9. DeJesus-Hernandez M, Mackenzie IR, Boeve BF, Boxer AL, Baker M, Rutherford NJ, Nicholson AM, Finch NA, Flynn H, Adamson J, et al. Expanded GGGGCC hexanucleotide repeat in noncoding region of C9ORF72 causes chromosome 9p-linked FTD and ALS. Neuron. 2011;72(2): 245–56.
10. Rzuczek SG, Park H, Disney MD. A toxic RNA catalyzes the in cellulo synthesis of its own inhibitor. Angewandte Chemie. 2014;53(41):10956–9.
11. Childs-Disney JL, Yildirim I, Park H, Lohman JR, Guan L, Tran T, Sarkar P, Schatz GC, Disney MD. Structure of the myotonic dystrophy type 2 RNA and designed small molecules that reduce toxicity. ACS Chem Biol. 2014;9(2): 538–50.
12. Park H, Gonzalez AL, Yildirim I, Tran T, Lohman JR, Fang P, Guo M, Disney MD. Crystallographic and computational analyses of AUUCU repeating RNA that causes spinocerebellar ataxia type 10 (SCA10). Biochemistry. 2015.
13. Zhang J, Ferre-D'Amare AR. Dramatic improvement of crystals of large RNAs by cation replacement and dehydration. Structure. 2014;22(9):1363–71.
14. Zhang J, Ferre-D'Amare AR. Co-crystal structure of a T-box riboswitch stem I domain in complex with its cognate tRNA. Nature. 2013;500(7462):363–6.
15. Klein DJ, Ferre-D'Amare AR. Structural basis of glmS ribozyme activation by glucosamine-6-phosphate. Science. 2006;313(5794):1752–6.
16. Battye TG, Kontogiannis L, Johnson O, Powell HR, Leslie AG. iMOSFLM: a new graphical interface for diffraction-image processing with MOSFLM. Acta Crystallogr D Biol Crystallogr. 2011;67(Pt 4):271–81.
17. Adams PD, Afonine PV, Bunkoczi G, Chen VB, Davis IW, Echols N, Headd JJ, Hung LW, Kapral GJ, Grosse-Kunstleve RW, et al. PHENIX: a comprehensive Python-based system for macromolecular structure solution. Acta Crystallogr D Biol Crystallogr. 2010;66(Pt 2):213–21.
18. Coonrod LA, Lohman JR, Berglund JA. Utilizing the GAAA tetraloop/receptor to facilitate crystal packing and determination of the structure of a CUG RNA helix. Biochemistry. 2012;51(42):8330–7.
19. Ferre-D'Amare AR, Zhou K, Doudna JA. A general module for RNA crystallization. J Mol Biol. 1998;279(3):621–31.
20. Klingl S, Scherer M, Stamminger T, Muller YA. Controlled crystal dehydration triggers a space-group switch and shapes the tertiary structure of cytomegalovirus immediate-early 1 (IE1) protein. Acta Crystallogr D Biol Crystallogr. 2015;71(Pt 7):1493–504.
21. Otwinowski Z, Minor W. Processing of X-ray Diffraction Data Collected in Oscillation Mod, Methods in Enzymology, vol 276. In: Carter CW Jr., Sweet RM, editors. Macromolecular Crystallography, part A. New York: Academic Press; 1997. p. 307–326.
22. Russi S, Juers DH, Sanchez-Weatherby J, Pellegrini E, Mossou E, Forsyth VT, Huet J, Gobbo A, Felisaz F, Moya R, et al. Inducing phase changes in crystals of macromolecules: status and perspectives for controlled crystal dehydration. J Struct Biol. 2011;175(2):236–43.
23. Dobrianov I, Kriminski S, Caylor CL, Lemay SG, Kimmer C, Kisselev A, Finkelstein KD, Thorne RE. Dynamic response of tetragonal lysozyme crystals to changes in relative humidity: implications for post-growth crystal treatments. Acta Crystallogr D Biol Crystallogr. 2001;57(Pt 1):61–8.
24. Bowler MW, Montgomery MG, Leslie AG, Walker JE. Reproducible improvements in order and diffraction limit of crystals of bovine mitochondrial F(1)-ATPase by controlled dehydration. Acta Crystallogr D Biol Crystallogr. 2006;62(Pt 9):991–5.
25. Rockland LB. Saturated salt solutions for static control of relative humidity between 5 and 40 C. Anal Chem. 1960;32(10):1375–6.

PERMISSIONS

The contributors of this book come from diverse backgrounds, making this book a truly international effort. This book will bring forth new frontiers with its revolutionizing research information and detailed analysis of the nascent developments around the world.

We would like to thank all the contributing authors for lending their expertise to make the book truly unique. They have played a crucial role in the development of this book. Without their invaluable contributions this book wouldn't have been possible. They have made vital efforts to compile up to date information on the varied aspects of this subject to make this book a valuable addition to the collection of many professionals and students.

This book was conceptualized with the vision of imparting up-to-date information and advanced data in this field. To ensure the same, a matchless editorial board was set up. Every individual on the board went through rigorous rounds of assessment to prove their worth. After which they invested a large part of their time researching and compiling the most relevant data for our readers.

The editorial board has been involved in producing this book since its inception. They have spent rigorous hours researching and exploring the diverse topics which have resulted in the successful publishing of this book. They have passed on their knowledge of decades through this book. To expedite this challenging task, the publisher supported the team at every step. A small team of assistant editors was also appointed to further simplify the editing procedure and attain best results for the readers.

Apart from the editorial board, the designing team has also invested a significant amount of their time in understanding the subject and creating the most relevant covers. They scrutinized every image to scout for the most suitable representation of the subject and create an appropriate cover for the book.

The publishing team has been an ardent support to the editorial, designing and production team. Their endless efforts to recruit the best for this project, has resulted in the accomplishment of this book. They are a veteran in the field of academics and their pool of knowledge is as vast as their experience in printing. Their expertise and guidance has proved useful at every step. Their uncompromising quality standards have made this book an exceptional effort. Their encouragement from time to time has been an inspiration for everyone.

The publisher and the editorial board hope that this book will prove to be a valuable piece of knowledge for researchers, students, practitioners and scholars across the globe.

LIST OF CONTRIBUTORS

Margaret M Pruitt and Clark R Coffman
Department of Genetics, Development and Cell Biology, Iowa State University, Ames, IA 50011, USA

Monica H Lamm
Department of Chemical and Biological Engineering, Iowa State University, Ames, IA 50011, USA

Bernhard Kuhle and Ralf Ficner
Abteilung für Molekulare Strukturbiologie, Institut für Mikrobiologie und Genetik, Göttinger Zentrum für Molekulare Biowissenschaften, Georg-August-Universität Göttingen, D-37077 Göttingen, Germany

Wei Qiu, Vladimir Romanov, Ashley Hutchinson, Andrés Lin and Maxim Ruzanov
Princess Margaret Cancer Center, University Health Network, Toronto, Ontario, M5G 2C4, Canada

Xiaonan Wang and Benjamin G Neel
Princess Margaret Cancer Center, University Health Network, Toronto, Ontario, M5G 2C4, Canada
Department of Medical Biophysics, University of Toronto, Toronto, Ontario, M5S 1A8, Canada

Emil F Pai
Princess Margaret Cancer Center, University Health Network, Toronto, Ontario, M5G 2C4, Canada
Departments of Biochemistry, Molecular Genetics, and Medical Biophysics, University of Toronto, Toronto, Ontario, M5S 1A8, Canada

Nickolay Y Chirgadze
Princess Margaret Cancer Center, University Health Network, Toronto, Ontario, M5G 2C4, Canada
Department of Pharmacology and Toxicology, University of Toronto, Toronto, Ontario, M5S 1A8, Canada

Kevin P Battaile
Hauptman–Woodward Medical Research Institute, IMCA-CAT, Advanced Photon Source, Argonne National Laboratory, Argonne, Illinois 60439, USA

Andrew M Ellisdon, Qingwei Zhang, Coral Ruby HP Law and James C Whisstock
Department of Biochemistry and Molecular Biology, Monash University, Clayton, VIC 3800, Australia

Travis K Johnson
Department of Biochemistry and Molecular Biology, Monash University, Clayton, VIC 3800, Australia
School of Biological Sciences, Monash University, Clayton, VIC 3800, Australia

Michelle A Henstridge and G Warr
School of Biological Sciences, Monash University, Clayton, VIC 3800, Australia

Seyed Majid Saberi Fathi
Department of Physics, Ferdowsi University of Mashhad, Mashhad, Iran

Jack A Tuszynski
Department of Physics, University of Alberta, Edmonton, Alberta, Canada

Abhishek Basu, Urmisha Das, Supratim Dey and Saumen Datta
Structural Biology and Bioinformatics division, Indian Institute of Chemical Biology, 4 Raja S.C. Mullick Road, Kolkata 700032 West Bengal, India

Kirill E Medvedev and Nikolay A Alemasov
Institute of Cytology and Genetics SB RAS, Prospekt Lavrentyeva 10, Novosibirsk 630090, Russia

Dmitry A Afonnikov
Institute of Cytology and Genetics SB RAS, Prospekt Lavrentyeva 10, Novosibirsk 630090, Russia
Novosibirsk State University, Pirogova str. 2, Novosibirsk 630090, Russia

Nikolay A Kolchanov
Institute of Cytology and Genetics SB RAS, Prospekt Lavrentyeva 10, Novosibirsk 630090, Russia
Novosibirsk State University, Pirogova str. 2, Novosibirsk 630090, Russia
NRC Kurchatov Institute, 1, Akademika Kurchatova pl., Moscow 123182, Russia

Yuri N Vorobjev
Institute of Chemical Biology and Fundamental Medicine SB RAS, Prospekt Lavrentyeva 8, Novosibirsk 630090, Russia

Elena V Boldyreva
Novosibirsk State University, Pirogova str. 2, Novosibirsk 630090, Russia

Institute of Solid Chemistry and Mechanochemistry, SB RAS, Novosibirsk 630090, Russia

Takeshi Kikuchi
Department of Bioinformatics, College of Life Sciences, Ritsumeikan University, 1-1-1 Nojihigashi, Kusatsu, Shiga 525-8577, Japan

Masanari Matsuoka
Department of Bioinformatics, College of Life Sciences, Ritsumeikan University, 1-1-1 Nojihigashi, Kusatsu, Shiga 525-8577, Japan
Japan Society for the Promotion of Science (JSPS), Tokyo, Japan

Laila Niiranen, Kjersti Lian, Kenneth A Johnson
The Norwegian Structural Biology Center (NorStruct), Department of Chemistry, UIT – the Arctic University of Norway, N-9037 Tromsø, Norway

Elin Moe
The Norwegian Structural Biology Center (NorStruct), Department of Chemistry, UIT – the Arctic University of Norway, N-9037 Tromsø, Norway
The Macromolecular Crystallography Unit, Instituto de Tecnologia Química e Biológica (ITQB), Universidade Nova de Lisboa, Oeiras 2780-157, Portugal

Soo Huei Tan, Yahaya M Normi, Adam Thean Chor Leow and Abu Bakar Salleh
Center for Enzyme and Microbial Biotechnology (EMTECH), Faculty of Biotechnology and Biomolecular Sciences, Universiti Putra Malaysia, Serdang, Selangor 43400, Malaysia

Roghayeh Abedi Karjiban
Center for Enzyme and Microbial Biotechnology (EMTECH), Faculty of Biotechnology and Biomolecular Sciences, Universiti Putra Malaysia, Serdang, Selangor 43400, Malaysia
Department of Chemistry, Faculty of Science, Universiti Putra Malaysia, Serdang, Selangor 43400, Malaysia

Mohd Basyaruddin Abdul Rahman
Center for Enzyme and Microbial Biotechnology (EMTECH), Faculty of Biotechnology and Biomolecular Sciences, Universiti Putra Malaysia, Serdang, Selangor 43400, Malaysia

Department of Chemistry, Faculty of Science, Universiti Putra Malaysia, Serdang, Selangor 43400, Malaysia
Malaysia Genome Institute, Ministry of Science, Technology and Innovation, Jalan Bangi, Kajang, Selangor 43000, Malaysia

Abdul Munir Abdul Murad
School of Biosciences and Biotechnology, Faculty of Science and Technology, Universiti Kebangsaan Malaysia, 43600 UKM, Bangi, Selangor, Malaysia

Nor Muhammad Mahadi
Malaysia Genome Institute, Ministry of Science, Technology and Innovation, Jalan Bangi, Kajang, Selangor 43000, Malaysia

Christopher B Stanley, Sudipa Ghimire-Rijal, Xun Lu, Dean A Myles and Matthew J Cuneo
Neutron Sciences Directorate, Oak Ridge National Laboratory, Oak Ridge, TN 37831, USA

Parthapratim Munshi
Neutron Sciences Directorate, Oak Ridge National Laboratory, Oak Ridge, TN 37831, USA
Department of Chemistry, Middle Tennessee State University, Murfreesboro, TN 37132, USA
Department of Chemistry & Center for Informatics, Shiv Nadar University, Dadri, Uttar Pradesh 203207, India

Sarin Chimnaronk, Jatuporn Sitthiroongruang, Monrudee Srisaisup, Albert J. Ketterman and Panadda Boonserm
Institute of Molecular Biosciences, Mahidol University, Salaya, Phuttamonthon, Nakhon Pathom 73170, Thailand

Kanokporn Srisucharitpanit
Faculty of Allied Health Sciences, Burapha University, Mueang District, Saen Sook, Chonburi 20131, Thailand

Renzhi Cao
Computer Science Department, University of Missouri, Columbia, Missouri 65211, USA

Jianlin Cheng
Computer Science Department, University of Missouri, Columbia, Missouri 65211, USA
Informatics Institute, University of Missouri, Columbia, Missouri 65211, USA

Christopher S. Bond Life Science Center, University of Missouri, Columbia, Missouri 65211, USA

Zheng Wang
School of Computing, University of Southern Mississippi, Hattiesburg, MS 39406-0001, USA

Boon Aun Teh and Nazalan Najimudin
School of Biological Sciences, Universiti Sains Malaysia, 11800 USM Pulau Pinang, Malaysia

Sy Bing Choi
School of Industrial Technology, Universiti Sains Malaysia, 11800 USM Pulau Pinang, Malaysia

Nasihah Musa, Abu Bakar Salleh and Yahaya M Normi
Enzyme and Microbial Technology Research Center (EMTECH), Faculty of Biotechnology and Biomolecular Sciences, Universiti Putra Malaysia, 43400 Serdang, Selangor, Malaysia

Few Ling Ling and See Too Wei Cun
School of Health Sciences, Health Campus, Universiti Sains Malaysia, 16150 Kubang Kerian, Kelantan, Malaysia

Habibah A Wahab
Malaysian Institute of Pharmaceuticals and Nutraceuticals, Ministry of Science, Technology and Innovation, Blok 5-A, Halaman Bukit Gambier, 11700 Pulau Pinang, Malaysia

Oliviero Carugo
Department of Structural and Computational Biology, Max F. Perutz Laboratories, Vienna University, Vienna, Austria
Department of Chemistry, University of Pavia, Pavia, Italy

Yu Zhang, Liangliang Wang, Xun Li and Fei Wang
College of Chemical Engineering, Nanjing Forestry University, Nanjing 210037, China
Jiangsu Key Lab of Biomass-Based Green Fuels and Chemicals, Nanjing 210037, China

Wenqian Li and Hao Shi
College of Chemical Engineering, Nanjing Forestry University, Nanjing 210037, China
Jiangsu Key Lab of Biomass-Based Green Fuels and Chemicals, Nanjing 210037, China
Department of Life Science and Chemistry, Huaiyin Institute of Technology, Huaian 223003, China

Xiangqian Li
Department of Life Science and Chemistry, Huaiyin Institute of Technology, Huaian 223003, China

Jianyun Guo, Xiaoling Jin, Ying Ji, Shengdi Fan and Chun-Shan Quan
Department of Bioengineering, College of Life Science, Dalian Nationalities University, Dalian 116600Liaoning, China

Yongbin Xu
Department of Bioengineering, College of Life Science, Dalian Nationalities University, Dalian 116600Liaoning, China
Laboratory of Biomedical Material Engineering, Dalian Institute of Chemical Physics, Chinese Academy of Sciences, Dalian 116023Liaoning, China

Jin-Sik Kim and Nam-Chul Ha
Department of Agricultural Biotechnology, College of Agriculture and Life Sciences, Seoul National University, Gwanak-gu, Seoul 151-742, Republic of Korea

Hisashi Mizutani and Hideaki Sugawara
Center for Information Biology, National Institute of Genetics, 1111 Yata Mishima, Shizuoka 411-8540, Japan

Kei Yura
Center for Information Biology, National Institute of Genetics, 1111 Yata Mishima, Shizuoka 411-8540, Japan
Graduate School of Humanities and Sciences, Ochanomizu University, 2-1-1 Otsuka, Bunkyo, Tokyo 112-8610, Japan
Center for Simulation Science and Informational Biology, Ochanomizu University, 2-1-1 Otsuka, Bunkyo, Tokyo 112-8610, Japan
School of Advanced Science and Engineering, Waseda University, 3-4-1 Okubo, Shinjyuku, Tokyo 169-8555, Japan

Ashley M. Buckle
The Department of Biochemistry and Molecular Biology, Biomedicine Discovery Institute, Monash University, Clayton, VIC 3800, Australia

Takeshi Sangawa
Laboratory of Protein Synthesis and Expression, Institute for Protein Research, Osaka University, Suita, Osaka 565-0871, Japan

Ken-ichi Miyazono, Jun Ohtsuka, Koji Nagata, Tomoki Shojima, Shohei Nosaki, Yuqun Xu, Delong Wang, Xiao Hu and Masaru Tanokura
Laboratory of Structural Biology and Food Biotechnology, Department of Applied Biological Chemistry, Graduate School of Agricultural and Life Sciences, The University of Tokyo, 1-1-1 Yayoi, Bunkyo-ku, Tokyo 113-8657, Japan

Dimitrios Georgios Kontopoulos
Department of Life Sciences, Imperial College London, Silwood Park Campus, Ascot, UK

Dimitrios Vlachakis
Bioinformatics & Medical Informatics Team, Biomedical Research Foundation, Academy of Athens, Athens, Greece

Georgia Tsiliki
School of Chemical Engineering, National Technical University of Athens, Athens, Greece

Sofia Kossida
IMGT®, The International ImMunoGeneTics Information System®, Université de Montpellier, Laboratoire d'ImmunoGénétique Moléculaire LIGM, UPR CNRS 1142, Institut de Génétique Humaine, Montpellier, France

Logan W. Donaldson
Department of Biology, York University, 4700 Keele Street, Toronto, ON M3J 1P3, Canada

Jamie J. Kwan
Department of Biology, York University, 4700 Keele Street, Toronto, ON M3J 1P3, Canada

McEwen Centre for Regenerative Medicine, Ontario Cancer Institute, 101 College Street, Toronto, ON M5G 1L7, Canada

HaJeung Park
X-ray Core Facility, The Scripps Research Institute, Scripps Florida, 130 Scripps Way, Jupiter, FL 33458, USA

Tuan Tran and Matthew D. Disney
Department of Chemistry, The Scripps Research Institute, Scripps Florida, 130 Scripps Way, Jupiter, FL 33458, USA

Jun Hyuck Lee and Hyun Park
Unit of Polar Genomics, Korea Polar Research Institute, Incheon 21990, Republic of Korea
Department of Polar Sciences, University of Science and Technology, Incheon 21990, Republic of Korea

Index